This book belongs to

S. Meech.

Standard Laboratory Equipment

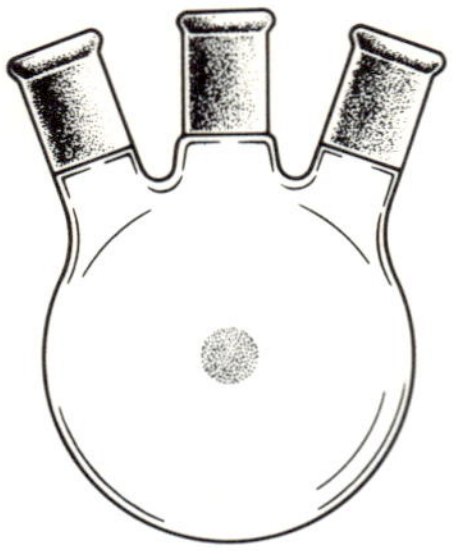

Three-Neck
Round Bottom Flask
Figure 20.2

Round Bottom
Flask
Figure 3.4

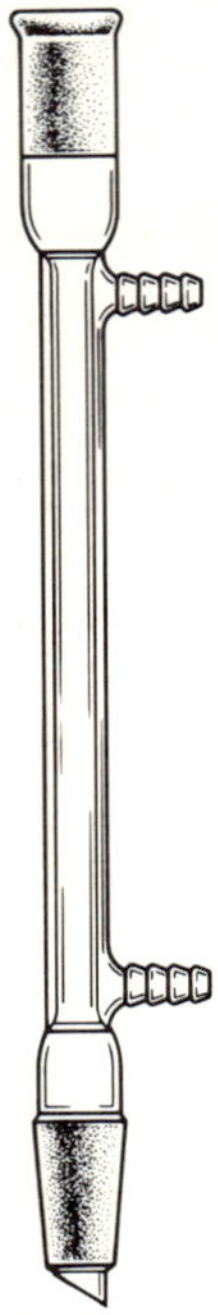

West
Condenser
Figures 3.4, 6.4

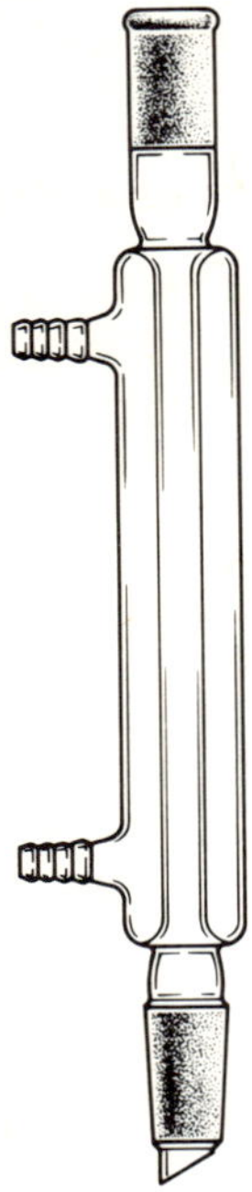

Distilling Column
(Liebig Condenser)
Figures 4.2, 5.3, 6.4

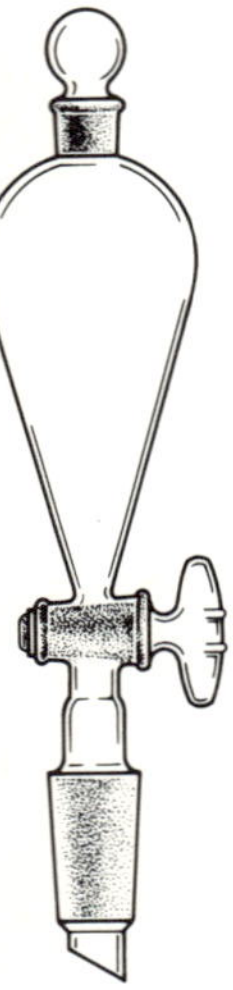

Separatory Funnel
(Addition Funnel)
Figures 1.2, 17.2, 20.2

Outlet Adapter
(Thermometer Adapter)
Figure 3.4

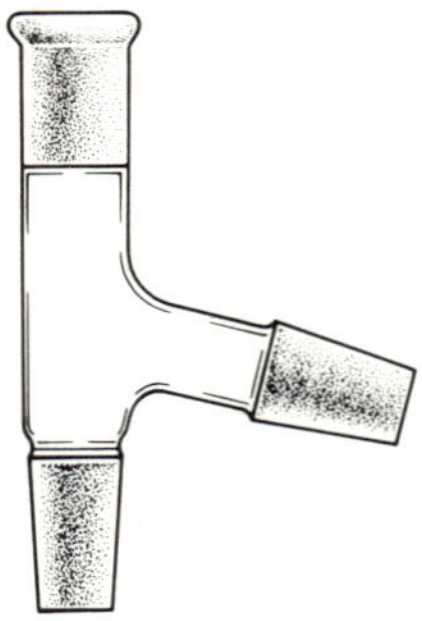

Connecting Adapter
(Distillation Head)
Figure 3.4

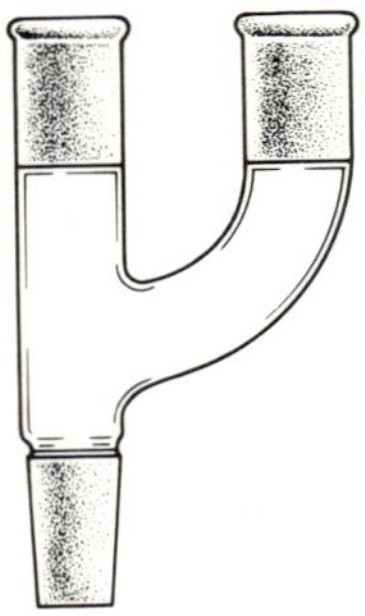

Claisen
Adapter
Figures 5.3, 14.3, 17.2, 20.2

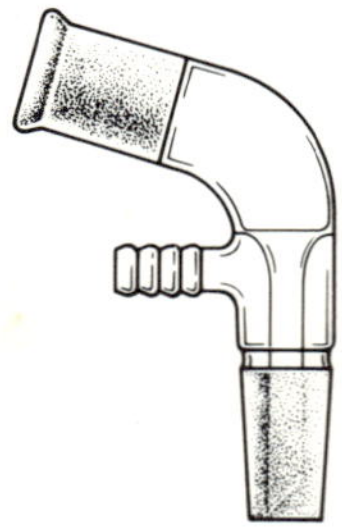

Vacuum
Adapter
Figures 3.4, 5.3, 20.2

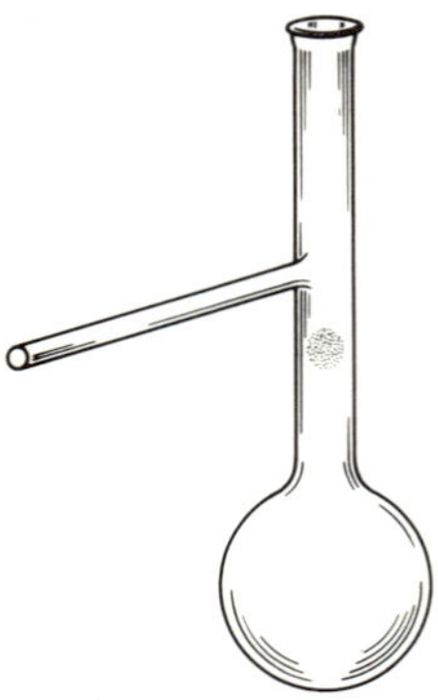

Side-Arm
Distilling Flask
Figures 11.1, 14.2, 17.3

Erlenmeyer
Flask
Figure 2.5

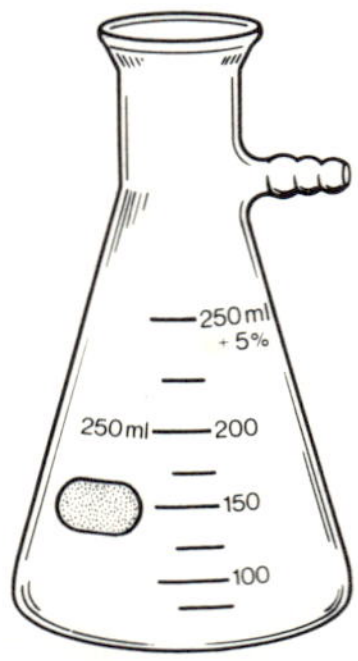

Filter Flask
Figures 1.7, 5.5, 14.2, 17.1

Büchner
Funnel
Figure 1.6

Useful Laboratory Information

FOR YOUR LABORATORY AND PERSONAL SAFETY — page

YOUR LABORATORY NOTEBOOK — page

LABORATORY TECHNIQUES — page

EXPERIMENTAL ORGANIC CHEMISTRY

EXPERIMENTAL ORGANIC CHEMISTRY

Michael P. Doyle
William S. Mungall

Hope College

John Wiley & Sons

New York
Chichester
Brisbane
Toronto

Library of Congress Cataloging in Publication Data:

Doyle, Michael P.
Experimental organic chemistry.

Includes index.
1. Chemistry, Organic—Laboratory manuals.
I. Mungall, William S., joint author. II. Title.
QD261.D67 547'.0028 79-18392

Printed in the United States of America

10 9 8 7 6 5 4 3 2 1

To Carol, Jennifer, and Teresa

Preface

We have written *Experimental Organic Chemistry* to complement the teaching of organic chemistry with an effective laboratory program that introduces students to the modern methods, techniques, and experimentation of organic chemistry. We believe that the combination of workable experiments with detailed experimental background and operational information will enhance students' interest in the laboratory and enable them to perform effectively and confidently. To fulfill this goal we have designed new, but thoroughly tested, experiments that can be performed within the natural constraints of health and safety standards, time and instrument availability, and laboratory economy. Laboratory operations are thoroughly described so that students can be expected to begin their experiments knowing what they are to observe and how to direct their experiment to a successful conclusion.

Experimental Organic Chemistry consists of three sections: the introduction to basic techniques and chemical transformations (Experiments 1-20), qualitative organic analyses (Experiments 21-26), and experiments in multistep syntheses, reaction mechanisms, and modern synthetic methods (Experiments 27-40). Basic laboratory techniques are introduced during realistic laboratory operations in Experiments 1-20 and serve as the foundation for subsequent experiments. Experiments 21-26 are sequential but may be employed at any time in the laboratory program after the introduction of basic techniques. Individual experiments in the section of Experiments 27-40 are not interdependent and can be rearranged to fit the needs of organic chemistry laboratory courses at a variety of schools.

In designing experiments for this laboratory textbook we have been attentive to the sequence of topics that are normally presented in the lecture portion of the organic chemistry course. Following experiments that explain basic laboratory techniques (Experiments 1-7), we proceed to introduce students to most of the important transformations of organic compounds. Reactions with aliphatic compounds are employed until Experiment 14 to correspond with the early emphasis on aliphatic systems in the lecture course. Although we have not included an early experiment on free radical halogenation reactions for reasons of health and safety, this textbook emphasizes substitution, elimination, addition, and oxidation transformations in correspondence with most modern lecture texts.

The vast majority of experiments found in *Experimental Organic Chemistry* will not be found in other laboratory textbooks. Our experiments have been selected from those that complement the lecture program so that students can be effectively introduced to the modern practices of organic chemistry with experiments that are attentive to laboratory safety and laboratory economy. Experiments 8–12, for example, use the reaction series substitution-elimination-addition to introduce students to the basic reactions of aliphatic organic compounds including instructions for the drying of organic liquids, choices of solvents for chemical reactions, and product analyses by gas-liquid partition chromatography and infrared spectroscopy. In addition, we have selected preparative experiments that can be expected to yield the desired products when performed by students in the introductory laboratory program. As an illustration, we have designed a new synthesis of benzocaine for this textbook (Experiment 28) because of students' frustrations with their inability to isolate this compound using standard procedures found in other laboratory textbooks.

One of the unique features of this book is its extensive use of sequential experiments whereby the product of one experiment is the reactant in the next experiment. Because each of the sequential experiments employs commercially available reactants or alternate experiments, student failure to produce the desired product in one experiment does not limit continuation of the sequence. Sequential experiments describe an ultimate design that cannot be identified in singular experiments and afford lower costs for laboratory operation.

We believe that an introduction to modern instrumental analyses including those performed through infrared or nuclear magnetic resonance spectroscopy and gas chromatography should be an integral component of the organic chemistry laboratory. We have therefore designed this laboratory textbook to expose students to instrumental analyses even when access to the necessary instrumentation is limited. Chemical instrumentation is introduced to provide students with some technical understanding of how specific instruments are operated and how to interpret the results obtained. However, each instrument-oriented experiment is designed to display actual experimental results without requiring actual student operation of the instrument. For example, Experiment 10 describes a chromatogram in the detail expected for correct gas chromatographic analysis; students may analyze this chromatogram or, when such instrumentation is available, follow the Experimental Procedure for an analysis of the experimentally obtained mixture. More than 60 spectra are included to promote student familiarity with spectral analyses of organic compounds.

Laboratory techniques are introduced in sufficient detail so that students can properly employ them to achieve the desired experimental result. In addition, these techniques are introduced at places in the text where they are expected to be used. For example, the introductions to distillation at reduced pressure (Experiment 14) and sublimation (Experiment 16) are given in experiments in which these laboratory operations are employed. Prelab and postlab questions test student understanding of experimental methods and techniques.

Important laboratory operations are reinforced and amplified in subsequent experiments.

Both traditional and modern methods for the structural identification of organic compounds are employed. In each experiment where an organic compound is isolated or synthesized, students are directed to identify the product by commonly used procedures. Experiments 13 and 21–26 provide detailed methods and procedures for the identification of unknown compounds. This section of experiments rapidly introduces students to the physical properties and chemical transformations of the major functional groups encountered in organic compounds. Appendix A lists the physical properties of more than 1400 organic compounds and provides melting points for their most useful derivatives. The selection of these compounds is based on their commercial availability and cost.

We have attempted to minimize the use of hazardous chemicals and procedures in the laboratory without minimizing student exposure to important techniques and procedures. The experiments selected meet reasonable health and safety standards. Cautionary statements regarding the handling of chemicals and the operation of laboratory procedures are provided along with instructions for the disposal of waste chemicals.

Laboratory economy is emphasized in two ways: by the use of sequential experiments and the employment of relatively inexpensive reagents. In addition, we have kept to a minimum the number of different reagents and solvents that must be employed in the laboratory program. Stock solutions prepared for the organic chemistry laboratory are standardized to minimize the number of different solutions that must be available in the laboratory. Finally, solvent recovery is promoted.

We have been accustomed to using thoroughly tested workable experiments in our laboratory program and the experiments we have developed for this textbook are no exception. Several have been used in our program for more than four years, and all the experiments have been class tested by Hope College students during two successive years. Their performance in this laboratory program has surpassed our expectations, and their helpful comments have led us to expect similar effectiveness and confidence from students in the future.

Our efforts in designing these laboratory experiments and in writing this laboratory textbook have been assisted by numerous individuals. We are grateful to Richard Paske who tested many of these experiments prior to their use in the organic chemistry laboratory. The organic chemistry classes who used the preliminary editions of this textbook receive our special thanks. Their enthusiasm and encouragement made this project worthwhile.

Furthermore, we are indebted to the people who reviewed the manuscript at different stages of its development: Steven Baldwin (Duke University), Edward L. Biersmith (Northeast Louisiana University), Kenneth G. Hancock (University of California, Davis), John R. Holum (Augsburg College), Evan Kyba (University of Texas, Austin), P. W. Le Quesne (Northeastern University), Anthony LoTempio (Broome Community College), John Meisenheimer

(Eastern Kentucky University), and Charles Stammer (University of Georgia.) Their advice, encouragement, and comments were of great help. We are especially grateful to Norma Plasman for typing the manuscript and preliminary editions and to David Wang for photographing the spectra included in this book.

The efforts of Wiley personnel, particularly our editor Gary Carlson, receive our sincere thanks.

Michael P. Doyle
William S. Mungall

Contents

EXPERIMENTAL ORGANIC CHEMISTRY

Introduction

The laboratory portion of the organic chemistry course lays the foundation upon which the basic principles of this experimental science can be understood. Organic chemistry has developed from experimental observations made in the laboratory. These observations have been summarized, tested, and correlated with related experimental information to form the basis of chemical theories and principles. By following the procedures in this manual and *observing the chemical changes and operations* involved, you will participate in the development of organic chemistry and better understand the basic principles of this experimental science.

Organic compounds have characteristic physical properties. They are gases, liquids, or solids. Some are classified as acids or bases. Certain organic compounds are soluble in water, but most are insoluble in this most common solvent. Because of the wide spectrum of their physical properties, organic compounds require a variety of techniques both for their isolation and purification and for their chemical transformations. Your success in this laboratory is dependent on your awareness of the physical properties of the compounds with which you are working, particularly their melting and boiling points, solubilities, densities, colors, and odors. *Laboratory techniques* for organic chemistry that include extraction, crystallization, distillation, reflux, and chromatography are based on the physical properties of organic compounds and are integral components of this laboratory program.

You will be required to identify the compounds you isolate or prepare. There are currently several million known organic compounds. The exact identification of any one of these compounds requires specific information concerning its source and its chemical and physical properties. In this laboratory course you will be introduced to the methodology for the *characterization* of organic compounds and for their *structural identification* by spectral means.

Your ability to observe chemical changes and your understanding of the changes that occur will be tested in the organic chemistry laboratory. Consequently, *you must thoroughly read the laboratory experiment and fully understand the laboratory procedure before you enter the laboratory.* Your success and safety in the laboratory is dependent on your awareness of the procedures you intend to employ and on the materials you plan to use.

Safety in the Laboratory

Standard safety practices are an essential part of all laboratory operations. The chemicals you employ in this laboratory are usually flammable, some are irritating, and many possess known or as yet undetermined toxic characteristics. Although the experiments you perform in this laboratory manual have been designed for safe operation, you should be prepared for any eventuality. Accidents in the organic chemistry laboratory can be avoided if you enter the laboratory properly prepared for the experiment, if you use good sense in reacting to unexpected situations, and if you rigidly follow basic safety rules that are enforced to ensure your personal safety.

RULES FOR YOUR PERSONAL SAFETY

1. **Wear safety glasses or other eye protection in the laboratory at all times.** Normally eyeglasses with safety lenses can be used, but goggles or safety glasses with side-shields are the better protection. Contact lenses do not protect your eyes. In fact, wearing contact lenses in the laboratory may result in eye irritation due to fumes in the air, since the eye cannot rapidly cleanse itself when the contact lens is in place.
2. **Never work alone in the laboratory.** The presence of another person in your laboratory in the event of a serious accident may save your life. You may work in the laboratory only during authorized times.
3. **Acts of carelessness, including those done in jest, endanger the safety of laboratory participants and are strictly prohibited.** Be aware of what others around you are doing; your safety also depends on their care in performing laboratory operations.
4. **Eating, drinking, or smoking in the laboratory is prohibited.** Laboratory chemicals that may have toxic properties dissolve in foods kept in the laboratory. Smoking represents a serious fire hazard.
5. **Keep your work area neat and uncluttered. Cleanup chemical and water spills at once. Unplug electrical equipment and turn off water and gas outlets when not in use.** If a chemical spill occurs and you are unfamiliar with the safe cleanup procedure for that chemical, immediately contact your laboratory instructor for assistance.
6. **Learn the location and proper use of fire extinguishers, fire blankets, safety showers, and eyewashes.**
7. **Avoid skin contact with chemicals.** If you spill a chemical on your skin, immediately wash the affected area with soap and water. Do not touch your face after handling chemicals; wash your hands after coming into contact with chemicals and chemical containers; and always wash your hands before leaving the laboratory. Rubber gloves should be worn when handling skin irritants.
8. **Do not heat an assembly of laboratory apparatus that is closed to the**

atmosphere and do not close an assembly of laboratory apparatus in which a gas is being evolved. Pressure buildup can cause serious damage. Always check your laboratory apparatus before you begin an experiment to ensure that the system is properly vented.

9. **Do not use flames in an unventilated laboratory area and never produce a flame near containers of flammable compounds.** Although you may be required to use a Bunsen burner to heat organic liquids for distillation in several laboratory experiments, the use of a flame for heating is to be avoided unless absolutely necessary.
10. **Maintain a familiarity with the general physical properties, fire hazards, and toxicities of chemicals that are stored in the laboratory.** Be prepared to react immediately to accidents to yourself or to your neighbor. Always inform your laboratory instructor immediately of any accidents, unexpected occurrences in the laboratory, or physical irritation due to exposure to chemicals.

The most common injuries incurred in the laboratory include cuts from broken glassware, burns from hot glassware or metal, and chemical burns on the skin or in the eyes. *To avoid cuts from broken glass tubing or thermometers, you should lubricate the tubing or thermometer with a drop of glycerine or water before insertion into a rubber or cork stopper and protect your hands with a cloth towel.* In the event of personal injury from cuts or burns, your instructor should be notified immediately and the following procedures should be adopted:

Cuts. If the cut is not serious, wash the affected area with water and dilute soap solution. If bleeding is serious, direct pressure with a clean, preferably sterile, dressing should be applied, and a physician should be contacted.

Heat Burns. Minor burns, where the tissue is not charred, can be treated by flushing with cold water. Serious burns should be treated by a physician.

Chemical Burns. Chemicals on the skin or in the eyes should be flushed with large amounts of flowing water from an eyewash. A physician should be contacted immediately in the event of eye damage.

Strong acids and bases damage skin tissue and may cause severe chemical burns if not immediately washed from the skin with copious amounts of water. These chemicals should always be handled with extreme care, and suitable procedures for their use that include (a) pouring the reagent from the laboratory container into a beaker or flask for subsequent volumetric or weight measurement and (b) always pouring acid into water, never the reverse, should always be employed.

Lachrymators are chemical substances that produce eye irritation and generally cause the production of tears. These chemicals, which are also referred to as "tear gases" when used in aerosol sprays, must be used only in well-ventilated areas and, preferably, in a fume hood. Although a short exposure to small amounts of lachrymator does not generally cause eye damage, a burning sensation in the eyes and on the skin is normally experienced. Flushing the eyes and skin with water generally alleviates the irritation.

The rules and precautions described in this and the following sections should not deter you from working effectively in the laboratory. You should not fear chemicals or laboratory procedures, but you should retain a knowledgeable respect for the materials and equipment you are handling. Specific precautionary statements for the safe handling of chemicals are given in the laboratory experiments for this manual. By being prepared for laboratory experiments, taking suitable precautions, and using good sense, you and your colleagues will safely experience the organic chemistry laboratory.

Heating Apparatus and Fire Safety

Heating is required in most organic chemistry experiments and a variety of apparatus are employed for this purpose. *Steam baths* are routinely used to efficiently heat substances to temperatures as high as 85°C and are always the first choice for these operations. The steam bath is placed under the flask that is to be heated and steam is applied to the exposed surface of the flask to effect the desired rate of heating.

Gas flames from *Bunsen burners* are employed in many laboratory situations to heat a substance to temperatures above 85°C. Since most organic compounds are flammable, the use of gas flames presents definite fire hazards and rigidly enforced precautions must be observed:

1. **A gas flame is preferably used in a well-ventilated area that is free of all flammable liquids.**
2. **Ensure that all volatile liquids in your vicinity are in closed containers before lighting your burner.** Fumes of volatile liquids from an open container may be ignited even when the containers are several feet from the flame.
3. **Never heat a solvent that has a boiling point less than 85° with a flame.** Steam baths or hot water baths are employed in these cases.
4. **Organic liquids may be heated over a flame only when the flask that contains the liquid is attached to a tightly fitted condenser.** The condenser prevents the escape of flammable material from the flask. When Bunsen burners are used for distillations (Experiments 3 and 6), the collection flask must be attached to the vacuum adapter (e.g., Figure 3.4); never distill a flammable substance into an open beaker or flask.
5. **Turn on your heating apparatus only when the heating operation is to begin. Turn off your burner as soon as the heating operation is completed.**
6. **Do not wear clothes having loose fitting sleeves and cuffs. Long hair should be tied back for laboratory work,** particularly when flames are used. In addition, your work area should be free of paper and other substances that can be ignited.

Table 1 Flash points of common organic solvents

Solvent	Structural formula	Boiling point (C°)	Flash point (C°)
Ethyl ether	$(CH_3CH_2)_2O$	35	-40
Pentane	$CH_3(CH_2)_3CH_3$	36	-49
Acetone	CH_3COCH_3	56	-10
Hexane	$CH_3(CH_2)_4CH_3$	68	-23
Ethanol	CH_3CH_2OH	78	12
Benzene	C_6H_6	80	-11
Toluene	$C_6H_5CH_3$	111	7

Electrical heating devices, either as heating mantles, hot plates, or oil baths, are employed in many laboratories. These heating devices are usually safer than Bunsen burners since they do not employ a flame. However, care must also be used with these devices since fires due to electrical malfunction or to flammable vapor contact with the heating element may occur. The apparatus must be thoroughly checked for obvious defects (frayed wire and bare contacts, for example) before use. Oil baths must not be heated above the flash point or decomposition point of the oil that is employed. *A liquid must not be heated to dryness when heating mantles or hot plates are used*, and none of the electrical heating devices should be used to distill volatile liquids with boiling points that are less than 85°C.

The *flash point* of a compound, which is the temperature at which that compound can be ignited and sustain combustion, provides a measure of flammability. As indicated by the flash points of common organic solvents in Table 1, volatile organic compounds are highly flammable. Mineral oil, by comparison, is high boiling, nonvolatile, and has a flash point of 220°C; this material cannot be ignited until the temperature of 220°C is reached. Generally, the flash points of organic compounds increase as their volatility and carbon-hydrogen content decrease.

In the event of a fire in the laboratory, get out of the immediate area of danger at once. Use a safety shower or blanket to extinguish clothing fire. Your instructor should be made immediately aware of the fire and its source. Only carbon dioxide or dry chemical extinguishers should be used on chemical fires. Water extinguishers usually disperse chemical fires over larger areas and should not be used.

Health Effects of Organic Chemicals

All chemicals are toxic. However, the toxicity of these substances varies widely. Ethanol, for example, is poisonous only if ingested in large amounts, whereas inhalation of bromine vapor in small amounts can cause serious damage. In addition to toxicity, certain chemicals are lachrymators or skin irritants, and a growing list of chemicals have been implicated as causing cancer (carcinogens) in test animals. Most organic compounds, however, have not been subjected to rigorous testing procedures to determine their health effects and, consequently, you should treat all chemicals that you use as if they could affect your health. *Always avoid skin contact with the chemicals you use, never taste a chemical, and avoid inhalation of vapors.* If you are to determine the odor of a chemical, cautiously bring the stopper from the container of that chemical to your nose. Thoroughly wash any area of skin that has contacted chemicals with soap and water. Specific warnings and instructions regarding the use of toxic chemicals and carcinogens are provided in the individual laboratory experiments for this manual.

Laboratory Economy

The experiments you perform in the organic chemistry laboratory utilize substantial amounts of a wide variety of chemicals as reagents, catalysts, and solvents. Ideally, the material cost for each experiment should be only the reagents that are consumed in forming products and all other materials should be recovered and recycled. Experiments in this laboratory manual are designed so that you can approach this goal. *Solvents are reclaimed by distillation or filtration and transferred to appropriately labeled solvent containers for further purification and re-use.* In several experiments you will employ the product from one reaction as the starting material for a subsequent reaction. *Final organic products obtained from laboratory experiments are placed in sample bottles or tubes, labeled,* and submitted to your instructor.* These materials can eventually be employed by yourself or others for additional experiments. By practicing good laboratory economy you not only save material costs, but, also, augment health safety in the laboratory and avoid despoiling the laboratory environment.

In addition to the general procedures that have been described, the following specific laboratory procedures should be employed in your work:

1. *Obtain only the amounts of chemicals that are required for an experiment.*
2. *Close laboratory containers of reagents and solvents immediately after you have measured out the desired material.* Open containers are health and fire hazards. In addition, certain reagents and solvents become contaminated when exposed to the air.
3. *Do not discard organic chemicals in the sink.* Aqueous solutions may be washed down the sink with large volumes of cold tap water. If you have any questions concerning the disposal of a chemical, consult your instructor.
4. *In sequential experiments, if you have only a fraction of the amount of the starting material, reduce the amounts of reagents and solvents to be used by a proportional amount.*

*The label should include the name of the compound and its structure, the melting point or boiling point of the compound, the weight of the sample, your name and notebook reference, laboratory section number, and the date.

Your Laboratory Notebook

The record that you maintain of your work in the laboratory is as much a part of chemical experimentation as the actual experiment. Your laboratory notebook contains a permanent record of your experimental observations, experimental results, and conclusions for experiments that you have performed. This experimental record is essential for scientific advancement and may be used as a legal document. Consequently, the development of suitable methods for keeping a clear and accurate experimental record in your laboratory notebook is an integral part of your experience in the organic chemistry laboratory.

The laboratory notebook is a bound notebook with a hard cover and is used solely for experimental data in this course. Your laboratory notebook is to be with you at all times while you are in the laboratory. Experimental observations and results are written into the notebook while the experiment is being performed.

The notebook is kept in ink, not pencil. Changes in the written record are made by drawing a single line through each incorrect word and then substituting the correct words. Three pages at the beginning of the notebook should be set aside for a Table of Contents, which should be kept current. The right-hand pages of the notebook are used for your experimental record; these pages are numbered consecutively. Left-hand pages are used for all informal records, including tare, gross, and net weights of materials, procedural outlines, and incidental notes or reminders. Separate items such as spectra or graphs, when added to the notebook, should be attached to a separate notebook page. The experimental writeup should be sufficiently clear and complete to enable the reader to reproduce the experiment exactly as it was performed by you and obtain the same results. *Your name and the date are recorded at the top of the right-hand page before each experiment is begun.* The following outline is a useful guide for your notebook description of experiments.

1. *Title of experiment.*
2. *Purpose of experiment*
   Why is the experiment being undertaken and what is to be achieved?
3. *Chemical process to be investigated*
   Usually an equation, such as a balanced chemical reaction, that describes the process.
4. *Physical properties of chemicals used in experiment*
   Includes molecular weights, melting points or boiling points, densities (if known), and other pertinent physical data.
5. *Tabular listing of chemicals required for the experiment*
   This table includes all chemicals that are to be used in the experiment and the source of these chemicals (chemical company, notebook reference, or, in the case of standard solutions, the chemistry stockroom). The weight or volume amounts of materials are listed as are the number of moles of these compounds.
6. *Flow diagram for all experimental operations*

7. *Brief description or listing of required equipment*
   Do not list beakers, flasks, or cylinders.
8. *Theoretical yield* (if applicable)
9. *Experimental procedure*
   Literature reference to procedure
   Experimental observations with description of procedure
   Product analysis and isolated product yield (if applicable)
   The experimental procedure should be written neatly in intelligible sentences and should describe what you have done, how the experiment was performed, and what was observed. A laboratory notebook description employs impersonal scientific language; the use of "I" (first person) should be avoided. Experimental observations include such items as construction of apparatus, solution preparations, order of addition of reagents, special manipulations, changes in temperature, color, phases, and times for reactions. The physical properties of products are fully described.
10. *Summary of experimental data and evaluation of experimental methods*
    Includes physical properties of products with listing of the same properties for the authentic material. The literature reference to the physical properties of the known material should also be given. The amount of product in grams and moles is given as is the percentage yield. A discussion of problems encountered in the experiment is often useful as are estimations of methods to improve the experiment.

Parts 1 to 8 should be completed prior to the start of your laboratory period. An example of such a prelab writeup is given for Experiment 1:

SEPARATIONS OF ORGANIC COMPOUNDS

Purpose: Isolation of benzoic acid from a mixture of benzoic acid and toluene by extraction methods.

$$C_6H_5CH_3, C_6H_5COOH \xrightarrow{\text{extraction}} C_6H_5COOH + C_6H_5CH_3$$

	Benzoic acid	*Toluene*
	m.p. = 122°C	b.p. = 111°C
	M.w. = 122 g/mol	M.w. = 92 g/mol
	K_α = 6.3 × 10^{-5}	insol. in H_2O
	sol. = 0.34 g/100 ml H_2O	

Reactants (Source)	Weight or volume	Moles
Benzoic acid in toluene (Stockroom)	30 ml	—
Sodium hydroxide (Stockroom, 10% aqueous solution, 2.5*M*)	2 × 15 ml	2 × 0.037
Hydrochloric acid (Stockroom, 10% aqueous solution, 1.2*M*)	2 × 35 ml	2 × 0.042

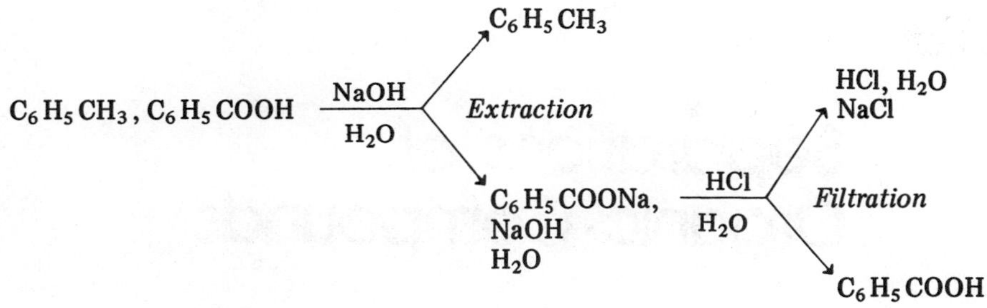

Equipment: separatory funnel, Büchner funnel.

Expected yield: Dependent on concentration of benzoic acid in toluene.

Your laboratory instructor may have alternate or more specific instructions concerning the format of your laboratory notebook. These instructions should be followed closely as you learn to develop formats appropriate to the different types of experiments that are encountered in this laboratory manual. The format for describing chemical reactions in your laboratory notebook is given in Experiment 4.

Yield Calculations

Organic chemistry processes are often complicated. More than one kind of chemical transformation takes place under a given set of reaction conditions with many organic compounds. In typical organic chemical preparations the desired product must be separated from undesired contaminants.

The efficiency of a chemical process in producing a desired compound is expressed as the *percentage yield.* For example, if a reaction is said to give an 85% yield of a compound, this means that the weight of product is only 85% of the amount that could theoretically have been formed (*theoretical yield*) in the absence of all competing processes if all of the starting material is converted to the desired compound.

The percentage yield is calculated for the purified product. However, since part of the product may be lost during purification, the yield of the crude material is reported as well. Samples must, of course, be dry when they are weighed. The percentage yield is reported to the nearest whole number. Since the actual laboratory operations that lead to the final product are generally not quantitative, the use of more than two significant figures for percentage yield is not appropriate.

Experiment One

Separations of Organic Compounds

The evolution of organic chemistry into a modern experimental science has occurred primarily because pure organic compounds can be isolated from the rich sources of organic compounds found in nature. Natural sources such as plants, coal, and petroleum contain complex mixtures of organic compounds. The separation of these complex mixtures into their individual components is performed through the use of techniques based on differences in physical properties of organic compounds. Physical properties such as solubilities and boiling points are translated into viable procedures for the selective isolation of particular materials from their environment.

The separation methods that are of greatest importance in organic chemistry are extraction, crystallization (filtration), distillation, and chromatography. Extraction and crystallization techniques depend on the solubility characteristics of organic compounds. Separations by distillation rely on differences in boiling points among the components of a mixture. Differences in the abilities of chemicals to adhere to surfaces (*adsorption*) are basic to chromatographic separations. Each of these methods is built upon a distinguishable physical

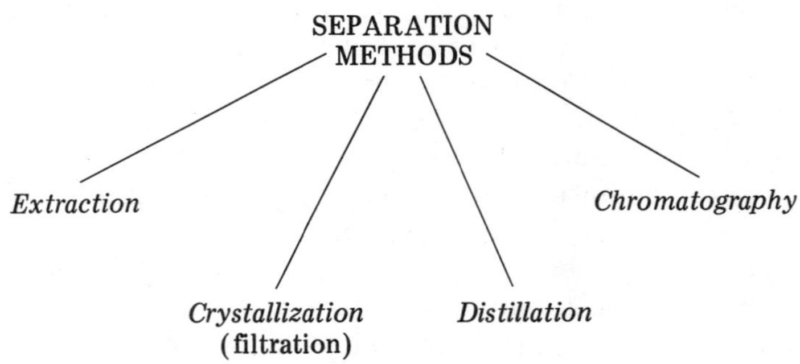

difference among the components of a mixture. We can separate compounds that differ in their solubility properties, boiling points, or adsorptivities by using separation techniques that are based on these physical differences. Frequently chemical reactions, particularly acid-base reactions, are employed to produce dramatic physical differences between components of mixtures.

In this experiment you will isolate benzoic acid from a mixture of benzoic acid and toluene by taking advantage of the physical and chemical properties of benzoic acid. Benzoic acid is produced on an industrial scale from toluene by oxidation at high pressure using a metal catalyst:

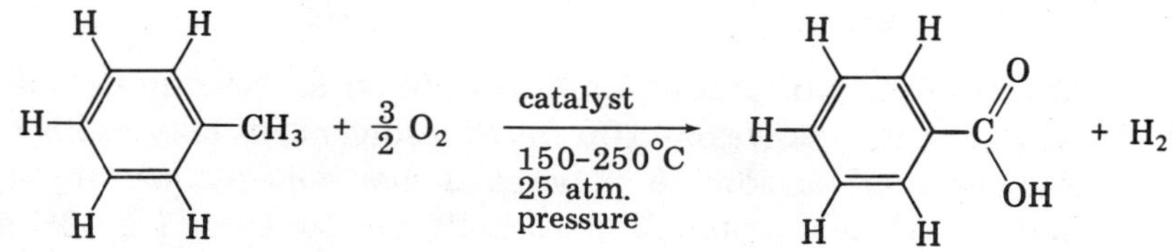

Toluene
b.p. = 111°C
m.p. = −95°C
sol. in H_2O = insoluble

Benzoic acid
b.p. = 249°C
m.p. = 122°C
sol. in H_2O = 0.34 g/100 ml H_2O
$K_a = 6.3 \times 10^{-5}$

Although the physical properties listed below the structures of toluene and benzoic acid indicate that separations based on boiling point or melting point differences are possible, you will use solubility differences for the separation of benzoic acid from toluene. Since benzoic acid is a stronger acid than water ($K_a = 1.8 \times 10^{-16}$), the conversion of benzoic acid to its conjugate base can be made to occur by treating benzoic acid with an aqueous solution of sodium hydroxide:

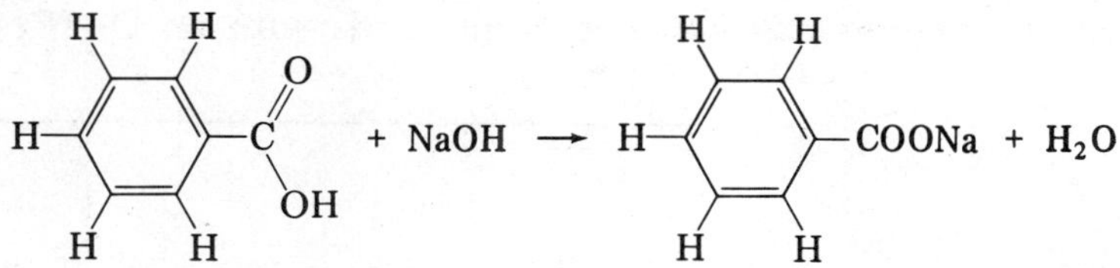

Stronger acid ($K_a = 6.3 \times 10^{-5}$) — *Stronger base* — *Weaker base* — *Weaker acid* ($K_a = 1.8 \times 10^{-16}$)

Through this acid-base conversion the solubility difference between toluene and benzoic acid (as sodium benzoate) is greatly enhanced, and the separation

Nonpolar covalent compound

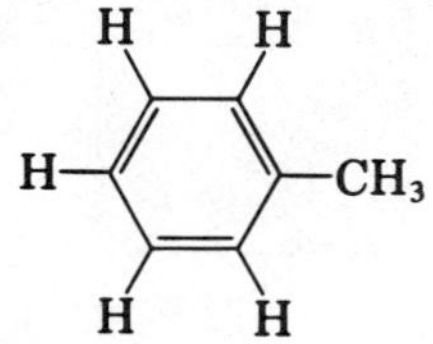

Toluene
(insoluble in water)

Polar ionic compound

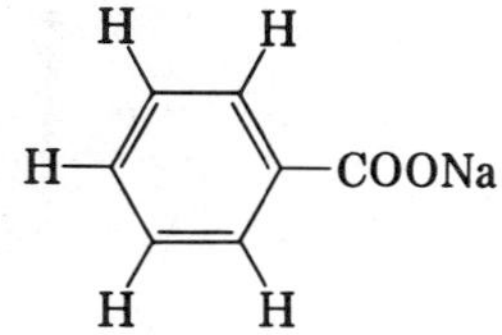

Sodium benzoate
(soluble in water)

technique that you will employ becomes the extraction of a polar ionic compound from a nonpolar covalent compound.

Solubilities of Organic Compounds

The solubility of a compound is defined as the number of grams of that compound that dissolves in 100 ml of a solvent, usually at 25°C. When we say that a compound is soluble in a particular solvent, we imply that a measurable amount of that compound (usually greater than 1 g/100 ml solvent) will dissolve in the specified solvent. On the other hand, when a compound is said to be insoluble in a particular solvent, we intimate that no appreciable amount of that compound (usually less than 1 g/100 ml solvent) will dissolve in the solvent. A compound will be soluble in a solvent if the attractive forces between molecules of the compound are like those that exist between solvent molecules; that is, "like dissolves like."

There is a spectrum of attractive forces that can exist between organic molecules. Basically, however, compounds that exhibit these attractive forces fall into four distinct categories based on the polarity of the bonds that constitute the molecular structure. *Nonpolar compounds* such as the hydrocarbons have an elemental composition (carbon and hydrogen) characterized by a low electronegativity difference. *Polar compounds* possess polar bonds (for example, C—O, C—N, C—Halogen) and are capable of stronger interactions than nonpolar compounds. Organic compounds with an O—H or N—H bond form

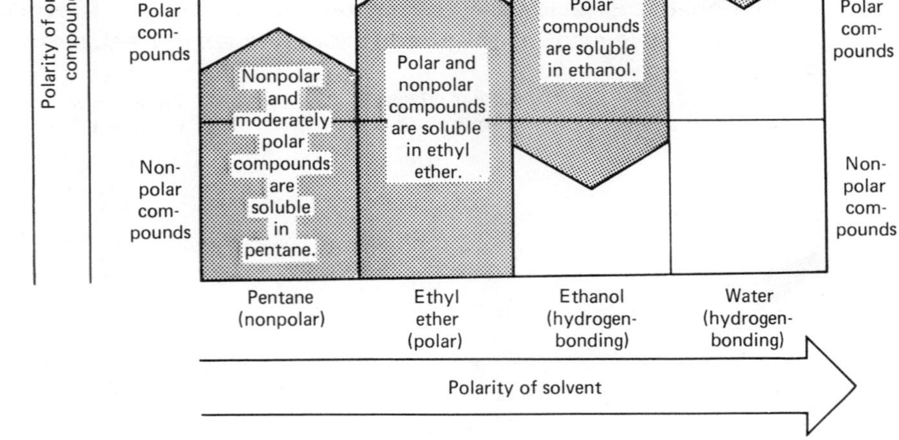

Figure 1.1 *Solubilities of organic compounds in nonpolar, polar, and hydrogen-bonding solvents.*

Increasing polarity of bonds in molecule →

Nonpolar compounds	Polar compounds	Hydrogen-bonding compounds	Ionic compounds

Increasing strength of attractive forces between molecules →

a special class of polar compounds called *hydrogen-bonding compounds*. The strongest attractive forces between molecular units are those of *ionic compounds*. By the rule of "like dissolves like" we mean that nonpolar compounds are most soluble in solvents that are also nonpolar and that polar compounds are most soluble in polar solvents.

The solvents you will find most useful for the experiments in this laboratory manual span the spectrum of solvent polarity and include pentane, ethyl ether, ethanol, and water. The solubilities of organic compounds in these solvents reflect their mutual compatibility in polar attractions (Figure 1.1). Pentane is the solvent of choice for nonpolar organic compounds, whereas ionic compounds are preferentially dissolved in water. Table 1.1 lists structural information and useful physical properties for the solvents that are included in Figure 1.1. Table 1.2 presents similar information for the solvents that are commonly used for extractions of organic compounds.

Extraction

Extraction is the separation method that involves the transfer of a substance from one material phase into a second phase. When the two phases are immiscible liquids, this method is known as a *liquid-liquid extraction*. In liquid-liquid extractions a compound is partitioned between two solvents. The success of the separation depends on the difference in the solubilities of the compound in the two solvents. Generally the compound that is to be extracted is insoluble or only partially soluble in one solvent but is very soluble in the other solvent.

Water is used as one of the two solvents in liquid-liquid extractions since most organic compounds are immiscible with water and because water dissolves ionic and very polar compounds. The solvents that are compatible with water for the extraction of organic compounds are generally chosen from those listed in Table 1.2. Each of these solvents meets the necessary criterion of relative

Table 1.1 Physical Properties of Common Solvents

Solvent	Structure	Molecular weight (g/mol)	Boiling point (°C)	Solubility (g/100 ml H_2O) at 20°C	Density at 20°C (g/cm^3)
Pentane	$CH_3CH_2CH_2CH_2CH_3$	72	36	0	0.626
Ethyl ether	$CH_3CH_2OCH_2CH_3$	74	35	7.5	0.714
Ethanol	CH_3CH_2OH	46	78	∞	0.789
Water	H_2O	18	100		1.000

Table 1.2 Physical Properties of Common Extraction Solvents

Solvent	Structural formula	Molecular weight, (g/mol)	Boiling point, (°C)	Density at 20°C (g/cm^3)	Comment
Ethyl ether	$CH_3CH_2OCH_2CH_3$	74	35	0.714	The most widely used extraction solvent. Upper layer in extraction with water. Flammable liquid.
Pentane	$CH_3CH_2CH_2CH_2CH_3$	72	36	0.626	Useful for extractions of nonpolar compounds. Upper layer in extractions with water. Flammable liquid.
Methylene Chloride[a]	CH_2Cl_2	85	41	1.335	Useful for extractions of polar compounds. Usually lower layer in extractions with water.
Chloroform[a]	$CHCl_3$	119	61	1.492	Useful for extractions of polar compounds. Lower layer in extractions with water.
Hexane	$CH_3(CH_2)_4CH_3$	86	68	0.659	Same as pentane for extractions. Flammable liquid.
Carbon tetrachloride[a]	CCl_4	154	77	1.594	Used for extractions of nonpolar compounds. Lower layer in extractions with water.
Benzene[b]	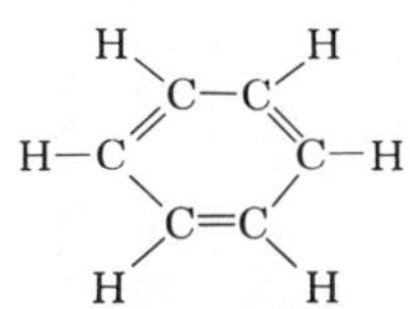	78	80	0.879	Used for extractions of aromatic compounds. Upper layer in extractions with water. Flammable liquid.
Toluene	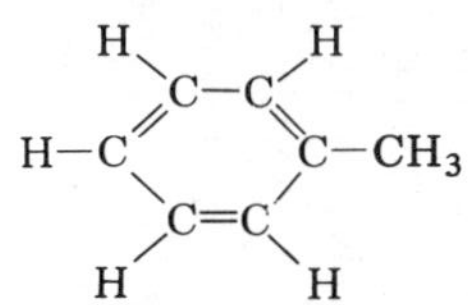	92	111	0.867	Same as benzene for extractions. Flammable liquid.

[a] Skin contact with chlorocarbon solvents and the breathing of their vapors is hazardous to your health. Carbon tetrachloride and chloroform have been linked to cancer in test animals and have been banned by FDA from use in drug, cosmetic and food packaging products.
[b] Breathing benzene vapors and absorption of benzene through the skin must be avoided. Inhalation of benzene causes irritation of mucous membranes. Benzene has been linked to cancer in test animals.

insolubility in water; water and any one of these solvents form two distinct phases. In extractions with water and an organic solvent the water phase is referred to as the *aqueous phase* and the organic solvent is called the *organic phase.*

In addition to the criterion of immiscibility for the aqueous and organic phases, the compound that is to be extracted must be separable from the phase in which it is concentrated. The organic solvent that is chosen generally has a boiling point that is much lower than that of the extracted compound and is usually selected from inexpensive and nontoxic organic compounds that have boiling points less than 100°C.

The extraction method is based on the distribution of a compound between two liquid phases at equilibrium. This equilibrium distribution (or partitioning) depends on the solubility of the compound in each of the two liquid phases. Consider the case of benzoic acid distribution in toluene and water as an example. Benzoic acid is soluble in water to the extent of 0.34 g/100 ml at 25°C and dissolves in toluene to the limit of 11 g/100 ml at the same temperature. If 5.0 g of benzoic acid that is dissolved in 50 ml of toluene is added to 100 ml of water and the resulting two-phase combination is thoroughly mixed, benzoic acid will be partitioned between the two liquid phases according to the equilibrium expression:

$$(\text{Benzoic acid})_{C_6H_5CH_3} \overset{K}{\rightleftharpoons} (\text{Benzoic acid})_{H_2O}$$

$$K = \text{Sol}_{H_2O}^{\text{Benzoic acid}} \Big/ \text{Sol}_{C_6H_5CH_3}^{\text{Benzoic acid}}$$

where K is the *equilibrium partition coefficient* and $\text{Sol}^{\text{Benzoic acid}}$ is the amount of benzoic acid that is dissolved in the solvent. The equilibrium partition coefficient is the solubility ratio which for benzoic acid in toluene and water is equal to

$$K = \frac{0.34 \text{ g/100 ml } H_2O}{11 \text{ g/100 ml } C_6H_5CH_3} = 0.031$$

Using this value for the equilibrium partition coefficient we can calculate the amount of benzoic acid that will be partitioned into 100 ml of water (X/100 ml) from the original 5.0 g that is dissolved in 50 ml of toluene [(5.0 g - X)/ 50 ml]:

$$K = 0.031 = \frac{X/100 \text{ ml}}{(5.0 \text{ g} - X)/50 \text{ ml}}$$

$$X = 0.29 \text{ g}$$

That is, 0.29 g of benzoic acid is in the water phase and the remaining 5.0 g - 0.29 g = 4.7 g of benzoic acid is in the toluene phase. This example indicates that organic compounds such as benzoic acid would be more effectively extracted from water by toluene (K = 32) than from toluene by water (K = 0.031). Indeed, extraction methods are generally used to separate organic compounds from water and from compounds that are soluble in water.

Acid-Base Extraction

Benzoic acid is significantly more soluble in toluene than in water. Consequently, unless the toluene solution of benzoic acid is extracted with extremely large volumes of water or unless multiple extractions with water are employed, benzoic acid will remain in the toluene phase. However, benzoic acid is a moderately strong acid that will react with aqueous sodium hydroxide to form the ionic salt sodium benzoate. Sodium benzoate, unlike benzoic acid, is insoluble in toluene but is appreciably soluble in water. If the toluene solution of benzoic acid is extracted with an aqueous solution of sodium hydroxide rather than water, benzoic acid will be converted to its sodium salt and the sodium salt will dissolve in the aqueous phase (Figure 1.2). The toluene phase can then be separated from the aqueous phase that now contains sodium benzoate.

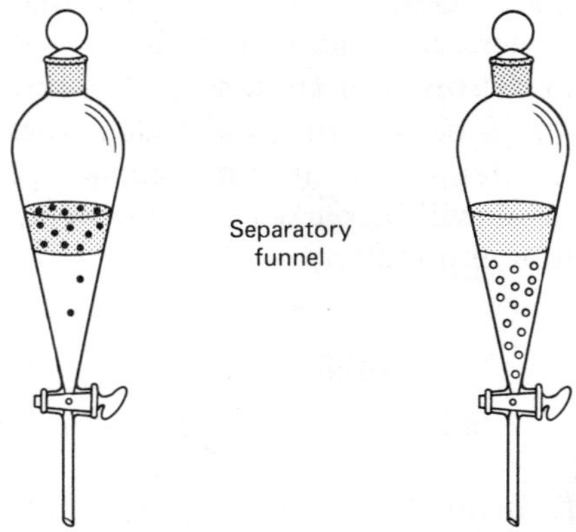

Figure 1.2 *Extraction of benzoic acid* (*a*) *by water* (● = C_6H_5COOH) *and* (*b*) *by aqueous sodium hydroxide* (○ = $C_6H_5COO^-Na^+$).

Benzoic acid is isolated by adding an aqueous mineral acid to the basic aqueous phase until the solution *p*H is less than 4. Since benzoic acid is a solid that is only slightly soluble in water, crystals of benzoic acid will precipitate from the acidic solution. Filtration of these crystals represents the final step in this isolation procedure. The following flow diagram outlines the separation procedure for this experiment:

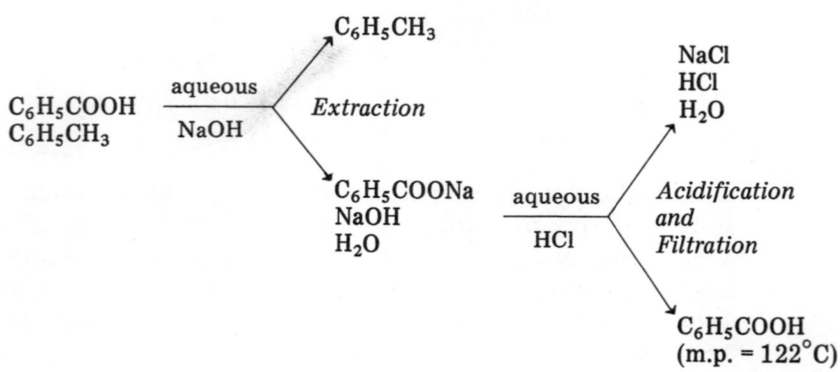

Base extractions are generally employed for the isolation of covalent acids from organic mixtures by procedures that are similar to the one you will use to isolate benzoic acid. Similarly, covalent bases are isolated from organic mixtures by extractions with aqueous mineral acids; the base is isolated after treating the aqueous solution that contains the conjugate acid of the covalent base with aqueous sodium hydroxide.

Use of the Separatory Funnel

The separatory funnel is designed for the multiple operations involved in liquid-liquid extractions: thorough mixing of the liquid phases and separation of the liquid phases. Thorough mixing of the two-phase combination is required to achieve partition equilibrium; each molecule of the compound that is to be extracted must be in contact with both liquid phases during extraction. The proper handling procedure for the separatory funnel during extractions is illustrated in Figure 1.3.

The identity of the upper and lower liquid layers in the separatory funnel is a function of the densities of the liquid solutions (Table 1.2). The solution that has the greater density will be the lower layer.

Filtration

Filtration is the laboratory operation that results in the separation of a solid material from a liquid. For the experiments in this laboratory manual you will employ two basic filtration methods: *gravity filtration* and *suction filtration.* Although the two methods employ different apparatus, they both involve the physical separation of a solid from a liquid.

Gravity filtration is commonly used for the collection of a solid material that is insoluble in the liquid with which it is associated. The apparatus that is

Figure 1.3 *Use of the separatory funnel.* (*continued on page 18*).

Before starting, remove from the area around you any flames and chemicals that will react with either of the solvents that you employ. Be certain that the stopcock of your separatory funnel is closed. Add the solution that is to be extracted and then add the extraction solvent. Do not fill the separatory funnel to more than three-fourths of its total volume to ensure thorough mixing in the extraction process. The following procedure requires both of your hands—one hand at the base of the separatory funnel to manipulate the stopcock and the other hand at the top of the separatory funnel to hold the stopper tightly and to support the separatory funnel.

Figure 1.3 *(continued)*

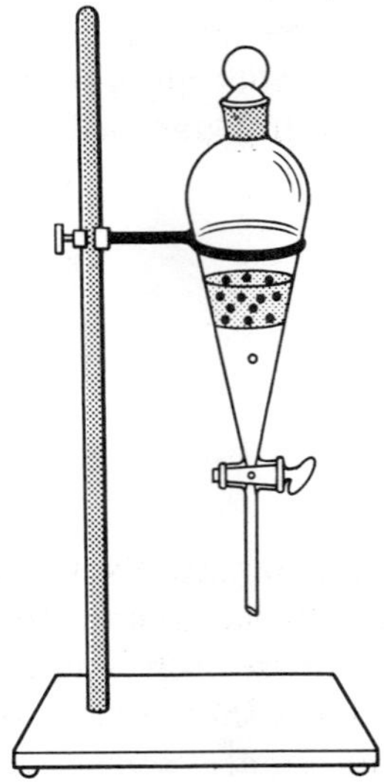

Stopper the separatory funnel, place your index and middle fingers over the stopper, and invert the separatory funnel to a tilted position. Now open the stopcock to release any pressure that has built up within the closed separatory funnel. Point the separatory funnel away from yourself and from any person near you.

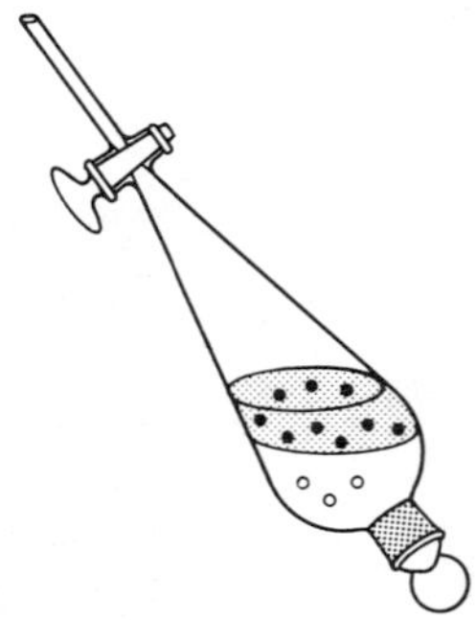

Close the stopcock. While the separatory funnel is in the tilted position gently shake the contents of the separatory funnel and then again release internal pressure through the stopcock. This cautious process allows you to sense the amount of pressure that will be built up in the separatory funnel by vigorous shaking. If you sense little pressure buildup, proceed to the next step; if there is substantial pressure buildup, repeat this step.

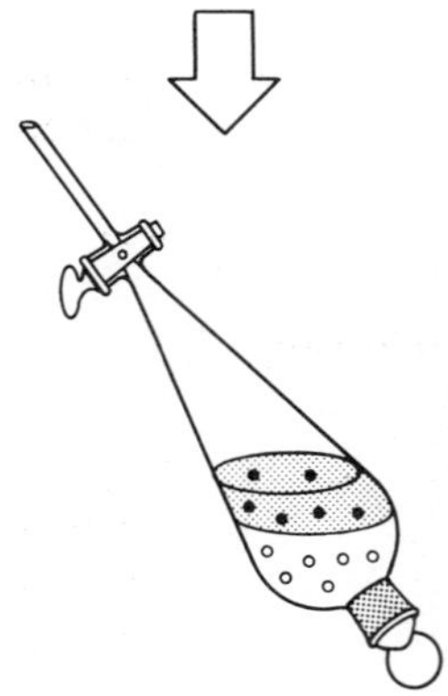

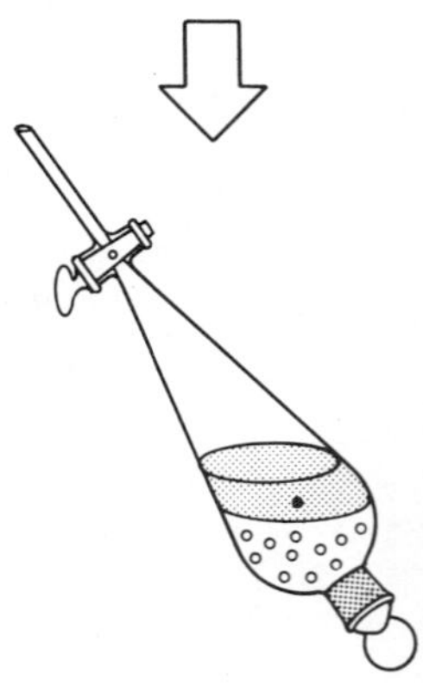

Again close the stopcock and now shake the contents of the separatory funnel vigorously so that the liquids are thoroughly mixed. Continue shaking the separatory funnel for more than 30 seconds. Open the stopcock to release internal pressure, then close the stopcock and return the separatory funnel to its upright position. Replace the separatory funnel in its iron ring support and remove the stopper.

When the two phases are visually separate, draw off the lower liquid layer through the stopcock and into a receiving flask.

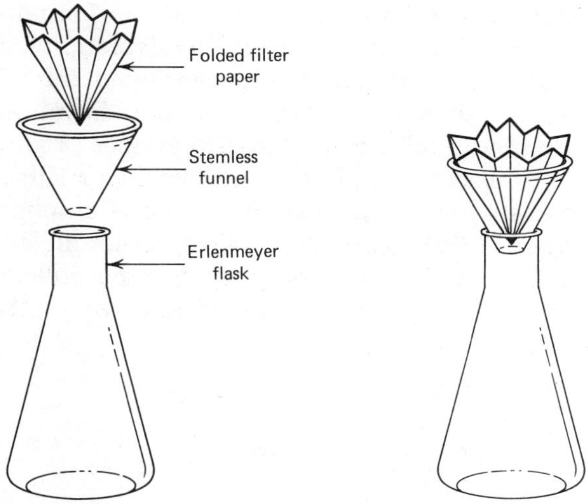

Figure 1.4 *Apparatus for gravity filtration.*

employed in this filtration procedure consists of an Erlenmeyer flask, a stemless funnel, and a fluted filter paper. The Erlenmeyer flask supports the stemless funnel, and the fluted filter paper is inserted into the funnel as shown in Figure 1.4. A stemless funnel is used instead of a long-stem funnel to avoid clogging of the stem by solids that may precipitate from the liquid during filtration. (The long-stem funnel is employed to transfer liquids between containers.) The filter paper is fluted so that the entire surface of the paper can efficiently function as a membrane to separate the solid from the liquid.

To construct a fluted filter, fold a circular piece of filter paper in half and then fold the half-circle into eight parts (seven folds) in the manner of constructing a fan: alternate the direction of the paper at each fold like those in the bellows of an accordion. Open the folded paper to the fluted construction that is described in Figure 1.5. The size of the filter paper you use for gravity

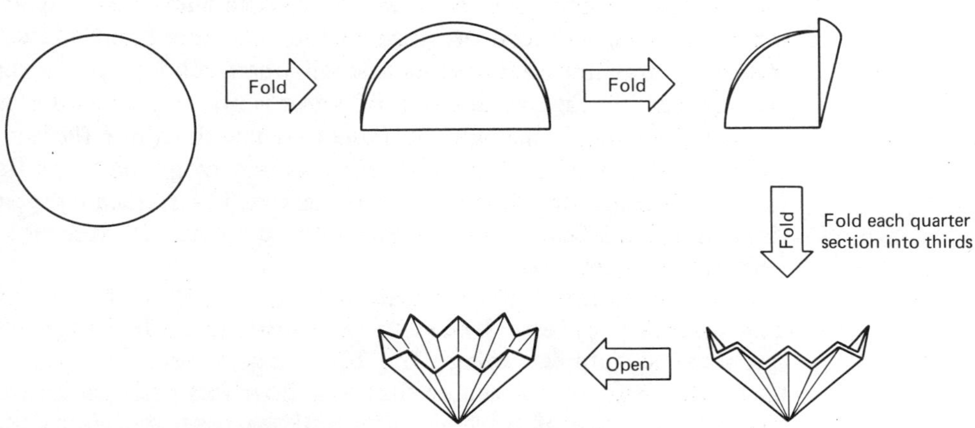

Figure 1.5 *Construction of a fluted filter.*

filtration is related to the size of the funnel: the diameter of the filter paper must be approximately twice the diameter of the top of the stemless funnel.

The method commonly employed for the collection of a solid that has crystallized from a solvent is suction filtration. Suction from a water aspirator pulls the liquid through the filter paper (Figure 1.6*a*). The filter paper is placed on the perforated plate of a Büchner or Hirsch funnel and the funnel is firmly attached to a filter flask by means of a neoprene adapter or a rubber stopper (Figure 1.6*b*). The filter paper covers all of the perforations of the funnel. The Hirsch funnel is designed for the collection of small quantities of solids (usually <1 g); the Büchner funnel will collect gram and larger quantities of solids.

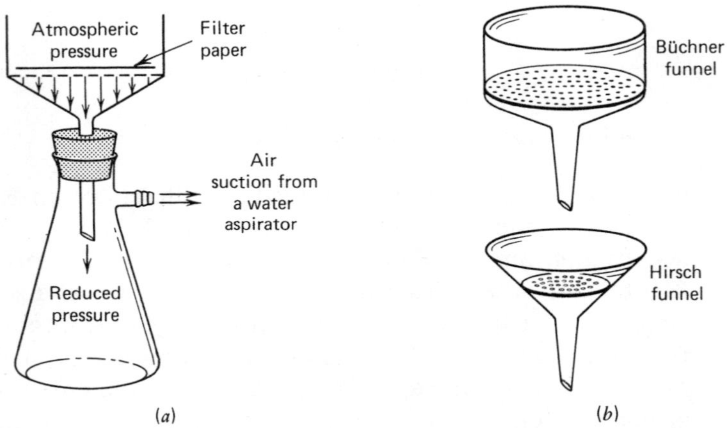

Figure 1.6 *(a) Operation of suction filtration. (b) Apparatus for suction filtration.*

Effective operation of the suction filtration technique depends on the extent of pressure reduction within the filter flask. A water aspirator is generally used to provide this pressure reduction. Water that is rapidly flowing from a faucet exerts a considerable reduced pressure within the faucet. A water aspirator releases the reduced pressure within the faucet and allows a faster flow of water by drawing air through a small open tube that is perpendicular to the water drain. So long as the flow of water is not constrained near the open tube, air will be sucked through the faucet. When the filter flask is attached to the aspirator for suction filtration, air is drawn from the filter flask and liquid is forced through the filter paper to relieve the reduced pressure in the filter flask (Figure 1.6*a*). Pressures as low as 20 Torr (20 mm Hg) can be achieved with water aspirators.

The setup for suction filtration (Figure 1.7) consists of the suction filtration assembly (Figure 1.6*a*) attached through a filter trap to the aspirator by thick-walled rubber tubing. The filter trap is used as a cautionary collection flask; the trap collects water that will flow from the aspirator if there is a decrease in the rate of flow of water and the pressure reduction within the filter flask is greater than the pressure reduction in the aspirator tube. The walls of

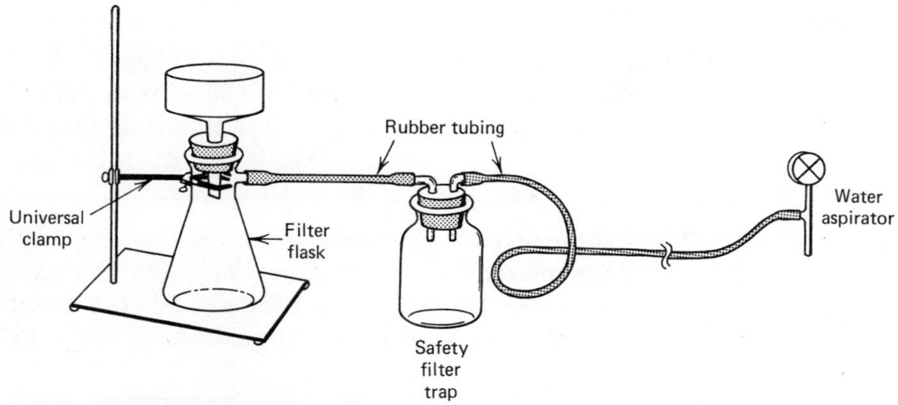

Figure 1.7 *Setup for suction filtration.*

To confirm that the aspirator and your apparatus are functioning properly, turn on the water for maximum water flow and place your thumb over the open tube that is perpendicular to the water drain. If you detect suction at that point, connect your apparatus to the aspirator and place your thumb over the end of the rubber hose that leads into your filter flask. If there is suction at this position, suction filtration will be possible.

the connecting rubber tubing must be thick enough so that no section of the tubing will collapse under the reduced pressure. The filter flask and filter trap are firmly supported by clamps to a ringstand. After filtration is complete remove the rubber tubing from the water aspirator before turning off the water to avoid a pressure change in the filtration assembly; such a pressure change causes water backup into the system.

Prelab Questions

1. The experimental procedure specifies that you use 30 ml of a stock toluene-benzoic acid mixture. What is the maximum weight of benzoic acid that could be isolated from a saturated solution of benzoic acid in 30 ml of toluene?

2. Calculate the number of moles of sodium hydroxide in 1.0 l of 10% aqueous sodium hydroxide (weight NaOH/volume H_2O). What minimum volume of 10% aqueous sodium hydroxide is required to convert the benzoic acid in 30 ml of a saturated toluene solution (Question 1) into its sodium salt?

3. What volume of 10% aqueous hydrochloric acid (volume concentrated HCl/total volume) is required to neutralize 30 ml of 10% aqueous sodium hydroxide? (The molar concentration of concentrated hydrochloric acid is 12*M*.)

4. What volume Erlenmeyer and suction filter flasks are suitable for this experiment? Could you use a 50-ml filter flask?

5. What effect does the porosity of a filter paper have on the efficiency and speed of filtration?

EXPERIMENTAL PROCEDURE

Certify that your separatory funnel does not leak either at the bottom stopcock or at the top stopper by filling your funnel half-full with water and then following the procedure that is described in Figure 1.3.

CAUTION: Toluene is a flammable hydrocarbon; be sure that there are no flames in your laboratory area before you begin this experiment. Although toluene is less toxic than benzene (Table 1.2), the breathing of its vapors should be avoided. If possible, work with toluene in the fume hood.

Pour 30 ml of the stock toluene-benzoic acid mixture into a graduated cylinder. Record the exact volume of the solution (±0.5 ml) and then transfer this solution to a separatory funnel. Add 15 ml of 10% aqueous sodium hydroxide to the separatory funnel; you will see two distinct liquid phases. (Which layer is the aqueous phase?) Follow the extraction procedure that is described in Figure 1.3 and draw off the lower liquid layer into an Erlenmeyer flask. A small fraction of the lower layer (<0.5 ml) may be allowed to remain in the separatory funnel; you should be careful to exclude any part of the upper layer from entering the Erlenmeyer flask (why?). Extract the organic phase with a second 15-ml portion of 10% aqueous sodium hydroxide and draw off the aqueous phase into another Erlenmeyer flask. Dispose of the organic phase in the container provided for waste toluene. Normally the aqueous phases from the two extractions are combined and further treated as one solution. In this experiment, however, keep the aqueous phases separate and label the Erlenmeyer flasks that contain these phases according to the order of extraction. Add 35 ml of 10% aqueous hydrochloric acid to each of the 15-ml aqueous phases in order to acidify these solutions. Confirm that the *p*H of the resulting solution is less than 2 using *p*H indicator paper; add additional 10% aqueous hydrochloric acid if necessary. You should observe a white precipitate in at least one of the Erlenmeyer flasks. Construct your filtration apparatus for suction filtration according to Figure 1.7. Turn on the water aspirator for maximum water flow and confirm that there is suction into the filter flask. Wet the filter paper with a small amount of water so that the filter paper will adhere to the Büchner funnel while you collect benzoic acid. Now filter the solid precipitate that is in the Erlenmeyer flask from your first extraction: swirl the contents of the Erlenmeyer flask so that the solid is evenly distributed in the liquid and then slowly pour the mixture onto the center of the filter paper. If some of the solid adheres to the Erlenmeyer flask, transfer a small portion of the *filtrate* (the filtered liquid solution) back into the Erlenmeyer flask, swirl the mixture, and filter. Allow the collected solid to air dry in the Büchner funnel for 10 minutes and then transfer it to a clean, dry, and previously weighed Erlenmeyer flask. Weigh the Erlenmeyer flask and calculate the weight of the solid; weigh the flask again at the start of your next laboratory period. If a precipitate was obtained in the Erlenmeyer flask from your second extraction, filter and weigh this solid separately. Determine the weight of the dry solid that you obtained from each extraction, and from the combined weight of solids calculate the number of grams of benzoic acid per milliliter of toluene that you recovered.

Postlab Questions

1. Excluding spillage and incomplete transfer of the solid from the filter paper to the flask, describe the operations in the extraction procedure that could account for a reduced recovery of benzoic acid.

2. Why is the filtrate and not fresh water used to wash adhering crystals of benzoic acid from the Erlenmeyer flask to the Büchner funnel?

3. The amount of benzoic acid that you obtained from your second extraction indicates how well you operated the extraction procedure. Suggest the causes for the extraction of benzoic acid in the second washing with aqueous base and describe how you would minimize the amount of benzoic acid in this extraction.

4. If the solid that you filter from the aqueous acid solution contains water, what will be the effect of the water contaminant on your calculation of g C_6H_5COOH/ml $C_6H_5CH_3$? How will you know when your solid precipitate does not contain water?

5. Using a flow diagram describe how you would isolate sodium benzoate (as benzoic acid) from an aqueous solution that contains sodium benzoate, sodium chloride, and undissolved solid particles.

6. The solubility of an organic material in water is 8.0 g/100 ml H_2O and its solubility in ethyl ether is 40 g/100 ml ether. If 8.0 g of this material is dissolved in 100 ml of water, which extraction procedure would remove more of the material from water: one extraction with 100 ml of ether or two extractions each with 50 ml of ether? Calculate the weight of the organic material that is extracted into 100 ml of ether for each procedure.

Experiment Two

Recrystallization and Melting Point Ranges

In Experiment 1 you obtained a solid material by filtering the precipitate that crystallized from water. Although certain physical characteristics of this solid are visually evident, you do not know yet if the solid is a single compound or a mixture of compounds that have similar physical properties. Similarly, you have yet to obtain evidence that the isolated compound is actually benzoic acid. The purpose of this experiment is to purify the solid that you obtained in Experiment 1 and to identify the structure of the purified material by a physical comparison with an authentic sample of benzoic acid.

After a material is isolated from its source, the desired component of the isolated material is separated from undesirable compounds by procedures that are refinements of the separation methods outlined in Experiment 1. The ultimate goal of a purification method is to refine a mixture so that only the desired compound is obtained. In this experiment the method of recrystallization will be employed to purify the solid you obtained in Experiment 1.

Recrystallization

The most widely used method for the purification of a solid is *recrystallization: the selective crystallization of one compound from a mixture of solid compounds.* The recrystallization method operates on the principle that the undesired compounds in the mixture will have solubility properties that differ from those of the desired compound in a selected solvent. Some of the undesired contaminants may be insoluble in the solvent after the desired compound has dissolved; others may remain dissolved in the solvent after the desired compound has crystallized. The following example illustrates this principle.

Benzoic acid and phthalic acid are both partially soluble in water. Their solubility increases as the temperature of water increases and is at a maximum when the solvent is heated at its boiling point. The solubility characteristics of benzoic acid and phthalic acid in water are described in Figure 2.1. Now suppose that you are given 6.0 g of a mixture that contains 90% benzoic acid, 10%

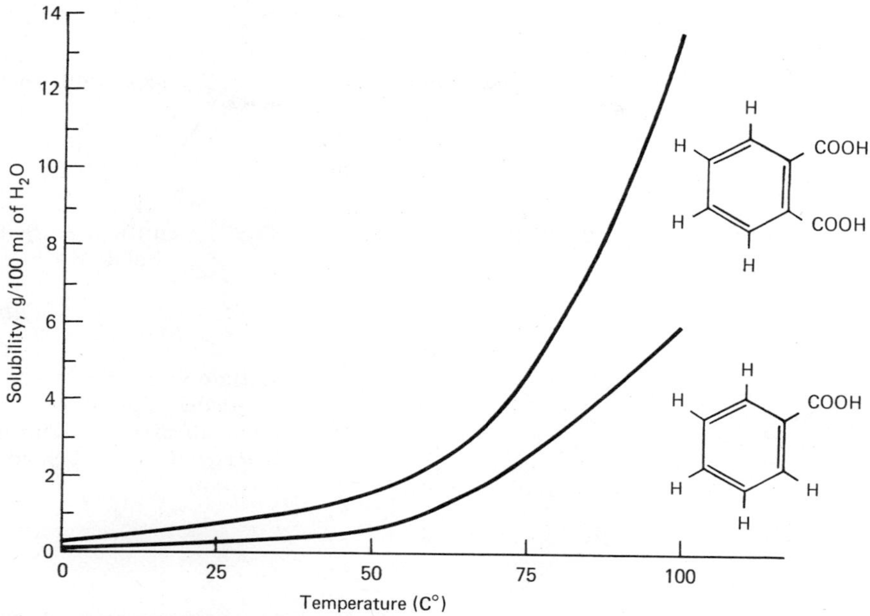

Figure 2.1 *Solubility characteristics of benzoic acid and phthalic acid in water at temperatures between the boiling point and freezing point of water.*

phthalic acid and asked to isolate pure benzoic acid. The minimum amount of water that can be used to dissolve all of the mixture is 100 ml; thus 6.0 g of the solid mixture will dissolve in 100 ml of water at the boiling point of water, 100°C. *A minimum amount of solvent is used in recrystallization procedures so that the greatest amount of solid can be recovered upon cooling the heated solution.*

Upon cooling the solution of benzoic acid and phthalic acid in water from 100°C to 25°C, benzoic acid crystallizes in an amount that is equal to the quantity that was dissolved at 100°C (5.4 g) minus the quantity that is soluble in 100 ml of water at 25°C (0.3 g):

$$(\text{Benzoic acid})_{\text{Crystalline}} = 5.4\text{ g} - 0.3\text{ g} = 5.1\text{ g}$$

Likewise, phthalic acid will crystallize in an amount equal to the quantity dissolved at 100°C (0.60 g) minus the quantity soluble in 100 ml of water at 25°C (0.70 g). In this case all of the phthalic acid contaminant is retained in the water solvent and pure benzoic acid can be filtered from the aqueous solution.

Recrystallization has three operations: (1) the dissolution of the solid mixture in a minimal volume of a solvent, usually at the boiling point of the solvent, (2) selective crystallization of the desired compound by cooling the solution, and (3) filtration of the purified solid from the liquid solution. Filtration has been discussed in Experiment 1. Let us now turn our attention to the individual operations of dissolution and crystallization.

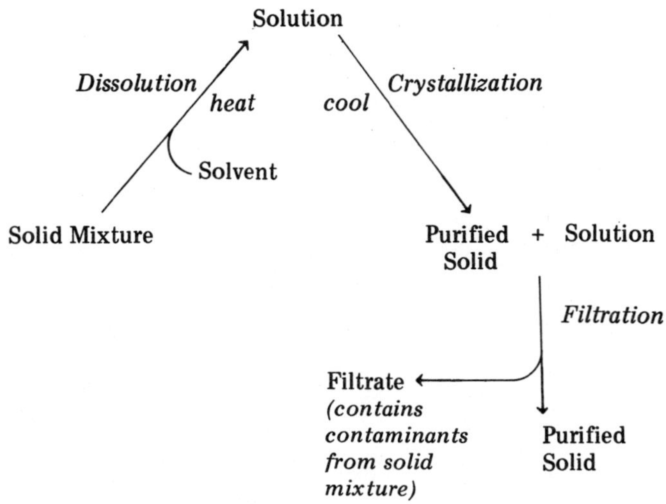

Dissolving Solids

When a solid compound is dissolved in the minimum amount of a solvent, that solution is said to be *saturated* at the temperature of dissolution. No more of the solid compound will dissolve in the saturated solution and cooling will cause precipitation of the solid compound. Dissolution of a solid in a solvent is optimum at equilibrium:

$$(\text{Compound})_{Solid} + n\ \text{Solvent} \rightleftharpoons (\text{Compound})_{Soln}$$

As with extractions, thorough mixing of the two phases is required to attain the equilibrium condition.

Energy is required to dissolve solid compounds. The molecules of a crystal are held together by intermolecular attractive forces that require a specific packing of the individual crystalline units. To break off a structural unit from the crystal requires an amount of energy that is equal to the attractive energy between the molecular crystalline unit and the crystal (*lattice energy*). This is why many solutions cool as solids dissolve in a solvent; when external heat is not applied, heat energy from the solvent is expended in breaking the lattice energy of the solid.

As a practical matter, a major difficulty encountered in recrystallizations is the use of too much of the recrystallization solvent. This problem results from either not allowing sufficient time for the solid to dissolve or from attempting to completely dissolve a solid mixture that includes a component insoluble in the recrystallization solvent. The use of too much solvent to dissolve a solid decreases the amount of material that can be recovered in the crystallization step.

Crystallization

The process of crystallization reverses the operation of dissolution. Like molecules of a dissolved solid (the *solute*) associate when the attraction between these molecules is greater than the attraction between solute molecules and solvent molecules. As solute molecules aggregate, they tighten their attraction by expelling solvent molecules associated with solute molecules (Figure 2.2). A lattice of oriented solute molecules is constructed and grows into a crystalline

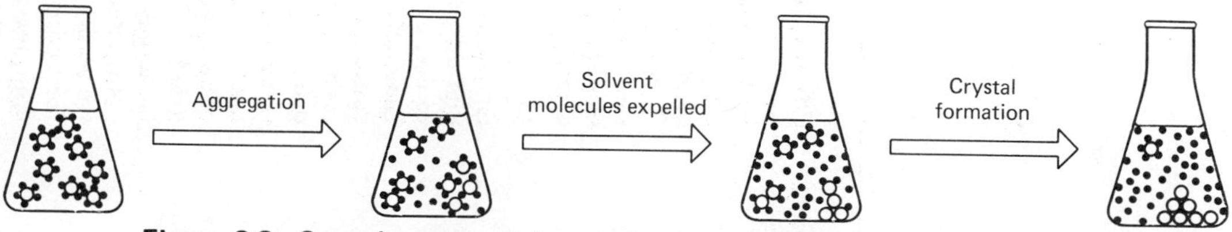

Figure 2.2 *Crystal growth. Solute molecules = ○; solvent molecules associated with solute molecules = ●.*

form that spreads through the solution. Energy is released in the formation of the crystalline lattice.

Molecules are oriented in a crystalline lattice to maximize intermolecular attractive forces. Molecules that are structurally identical and have the same volume can attain the closest fit and the greatest attraction. Unlike molecules disrupt the crystalline lattice and thereby decrease the lattice energy. However, molecules of similar but not identical structure and volume combine during crystallization with a minimum disruption of the lattice; compounds represented by such molecules also have similar solubility and physical properties.

The crystallization of a pure compound is optimum when the system is at equilibrium. Unlike the dissolution process, however, the crystallization process is most effective when the crystalline medium is minimally disturbed. Swirling of a medium in which crystals are beginning to form physically shears the crystals into smaller aggregates. Rapid cooling forces the solution into a condition of supersaturation and leads to smaller and less pure crystalline forms.

Crystals of solid organic compounds form into numerous shapes that are dependent on the structure of the compound and on the procedure employed for crystallization. The most common of these crystalline forms are described in Figure 2.3. Whenever you obtain a crystalline material you should describe the form of the crystals according to this selection.

Recrystallization Solvents

Solvents that are most commonly used for recrystallization are chosen from liquids that are relatively inexpensive and unreactive, that will dissolve organic solids, and that can be conveniently vaporized. Recrystallization solvents that fulfill these selection criteria are listed in Table 2.1 along with their physical properties. The choice of a solvent for recrystallization generally is made through consideration of the following requirements:

Table 2.1 Physical Properties of Common Recrystallization Solvents

SOLVENT POLARITY

Solvent[a]	Structural formula	Boiling Point (C°)	Freezing Point (C°)	Flammability[b]	Comments
Water	H_2O	100	0	I	Particularly useful for recrystallization of salts and very polar solids. Crystallized solids dry slowly.
Methanol	CH_3OH	65	-198	F	Generally useful for polar compounds. Completely miscible with water.
Ethanol	CH_3CH_2OH	78	-117	F	Generally useful for polar compounds. Completely miscible with water.
Acetone	$CH_3\overset{\overset{O}{\parallel}}{C}CH_3$	56	-95	F	Generally useful for polar compounds. Should be dried (water removed) prior to to use. Completely miscible with water.
Methylene chloride[c]	CH_2Cl_2	40	-97	I	Generally useful for polar and nonpolar compounds; particularly good for low-melting compounds. Crystals are easily dried.
Ethyl ether	$(CH_3CH_2)_2O$	35	-116	F	Generally useful for polar and nonpolar compounds; particularly good for low-melting compounds. Crystals are easily dried.
Chloroform[c]	$CHCl_3$	61	-64	I	Generally useful for polar and nonpolar compounds.
Benzene[d]	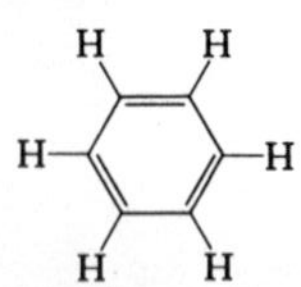	80	6	F	Generally useful for aromatic compounds, for hydrocarbons, and for molecular complexes.
Carbon tetrachloride[c]	CCl_4	76	-23	I	Generally useful for nonpolar compounds.

Ligroin	Mixture of C_7 and C_8 hydrocarbons	90-115		F	Generally useful for nonpolar compounds.
Hexane	$CH_3(CH_2)_4CH_3$	68	-195	F	Generally useful for nonpolar compounds.
Petroleum ether	Mixture of C_5 and C_6 hydrocarbons	35-60		F	Generally useful for nonpolar compounds; particularly good for low-melting compounds. Crystals are easily dried.
Pentane	$CH_3(CH_2)_3CH_3$	36	-130	F	Generally useful for nonpolar compounds; particularly good for low-melting compounds. Crystals are easily dried.

[a] Solvents are listed in approximate order of polarity.
[b] F = flammable, I = not flammable.
[c] Skin contact with chlorocarbon solvents and the breathing of their vapors is hazardous to your health. Carbon tetrachloride and chloroform have been linked to cancer in test animals and have been banned by the FDA from use in drug, cosmetic, and food packaging products.
[d] Breathing benzene vapors and absorption of benzene through the skin must be avoided. Inhalation of benzene causes irritation of mucous membranes. Benzene has been linked to cancer in test animals. Toluene is often substituted for benzene for use as a recrystallization solvent.

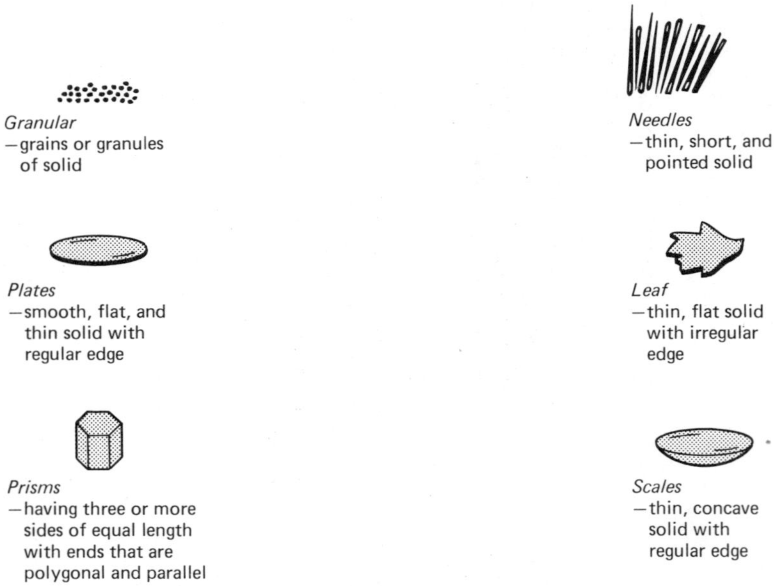

Figure 2.3 *Common crystalline forms of organic solids.*

1. *The solvent should not react with the solid that is to be recrystallized.* Although most organic compounds are unreactive with the solvents listed in Table 2.1, water should be separately considered because of its reactivity and its ability to combine with polar and ionic compounds as "hydrates." Compounds that decompose in water will also react with methanol and ethanol.
2. *The solid to be recrystallized should be only partially soluble in the recrystallization solvent and relatively insoluble in that solvent at the temperature used for filtration.* If the solubility of the solid is appreciable at the low temperature (>10 g/100 ml solvent), less recrystallized solid is recovered from a given amount of solvent and less solvent is available to dissolve the undesirable impurities.
3. *The solid should be reasonably soluble in the recrystallization solvent at the boiling point of the solvent.* Generally, if a solid is less soluble than 5 g/100 ml at the boiling point of the solvent, the recrystallization process becomes uneconomical and inefficient. Physical losses of solid often occur in the filtration step.
4. *The boiling point of the recrystallization solvent should not exceed the melting point of the solid.* If the solid melts before it dissolves in the solvent, the melted solid often separates from the solvent as an oil during the crystallization process. Crystallization then occurs in the separated oil rather than in the solvent solution. Minimal purification occurs because impurities that are more soluble in the oil than in the solvent are included in the crystalline lattice.

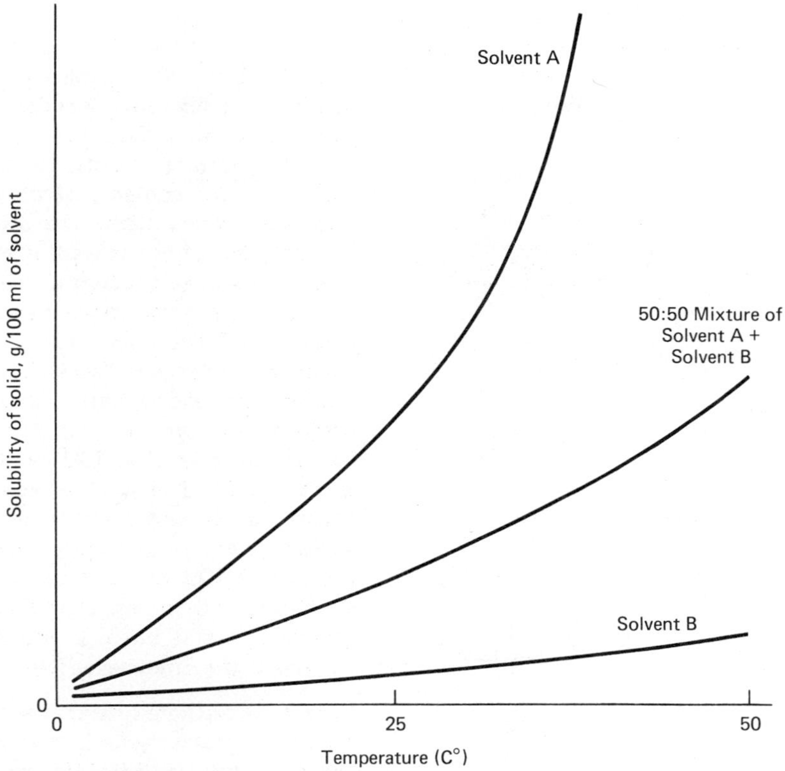

Figure 2.4 *Use of solvent pairs for recrystallization.*

The solid to be recrystallized is very soluble in Solvent A but is practically insoluble in Solvent B. However, an equal volume mixture of the two solvents forms a solvent system from which the solid can be effectively recrystallized.

Solubility data for the compounds you isolate or prepare are not usually available in the literature. If a recrystallization solvent is not specified, you should determine solubilities of the solid you wish to purify in a representative series of solvents that have different solubility characteristics: pentane, ethyl ether, ethanol, and water (see Figure 1.1).

The solvents that are listed in Table 2.1 do not always meet the previously described requirements. Often a solid is found to be too soluble in one solvent and too insoluble in another recrystallization solvent. In such cases compatible *solvent pairs* are used—the solid is recrystallized from a miscible mixture of two solvents: in one solvent the solid readily dissolves and in the other solvent the solid is relatively insoluble. A wide range of solubility characteristics exists for solvent pairs between the solubility extremes for the individual solvents (Figure 2.4).

The two solvents of a solvent pair are generally chosen from the list in Table 2.1. They are completely miscible in each other and usually have different characteristics of polarity and structure. Often the two solvents are chosen to have similar boiling points so that they will distill together and remain at

Figure 2.5 *General procedure for recrystallization*

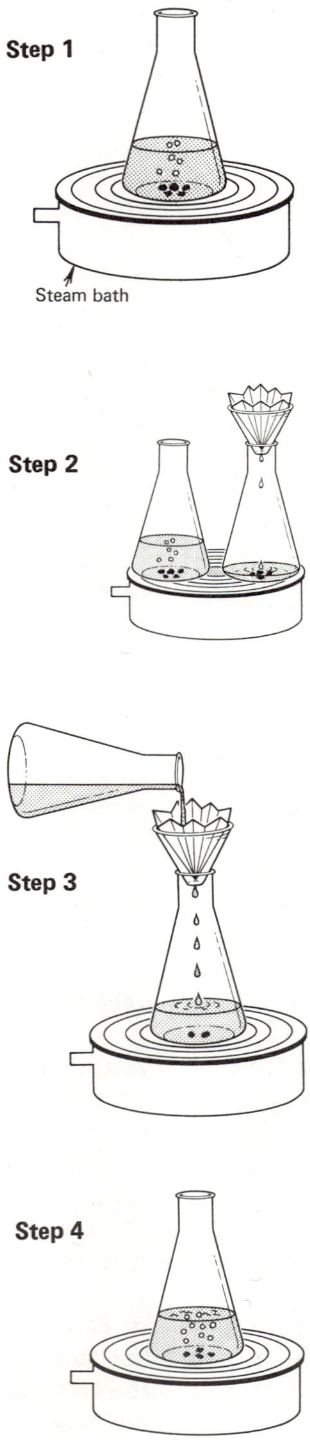

Step 1. Dissolve the solid in the minimum amount of solvent by adding small portions of the solvent to the solid in an Erlenmeyer flask. Crush large particles of solid prior to adding solvent in order to increase the rate for dissolution. Swirl the contents of the flask for thorough mixing and heat (steam bath) the solvent-solid mixture to the boiling point of the solvent after each addition of solvent. Add 2 to 3 boiling stones to the Erlenmeyer flask in order to maintain constant boiling and to avoid superheating and bumping of the liquid solution. Observe the extent to which the solid has dissolved after each solvent addition and estimate the minimum amount of solvent that should dissolve all of the solid. (If 50 ml of solvent dissolves one-half of the solid, then 100 ml of solvent should dissolve all of the solid.) After you have added the estimated minimum volume of solvent, add an additional small portion of solvent to ensure that no more solid dissolves (usually not more than 10% of the total volume of solvent). The undissolved solid is separated from the heated solution by gravity filtration (Steps 2 and 3). If no undissolved solid remains in the Erlenmeyer flask, proceed to Step 4.

Step 2. A separate Erlenmeyer flask containing a sufficient amount of the recrystallization solvent to fill the bottom of the flask and 2 or 3 boiling stones is fitted with a stemless funnel and a fluted filter paper for filtration of undissolved solid. This flask is placed over a steam bath and the solvent is heated to boiling so that its vapors warm the flask, funnel, and filter paper. Heating the filtration apparatus with the boiling recrystallization solvent prevents crystallization during filtration that would occur if the recrystallization solvent is cooled.

Step 3. Using a towel or clamp to grasp the flask, remove the flask containing the dissolved solid from the steam bath, swirl the contents of the flask to ensure mixing, and carefully pour the hot solution onto the solvent-warmed filter paper. Do not allow the contents of either flask to cool. If crystallization occurs on the filter paper or in the flask from Step 1, heat a small volume of recrystallization solvent in the Erlenmeyer flask and pour this solution onto the filter paper. After the solvent has passed through the filter paper remove the stemless funnel from the Erlenmeyer flask.

Step 4. In Steps 1 and 3 solvent was added to prevent saturation. This excess of solvent is now removed by evaporation through boiling the recrystallization solution on a steam bath. If Steps 1 and 3 were performed properly, volume reduction by only 10% is required to reach saturation.

Step 5

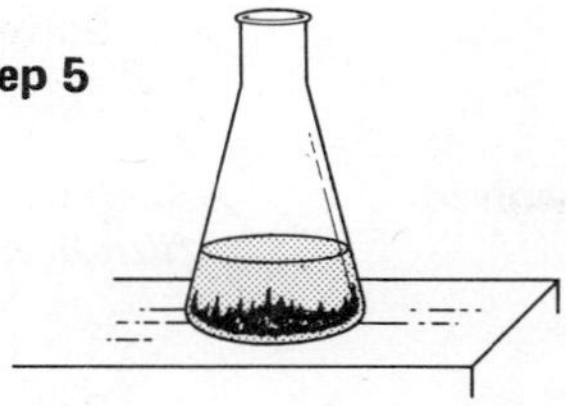

Step 5. Remove the Erlenmeyer flask from the steam bath and allow the recrystallization solution to cool to room temperature. After crystal growth is no longer visible at room temperature, the flask is often cooled in an ice bath so that additional crystallization will occur.

Step 6

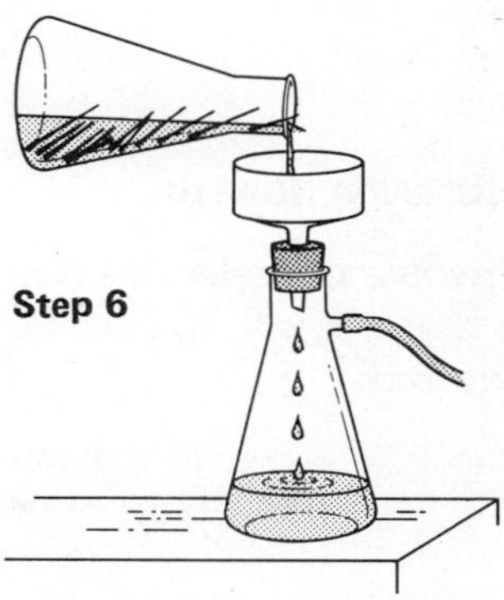

Step 6. The crystallized solid is now filtered through a Büchner or Hirsh funnel under reduced pressure (see Figure 1.7). A glass rod is used to break up the crystalline mass and to mix the contents of the Erlenmeyer flask so that the mixture can be poured onto the funnel. A small volume of cold solvent is first poured through the filter paper to ensure that the filter paper fits tightly against the holes of the funnel. The contents of the Erlenmeyer flask are swirled and then rapidly poured onto the center of the filter paper. A small amount of cold solvent is poured into the Erlenmeyer flask to aid in the transfer of solid that adheres to the walls of the flask and to wash the solid in the filter flask. This washing prevents impurities present in the solvent that adhere to the filtered solid from depositing on the surface of the purified crystals. Suction is continued until the solvent has been drawn through the solid and the crystals are dry.

The crystalline solid that is obtained in Step 6 is referred to as the *first crop* of crystals. If solvent is evaporated from the filtrate and Steps 4–6 are repeated, a *second crop* of crystals can also be obtained. A third crop of crystals is usually insignificant in amount when compared to the weighed amount of the first two crops of crystals.

Figure 2.5 (*continued*)

the same composition in the recrystallization solution. The most commonly used solvent pairs are ether-pentane, benzene-hexane, and ethanol-water.

Recrystallization Procedures

The procedure you employ to recrystallize an impure solid depends on whether the impurity is more soluble or less soluble in the recrystallization solvent than is the compound to be isolated, and on whether you employ one solvent or a solvent pair for recrystallization. Each of these procedures will be separately described within the general scheme for recrystallizations (Figure 2.5).

Impurity is Less Soluble in the Recrystallization Solvent

The procedure that you employ should involve the selective removal of the impurity by filtration prior to crystallization. In this case Steps 1 through 6 of the recrystallization procedure (Figure 2.5) must be followed in order to obtain a purified crystalline solid.

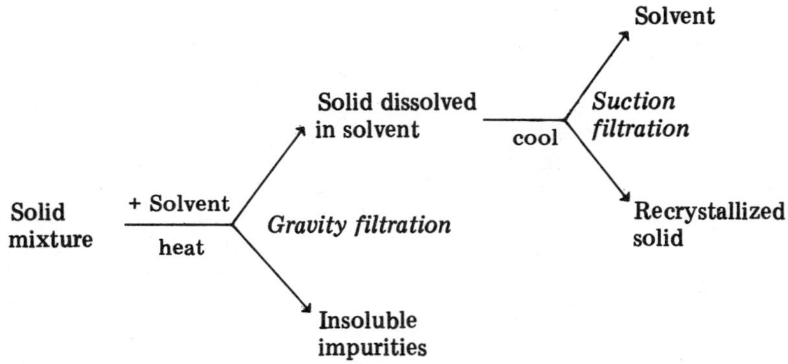

Impurity is More Soluble in the Recrystallization Solvent

The procedure that you employ should involve the selective removal of the impurity in the crystallization step. In this case Steps 2, 3, and 4 of the recrystallization procedure (Figure 2.5) are not required.

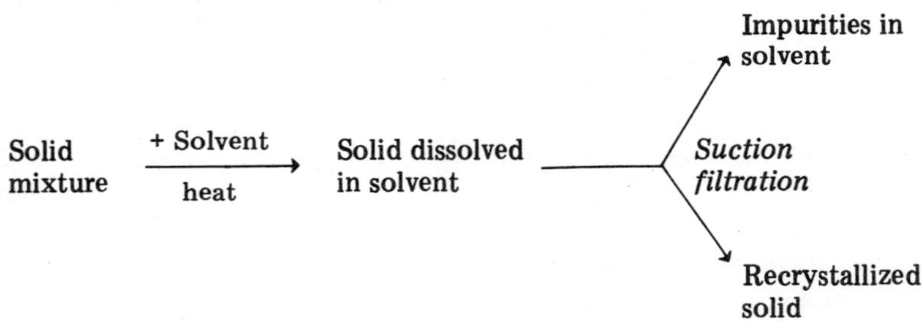

Use of Solvent Pairs

Step 1 of the recrystallization procedure (Figure 2.5) is modified when solvent pairs are employed. The solid is dissolved at room temperature in a minimum amount of the solvent in which the solid is more soluble. The resulting solution is then heated to its boiling point and the cosolvent is slowly added with continued heating until the solution becomes slightly cloudy. The dissolved compound is saturated in the solvent pair at the point cloudiness in the solution just appears. No more than an equal volume of cosolvent should be added to the first solvent to ensure effective recrystallization from the solvent pair. A small amount of the first solvent can be added to clarify the solution. Steps 5 and 6 are followed to complete the recrystallization process. In Step 6 the solvent in which the solid is less soluble is used to wash the filtered crystals.

Melting Points

The melting point of a pure compound is the temperature at which the liquid and solid phases of that compound are in equilibrium at 1 atm pressure. When

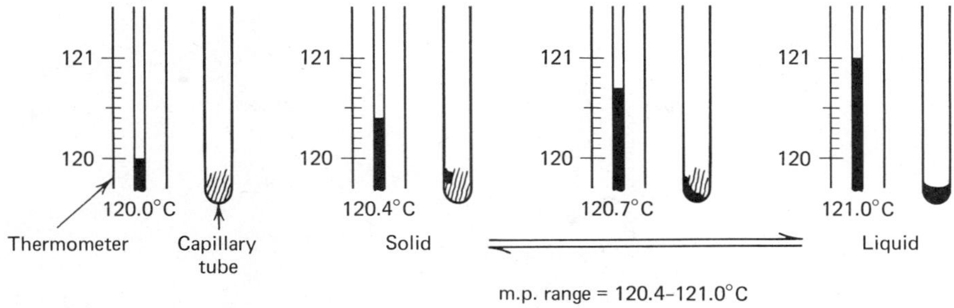

Figure 2.6 *Observing the melting point range.*

the thermal energy that is exerted on a pure solid is equal to the lattice energy

$$\text{Solid compound} \underset{}{\overset{\text{Temp = m.p.}}{\rightleftharpoons}} \text{Liquid compound}$$

that binds together the molecular units of the crystal, molecules of the crystalline lattice escape from their highly ordered environment. The temperature required for the transition from the ordered molecular arrangement in the crystalline lattice (solid) to the disordered condition of the fluid (liquid) is an invariant physical characteristic of a pure solid.

The melting point (or *freezing point* if you are considering the temperature at which the solid begins to form from the liquid) is a measure of intermolecular attractive forces between molecules; the higher the melting point, the greater the attraction. Among compounds having the same molecular weight, the more polar the compound and the more symmetrical its molecular structure, the higher the melting point. Thus the melting point of a compound provides information about one physical dimension of molecular structure.

The melting point of a pure compound is determined by observing the temperature at which the transition solid ⇌ liquid occurs. *A representative small amount of solid is placed at the sealed end of a thin-walled glass capillary tube and heat is uniformly applied to the capillary tube.* The temperature at which liquid is first observed and the temperature at which solid is no longer visible represent the temperature extremes (*melting point range*) for the solid-liquid transition (Figure 2.6). The melting point reported in the literature is usually taken as the upper temperature extreme of the melting point range.

A representative sample of the solid is used to obtain the melting point. After filtering and drying, the solid is thoroughly mixed and the open end of the capillary tube is pressed into the solid until solid enters the tube. The capillary tube is gently tapped on a hard surface until the solid is packed onto the sealed end. The solid should extend no higher than 5 mm from the sealed end of the capillary tube.

Apparatus commonly used to obtain melting points are described in Figure 2.7. Each of these systems is designed to uniformly heat the melting point sample and to provide an adequate window to view the sample. The Thiele tube is an oil bath that requires external heating with a Bunsen burner; the tube is constructed with a side-arm that provides circulation of the heated oil for uni-

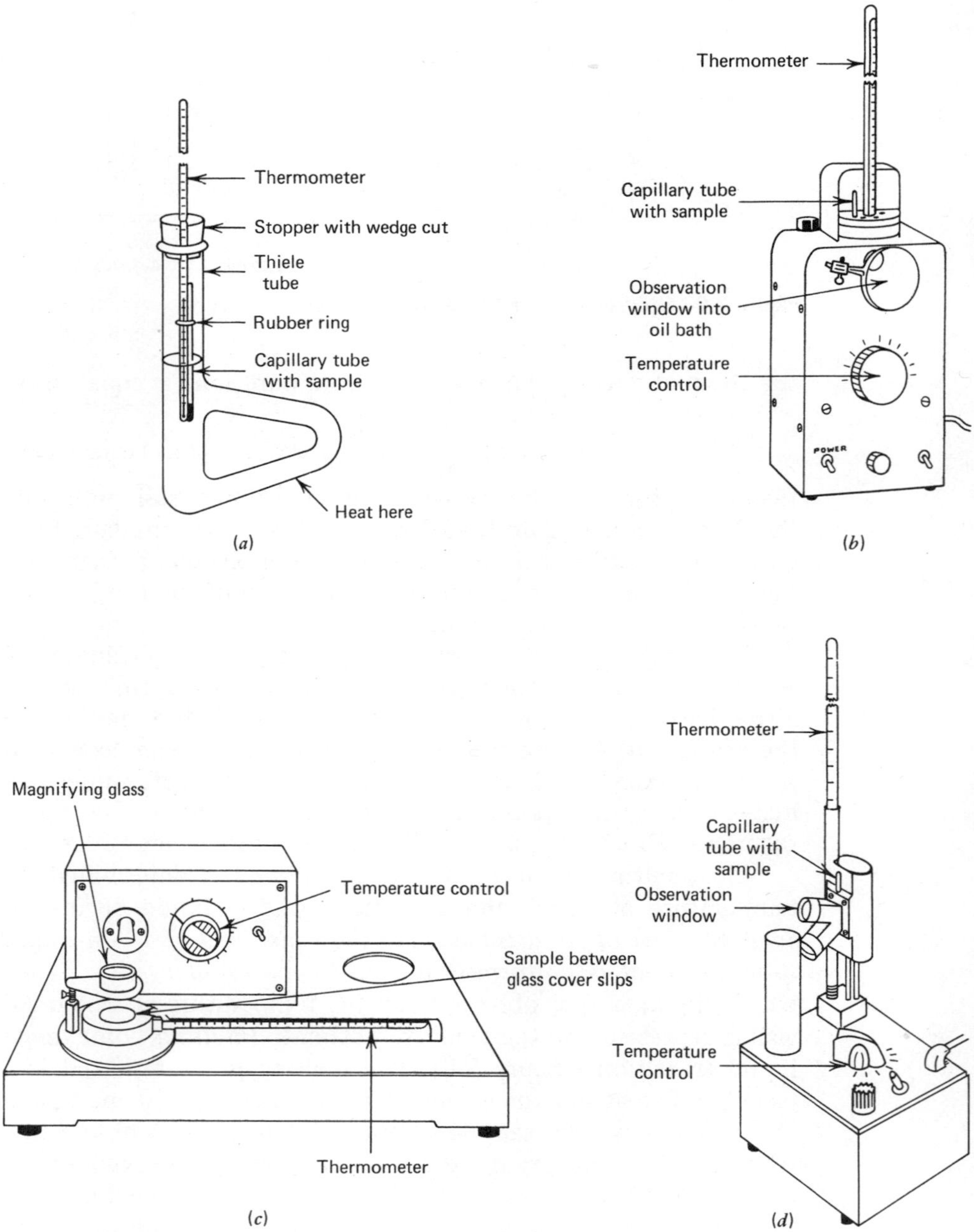

Figure 2.7 *Commonly used melting point apparatus.*

(*a*) Thiele tube; temp. range: 25–180°C (mineral oil bath). (*b*) Thomas-Hoover apparatus; temp. range: 25–300°C (silicone fluid bath). (*c*) Fisher-Johns apparatus; temp. range 25–300°C (heating block). (*d*) Mel-Temp; temp. range: 25–400°C (heating block).

form temperature changes. If mineral oil is employed as the bath oil, the temperature of the oil bath should not exceed 180°C. A silicon fluid can be used as the bath oil if temperatures up to 300°C are required.

The Thomas-Hoover melting point apparatus uses an electrically heated oil bath and is constructed with a window through which the sample in a capillary tube can be viewed; the mercury level of the thermometer is observed through a periscope for simultaneous viewing of the temperature and melting crystals. The Fisher-Johns and Mel-Temp melting point apparatus have electrically heated hot plates (the heating block) that uniformly heat the sample in a capillary tube (Mel-Temp apparatus) or between microscope slides (Fisher-Johns apparatus). A variable voltage transformer is used in these electrically heated devices to change the rate of heating; a low voltage provides a slow rate of heating. The electrical current to these devices must always be turned off after obtaining a melting point.

The melting point and melting point range for a solid depends on the rate of heating and the accuracy of your thermometer as well as on the nature of the sample. The rate of heating should be controlled so that the rate of temperature increase within 5° of the melting point is approximately 2°C per minute; a faster rate of heating will not allow the sample within the capillary tube to attain thermal equilibrium with the heating surface. Your thermometer should be corrected for accuracy prior to use by measuring the temperature and temperature range at which known pure samples melt (see Experimental Procedure). When these technical procedures are followed the melting point that you measure will correspond to the melting point of the sample.

Melting Point Range as a Criterion for Purity

Samples of a pure compound usually contain only one crystalline form and therefore melt within a narrow range of temperatures, usually less than 1°C. A

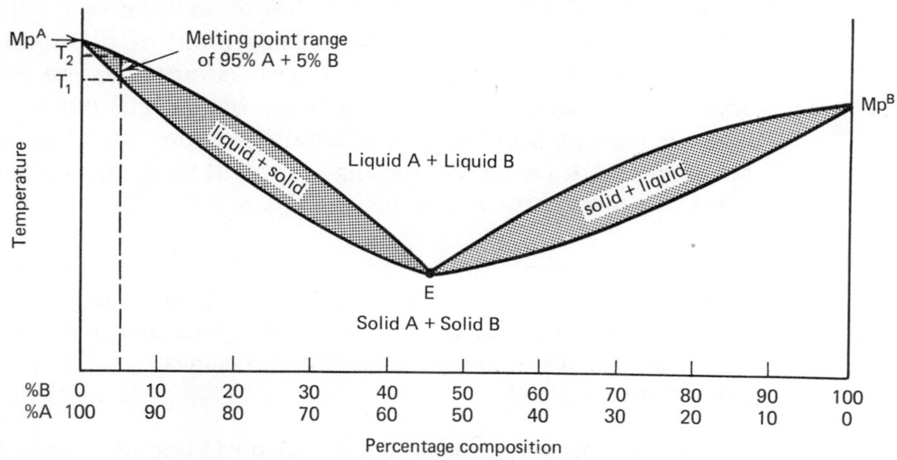

Figure 2.8 *Typical liquid-solid phase diagram for a two-component solid mixture.*

broad melting point range of greater than 2°C usually indicates the presence of an impurity in the sample. A solid mixture exhibits a very different melting point behavior than would be expected of its pure components. *Impurities generally cause a depression of the melting point and a broadening of the melting point range.*

The liquid-solid *phase diagram* (plot of temperature versus composition) that is shown in Figure 2.8 describes the phase behavior that is typical of a two-component solid mixture. Pure A and Pure B have sharp melting points that do not change upon repeated recrystallizations. On the other hand, a mixture of 95% A + 5% B has a broad melting point range and a melting point that is lower than that of pure A; this solid mixture begins to melt at T_1 and between T_1 and T_2 the solid mixture is in equilibrium with the liquid. Recrystallization of this 95% A/5% B mixture that changes the percentage composition of B in A also changes the melting point and melting point range. Only at the composition represented by point E (*eutectic composition*) in Figure 2.8 will the melting point of a solid mixture be sharp; however, changing the percentage composition of the eutectic mixture by recrystallization will also change the melting point behavior.

The melting behavior of a sample can be used as a criterion for purity if *a pure compound is recognized as having a sharp melting point that does not change upon repeated recrystallizations.*

Mixed Melting Points in Structure Identification

A pure compound whose structure is unknown can be identified by comparison of its melting point with that of known compounds. The melting point of an unknown compound narrows its structural identification to only those compounds that have that physical property. Although many known compounds have a melting point identical with that of an unknown, only one of the known compounds will not depress the melting point of the unknown when the two compounds are mixed together. *If two compounds are identical, the melting point of the mixture of the two compounds will not be depressed from the separate melting points of the individual compounds.* If the two compounds are not identical, their mixed melting point will be depressed and the melting point range of the mixture will be broad (Figure 2.8).

Prelab Questions

1. Calculate the amounts of benzoic acid and phthalic acid obtained in a second crop of crystals if the volume of filtrate from the 90% benzoic acid and 10% phthalic acid recrystallization example is reduced from 100 ml (containing 0.3 g benzoic acid and 0.6 g of phthalic acid) to 10 ml.

2. Comment on the factor or factors that would lead to a sample having a melting point that is lower after recrystallization than before recrystallization.

3. Explain why "wet" solids—those that contain solvent—generally have broad melting point ranges and low melting points.

4. Explain in terms of intermolecular attractions how crystal growth of one compound from a solution that contains that compound and impurities can be a selective process. What effect on crystal growth and crystal purity would be expected if the process of crystallization is physically disturbed?

5. Describe the flow diagram for the recrystallization procedure that should be used if the solid mixture contains (a) impurities that are less soluble and (b) impurities that are more soluble in the recrystallization solvent.

6. Describe a flow diagram for the recrystallization procedure that employs a solvent pair.

EXPERIMENTAL PROCEDURE

To correct your thermometer for accuracy obtain the melting points for selected pure compounds with the thermometer that you will use throughout this laboratory. A selection of compounds that spans a wide range of melting points includes *p*-dichlorobenzene (m.p. = 53°C), benzoic acid (m.p. = 122°C), sulfanilamide (m.p. = 166°C), and *p*-nitrobenzoic acid (m.p. = 242°C). Plot the expected melting point verses the difference between the expected melting point and the melting point that you observe. Use this graph to correct the melting points that you take in this and subsequent experiments.

Obtain the dry weight of the solid compound that you isolated in Experiment 1 and take its melting point. Transfer this solid to a clean 125-ml Erlenmeyer flask and add 15 ml of water per gram of solid. After adding 2 or 3 boiling stones, place the Erlenmeyer flask on a wire screen that is supported on a ring stand and heat the water mixture to boiling using a Bunsen burner. Add water in 5-ml portions to the boiling mixture until the solid is dissolved. Complete the recrystallization process according to Steps 5 and 6 of Figure 2.5. Record the color and crystalline form of the filtered solid. Determine the weight of the dry solid after removal of the boiling stones and calculate the weight percentage of recovered solid (100 × wt. recrystallized solid/wt. initial solid). If the initial solid contains only a small amount of impurities, the weight percentage of recovered solid is related to the effectiveness of the recrystallization procedure for the isolation of purified product.

Obtain a melting point for the dry recrystallized solid and take a mixed melting point of this compound with pure benzoic acid. For a mixed melting point equal amounts of the two solids are thoroughly ground together. Use a fire-polished glass rod and a small test tube to crush and grind the two solids together.

Postlab Questions

1. Describe the operations in the recrystallization procedure that could account for a low weight percentage of recovered solid. Describe simple experiments that would test your hypotheses regarding losses of solid.

2. Describe the operations by which you could increase the weight percentage of recovered solid.

3. Describe how your experimental data suggests that the solid isolated in Experiment 1 is benzoic acid.

4. Suggest a simple test that would show that the isolated solid is an acid. Can you suggest a simple test that would show that the isolated compound contains carbon and hydrogen?

Experiment Three

Separation Techniques: Isolation of Trimyristin from Nutmeg

Solid-Liquid Extraction

In Experiment 1 you were introduced to the extraction technique for the separation of a substance in a liquid phase mixture by the transfer of this substance to an immiscible solvent: *liquid-liquid extraction.* In this experiment a compound in a solid phase mixture will be separated by extraction with a solvent. Extractions of this type, called *solid-liquid extractions*, are important in the chemistry laboratory since they involve the selective separation of individual compounds from mixtures. The solvent employed in solid-liquid extractions selectively dissolves only a portion of the compounds in the solid mixture (Figure 3.1).

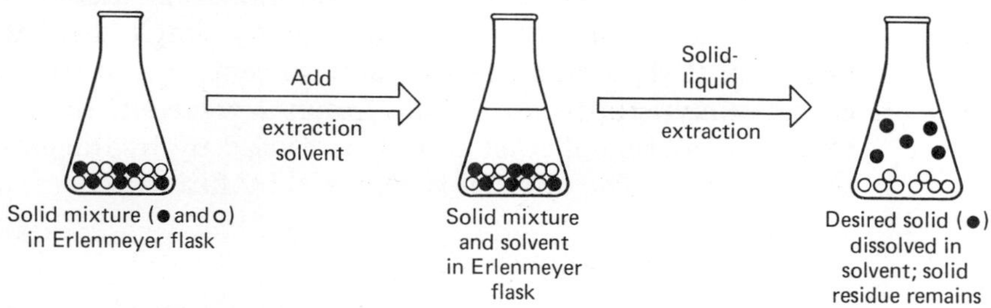

Figure 3.1 *Solid-liquid extraction.*

Solid-liquid extractions are common in our everyday experience. For example, the brewing of tea involves the extraction of compounds from tea leaves by hot water. The solid residue from the leaves that remains undissolved is removed from the aqueous solution by filtration (the tea bag) and the aqueous extract is then used as a beverage.

In the solid-liquid extraction process the efficiency of the separation depends on the solubility of the desired compound in the extraction solvent at equilibrium, on the volume of the solvent that is used in the extraction, and on the number of times that the extraction process is repeated. Extraction efficiency is measured as the percentage of the desired compound in the mixture that is isolated,

$$\textit{Extraction efficiency} = \frac{\text{wt. of compound isolated}}{\text{wt. of compound in mixture}}\ 100\%$$

Factors that cause non-equilibrium conditions and decrease the solid-liquid extraction efficiency are: large particle size of the solid mixture, insufficient contact time of the solvent with the solid, and inefficient mixing of the solvent and solid. Although extraction is a valuable separation method, it is often not possible to obtain a compound in pure form by extraction since more components of the solid mixture than the desired compound may be soluble in the solvent. Other separation techniques such as recrystallization (Experiment 2) or fractional distillation (Experiment 6) are usually required to purify the compound after extraction.

In this experiment you are directed to isolate the natural fat trimyristin from ground nutmeg by solid-liquid extraction. The solvent of choice for this particular separation is ethyl ether. Trimyristin, like other common fats, is a nonpolar compound that will readily dissolve in low polarity organic solvents

$CH_3(CH_2)_{12}CO_2CH_2$
$CH_3(CH_2)_{12}CO_2CH$
$CH_3(CH_2)_{12}CO_2CH_2$

Trimyristin
b.p. = 311°C
m.p. = 56°C

$CH_3CH_2OCH_2CH_3$

Ethyl ether
b.p. = 35°C

such as ether. After extraction the ether extract is filtered to remove the insoluble nutmeg residue and then concentrated by simple distillation. The addition of methanol (CH_3OH), a polar organic solvent, to the concentrated ether extract causes trimyristin to precipitate from the solution. The insolubility of trimyristin in the methanol-ether solvent is caused by hydrogen bonding of methanol to ethyl ether. The product is collected by filtration.

Isolation of Organic Compounds from Natural Sources

Organic compounds isolated from plant or animal material are called *natural products.* From earliest recorded history to the present time natural products have provided man with drugs to cure disease, dyes to color clothing or to paint pictures, flavoring agents for food and drink, perfumes, and a myriad of materials for other purposes. Natural products are isolated by a number of different techniques including crystallization (Experiment 2), extraction (Experiment 3), steam distillation (Experiment 5), fractional distillation (Experiment

6), sublimation (Experiment 16), and chromatography (Experiments 7, 10, and 19). We will discuss and use these techniques in this manual for the organic laboratory.

Scarcity of an important natural product has encouraged chemists to investigate ways to obtain the material by synthesis. These investigations have led to a number of significant scientific discoveries. One interesting example of this is the history of the drug quinine.

When the Spanish first arrived in Peru they found that the natives could effectively treat malaria with an extract from the bark of the cinchona tree. In 1827 two French chemists isolated pure quinine from the cinchona bark extract and showed that this compound was the active principle in the extract. The demand for quinine in the nineteenth century was so high that the material became very scarce. Large prizes were offered to any scientist who could synthesize quinine. William Perkin, a young British chemist, attempted to prepare quinine by the oxidation of aniline. Perkin never synthesized quinine, but the prize that he won was much larger than he ever expected. Perkin's experiments with aniline led to the production of the first synthetic dye (1856) and to the subsequent growth of a new chemical industry. Although quinine was eventually synthesized in 1944 by R. B. Woodward and W. von E. Doering, the drug is still most economically obtained by extraction of the bark of the cinchona tree.

Introduction to Simple Distillation

Distillation is a widely used technique for the separation and purification of compounds. The separation depends upon differences in vapor pressures of the components in the mixture. The vapor pressure (P_{vp}) of a liquid is a measure of the tendency of molecules on the surface of the liquid to pass into the vapor state. At a given temperature the vapor pressure is a constant that is independent of the total pressure of the atmosphere (P_{atm}) on the surface of the liquid. As the temperature of the liquid is raised, the vapor pressure of the liquid increases until it is equal to the pressure of the atmosphere on the liquid. At that temperature the liquid boils. We define the temperature at which $P_{atm} = P_{vp}$ as the boiling point (b.p.). The boiling point of a pure liquid is a constant at constant atmospheric pressure. For example, pure water has a boiling point of 100°C at one atmosphere pressure. Liquids that have a high vapor pressure at room temperature will have a lower boiling point than do liquids with a low vapor pressure at room temperature.

When a liquid phase mixture is boiling, the vapor above the liquid will not have the same composition as the liquid, but will be enriched in the more volatile or lower boiling component (Figure 3.2). If the vapor above the liquid is collected and cooled, the liquid that condenses will have the same composi-

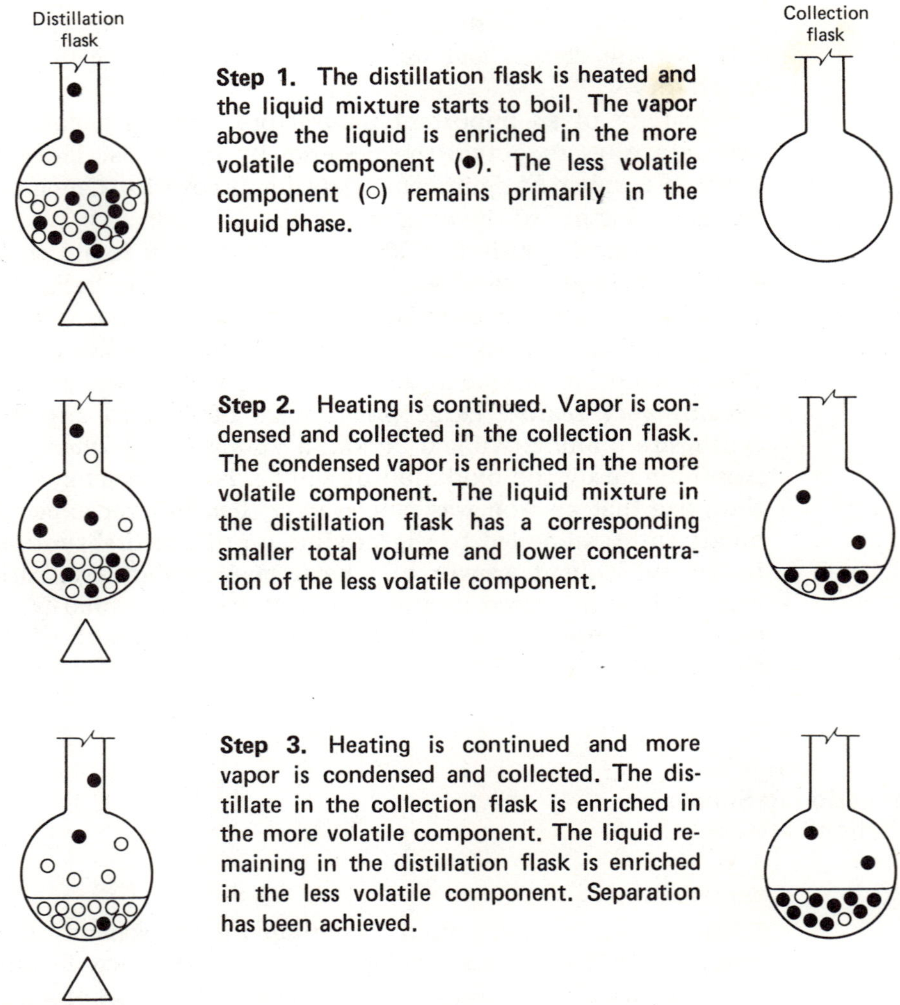

Figure 3.2 *The simple distillation process.*

tion as the vapor and thus be enriched in the more volatile component. The boiling mixture will then have a correspondingly lower concentration of the more volatile component, and the boiling point of the mixture will be slightly higher now because of this change in composition. If we continue to boil the mixture, collect the condensed vapor (*distillate*) and measure both the volume of distillate and the boiling point at which the distillate is collected, we can translate this data into a graph of boiling point versus volume of distillate collected. This graph will resemble the graph in Figure 3.3 if the boiling points of the two components in the mixture differ by at least 100° (Δ b.p. > 100°C) and equal volumes of the two components were present in the original mixture. The distillates that are collected while the temperature is relatively constant (level regions of the curve) will be pure samples of the components of the origi-

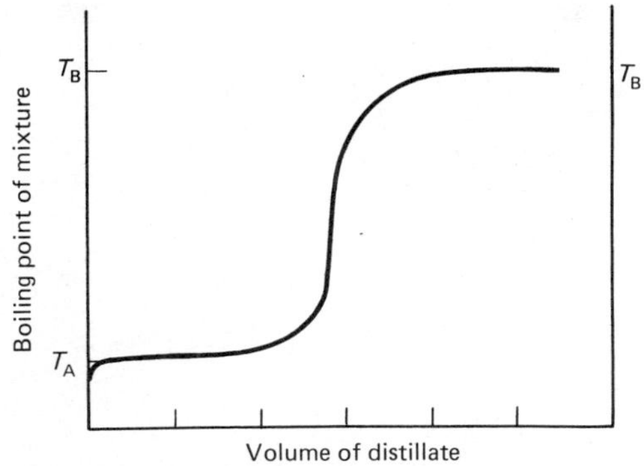

Figure 3.3 *Distillation curve.*

T_A = boiling point of component A.
T_B = boiling point of component B.

nal mixture. If the boiling points of the two components are close to each other, the separation of the two components in the mixture will not be adequate and the distillation curve will not show regions of relatively constant temperature.

The distillation apparatus shown in Figure 3.4 is designed to allow the efficient heating of a liquid to boiling in the distillation flask, the cooling of the hot vapor in the condenser, and the collection of the condensed vapor (*distillate*) in the collection flask. The thermometer is positioned to measure the temperature of the vapor before it enters the condenser. The theory and practice of distillation is further discussed in Experiments 5 and 6.

The boiling point that you record for the distillate is dependent on operational factors that are related to the equipment that you employ and to your procedure for distillation. Your understanding of these factors is essential if you are to use the distillation procedure properly.

(*a*) *The thermometer that you use to measure boiling point is uncorrected. To ensure that your thermometer is accurate, measure the boiling point of water (100°C at atmospheric pressure) and the boiling point of ethyl ether (35°C at atmospheric pressure). The thermometer correction using the boiling point of ethyl ether can be determined in this experiment.*

(*b*) *To record a boiling point from your thermometer, the vapors of the liquid must envelop the thermometer bulb and drops of liquid should condense from your thermometer.*

(*c*) *When the vapors of a boiling liquid rise to the thermometer, the temperature indicated on your thermometer increases until the thermometer is in thermal equilibrium with the vapor and records the actual temperature of the vapor. The initial increase in temperature on your thermometer does not indicate that the temperature of the vapor is also increasing.*

(*d*) *When the volume of liquid in the distilling flask becomes so low that vapors from the liquid no longer envelop the thermometer, the tempera-*

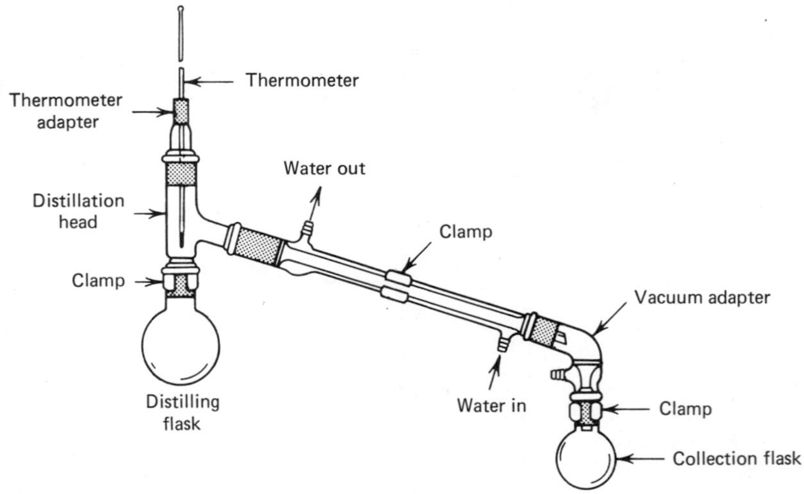

Figure 3.4 *Simple distillation apparatus.*

Assembly Instructions

1. Lightly grease the male joints on the standard vapor glassware.
2. Clamp the 100-ml round-bottom flask to a ring stand directly over a steam bath. The steam bath should be setting on the base of the ring stand so that the weight attached to the ring stand is directly over its base. This will help to prevent the apparatus from tipping over.
3. Insert the distillation head in the top of the distillation flask. Adjust the clamp on the flask if necessary so that the apparatus is exactly vertical.
4. The condenser is now attached to the second joint of the distillation head and secured with a clamp to another ring stand. The condenser will be at the correct angle to allow the distillate to run down to the collection flask.
5. The vacuum take-off adapter and the collection flask are attached and clamped in place as shown. This clamp can be fixed to the ring stand used to hold the condenser or to a third ring stand. It is often advantageous to use a rubber band around the vacuum adapter and the water inlet tube on the condenser to hold the adapter in place when the collection flask is changed.
6. The thermometer is carefully inserted into the thermometer adapter (a very small amount of lubricant may be necessary on the rubber seal). The height of the thermometer is adjusted so that the bulb at the bottom of the thermometer is just below the side-arm of the distillation flask. With this arrangement, the thermometer will respond to the temperature of the hot vapor as it enters the condenser.
7. Attach the rubber tubing to the inlet (bottom) and outlet (top) of the condenser. Be sure the tubing is secure. Turn the water on slowly. Check to be sure that the water is flowing through the condenser in a slow stream by looking at the water flowing out the outlet tubing into the sink. Adjust the rate if necessary.
8. Check to be sure that all joints fit tightly and that the parts are securely clamped.

ture will drop. Alternatively, in the distillation of a mixture the temperature will drop after a low boiling component is distilled if sufficient heat is not applied to the distilling flask so that the boiling point of a second component is maintained.

Your use of a distillation apparatus such as that described in Figure 3.4 does not allow you to know the boiling point of the distillate until the vapors envelop the thermometer. A finite amount of time is required for the thermometer to reach thermal equilibrium with the vapor. Consequently, the boiling point of the first drops of distillate is not accurately known. In addition, changes in pressure, heating rate, and amount of liquid in the distilling flask, among others, affect the recorded temperature at which the distillate is collected. Therefore, the distillate is collected over a range of temperatures and the boiling point that is obtained is actually a range of boiling points. *For all separations by distillation the* **boiling point range** *is recorded as the minimum to maximum temperature observed for the distillate fraction that is collected.*

The mixture that you will distill consists primarily of ethyl ether (b.p. = 35°C) and trimyristin (b.p. > 300°C). Since the ether has a very high vapor pressure (low boiling point) compared to trimyristin, the distillate will consist almost entirely of ethyl ether until the ether has been distilled from the flask. At that point, a large increase in temperature would be necessary to cause the remaining trimyristin to boil. Since heating the impure trimyristin to its boiling point may cause decomposition of the product, we will not attempt its purification by distillation. Trimyristin has a melting point that is above room temperature; consequently, we will be able to crystallize trimyristin from an ether-methanol mixture. The separation of trimyristin from nutmeg thus involves three laboratory operations: solid-liquid extraction, simple distillation, and filtration (crystallization). The following flow diagram outlines the separation procedure:

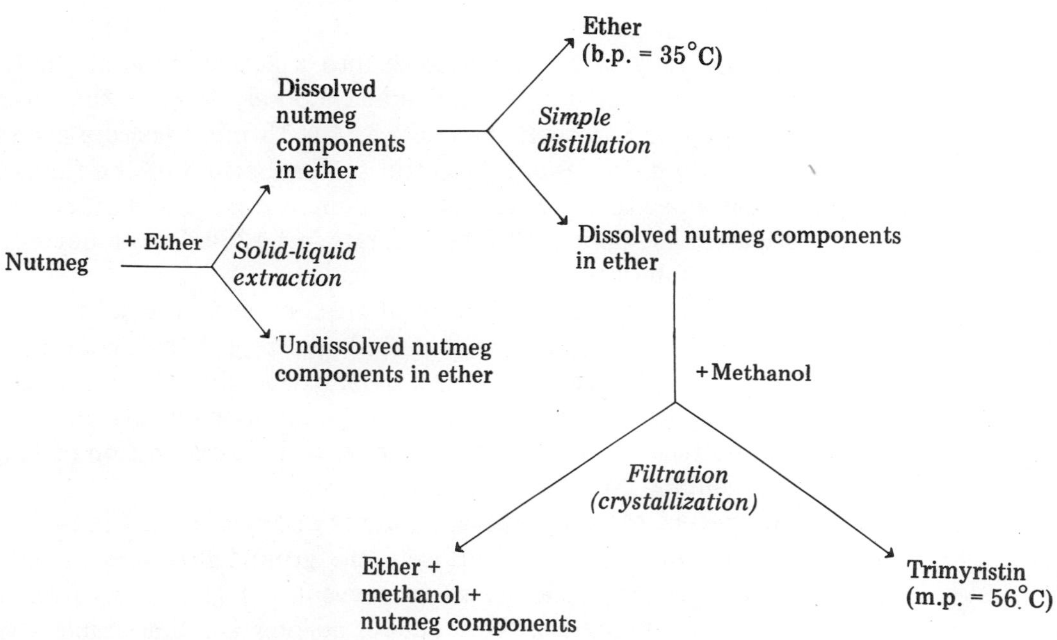

Prelab Questions

1. What other common examples of extraction are you familiar with besides brewing tea?

2. Will the extraction of a solid mixture with a hot solvent be more or less efficient than an extraction with the same solvent at room temperature? Why?

3. Would it be more efficient to use ground nutmeg than to use whole nutmeg for this experiment? Why?

4. Methanol boils at 65°C (760 Torr pressure) and water boils at 100°C at the same pressure. Which liquid has the highest vapor pressure at 25°C? At 65°C? At their respective boiling points?

5. Describe how the following factors affect the recorded boiling point: (a) the pressure of the atmosphere, (b) the rate of heating of a liquid in a distilling flask, and (c) the use of an uncorrected thermometer to record temperatures.

EXPERIMENTAL PROCEDURE

CAUTION: Ethyl ether is a highly volatile and extremely flammable solvent. Be sure that there are no flames in your area before you handle ethyl ether. If possible, always work with ethyl ether and other volatile solvents in the fume hood and not on the open lab bench. Solvents such as ether that are flammable, volatile, or insoluble in water should not be poured into sinks. These solvents should be transferred to special waste solvent containers that are provided in the laboratory.

Weigh 15 g of ground nutmeg into a 250-ml Erlenmeyer flask and carefully add 30 ml of ethyl ether. Loosely stopper the flask with a cork stopper and swirl the contents for 15 min to ensure good mixing. Allow the solid residue to settle to the bottom of the flask. Quickly decant the ether extract into a stemless funnel and filter by gravity (Figure 1.4). Collect the ether filtrate in a 125-ml Erlenmeyer flask and save this solution.

Add a second 30-ml portion of fresh ether to the solid nutmeg residue in the 250-ml Erlenmeyer flask and repeat the extraction process for 15 min. Decant the second ether extract from the solid residue and filter through the same fluted filter paper used for the filtration of the first extract. Rinse the filter paper with a 10-ml portion of ethyl ether and combine the ether filtrates.

Construct the distillation apparatus as described in Figure 3.4 using a 250-ml distilling flask. Lubricate the ground-glass joints with a *small amount* of stopcock grease to prevent the glass parts from freezing together. Remove the thermometer adapter and thermometer from the distillation apparatus, and carefully pour the ether extract into the distillation flask with the aid of a long-stemmed funnel. Add several (3 to 5) boiling stones to the distillation flask. Replace the thermom-

---

CAUTION: The boiling stones are essential to maintain constant and even boiling. The stones are small pieces of carborundum or other porous material. Vapor bubbles readily form in the pores of the stones when the liquid is heated. These small bubbles provide a steady evolution of vapor and help to prevent superheating and bumping of the liquid. If the temperature of the boiling liquid should drop and boiling ceases, the pores in the stones will fill with liquid and new stones must be added before the liquid is heated to boiling again.

---

eter and thermometer adapter and heat the liquid in the distillation flask by means of a steam bath. Concentrate the ether solution to approximately one-half of its original volume; you should collect about 35 ml of distillate. Record the boiling range and the volume of the distillate collected and then dispose of the collected ether in the container provided for waste ether. Pour the concentrated ether solution from the distilling flask into a 250-ml Erlenmeyer flask. The distilling flask should be rinsed with a small portion of ether (5 ml) to ensure complete transfer. Measure 70 ml of methyl alcohol in a graduated cylinder and slowly add this solvent to the concentrated ether solution. Continually stir the solution in your Erlenmeyer flask while you add the methanol solvent. A solid will begin to form during the methanol addition. Allow the resulting precipitate to settle for a period of 5–7 min while you prepare your small Büchner funnel for suction filtration (Figure 1.7).

Collect the precipitated trimyristin by suction filtration and wash the precipitate with a 10-ml portion of ether-methanol (1 : 1). The trimyristin may be dried by drawing air through the Büchner funnel for several minutes after all of the solvent has been collected in the filter flask. Weigh the trimyristin and report the isolated yield as a weight percent based on the amount of ground nutmeg used for the extraction. Save the trimyristin in a small open flask for use in the next experiment and discard the ether-methanol filtrate in the proper container. Let the trimyristin dry in the open flask until the next laboratory period and then take a melting point of your product. Save the isolated trimyristin for use in Experiments 4 and 7.

The melting point of trimyristin is reported in the literature to be 56°C. However, trimyristin may exist in more than one crystalline structure: a *polymorphic compound.* Two other solid forms of trimyristin melt at 32 and 44°C. These forms are not as stable as the form that melts at 56°C and slowly convert into the more stable crystalline structure.

Postlab Questions

1. How could you have improved the efficiency of the extraction of trimyristin?

2. What experimental evidence do you have that the distillate obtained in this experiment is ethyl ether?

3. What other solvents could be used to extract trimyristin from nutmeg? Would methyl alcohol be a good choice for an extraction solvent?

4. Why is ethyl alcohol commonly used to extract the flavoring agents that are used in cooking rather than solvents such as ether or chloroform?

5. Explain how the ethyl ether and methanol used in this experiment could be separated, recovered, and used again.

Experiment Four

The Preparation and Purification of Myristic Acid

Trimyristin, which you isolated in Experiment 3, is one member of a class of organic compounds found in animal "fatty" tissues and vegetable or marine oils and known as *triglycerides.* Triglycerides are triesters of glycerol and *fatty acids*: long-chain aliphatic carboxylic acids that have an even number of carbon atoms. Myristic acid, which will be prepared from trimyristin in this experiment, is one such fatty acid.

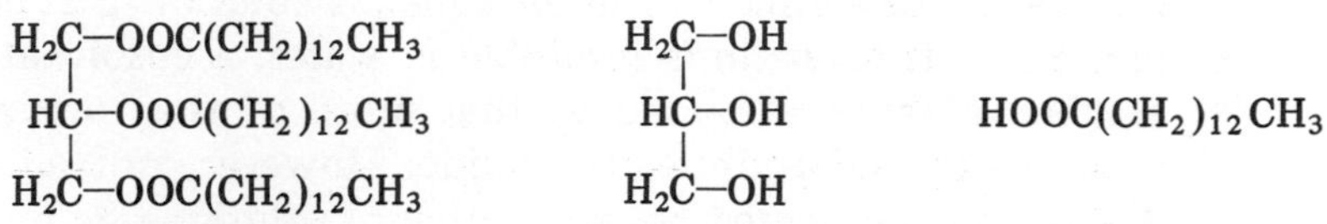

Trimyristin *Glycerol* *Myristic acid*

In this first experiment with chemical reactions of organic compounds you will investigate one of the oldest and most important organic reactions—hydrolysis. A hydrolysis reaction involves the cleavage of a polar bond in an organic compound (Y—Z) and an H—O bond in water with the formation of a bond between the more electronegative atom (Y) of the polar compound and hydrogen and the formation of a bond between the less electronegative atom (Z) of the polar compound and hydroxyl:

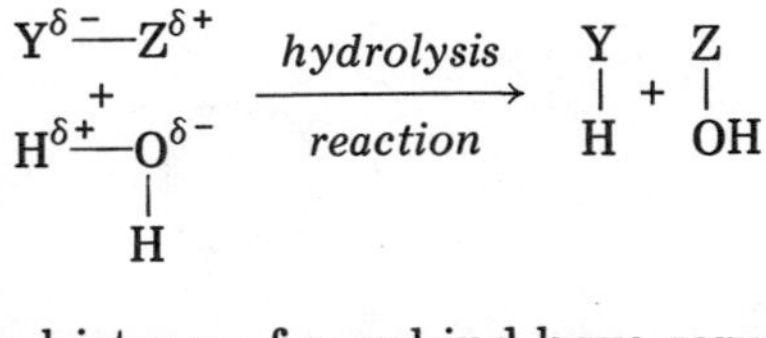

Few reactions in the history of mankind have served a more useful purpose than the hydrolysis of triglycerides. Natural fats such as tallow from cattle and sheep are hydrolyzed by treatment with lye (a strongly alkaline solution obtained by leaching wood ashes and mainly composed of aqueous sodium hydroxide). The residue that is obtained from this reaction is processed into *soap.* Although the soaps that we use today contain additional ingredients that improve their appearance and odor or texture, the active ingredients in soap are the products from the basic hydrolysis (saponification*) of naturally occurring fats—the sodium salts of fatty acids.

$$\begin{array}{l} H_2C\text{—}OOCR \\ \quad | \\ \;HC\text{—}OOCR' \\ \quad | \\ H_2C\text{—}OOCR'' \end{array} \quad + \quad 3\ NaOH \quad \xrightarrow{H_2O} \quad \begin{array}{l} H_2C\text{—}OH \\ \quad | \\ \;HC\text{—}OH \\ \quad | \\ H_2C\text{—}OH \end{array} \quad + \quad \begin{array}{l} RCOONa \\ R'COONa \\ R''COONa \end{array}$$

Triglycerides (R,R′,R″ = long chain hydrocarbon) — *Sodium hydroxide* — *Glycerol* — *Sodium salts of fatty acids*

$$\text{Natural fats} + \text{Lye} \xrightarrow{H_2O} \text{Glycerol} + \text{Soap}$$

The cleansing action of soap is due to the ability of the fatty acid salts to attract both water and hydrocarbonlike molecules. The ionic portion of the acid salt renders the compound soluble in water (*hydrophilic* attraction) whereas the aliphatic hydrocarbon portion of the acid salt attracts nonpolar molecules (*hydrophobic* attraction). The net result is that water-soluble fatty acid salts solubilize nonpolar compounds that are by themselves insoluble in water (Figure 4.1).

The hydrolysis of trimyristin involves heating the triglyceride that you isolated in Experiment 3 in an aqueous solution that contains sodium hydroxide. Since trimyristin is insoluble in water, a cosolvent in which it is partially soluble—ethanol—is added so that more of the trimyristin will be in contact with water and sodium hydroxide. However, not all of the triglyceride will dissolve in the heated aqueous ethanol solution. As a consequence, the extent of reaction can be followed by observing the amount of undissolved trimyristin since the products of the hydrolysis reaction are soluble in the aqueous solution. The isolation of myristic acid is completed through acidification of the basic solution after complete hydrolysis. Myristic acid is then purified by recrystallization from methanol.

*Although "saponification" is the term originally used to designate the process for making soap, it is now used to describe the basic hydrolysis of esters.

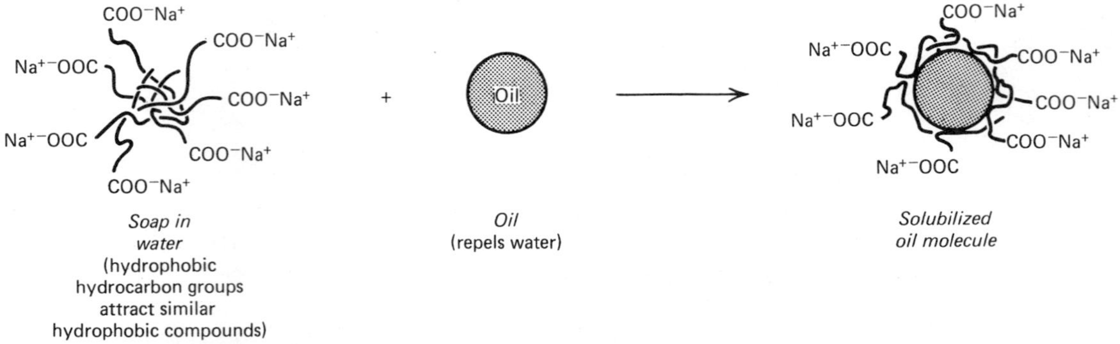

Figure 4.1 *The cleansing action of soap.*

Describing Chemical Reactions in your Laboratory Notebook

The description of experimental processes in your laboratory notebook is an essential part of your involvement in the chemistry laboratory. Your writeup for laboratory experiments is a measure of your understanding of the experimental processes and therefore must be a complete record of your experimental design and all experimental observations. Experiments that involve chemical reactions require a more detailed notebook description than that previously described in the Introduction because such experiments usually involve several experimental operations. The following outline is a useful guide for your notebook description of these experiments:

1. *Title of experiment*
2. *Purpose of experiment*
3. *Balanced chemical reaction*
4. *Physical properties of reactants and products*
5. *Table of reactants*
   This table should include all reactants that are to be used in the experiment and the source of these chemicals (chemical company, notebook reference, or, in the case of standard solutions, the chemistry stockroom). The listing of chemical sources provides information concerning the quality of the reagents that you employ. The amount of each reagent in weight or volume and moles completes this table.
6. *Flow diagram for all experimental operations*
7. *List of competing reactions*
   Reactions that compete with the desired process lower the yield of the expected product. Prior knowledge of competing reactions will often allow you to change reaction conditions to those that favor the formation of the desired product.
8. *Theoretical yield of product*
   The amount of the desired product in grams that is expected if the limiting reagent is quantitatively converted to that product in the theoretical yield:

$$\begin{pmatrix}\text{Mol of}\\ \text{limiting}\\ \text{reagent}\end{pmatrix} \times \begin{pmatrix}\text{Stoichiometric}\\ \text{ratio of product}\\ \text{to limiting reagent}\end{pmatrix} \times \begin{pmatrix}\text{Molecular}\\ \text{weight of}\\ \text{product}\end{pmatrix} = \text{Theoretical yield}$$

9. *Experimental procedure*
   Literature reference to procedure
   Description of chemical reaction
   Description of product isolation and purification
   Product analysis and isolated product yield

Isolated product yield is the percentage yield of product based on the theoretical yield.

10. *Summary of experimental data and evaluation of experimental methods.*

Parts 1–8 should be completed prior to the start of your laboratory period. An example of such a prelab writeup is given for this experiment.

THE PREPARATION AND PURIFICATION OF MYRISTIC ACID

Purpose: The preparation of myristic acid by alkaline hydrolysis of trimyristin. Isolation and recrystallization of myristic acid.

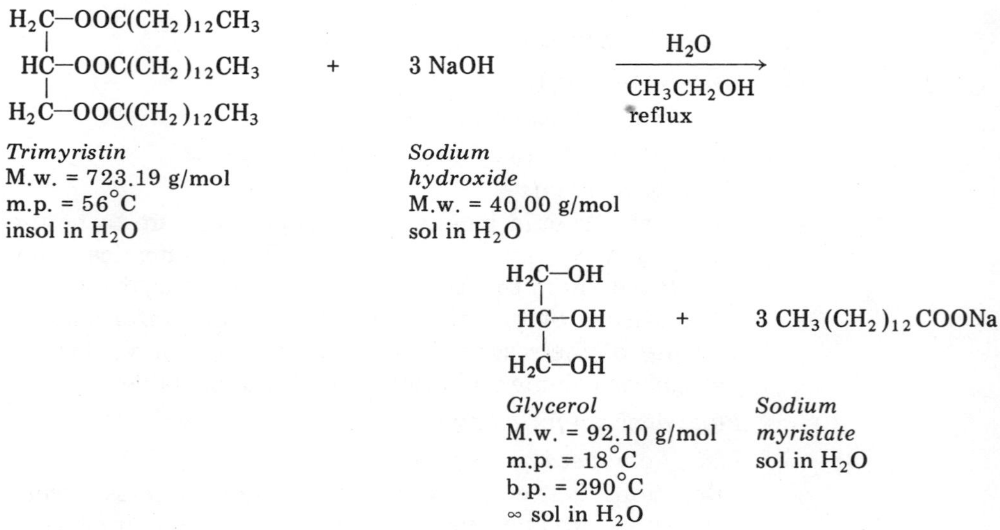

$$CH_3(CH_2)_{12}COONa + HCl \xrightarrow{H_2O} CH_3(CH_2)_{12}COOH + NaCl$$

Hydrogen chloride M.w. = 36.5 g/mol sol in H_2O	*Myristic acid* M.w. = 228.36 g/mol m.p. = 54°C insol in H_2O

Reactants (Source)	Weight or Volume	Moles
Trimyristin (Exp. 3, Notebook I-17)	1.50 g	2.07 mmol
Sodium hydroxide (Stockroom, 10% solution, 2.5*M*)	5.0 ml	12.5 mmol
Hydrochloric acid (Stockroom, 10% solution, 1.2*M*)	20 ml	24 mmol

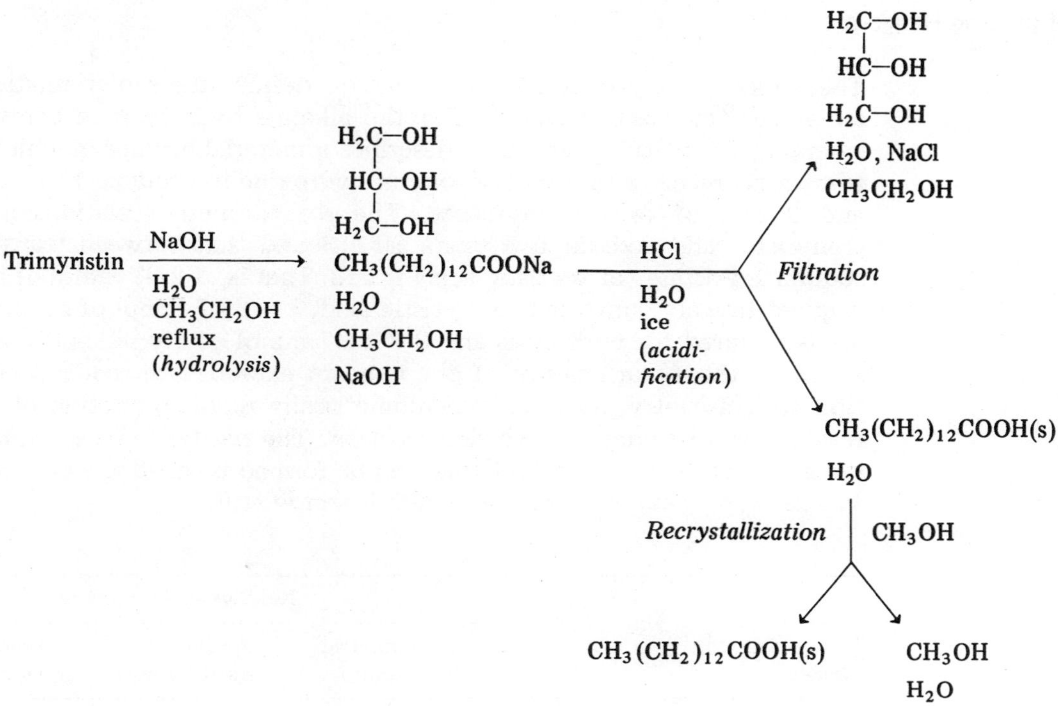

Competing reactions: **Partial hydrolysis—side products filtered with myristic acid**

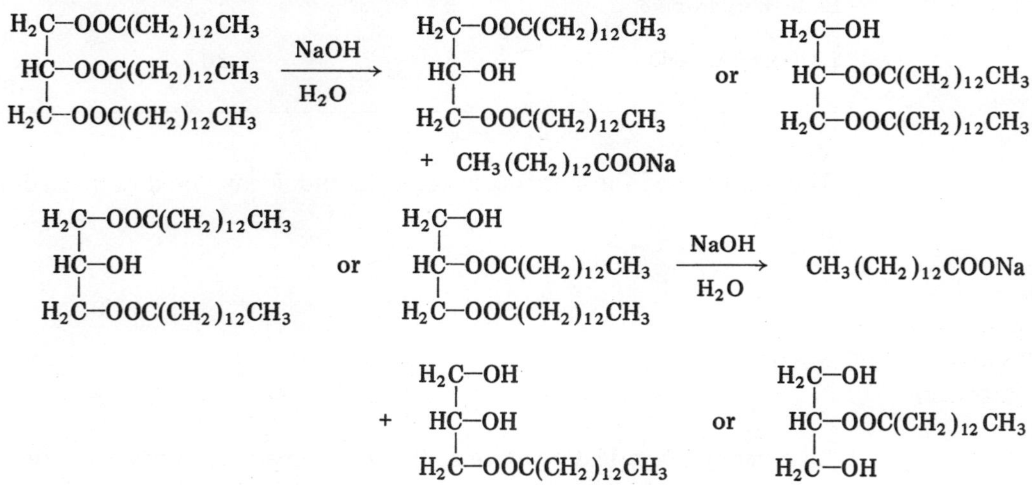

Theoretical yield:

$$2.07 \text{ mmol Trimyristin} \times \left(\frac{3 \text{ mmol Myristic acid}}{\text{mmol Trimyristin}}\right) = 6.21 \text{ mmol Myristic acid}$$

$$6.21 \text{ mmol Myristic acid} \times \left(\frac{1 \text{ mol}}{1000 \text{ mmol}}\right) \times 228 \text{ g/mol} = 1.42 \text{ g Myristic acid}$$

Limiting Reagent

The stoichiometry of a chemical reaction defines the molar relationship between reactants and products. For the alkaline hydrolysis of trimyristin, for example, the balanced equation describes a material balance in which 1 mol of trimyristin reacts with 3 mol of sodium hydroxide to produce 1 mol of glycerin and 3 mol of sodium myristate. The stoichiometric relationship between trimyristin and sodium hydroxide is 1:3 and that between trimyristin and sodium myristate (or myristic acid) is 1:3. That is, if 2.07 mmol of trimyristin is quantitatively converted to myristic acid, 3 × 2.07 mmol of sodium hydroxide is required for hydrolysis and 3 × 2.07 mmol of myristic acid will be produced. If the molar amount of one reactant exceeds that required by the reaction stoichiometry, only the stoichiometrically required portion of the excess reactant will be consumed during reaction. The reactant whose molar amount limits the amount of product that will be formed is called the *limiting reagent.* In this experiment the limiting reagent is trimyristin:

	Relative to Trimyristin		
Reagent	Stoichiometric relationship	Amount used (mmol)	Amount required (mmol)
Trimyristin (*limiting reagent*)	1	2.07	2.07 (1 × 2.07)
Sodium hydroxide	3	12.5	6.21 (3 × 2.07)
Hydrochloric acid	3	24	6.21 (3 × 2.07)

The amount of the limiting reagent should always be determined accurately by weight.

Solvent Reflux—Temperature Control for Reactions

Like many chemical reactions, the hydrolysis of trimyristin occurs slowly at room temperature. In order to increase the rate of reaction so that products will be formed within a reasonable time period, the reaction solution must be heated. Since reactants are normally less volatile than the solvent employed for a chemical reaction, the boiling point of the solvent restricts the temperature range that can be used in heating a reaction solution. At atmospheric pressure the boiling point of the solvent is the upper temperature limit to which a reaction solution can be heated. However, it is precisely this limitation that is used advantageously in controlling the temperature for reactions.

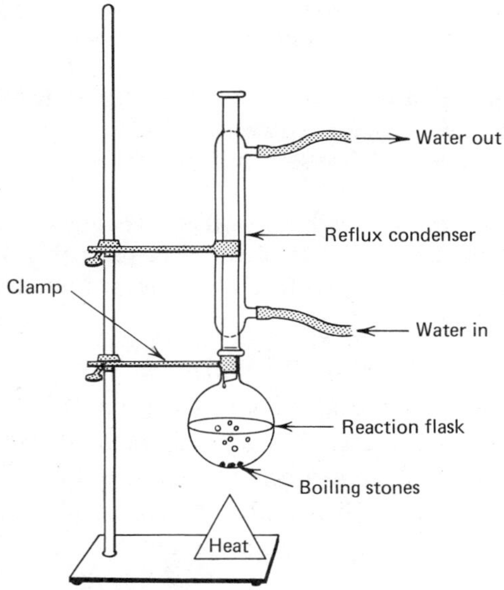

Figure 4.2 *Apparatus for heating with reflux.*

Organic compounds boil at a constant, well-defined temperature. Consequently, heating a volatile solvent at its boiling point is a convenient procedure for maintaining a constant, well-defined reaction temperature. Boiling the solvent chosen for this experiment (80% ethanol; 20% water) will allow you to maintain a constant reaction temperature of 80°C. The boiling solvent will also agitate the solution to ensure thorough mixing of reactants throughout the reaction.

Heating a solution containing a volatile solvent will result in the loss of this compound unless the hot vapor is condensed to the liquid phase and returned to the reaction solution. This process is called *heating under reflux*, or *refluxing*, and is a widely employed technique in organic chemistry. Figure 4.2 illustrates the apparatus set up for refluxing a solution. Cold water flowing through the vertical condenser cools the organic vapors. The condensed liquid then drains back into the boiling solution. The rate of heating is controlled so that liquid condensing on the inner walls of the vertical condenser reaches a level that is one-third the distance up the condenser but no higher. More rapid heating does not change the temperature of the refluxing solution but may cause vapor to escape through the top of the condenser.

Prelab Questions

1. Write a flow diagram for this experiment that includes the products from the competing reactions and describe how these compounds could be separated from myristic acid.

2. A quantity of 1.24 g of myristic acid was isolated from the trimyristin hydrolysis that is described in the sample prelab writeup. Calculate the percentage yield of myristic acid that was isolated.

3. Describe a procedure for the isolation of glycerol in this experiment. (Glycerol is a major component of many cosmetic soaps.)

4. The apparatus for refluxing a solution is always left open to the atmosphere during heating and is never stoppered. Describe what change occurs in a closed system when the temperature of that system is increased. Why is heating a closed system an unsafe procedure?

5. Tristearin is a triglyceride that is abundant in the depot fat of cattle and sheep. Calculate the theoretical yield of stearic acid (M.w. = 284.49 g/mol) if 1.50 g of tristearin (M.w. = 891.45 g/mol) is saponified in aqueous sodium hydroxide.

EXPERIMENTAL PROCEDURE

Weigh one-half of the trimyristin* that you isolated in Experiment 3 (approximately 1.0 g) and quantitatively transfer the weighed solid into a 100-ml round-bottom flask. Add 20 ml of ethanol to the round bottom flask followed by 5.0 ml of 10% aqueous sodium hydroxide.

CAUTION: Sodium hydroxide and solutions of sodium hydroxide are caustic; they can destroy living tissue by chemical reactions. Handle such basic substances carefully and, if contact is made with your person, thoroughly wash the affected area with water.

Caustic solutions can also damage glassware. Avoid contact of such solutions with ground glass joints. Caustic causes ground glass joints to tightly fasten together (to "freeze"). Use a long-stem funnel to transfer a caustic solution into a flask.

Construct your apparatus for refluxing as described in Figure 4.2. Lubricate the ground glass joint of your 100-ml reaction flask and the outer joint of your reflux condenser with a small amount of stopcock grease to prevent the glass joint from freezing together. Turn on the water to the condenser after you determine that all rubber hose connections are tightly fastened and that the water return hose is placed into the drain of your sink. After adding 2 or 3 boiling stones, heat the reaction solution to boiling on a steam bath and reflux this solution for 1 hr or, if an oily trimyristin layer is visible, until all of the triglyceride has been converted to soluble products.

After hydrolysis is complete turn off the steam and allow the reaction mixture to cool to room temperature. Prepare a cold acidic solution by adding 20 ml of 10% aqueous hydrochloric acid to an equal volume of crushed ice and then pour the reaction solution into the cold acidic solution while vigorously stirring the acidic solution with a glass stirring rod. Continue stirring for several minutes after combining the two solu-

*Save the remaining trimyristin. This compound will be used again in Experiment 7.

tions to allow complete precipitation of the solid that forms in the acidification process. Further dilute the acidic solution with 50 ml of distilled water and then collect the precipitate by suction filtration using a Büchner funnel. Wash the collected solid with two 5-ml portions of distilled water to remove any mineral acid that adheres with water to the solid. With your spatula press the solid in the funnel to aid the removal of water and draw air through the funnel until the solid is visually dry (water is no longer visible on the surface of the solid).

Recrystallize the collected solid from methanol-water by dissolving the product in a minimal volume of methanol (15–20 ml) at room temperature. Follow Steps 2 and 3 of the recrystallization procedure in Figure 2.5. After gravity filtration of the undissolved solid, heat the methanol solution to boiling on a steam bath, and then add water dropwise until the solution turns slightly cloudy. Add a small amount of methanol dropwise to clarify the solution. Remove the recrystallization flask from the steam bath and allow the solution to cool to room temperature. When crystallization is visually complete at room temperature cool the solution in an ice bath to precipitate more of the product. Collect the crystals by suction filtration. Allow the collected solid to air dry on a filter paper until your next laboratory period. At that time weigh the solid and obtain its melting point. Calculate the percentage yield of the isolated product. Save this product for further use in Experiment 7.

Postlab Questions

1. In addition to a comparison of melting points between your product and myristic acid, how can you certify that your reaction product is actually myristic acid? Trimyristin and myristic acid have similar melting points.

2. What evidence do you have from your experimental procedure that suggests that the isolated product is myristic acid and not trimyristin or a product from competing reactions?

3. What would have been the effect on the product isolation procedure if the 10% aqueous hydrochloric acid solution actually contained only 1% hydrochloric acid?

4. Esters can be hydrolyzed in acid-catalyzed reactions:

$$RCOOR' + H_2O \xrightarrow{H^+} RCOOH + R'OH$$

Describe the advantages of the saponification reaction over the acid-catalyzed reaction for the hydrolysis of trimyristin.

5. Why is methanol, and not ethanol, the preferred solvent for recrystallization of myristic acid?

6. Describe the changes in your experimental procedure that would have allowed you to isolate the sodium salt of myristic acid.

7. Soap solutions foam when agitated by shaking, and deposit an insoluble precipitate when treated with iron or calcium salts, the principal mineral content of hard water. Describe your observations from this experiment that indicate that sodium myristate possesses the qualities of a soap. In the preparation of solutions for this experiment, why is the use of distilled water rather than tap water important?

Experiment Five

Steam Distillation: Isolation of Limonene

The most widely used method for the isolation and purification of liquid organic compounds is *distillation*. Three distillation techniques—simple distillation, steam distillation, and fractional distillation—are common operations of this separation method in the organic chemistry laboratory. As you have already observed in Experiment 3, a simple distillation will separate a volatile liquid from nonvolatile materials. Steam distillation, which is the laboratory operation introduced in this experiment, is a method for codistillation of water and volatile organic compounds that are immiscible with water; steam distillations

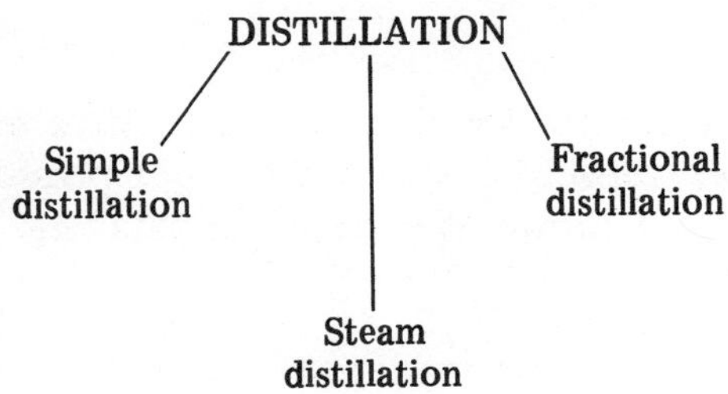

occur at temperatures less than 100°C. Fractional distillation, which is discussed in the following experiment, is used when the separation of volatile liquids that are miscible is desired. In this experiment you will isolate limonene, a naturally-occurring cycloalkene, by steam distillation.

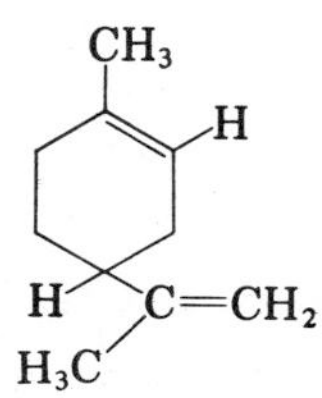

Limonene

Distillation and Raoult's Law

A liquid exerts pressure on its surroundings. This pressure, the *vapor pressure*, increases with increasing temperature in the manner described for pure limonene and for pure water in Figure 5.1. As the temperature of the heated liquid increases, the vapor pressure of the liquid also increases until the vapor pressure (P_{vp}) equals the pressure of the atmosphere on the liquid (P_{atm}). When $P_{atm} = P_{vp}$ the liquid boils and the temperature at which boiling occurs is defined as the boiling point.

For a pure liquid the temperature of the boiling liquid (*pot temperature*) is identical to the temperature of the vapor prior to condensation in the condenser (*head temperature*). The temperature measured by the thermometer (Figure 3.4) is equal to the temperature of the boiling liquid and remains constant throughout the distillation. This temperature is the boiling point of the pure liquid.

The boiling point of a liquid solution containing two or more miscible components also occurs when $P_{vp} = P_{atm}$, but in this case P_{vp} is the sum of the individual vapor pressures of all volatile components (*partial vapor pressures*):

$$P_{vp} = P_1 + P_2 + P_3 + \cdots + P_n$$

The partial vapor pressure of any one volatile component (P_n) depends on the mole fraction of that component in the liquid solution (Xn) and is related to the vapor pressure of the pure liquid (P_n^0) by the equation

$$P_n = P_n^0 X_n$$

This equation, which is a quantitative expression of *Raoult's Law*, describes the relationship between the vapor pressure and the composition of the homogeneous solution at a given temperature. The mole fraction of component *n* in

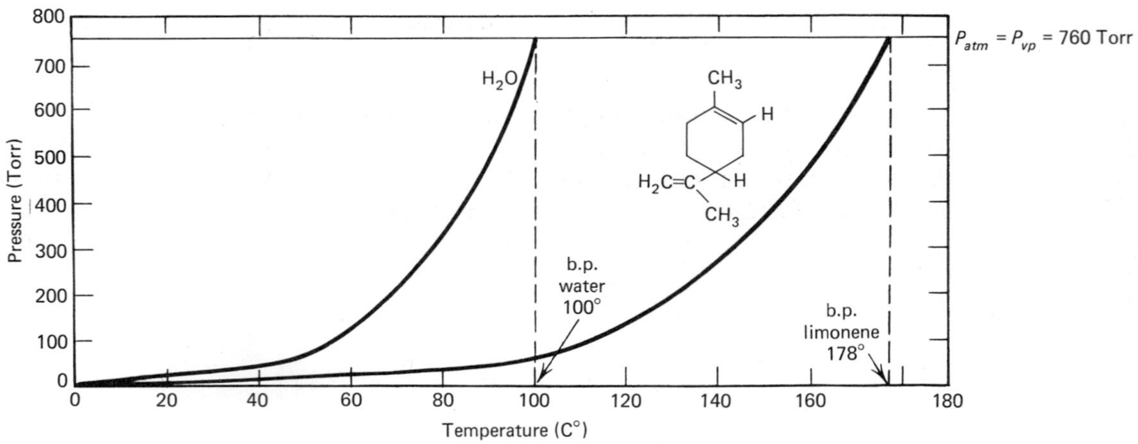

Figure 5.1 *Relationship between vapor pressure and temperature for water and for limonene.*

the solution (X_n) is that fraction of the total number of moles of all solution components that is component n:

$$X_n = \frac{\text{mol}_n}{\text{mol}_1 + \text{mol}_2 + \text{mol}_3 + \cdots + \text{mol}_n}$$

Consider the case in which a nonvolatile compound is dissolved in a volatile solvent—a situation similar to trimyristin dissolved in ethyl ether (Experiment 3). Since trimyristin is relatively nonvolatile, its partial pressure is negligible at the temperatures used to distill the ether. However, by Raoult's Law the temperature required to maintain boiling of the ether solution must be higher than the boiling point of ethyl ether since the mole fraction of ethyl ether in the solution is less than 1.0. On the other hand, if the vapor released from the boiling liquid is allowed to maintain a liquid-vapor equilibrium as it traverses the distance from the boiling liquid to the distillation head, the head temperature will be the boiling point of the pure ethyl ether. The liquid-vapor equilibrium that is observed along the walls of the distillation apparatus and at the thermometer in a distillation process is not the liquid-vapor equilibrium that occurs at the surface of the boiling solution. The ether vapor released from the boiling solution condenses into the liquid phase as the vapor advances to the distillation head. The condensed liquid is pure ethyl ether and therefore requires a lower boiling temperature—the boiling point of the pure liquid.

Raoult's Law relates vapor pressure and composition for *ideal solutions*—those in which interactions between like molecules are the same as those between unlike molecules. Deviations from Raoult's Law occur when the attraction between unlike molecules is either stronger or weaker than the attraction between molecules of the same compound. For example, the temperature required to boil ether from an ether solution containing myristic acid would be slightly greater than that predicted from application of Raoult's Law (nonideal behavior) since myristic acid forms hydrogen bonds to ethyl ether. With the exception of solutions that contain compounds capable of hydrogen bonding, however, most organic solutions approximate the ideal behavior of Raoult's Law.

Steam Distillation

The distillation technique that involves the codistillation of a mixture of water and volatile organic compounds that are immiscible with water is known as steam distillation. Steam distillation is a valuable technique for the separation of volatile compounds from nonvolatile compounds. One advantage of this technique, particularly for compounds that decompose at their boiling point, is that distillation temperatures that are less than 100°C, the boiling point of pure water, are used.

A volatile organic compound in the steam distillation process is one that has a discernible vapor pressure at 100°C. Generally, a vapor pressure greater than 10 Torr at 100°C for the volatile compound ensures an effective steam distillation. A nonvolatile compound is usually considered to have a vapor pressure that is at least an order of magnitude less than that of the volatile compound.

Since water and the volatile organic component that codistills with water are immiscible, their partial vapor pressures do not depend on their composition in the mixture; that is, $P_n = P_n^0$. Therefore, regardless of the relative amount of each component, the total vapor pressure (P_{tot}) of the liquid mixture is the sum of the vapor pressures of the individual immiscible components at a given temperature:

$$P_{tot} = P_{H_2O}^0 + P_{organic}^0$$

In contrast to the behavior of liquid solutions composed of miscible components that is described in the previous section, the total vapor pressure of liquid mixtures composed of immiscible components is always greater than the vapor pressure of any one of the mixture components. *The boiling point of the immiscible mixture of volatile compounds must therefore be lower than the boiling point of any one of the mixture components.*

In the case of an immiscible mixture of water (b.p. = 100°C) and limonene (b.p. = 178°C),

$$P_{tot} = P_{H_2O}^0 + P_L^0$$

where $P_{H_2O}^0$ is the vapor pressure of pure water and P_L^0 is the vapor pressure of pure limonene. When $P_{tot} = P_{atm}$ boiling will occur. At $P_{atm} = 760$ Torr the temperature that is required to boil this mixture can be determined from Figure 5.1 as 98°C (Figure 5.2). At this temperature, which is 2°C less than the boiling point of the more volatile component, water, both water and limonene codistill.

The composition of the vapor as it condenses can be determined by approximating the behavior of each vapor component as that of an ideal gas:

$$P_n^0 V_n = N_n RT$$

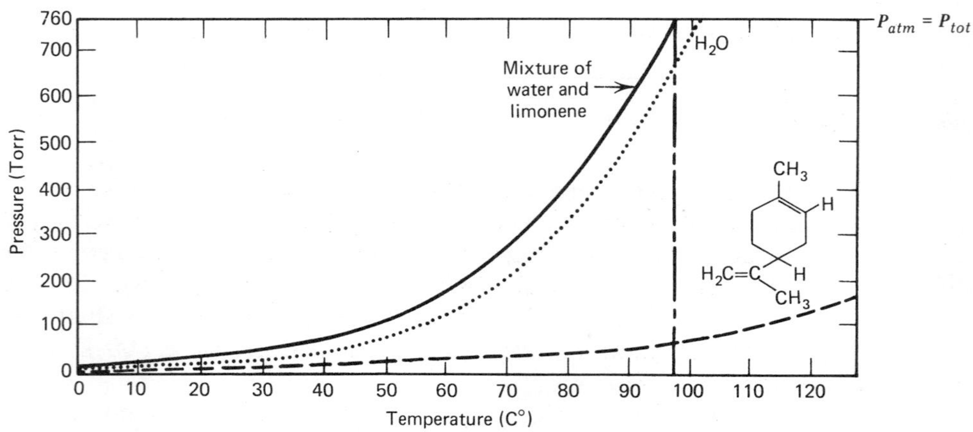

Figure 5.2 *Steam distillation of limonene. Relationship between vapor pressure and temperature.*

where P_n^0 is the vapor pressure of the pure liquid, V_n is the volume occupied by the vapor, N_n is the number of moles of component n ($N_n = g_n/\text{M.w.}_n$), R is the universal gas constant, and T is the absolute temperature. From this relationship the weight ratio of the two volatile components, water and limonene in the present experiment, can be calculated:

$$\frac{P_L^0 V_L}{P_{H_2O}^0 V_{H_2O}} = \frac{(g_L/\text{M.w.}_L)RT}{(g_{H_2O}/\text{M.w.}_{H_2O})RT}$$

Since the volume that contains the water and limonene vapors is the same ($V_L = V_{H_2O}$) and since both components distill at the same temperature, the above expression becomes

$$\frac{g_L}{g_{H_2O}} = \left(\frac{P_L^0}{P_{H_2O}^0}\right)\left(\frac{\text{M.w.}_L}{\text{M.w.}_{H_2O}}\right)$$

The weight ratio of two immiscible components in the distillate from a steam distillation depends on the ratios of their partial vapor pressures and their molecular weights. At 98°C $P_{H_2O}^0 = 700$ Torr and $P_L^0 = 60$ Torr; therefore, the weight ratio of limonene to water is

$$\frac{g_L}{g_{H_2O}} = \left(\frac{60 \text{ Torr}}{700 \text{ Torr}}\right)\left(\frac{136 \text{ g/mol}}{18 \text{ g/mol}}\right) = 0.65$$

That is, the distillate from steam distillation of limonene will contain 0.65 g of limonene for every gram of water that codistills. Even though limonene has a vapor pressure that is an order of magnitude less than that of water at the distillation temperature, the weight yield of the less volatile organic compound is raised due to the low molecular weight of water.

The apparatus for steam distillation is shown in Figure 5.3. Steam from an external source is passed into the distillation flask through a glass tube to a level below the upper level of the organic liquid. The flask may be heated on a steam bath or by gentle heating with a Bunsen burner to avoid condensation of a substantial volume of water from the external source of steam. The Claisen head that is fitted onto the distillation flask helps to prevent the liquid from splattering into the condenser during the distillation.

If the organic compound has a sufficiently high vapor pressure and molecular weight so that the weight ratio g_n/g_{H_2O} is expected to be greater than 0.1, water can be added directly to the distillation flask and heated to produce steam internally. This alternate procedure avoids the use of an external source of steam, but cannot be used when large quantities of water are required to steam distill the volatile organic compound. To employ this procedure the adapter and glass tube for external steam attached to the Claisen head (Figure 5.3) are replaced by a stopper.

External steam is obtained from a laboratory steam line or may be generated by boiling water in an apparatus such as that shown in Figure 5.4. If steam from a laboratory steam line is used, a water trap (Figure 5.5) is normally inserted between the steam line and the distillation apparatus to allow removal of water and impurities that are present in the steam line.

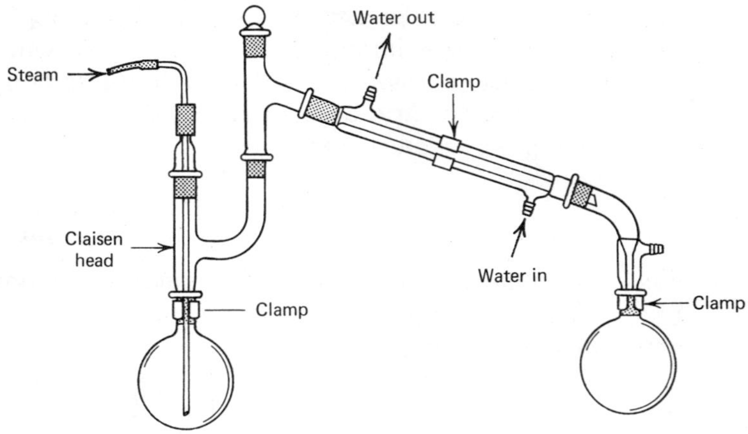

Figure 5.3 *Steam distillation apparatus.*

Assembly Instructions

1. Lightly grease the male joints of the standard taper glassware.
2. Clamp a 250-ml round-bottomed flask to a ring stand directly over a steam bath or Bunsen burner.
3. Insert the Claisen head and assemble the distillation head, condenser, vacuum take-off adapter, and collection flask as shown in the above figure and described in Figure 3.4.
4. Insert a piece of long glass tubing through the thermometer adapter and down through the Claisen head into the bottom of the distilling flask. Mark the level of the tubing above the thermometer adapter, remove the tubing and bend it to a 90° angle by heating the marked section with a Bunsen burner. Now reinsert the bent tubing into the distillation apparatus.
5. Complete the assembly by connecting the external steam line to the water trap (Figure 5.5) and the water trap to the bent glass tube with short pieces of rubber tubing. If a laboratory steam line is used, condensed water should be drained from the steam line before connecting the steam line to the water trap.
6. Do not fill the distilling flask more than one-third full to ensure that water condensing from the steam will not fill the flask.

Isolation of Volatile Organic Compounds from Natural Sources

The volatile components of plants constitute an enormous wealth of organic chemicals that have served mankind as pharmaceuticals, perfume ingredients, preservatives, flavoring agents, and in many other capacities. Volatile components are generally isolated from a plant by steam distillation at atmospheric pressure. The plant is usually finely divided or blended to release its volatile components from the plant's cellular structure and steam is then introduced into the mixture to codistill the volatile organic compounds with water. If the plant source is easily permeated, as is the case with leaves, flower petals and buds, grasses, and herbs, steam distillation is carried out without pretreatment

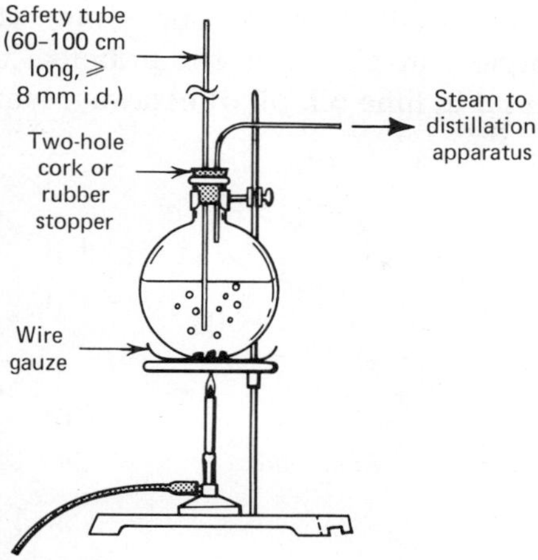

Figure 5.4 *Laboratory steam generator.*

Fill the round-bottom flask half-full with water and, after adding several boiling chips, heat the water to boiling. The straight glass tube that is inserted into the water provides a pressure release if steam is generated at too rapid a rate for its removal through the bent glass tube.

of the vegetable material. The mixtures of volatile organic compounds that are obtained from plants and are immiscible with water are known as *essential oils*. The odors of plants are due to these essential oils.

Essential oils have been isolated from a wide variety of plants but constitute only a small percentage of the total plant composition. Oil of caraway, for example, is obtained from caraway seeds but constitutes only 3 to 7% of the total weight of the seed; rose oil is obtained in only 0.04% weight yield from the petals and buds of roses. Both rose oil and oil of caraway are used as perfume ingredients. Orange oil, which is employed in this experiment and is obtained from orange peels, constitutes as much as 1% of the weight composition of the peels from the Valencia variety of the fruit.

The major constituents of the essential oils are terpenes and their oxygenated derivatives. Terpenes are structurally composed of isoprene units combined

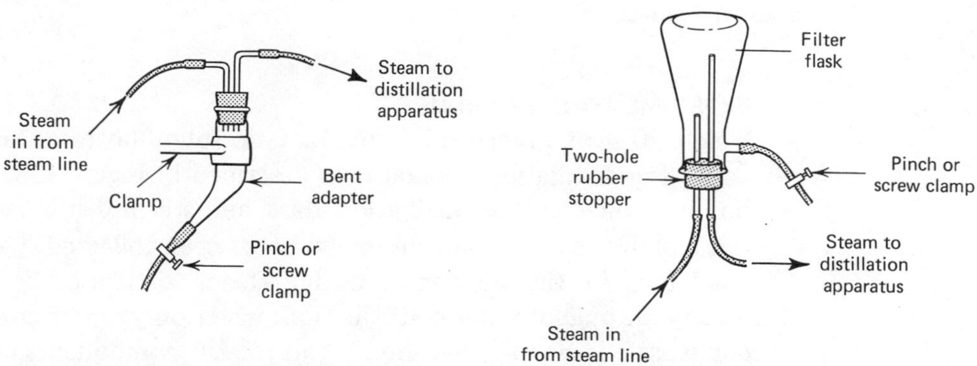

Figure 5.5 *Water traps for use in steam distillations that employ an external source of steam.*

in a "head-to-tail" fashion. Limonene, which is the most widely distributed cyclic terpene in plants, is the principal ingredient of orange oil (90 to 95%) and is found in lime oil, oil of caraway, lemon oil, and many other essential oils.

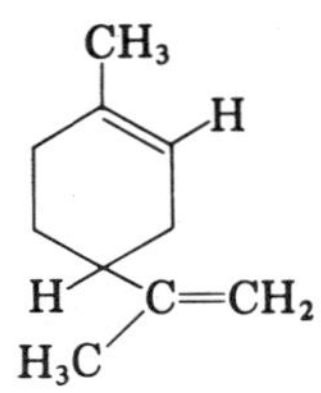

Limonene

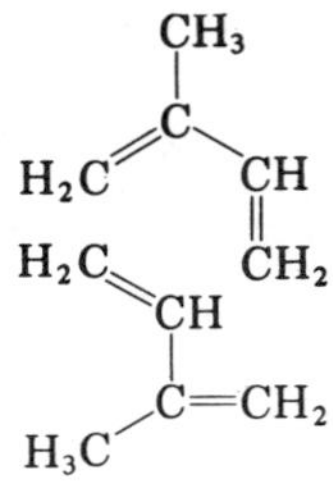

Two isoprene units arranged in "head to tail" fashion

Prelab Questions

1. Calculate the vapor pressure of the liquid solution composed of 74 g of ethyl ether and 36 g of trimyristin at the boiling point of pure ethyl ether. Assume Raoult's Law is obeyed and that trimyristin is nonvolatile. To distill ethyl ether will you have to raise or lower the pot temperature from the boiling temperature of pure ethyl ether?

2. Calculate the ratio of number of moles of limonene to number of moles of water for the steam distillation of limonene at 98°C. How does this ratio compare with the ratio of partial vapor pressures, $P^0_L/P^0_{H_2O}$?

3. What is the minimum volume of water that is required to steam distill 20 g of limonene? Assume that the density of water is 1.0 g/ml.

4. Water and ethyl ether (b.p. = 35°C) are immiscible. What is the maximum predictable temperature at which ethyl ether will codistill with water?

5. Describe a flow diagram for the experimental procedure that you will employ in this experiment.

6. Will the organic components of the distillate from the steam distillation form the upper or lower liquid layer?

EXPERIMENTAL PROCEDURE

Weigh 20 g of orange oil into a 250-ml round-bottom flask and construct the distillation apparatus as described in Figure 5.3. Add several boiling stones to the distillation flask and steam distill the orange oil until all of the volatile organic material has been collected. The vapor that condenses in the condenser during steam distillation is discernably cloudy throughout the codistillation; when only water condenses, the condensate is clear. To ensure that all volatile components of the orange oil have been distilled, collect 1 ml of the condensate in a 10-ml graduated cylinder or a small test tube and observe if only one liquid layer is present; if two liquid layers are observed, pour the contents of the graduated cylinder or test tube into the receiving flask and continue the steam distillation.

After steam distillation is complete, pour 30 ml of hexane into the receiving flask. Pour this liquid mixture into a separatory funnel and draw off the water layer. Transfer the organic layer to a 100-ml round bottom flask without entrapping any of the water layer. Save the organic solution for fractional distillation in the next experiment.

Alternate Experiment: Isolation of Oil of Nutmeg

Weigh 20 g of ground nutmeg into a 250-ml round-bottom flask and construct the distillation apparatus as described in Figure 5.3. Add 50 ml of water and several boiling stones to the distillation flask and steam distill the volatile components of nutmeg using an external source of steam. Heat the distillation flask on a steam bath or by gentle heating with a Bunsen burner during steam distillation to minimize condensation of water from the steam line. Continue the steam distillation for 90 min or until the condensate is clear, whichever time is shorter. After steam distillation is complete, add 50 ml of ethyl ether to the distillate and thoroughly mix the ether and aqueous layers. Transfer the ether-water mixture from the receiving flask into a separatory funnel. Rinse the receiving flask with 5 ml of ether and then pour this solution into the separatory funnel. Carefully separate the ether layer from the water layer and transfer the ether layer into a 50-ml round bottom flask. Remove the ether by simple distillation as is described in Experiment 5. Obtain the weight of the residue after all of the ether has been distilled and calculate the weight percentage of oil of nutmeg in ground nutmeg. Ground nutmeg contains terpineol, borneol, pinene, camphene, safrole, eugenol, linalool, and limonene, in addition to a small portion of the trimyristin that you isolated in Experiment 3.

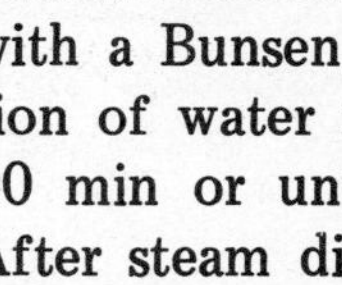

Terpineol
b.p. = 220°C
m.p. = 35°C

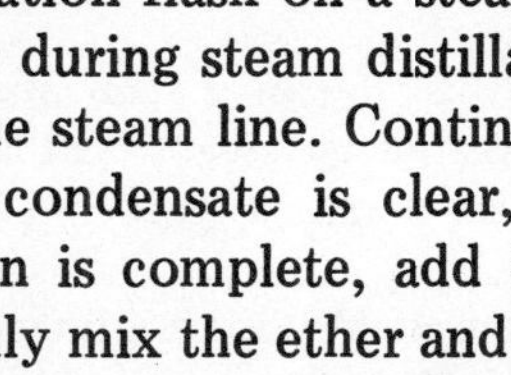

Camphene
b.p. = 158°C
m.p. = 45-6°C

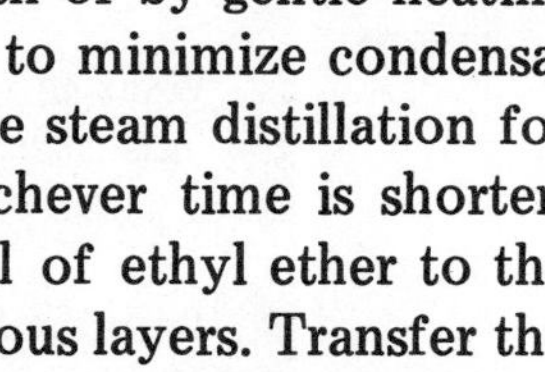

α-Pinene
b.p. = 156°C

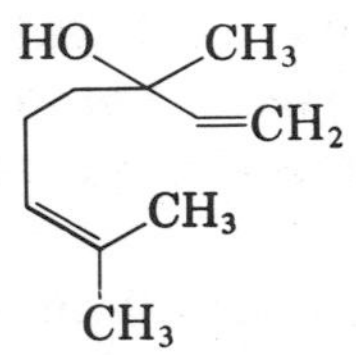

Linalool
b.p. = 198–200°C

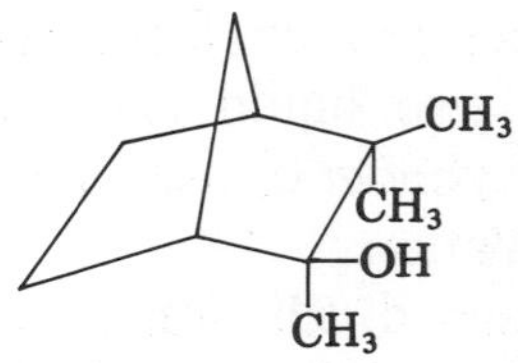

Borneol
b.p. = 212°C
m.p. = 208°C

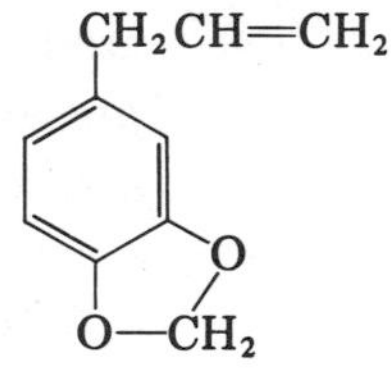

Safrole
b.p. = 234°C

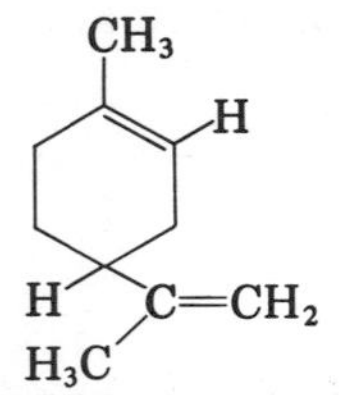

Limonene
b.p. = 178°C

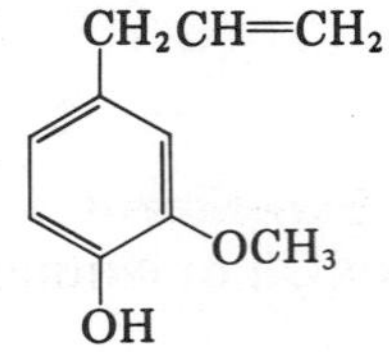

Eugenol
b.p. = 255°C

Postlab Questions

1. Explain why the vapor that condenses during steam distillation is cloudy.

2. What should be the odor of a plant after steam distillation? How are the odors of essential oils that are isolated by steam distillation related to the original odor of the plants from which the oils are obtained?

3. Describe how you could impart a lime odor to soap. (Compare the odors of orange oil and myristic acid and relate the process for making perfumed soap.)

4. Describe the vapor pressure relationship between P_{tot}, $P^0_{H_2O}$, and the partial pressures of two miscible organic liquids that are immiscible with water. What is the partial pressure of each of the organic compounds relative to the partial pressure of the pure compound?

5. Comment on the feasibility of steam distillation of (a) a volatile compound that is a solid at room temperature, (b) a volatile compound that reacts with water at temperatures near the boiling point of water, and (c) a volatile compound that is miscible with water.

Experiment Six

Fractional Distillation: Purification of Limonene

Fractional distillation is employed for the separation of volatile miscible liquids. In this distillation method selective separation of a volatile compound occurs when the boiling liquid is maintained at a sufficiently high temperature, and the pathway between the boiling liquid and the distillation head is appropriately designed to allow the vapors issued from the boiling liquid to condense and redistill from the condensed liquid. Condensation and redistillation from the condensed liquid continues along the pathway to the distillation head. The effectiveness of the separation, which is determined by the purity of the distillate, is a function of the nature of the liquid components and of the pathway to the distillation head. In this experiment the solution of hexane and limonene that you obtained in Experiment 5 will be fractionally distilled in order to separate limonene from hexane.

Purification by Fractional Distillation

As predicted by Raoult's Law, the composition of the vapor above a boiling ideal solution differs from the composition of the liquid. For example, the composition of hexane and limonene in a solution that consists of only these two components is given by

$$1.00 = X_{\text{Hex}} + X_{\text{Lim}}$$

where X_{Hex} is the mole fraction of hexane and X_{Lim} is the mole fraction of limonene in solution. In the vapor, however, the relationship between total pressure (P_{tot}) and the mole fraction of each solution component is

$$P_{\text{tot}} = P^{\circ}_{\text{Hex}} X_{\text{Hex}} + P^{\circ}_{\text{Lim}} X_{\text{Lim}}$$

That is, the composition of the vapor is dependent on the mole fraction of the solution components as well as on the partial pressures of the pure compounds.

The more volatile component is concentrated in the vapor phase relative to its molar composition in the liquid phase: since $P^\circ_{Hex} > P^\circ_{Lim}$, then

$$\underset{\textit{Mole ratio in vapor}}{\frac{P^\circ_{Hex} X_{Hex}}{P^\circ_{Lim} X_{Lim}}} > \underset{\textit{Mole ratio in liquid}}{\frac{X_{Hex}}{X_{Lim}}}$$

The relationship between the boiling point and the composition of the liquid and vapor phases from a binary solution of hexane and limonene is described in Figure 6.1. The lower curve represents the boiling points of all mixtures of hexane and limonene. The upper curve describes the composition of the vapor phase that is in equilibrium with the boiling liquid phase at the same temperature. For example, at a percentage molar composition of 40% hexane–60% limonene the liquid will boil at 118°C and the vapor phase in equilibrium with the boiling liquid phase has a composition of 73% hexane–27% limonene. The vapor phase is enriched in hexane. If the vapor phase that is in equilibrium with the boiling liquid phase is condensed, the condensate will be composed of 73% hexane–27% limonene.

The fundamental operation of fractional distillation can be understood by taking a binary solution through a series of consecutive distillations (Figure 6.2). For example, condensation of the vapor that is in equilibrium with the boiling liquid (40% hexane-60% limonene), distillation of the condensate (73% hexane–27% limonene) and collection of its vapors, and successive repetition of this process results in a final condensate that is composed of essentially pure hexane (Figure 6.3). Each successive distillation occurs at a lower boiling temperature to the limit that is the boiling point of pure hexane. In addition,

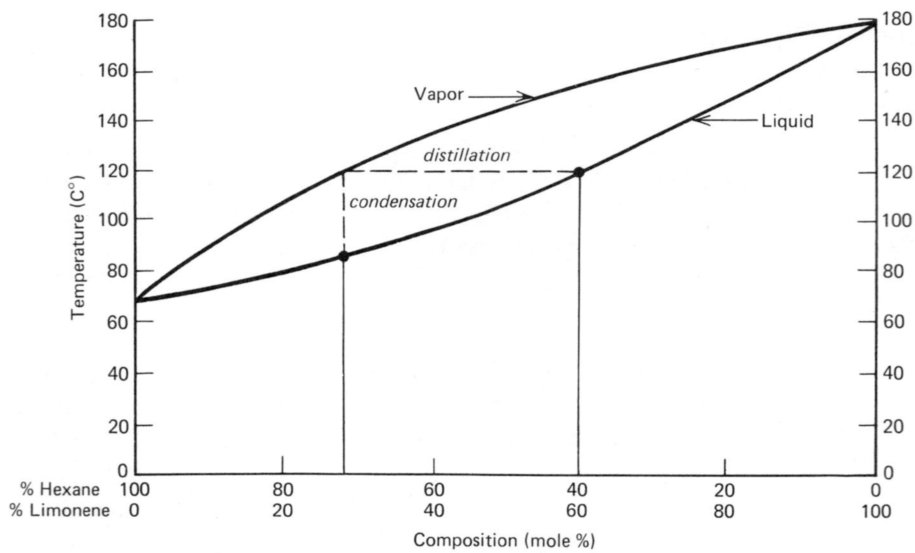

Figure 6.1 *Boiling point-composition diagram for the binary solution of hexane and limonene.*

Vapor condenses at each stage of the fractionating column and redistills to the next stage until vapor reaches the distillation head. The boiling point of the liquid at the lower stage is higher than the boiling point of the liquid at the higher stage. A lightweight plastic ball atop the reservoir of each stage in this fractionating column prevents condensed liquid from reentering the next lower stage or the distillation flask.

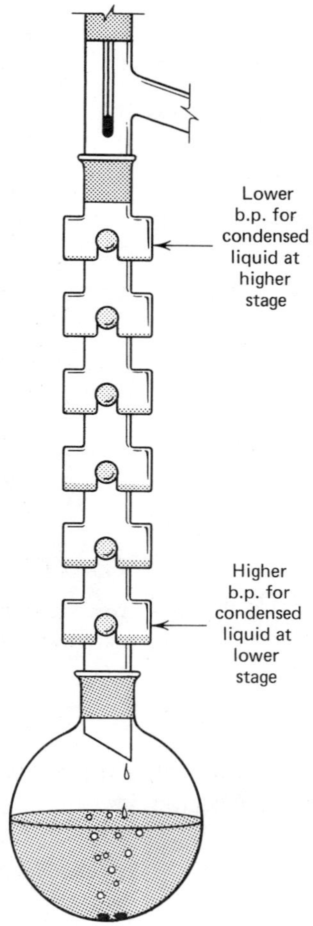

Figure 6.2 *Fractional distillation as a series of consecutive distillations.*

as hexane is selectively removed from the boiling solution the liquid becomes enriched in limonene and the temperature required to maintain boiling increases to the limit that is the boiling point of pure limonene.

The requirements for fractional distillation are represented in this analysis of consecutive distillations. The temperature of the liquid must be sufficiently high to maintain boiling. The vapor initiating from the boiling liquid condenses and redistills along the pathway between the boiling liquid and the distillation head, causing a temperature gradient to extend along this pathway from the higher pot temperature to the lower distillation head temperature. Thus, the design of the pathway between the distillation flask and the distillation head (the *fractionating column*) is an essential consideration in fractional distillation. In addition, as the vapor that is enriched in the more volatile component is condensed, the boiling liquid becomes enriched in the less volatile component and the temperature required to maintain boiling increases.

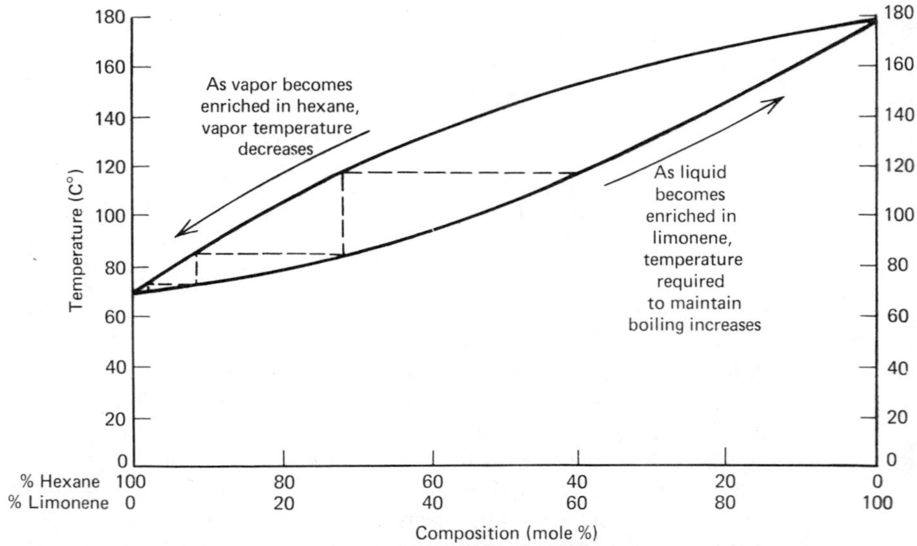

Figure 6.3 *Fractional distillation of the binary solution of hexane and limonene.*

Note that the smaller the boiling point difference between the two liquid components, the narrower the temperature gradient within the fractionating column for optimal separation.

Fractionating Columns and their Efficiencies

A fractionating column provides the vertical pathway through which the vapor must pass from the distillation flask to the distillation head. At the boiling point of the liquid, vapor rises into the column. The fractionating column initially cools the vapor so that you observe a line of condensing vapor rise from the distillation flask through the fractionating column. As the vapor rises along this vertical pathway, a temperature gradient that is due to the consecutive condensation and redistillation of the vapor issued from the boiling liquid forms in the column. The temperature at the lower end of the column is higher than the temperature in the upper portion of the column. The temperature gradient ensures the equilibrium condition for fractional distillation as the vapor condenses and redistills.

The fractional distillation apparatus described in Figure 6.4 is designed for the separation of volatile liquids whose boiling points differ by >50°C. A Liebig condenser, which is constructed with a spacious air pocket for insulation, serves as the fractionating column. A West condenser, which has a more confined inner column, is employed to condense the vapor. For applications in which separations of volatile liquids having boiling point differences of less than 50°C are required, the Liebig condenser can be filled with glass beads, glass helices, or stainless steel turnings to increase the surface area in the column for condensation-redistillation of the vapor. In these applications the column and distillation head are also insulated with a layer of glass wool or

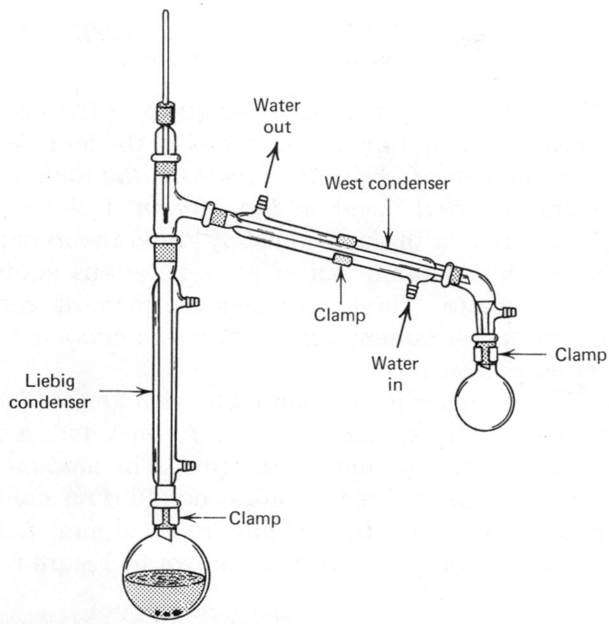

Figure 6.4 *Apparatus for fractional distillation.*

Assembly Instructions

1. All glassware must be clean and dry before assembly. Water in the outer column (*jacket* of column) of the Liebig condenser is to be avoided.
2. Lightly grease the male joints on the standard taper glassware. Grease on the joints of the vacuum adapter is not necessary for distillations at atmospheric pressure since vapors will not escape from the apparatus after they are condensed.
3. Assemble the glassware as described in Figure 6.4 according to instructions given in Figure 3.4. The Liebig condenser and distillation head are attached to the distilling flask so that they are exactly perpendicular to the laboratory benchtop.
4. Close the two openings to the jacket of the Liebig condenser with plugs of cotton or rubber bulbs so that the resulting dead air space will insulate the internal column against heat loss to the atmosphere.
5. The distilling flask used in the distillation should be not more than two-thirds full nor less than one-third full at the start of the distillation.

aluminum foil in order to maintain a narrow equilibrium temperature gradient within the column.*

The efficiency and practicality of a fractionating column is determined by the degree of separation that can be obtained and by the volume of liquid that is retained by the column. The efficiency of a column is measured by the *number of theoretical plates*, *n*, which is defined by the following equation for a two-component (A and B) liquid:

*Fractional distillation operates effectively as an *isothermal* process. That is, the temperature within the column is ideally controlled by the condensation-redistillation operation of the vapor within the column rather than by external cooling. Vacuum jacketed columns are employed to aid in maintaining isothermal conditions.

$$\frac{\text{Mole fraction ratio in condensed vapor}}{\text{Mole fraction ratio in liquid}} = \frac{X_A^{\text{Vap}}/X_B^{\text{Vap}}}{X_A^{\text{Liq}}/X_B^{\text{Liq}}} = \left(\frac{P_A^\circ}{P_B^\circ}\right)^{n+1}$$

The ratio P_A°/P_B° is often referred to as the enrichment factor since the vapor above a boiling liquid is enriched in the more volatile component by this ratio. *The number of theoretical plates is the number of condensation-redistillation operations that occur in the column.* Columns such as that shown in Figure 6.4 operate with approximately three theoretical plates. When filled with glass beads, this column can effect separations equivalent to as many as six theoretical plates. These estimates are made in reference to the standard 20-cm fractionating column; *the longer the column the greater the number of theoretical plates.*

Although a long column allows a greater degree of separation between the volatile components of a mixture than does a short column, the long column retains a greater volume of liquid. The amount of liquid required to maintain the condensation-redistillation equilibrium condition within the fractionating column is the *hold-up volume* of a column. A balance must be met with fractionating columns like that shown in Figure 6.4 so that an optimum amount

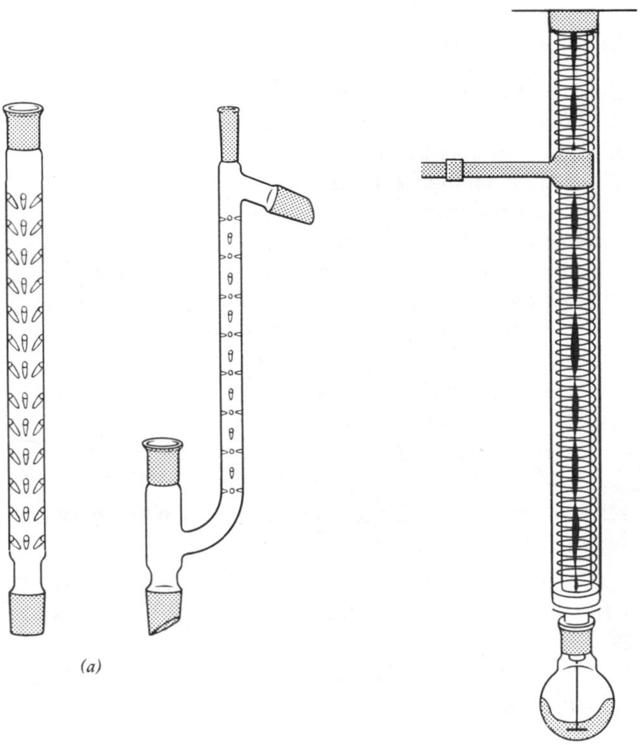

(a) (b)

Figure 6.5 *(a) Vigreux column: the glass of the column is indented at regular intervals to provide more surface area than in conventional straight columns. (b) Spinning band column: a motor-driven wire (or Teflon) band is rotated at high speed to increase liquid-vapor contact within the column without increasing hold-up volume.*

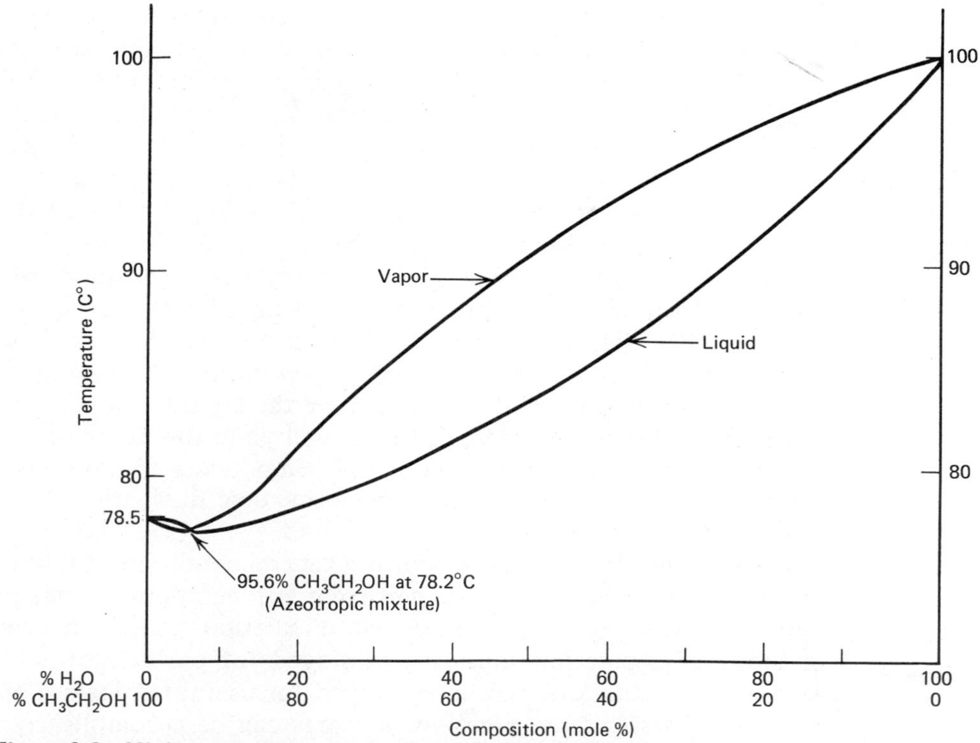

Figure 6.6 *Minimum-boiling azeotrope of ethanol and water.*

of liquid is delivered to the receiving flask with a satisfactory degree of component separation.

The apparatus described in Figure 6.4 is limited by a hold-up volume of nearly 3 ml. Adding glass helices to this column doubles the number of theoretical plates but also doubles the hold-up volume. The Vigreux column (Figure 6.5*a*) is designed to provide additional column surface area for increased condensed liquid-vapor contact but does not measurably increase the hold-up volume over that of a straight column. The most efficient column for fractional distillation, both in terms of the number of theoretical plates and in minimum hold-up volume, is the spinning band column (Figure 6.5*b*). Spinning band column assemblies having more than 100 theoretical plates and hold-up volumes of less than 0.5 ml are commercially available and are capable of efficient separations of compounds whose boiling points differ by less than 2°C.

Fractional Distillation of Nonideal Solutions

Solutions whose liquid-vapor behavior differs from that predicted by Raoult's Law are known as *nonideal solutions.* The most common nonideal behavior occurs with solutions that display a higher vapor pressure than that predicted by Raoult's Law. In such cases, the forces of attraction between molecules of different compounds are weaker than those between molecules of the pure compound. As a result of this weaker attraction, the total pressure of the two-component mixture is greater than the vapor pressure of the pure less volatile component in a certain composition range. Figure 6.6 illustrates this behavior in the boiling point–composition diagram for the system ethanol-water. At ethanol-water compositions between 100% ethanol and 95.6% ethanol–4.4% water, the boiling points of the liquid solution are less than the boiling points of either pure ethanol (b.p. = 78.5°C) or pure water (b.p. = 100°C). The minimum boiling composition has a constant boiling point; the liquid and vapor in equilibrium with the liquid have the same composition. Mixtures that boil at a constant boiling point with constant liquid and vapor compositions are known as *azeotropic mixtures.* Such mixtures cannot be separated by fractional distillation into their pure components.* Fractional distillation of these nonideal solutions will yield the azeotropic mixture as the separation limit; only after one of the components of the azeotropic mixture has been completely removed in the azeotropic distillation can the remaining component be distilled as a pure compound.

Nonideality in boiling behavior is a general condition of mixtures composed of polar compounds, and, in particular, of compounds that are capable of hydrogen bonding. When fractional distillation yields an azeotrope, other methods of separation must be employed for purification. Often, however, azeotropic distillations can be employed for useful purposes. For example, the removal of water from reaction mixtures can be accomplished through azeotropic distillation with toluene (b.p. of azeotrope = 85°C, 80% toluene–20% water) or with hexane (b.p. of azeotrope = 62°C, 94% hexane–6% water).

Prelab Questions

1. If the column you use for fractional distillation is capable of one theoretical plate and your liquid mixture is composed of 40% hexane–60% limonene, estimate the boiling point and mole percentage of hexane for the first few drops of distillate from Figure 6.3.

2. Assume that your fractionating column is capable of three theoretical plates. (a) What will be the head temperature while hexane is being collected as the distillate? Explain. (b) What will be the boiling temperature of the liquid in the distilling flask as the last drop of hexane is collected? Explain.

3. Explain the significance of a temperature range of 72 to 68°C for the boiling point of hexane that is distilled from a hexane-limonene mixture. Would you estimate that this fraction is composed of pure hexane?

*Anhydrous *absolute alcohol*, 99.9+%, may be prepared from the azeotropic mixture by distillation from a chemical drying agent such as calcium oxide. Alternatively, anhydrous ethanol may be obtained by adding benzene to the ethanol-water azeotrope, and making use of the fact that the three-component mixture will form an azeotrope boiling at 64.6°C having the composition of 18.5% alcohol, 7.4% water, and 74.1% benzene. Anhydrous ethanol is obtained after azeotropic distillation of water in the three-component azeotrope. However, absolute ethanol is often contaminated by trace amounts of benzene when this latter method is employed.

4. Predict the distillation curve (b.p. versus volume of distillate, similar to Figure 3.3) for the fractional distillation of the hexane-limonene mixture.

5. The head temperature often drops towards room temperature just before a high boiling fraction reaches the distillation head and just after a low boiling fraction has been collected. Would this observation be compatible with the distillation curve that you constructed for Question 4? Does a decrease in the head temperature necessarily correspond to a decrease in the boiling temperature of the liquid in the distilling flask?

6. Estimate the time required for distilling vapors to reach thermal equilibrium with a thermometer inserted into the distillation head. At what point in a distillation should you begin to record the temperature of the distillate?

EXPERIMENTAL PROCEDURE

CAUTION: Hexane and limonene are volatile and flammable organic compounds. Be certain that there are no flames in your area before you begin this experiment.

Construct the distillation apparatus as described in Figure 6.4 using as the distilling flask the 100-ml round-bottom flask containing the hexane-limonene mixture that you obtained in Experiment 5. For this experiment use a graduated cylinder as the receiving flask rather than a round-bottom flask so that you can measure the volume of distillate. Add several boiling stones to the distillation flask and begin heating the hexane-limonene solution. Describe your observations of the distillation process when boiling commences and the condensing vapor rises in the fractionating column toward the distillation head. Include in this description your visual observations as well as the approximate rate of heating and time required for vapor to reach the distillation head after boiling has commenced. The rate of heating should be sufficiently rapid so that vapor reaches the distillation head within 15 min after the liquid has begun to boil. Vapor should condense on the thermometer and fall in droplets into the distillation flask; the rate of heating is too rapid if condensation of vapor on the thermometer does not occur. Slowly increase the rate of heating as hexane distills and is collected in the receiving graduated cylinder.

Record the head temperature when the first drop of distillate is collected and the temperatures at each 2-ml interval for collection of distillate. When the temperature begins to rise above 85°C, transfer the collected distillate from the graduated cylinder into a receiving flask and save this fraction (Fraction 1) for further use in Experiment 7. Record the head temperature at smaller volume intervals (<0.5 ml) as the head temperature increases from 85° to the temperature at which pure limonene distills. When the head temperature has stabilized so that the rate of temperature change is less than 1°C per 0.5 ml of distillate, again transfer the distillate from the graduated cylinder into a receiving flask (Fraction 2). Begin the collection of the limonene fraction (Fraction 3) in a clean, dry graduated cylinder and record the distillation temperature at 2-ml intervals. Continue the distillation until less than 3 ml of liquid remains in the distillation flask. If the temperature drops from the boiling point of limonene during this stage of the fractional distillation, either not enough heat is being applied to the distillation flask or not enough liquid remains in the distillation flask and fractionating column to continue. Save Fraction 3 for further use in Experiment 7.

Record the barometric pressure, make any necessary thermometer corrections, and list the boiling point ranges for Fractions 1, 2, and 3. Plot boiling point versus volume of distillate and evaluate the efficiency of the hexane-limonene separation.

Postlab Questions

1. Why must the rate of heating be increased as hexane distills from the distillation flask?

2. The temperature when the first drop of distillate is collected is lower than the temperature after 1 ml of distillate is collected. However, the first drop of distillate has the same composition as the first ml of distillate. How do you account for the temperature difference?

3. Column efficiency for fractional distillation is lower when the rate of heating is too rapid. Assume that, due to rapid heating, your initial column efficiency is one theoretical plate, and that, after adjustment of the heating rate, the column efficiency is two theoretical plates. If you are distilling a 40% hexane-60% limonene mixture, what will be the boiling point of the initial distillate (first few drops) and the boiling point of the distillate after adjustment of the heating rate?

4. Water (b.p. = 100.0°C) and formic acid (b.p. = 100.7°C) form an azeotropic mixture that has a boiling point of 107.3°C (77.5% formic acid-22.5% water). Draw a boiling point-composition diagram for this maximum boiling azeotrope and describe this deviation from Raoult's Law.

5. What will be the composition of the first fraction from an efficient distillation of a solution that contains 90% formic acid-10% water (refer to Question 4)? What will be the composition of the last fraction from this distillation?

Experiment Seven

Observing Chemical Reactions: Solubility Tests, Chemical Characterization Tests, and Thin Layer Chromatography

The structure of an organic compound determines the chemical reactions the compound can undergo. Conversely, knowledge of the reactions that a compound undergoes provides information about the structure of that compound. Certain transformations are characteristic of particular functional groups. Under a uniform set of reaction conditions a reaction will occur with a compound that possesses a particular functional group and will not occur with another compound that does not possess that functional group. For example, the reaction of bromine at the carbon-carbon double bond of an alkene is a characteristic reaction of alkenes, but bromine does not react with an alkane under similar reaction conditions. Such characteristic reactions are used to classify organic compounds by functional groups and to assist in the structural identification of organic compounds.

The changes that occur during a chemical reaction are often directly observable. In Experiment 4, for example, the extent of hydrolysis of trimyristin could be followed by observing the decrease in the volume of the oil layer with time; when the oil layer of trimyristin was no longer visible, hydrolysis was assumed to be complete. If such observable reactions are specific to one or only a few functional groups and occur rapidly, they can be used to identify the functional groups that are present in an organic compound. In this experiment you will be introduced to methods for functional group identification through *solubility tests* for acids and bases and by rapid, observable chemical reactions known as *chemical characterization tests*.

Not all chemical reactions are visually observable. Many chemical reactions occur in a homogeneous solution and yield products that are visually identical with the reactants. In such situations the extent of reaction is conveniently followed by methods that separate the reactants from the products for subsequent visual identification. *Thin layer chromatography* (TLC) is introduced in this experiment as a method for the separation of reactants from products.

Solubility Tests for Organic Compounds

Organic compounds can be classified according to their solubility behavior in a select number of solvents and solutions. The characteristic solubilities of organic compounds in water and in ethyl ether that are described in Experiment 1 (Figure 1.1), for example, provide a method to distinguish ionic and very polar organic compounds from nonpolar and moderately polar compounds. In addition to these solubility properties, however, classification of acidic and basic organic compounds can be made by determining their solubilities in basic and acidic aqueous solutions. If an organic compound that is not soluble in water is soluble in 10% aqueous sodium hydroxide, that compound has undergone a chemical change to form a water soluble ionic compound in an acid-base reaction. Compounds that dissolve in 10% aqueous sodium hydroxide are more acidic than water and must therefore possess an acidic functional group. Like-

Solubility test for organic acids

$$HA + OH^- \longrightarrow H_2O + A^-$$

HA	OH^-	H_2O	A^-
Organic acid— insoluble in H_2O	*10% Aqueous sodium hydroxide*	*Water*	*Soluble in* H_2O

wise, an organic compound that is not soluble in water but is soluble in 10% aqueous hydrochloric acid must possess a basic functional group.

Solubility test for organic bases

$$B + H_3O^+ \longrightarrow H_2O + BH^+$$

B	H_3O^+	H_2O	BH^+
Organic base— insoluble in H_2O	*10% Aqueous hydrochloric acid*	*Water*	*Soluble in* H_2O

The solubilities of organic compounds in concentrated sulfuric acid also provide useful structural information for compounds that have been shown to be insoluble in water, 10% hydrochloric acid, and 10% sodium hydroxide. Organic compounds that possess functional groups with oxygen or nitrogen atoms are protonated by concentrated sulfuric acid since sulfuric acid is a stronger acid than most protonated organic compounds that possess such functional groups. In addition, unsaturated compounds such as alkenes undergo addition reactions in sulfuric acid to form polar products that dissolve in the acid medium.

Solubility test for neutral organic compounds

$$N + H_2SO_4 \longrightarrow HN^+ + HSO_4^-$$

N	H_2SO_4	HN^+	HSO_4^-
Neutral compound— insoluble in H_2SO_4	*Conc. sulfuric acid*	*Soluble in* H_2SO_4	*Hydrogen sulfate*

SOLUBILITY TEST PROCEDURE

3.0 ml of the solvent is slowly added in 1.0-ml portions to 0.1 g (0.1 ml) of the organic compound that is contained in a small test tube. The mixture is shaken vigorously after the addition of each portion of solvent to allow thorough mixing. Observe and record the change either in the volume of the liquid organic layer or in the amount of solid as solvent is mixed with the organic compound. The organic compound is soluble in the solvent if the organic compound dissolves completely.

Table 7.1 describes possible results obtained in solubility tests and their implications concerning the structures of organic compounds. Since compounds that are soluble in water will also be soluble in aqueous hydrochloric acid, aqueous sodium hydroxide, and concentrated sulfuric acid, these latter solubility tests are not performed on water soluble compounds. Likewise, since compounds that are soluble in either 10% aqueous hydrochloric acid or 10% aqueous sodium hydroxide will also be soluble in concentrated sulfuric acid, solubility tests in concentrated sulfuric acid are not carried out on these compounds. Color changes that are often observed when an organic compound is dissolved in concentrated sulfuric acid indicate the occurrence of a chemical reaction between the organic compound and sulfuric acid.

Chemical Characterization Tests

Certain chemical reagents react selectively with organic functional groups to produce distinct visual changes. These reactions are known as chemical characterization tests. A positive chemical characterization test—one in which the expected chemical reaction is observed to occur—indicates the presence of a particular functional group in the compound being examined. In this experiment you will investigate the use of chemical characterization tests for alkenes—chemical reactions that indicate the presence of a carbon-carbon double bond in an organic compound. Subsequent experiments will describe chemical characterization tests that are selective for other functional groups.

Reaction with Bromine in Carbon Tetrachloride

Alkenes react rapidly with molecular bromine to give the vicinal dibromide addition product in quantitative or nearly quantitative yield:

Bromine addition to alkenes

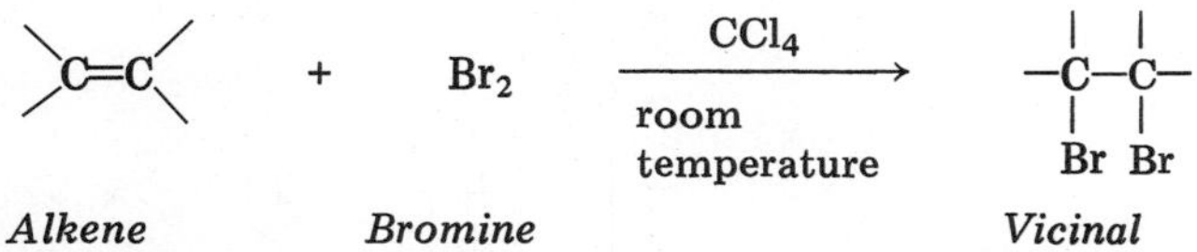

Alkene (usually colorless) | *Bromine* (red-brown) | *Vicinal dibromide* (colorless)

Since bromine is red-brown and the organic reactant and the dihalide product are colorless, decolorization of a bromine solution by bromine addition is a color test for the presence of a carbon-carbon double bond.

Table 7.1 Solubility Tests for Organic Compounds (+ = Soluble; – = Insoluble)

Solubility of Organic Compound in					Structural implication	Examples
H_2O	$(CH_3CH_2)_2O$	10% aq. HCl	10% aq. NaOH	H_2SO_4		
+	–				Ionic and very polar compounds	Ammonium salts, carboxylate salts
+	+				Very polar compound, usually having low molecular weight	Ethyl alcohol, acetone, acetic acid
–	+	+	–		Organic base	Amines
–	+	–	+		Organic acid	Carboxylic acids, phenols
–	+	–	–	+	Polar neutral compound	Ethers, alcohols, aldehydes, ketones, alkenes, alkynes
–	+	–	–	–	Nonpolar compound	Alkanes, alkyl halides

Bromine will also react with certain organic compounds to substitute a bromine atom for a hydrogen atom that is bonded to carbon:

Bromine substitution

$$\underset{}{RH} + \underset{\textit{Bromine}}{Br_2} \xrightarrow[\text{or light}]{\text{heat}} RBr + \underset{\textit{Hydrogen bromide}}{HBr}$$

The strong acid hydrogen bromide is produced in this substitution reaction. The absence of the strong acid hydrogen bromide differentiates bromine addition from bromine substitution reactions in which hydrogen bromide is produced. Thus, *if decolorization of the bromine solution occurs without concurrent formation of hydrogen bromide, this chemical characterization test indicates the presence of a carbon-carbon double bond.* Among the common families of organic compounds only alkenes and alkynes give a positive test for bromine addition; alkynes are selectively detected by alternative tests (Experiment 21). Although the bromine addition test does not differentiate alkenes from alkynes, this chemical characterization test does distinguish these two classes of unsaturated compounds from all other common classes of organic compounds.

PROCEDURE

In a small test tube dissolve 0.1 g of solid or four drops of the liquid organic compound in 1-2 ml of carbon tetrachloride. Avoid breathing the vapor of this chlorinated solvent. Add two drops of a 2% solution of bromine in carbon tetrachloride to the dissolved organic compound and thoroughly mix the reactants. Observe and record the color of the reactant solution during the addition and mixing process. If the bromine color fades, continue the dropwise addition until the bromine color persists and then test the vapor above the solution for hydrogen bromide by using a wet piece of litmus paper or *p*H test paper. Avoid skin contact with the bromine solution—bromine is an oxidant that can produce chemical burns.

Reaction With Potassium Permanganate

Potassium permanganate oxidizes alkenes to vicinal diols under mild conditions:

Permanganate oxidation of alkenes

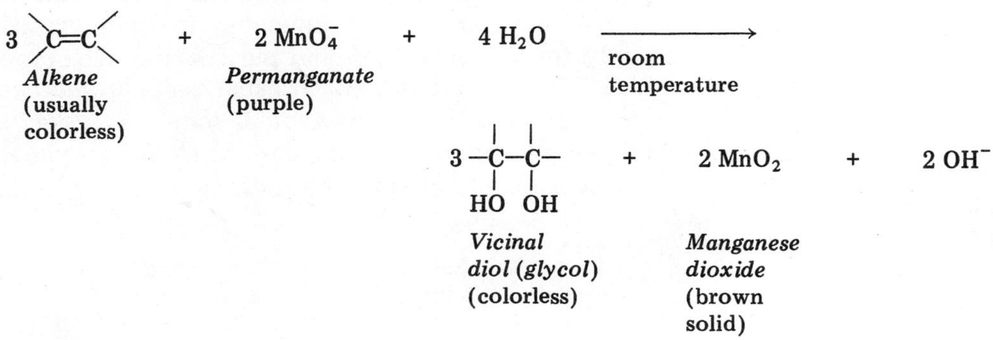

In this reaction the purple solution of potassium permanganate changes to dark brown and a brown insoluble precipitate of manganese dioxide is formed. Further oxidation of the glycol to ketone and carboxylic acid products may also occur, but these processes also produce decolorization of permanganate and manganese dioxide formation.

Although permanganate oxidation is an easily observable reaction, it is not a selective chemical characterization test for alkenes. All easily oxidized organic compounds, including alkynes, alcohols, aldehydes, and phenols, give a positive test with potassium permanganate. Indeed, a positive permanganate test indicates only that the organic compound is oxidized by this reagent, just as decoloration of bromine without hydrogen bromide evolution indicates only that the organic compound undergoes addition of bromine. However, the combination of several chemical characterization tests, each specifying the result from a particular reaction process, does limit the choice of functional groups that can be present in an organic compound whose structure is unknown. If an organic compound gives a positive test both for addition of bromine and for oxidation by permanganate, the organic compound can be classified as either an alkene or an alkyne with a high degree of confidence.

PROCEDURE

In a small test tube dissolve 0.1 g of the solid or four drops of the liquid organic compound in 1.0 ml of reagent grade acetone. Then add three drops of a 2% aqueous solution of potassium permanganate drop by drop with vigorous shaking of the test tube. Avoid contact of the permanganate solution with your skin. A positive permanganate oxidation test has occurred if more than one drop of the permanganate solution is decolorized and a brown precipitate of manganese dioxide is observed within 3 min.

Thin Layer Chromatography (TLC)

Developed over the past eighty years, the techniques of chromatography are relatively new to chemistry. Their value lies in their versatility—providing methods not only for the separation and purification of compounds but, also for product identification and quantitative analyses. *Chromatographic methods involve the reversible transfer of a compound that is adsorbed on a* **stationary phase** *into a* **mobile phase** *that is flowing past the stationary phase.* The exchange of the adsorbed compound between the stationary phase and the mobile phase is an equilibrium partitioning process:

$$\text{Compound on stationary phase} \overset{K}{\rightleftharpoons} \text{Compound in mobile phase}$$

The value of the equilibrium constant for this dynamic interchange is related to the difference in the attractive forces between the compound absorbed on the stationary phase and the compound dissolved in the mobile phase.

Chromatography is similar to extraction in that there is partition of a compound between two phases; but, unlike extraction, chromatography involves the transfer of a compound from a stationary phase into a mobile phase and back to the stationary phase. Since the mobile phase is flowing past the stationary phase, the time the compound remains dissolved in the mobile phase results in migration of the compound along the stationary phase in the direction of flow. The more strongly a compound is adsorbed onto the stationary phase (the lower the value of K), the less is its migration along the stationary phase. Selective separation of a mixture into its components by partition chromatography is due to differences in the migration of the individual components along the stationary phase.

Thin layer chromatography involves the partitioning of a compound between a solid adsorbent (the stationary phase) that is thinly coated on a flat plate of glass or rigid plastic and a solvent (the mobile phase) that flows through the solid adsorbent. As a technique of solid-liquid partitioning, TLC is one form of *adsorption chromatography* and the flowing solvent is known as the *developing* (*eluting*) *solvent.* TLC is primarily used for qualitative analysis of small amounts organic compounds—to determine the number of components in a mixture and to identify components through comparisons with structurally known compounds. In addition, TLC is often used to identify the optimum conditions for separation of mixture components in larger samples by another form of elution chromatography, column chromatography (Experiment 19).

In the procedure employed for TLC (Figure 7.1) a small amount of the sample to be analyzed is applied near one end of the adsorbent coated plate. The coated plate is then placed upright in a developing chamber that contains a shallow pool of the eluting solvent. The developing solvent rises along the coated surface of the plate by capillary action and carries with it the components of the sample. The components of a mixture move up the TLC plate at different rates, dependent on the solubility of the component in the solvent and the degree to which the component is adsorbed by the stationary phase. The result is a series of spots on a line perpendicular to the solvent level in the developing chamber.

A compound will move up the TLC plate at a rate relative to that of the solvent front. This relative mobility is known as the *R_f value* of the compound and is defined by

$$R_f = \frac{\text{Distance from origin traveled by compound}}{\text{Distance from origin traveled by solvent}}$$

where the origin is the midpoint of the spot of sample that is applied to the TLC plate (Figure 7.2). Under a defined set of conditions (adsorbent, solvent, layer thickness, temperature, and humidity) the R_f value is a characteristic property of a compound.

Figure 7.1 *Thin layer chromatographic analysis procedure.*

Spotting Procedure

Prepare a 5 to 10% solution of the sample to be analyzed in a volatile solvent. (Ethyl ether is commonly used.) A capillary tube is constructed by drawing out a melting point capillary tube to a very narrow constriction. Using a dull-pointed pencil lightly draw a straight line across the TLC plate at approximately 1 cm from the end of the plate. Now, with your capillary tube carefully apply the sample solution to one spot along the drawn line on the TLC plate. For application dip the capillary tube in the sample solution, and then just touch the end of the capillary tube to the plate for a fraction of a second to obtain a concentrated spot with a small diameter (<2 mm). For most samples 0.02 ml of the 5 to 10% sample solution is sufficient application for subsequent visual identification of sample components. Allow the solvent to evaporate from the plate prior to additional solution application to the same spot and prior to development of the TLC plate.

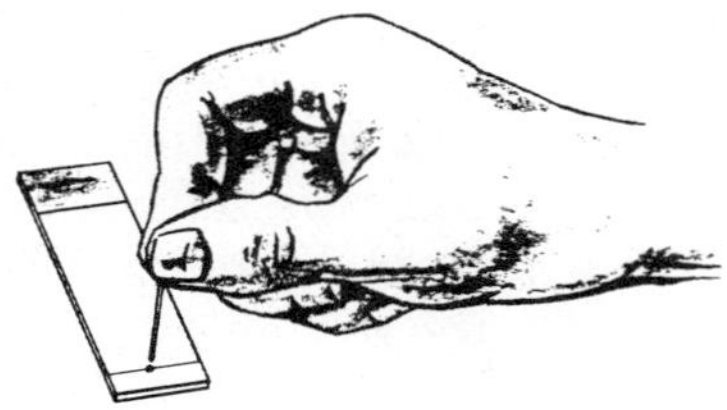

Developing Procedure

A closed wide-mouth container—screw-cap bottle, or an Erlenmeyer flask stoppered with a rubber stopper—is used as the developing chamber. Fill the developing chamber to a depth of 0.5 cm with the chosen developing solvent. Fit a piece of filter paper around the inside of the developing chamber to ensure rapid equilibration of liquid and vapor in the container. After the filter paper has been saturated with the elution solvent, place the TLC plate in the developing chamber (spotted end down) at an angle from the side of the container and then cover the container. Be certain that the solvent level in the developing chamber is below the level of the spot of sample. Allow the solvent to rise on the TLC plate to a level that is approximately 1 cm from the top of the plate (usually <10 min) and then remove the plate from the developing chamber. Immediately mark the solvent level on the plate and allow the solvent to evaporate from the TLC plate (<2 min).

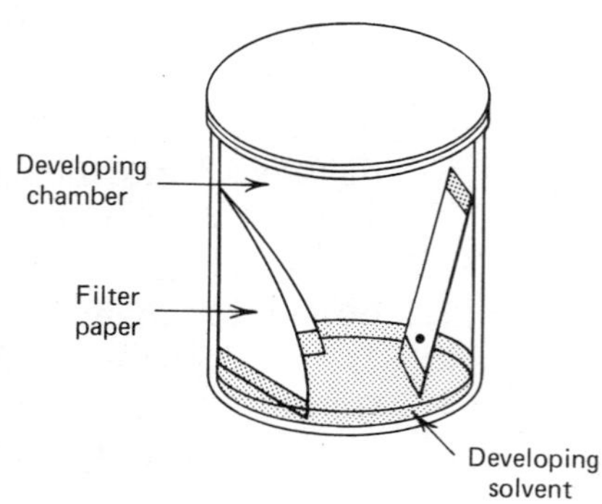

Figure 7-1 *(continued)*

Visualization Procedure:

When the compounds analyzed by TLC are colorless, a procedure for visual spot detection is required. Compounds that fluoresce can be visualized by irradiation of the plate with ultraviolet light (ultraviolet lamp) in the dark; such compounds emit absorbed light and appear as bright spots on the plate. If the solid adsorbent on the TLC plate is impregnated with a fluorescent indicator, the entire plate will brighten under a UV lamp except for those regions where compounds absorb some of the UV light; the presence of these compounds is indicated by dark spots on a bright background. Another common method for the visualization of organic compounds involves the formation of a molecular complex with iodine (except for alkanes and alkyl halides) and is shown in this figure. Several crystals of iodine are placed in a container similar to that used in the developing procedure. The developed TLC plate is then set into this covered chamber and allowed to remain until dark brown spots are visible. When the spots are sufficiently intense for visual identification, the TLC plate is removed from the container and the spots are immediately outlined with a pencil. Left in the open, iodine sublimes from the plates and the spots fade. Numerous alternate chemical procedures are also employed for the detection of specific organic compounds.

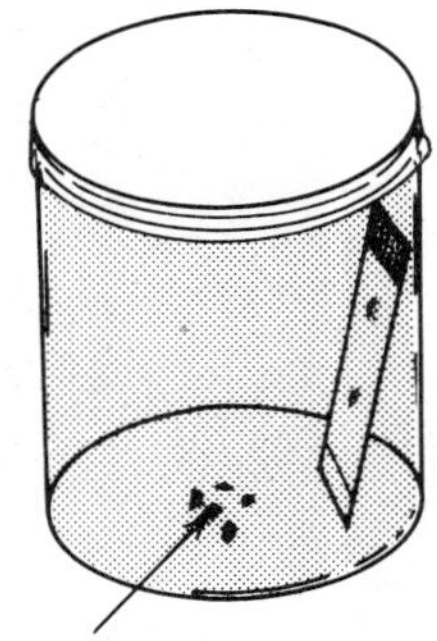

The most frequently used solid adsorbents for thin layer chromatography are alumina (Al_2O_3) and silica gel (SiO_2). Alumina is more polar than silica gel and is often referred to as being *more active* than silica gel. Alumina is more suitable for analyses of less polar compounds (hydrocarbons, ethers, aldehydes, ketones, alkyl halides) since polar compounds are strongly adsorbed on this adsorbent. TLC analyses of polar compounds on alumina generally result in low R_f values and minimal separations. Conversely, silica gel is the adsorbent of choice for analyses of polar compounds (carboxylic acids, alcohols, amines) since nonpolar compounds are weakly adsorbed on silica gel. TLC analyses of nonpolar compounds on silica gel generally results in high R_f values and minimal separations.

The nature of the developing solvent is also a major factor in determining the mobility of the components in a mixture. If the solvent is more polar than

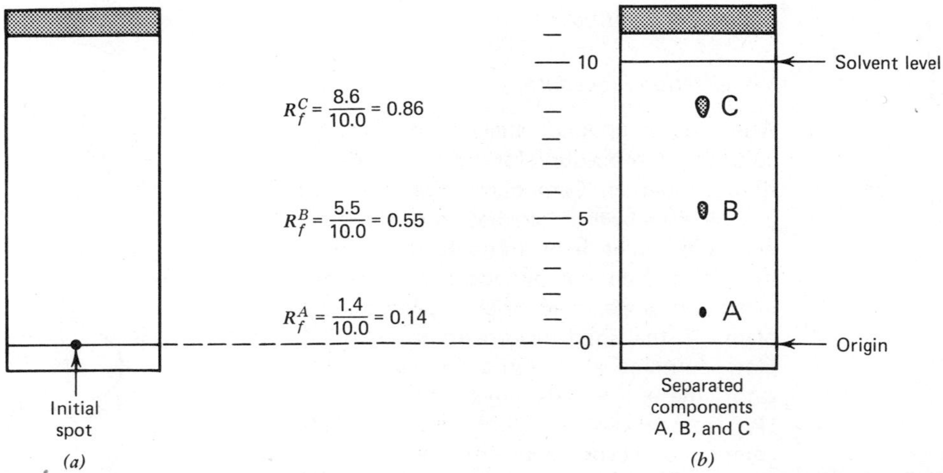

Figure 7.2 *Thin layer chromatographic plate (a) before and (b) after development in the solvent.*

a mixture component, the solvent will replace that component on the solid adsorbent and the component will remain almost entirely in the mobile liquid phase (high R_f value). If the solvent is much less polar than a mixture component, that component will adhere to the solid adsorbent and not be displaced by the solvent ($R_f = 0$). In general, the ability of the developing solvent to displace a compound from a solid adsorbent (*eluting power*) is related to solvent polarity and follows the order depicted in Table 7.2. Mixtures of two or more solvents are often employed as the developing solvents for TLC analyses. The polarity of a solvent mixture is intermediate between those of the pure solvent components.

Table 7.2 Common Eluting Solvents for Thin Layer Chromatography

INCREASING SOLVENT POLARITY ↓

Solvent	Boiling point (C°)
Hexane	68
Carbon tetrachloride	77
Toluene	111
Benzene	80
Methylene chloride	41
Ethyl ether	35
Chloroform	61
Ethyl acetate	77
Acetone	56
Ethanol	78
Methanol	65

INCREASING ELUTING POWER ↓

Preparation of TLC Plates

TLC sheets (20 m × 20 cm) with a uniform coating of alumina or silica gel on glass or plastic are commercially available. The plastic sheets use an organic polymer to bind the coating of solid adsorbent to the flexible plastic. These sheets are durable and can be cut by a scissors into strips for use in routine qualitative analyses. Each 20 × 20 cm sheet can be cut into either twenty-four 2.5 × 6.6 cm plates or sixteen 2.5 × 10 cm plates. The commercial TLC sheets may be obtained with a fluorescent indicator impregnated in the solid adsorbent; the fluorescent indicator aids the visual identification of compounds that absorb ultraviolet light when the developed slide is viewed under an ultraviolet lamp. TLC plates without the fluorescent indicator are used in this experiment.

TLC plates can be prepared in the laboratory at low cost by coating silica gel on glass microscopic slides. In a convenient dipping procedure, a slurry of 50 g of silica gel G (for TLC slide preparation) in 150 ml of chloroform* is prepared in a wide-mouth, screw-cap jar so that the depth of the slurry is at least 6 cm. Just before dipping the clean dry microscope slides into the slurry, vigorously shake the covered jar. Without allowing the slurry to settle, dip two microscope slides that are held uniformly and tightly together into the silica gel until all but 1 cm of the slides is immersed. Slowly remove the plates and allow the solvent to drain back into the jar. The coating of silica gel must be uniform across the plate and without any observable breaks in the surface. Wipe the excess silica gel from the edges of the plates, and after the solvent has evaporated carefully separate the two plates. Handle these plates carefully since, unlike the commercially available TLC sheets that contain a surface binder, the silica gel coatings on the laboratory-prepared plates easily chip and crack. Silica gel plates prepared by this procedure can be used immediately for TLC analyses. The chloroform solution can be stored for future use; approximately 20 TLC slides can be prepared from the stock slurry.

Qualitative Analysis By TLC

Thin layer chromatography is routinely used to identify organic compounds and monitor organic reactions. Although chromatographic methods do not provide direct structural information for organic compounds, identical R_f values for an unknown compound and an authentic sample of a known compound in several different developing solvents do indicate that the unknown and known compound are identical. When the identity of an unknown compound is limited

*Skin contact with chloroform and the breathing of its vapors is hazardous to your health. Work with chloroform in the fume hood and wear protective gloves during the preparation of TLC plates.

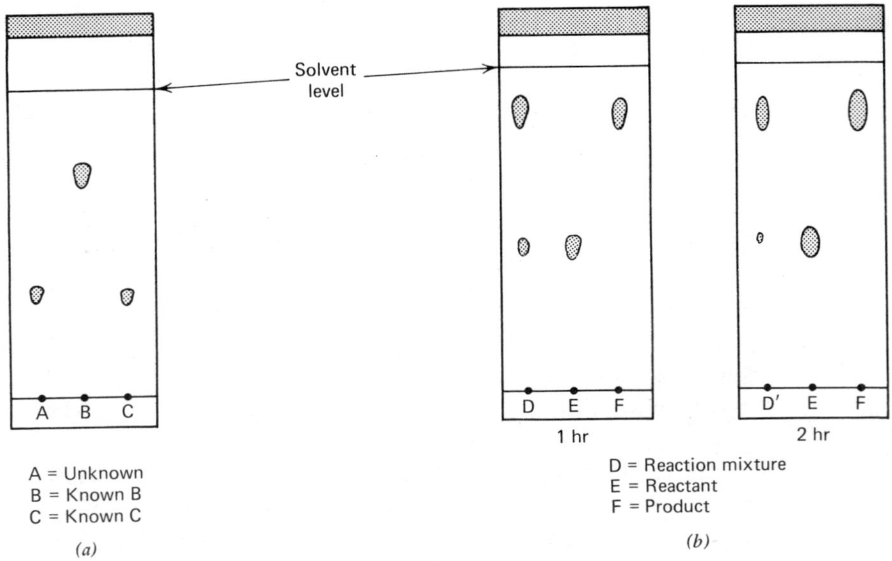

Figure 7.3 *Applications of thin layer chromatography: (a) Compound identification and (b) monitoring chemical reactions.*

to one of only a few known compounds and each of the known compounds is separable from the others by TLC, positive structural identification of the unknown can be obtained. To compare an unknown with known compounds, each compound is spotted on one TLC plate and the TLC plate is developed in a solvent suitable for the separation of the known compounds (Figure 7.3*a*). The identity of the unknown is deduced by comparing R_f values.

The progress of an organic reaction can be followed by using a similar TLC procedure (Figure 7.3*b*). In this latter application authentic samples of the reactant and, if available, expected product are spotted on the TLC plate along with the reaction mixture. If the reactant and product are separable by TLC, the progress of the reaction is followed by observing the increase in the size of the TLC spot for the product and a corresponding decrease in the size of the reactant spot with time.

Prelab Questions

1. In determining the solubility properties of a compound of unknown structure, a student determines that the compound is soluble in water, ether, 10% HCl, 10% NaOH, and conc. H_2SO_4. The student concludes that the unknown is a highly polar organic compound that contains both an acidic and a basic functional group. Criticize this conclusion.

2. Describe the expected change in color of a solution of 2% bromine in carbon tetrachloride upon adding two drops of that solution to 2 ml of carbon tetrachloride.

3. Predict the more suitable solid adsorbent (silica gel or alumina) for TLC analysis of the reaction product obtained from the hydrolysis of trimyristin (Experiment 4) and explain your choice.

4. A colorless compound is to be analyzed by TLC. In spotting a solution of that compound on the TLC plate how will you identify the exact position of the spotted compound if you intend to apply additional solution to that same spot?

5. Can thin layer chromatography be used to separate and identify very volatile compounds? Why?

6. An organic chemistry student did not observe a spot on a TLC plate that had been spotted with trimyristin and developed. The student knew from the results of other students in the laboratory that trimyristin could be visualized by the procedure that was employed and knew that the level of the developing solvent was below the applied spot. What could be the cause of this student's failure to observe a spot for trimyristin? What advice would you give this student that would ensure the completion of the experiment?

EXPERIMENTAL PROCEDURE

Perform solubility tests for trimyristin (Experiment 3), myristic acid (Experiment 4), limonene (Experiment 6), and hexane according to the procedure described in this experiment. Record your observations and deduce the structural implications (polarity, acid-base properties) of your results. Characterize this same set of organic compounds by chemical tests with bromine in carbon tetrachloride and with potassium permanganate according to the procedures described in this experiment. Describe your observations in detail for future reference.

Prepare three silica gel TLC plates according to the method designated by your instructor and constrict two capillary tubes for spotting these TLC plates. Prepare separate solutions of trimyristin (Experiment 3) and myristic acid (Experiment 4) by dissolving 0.1 g of each of these compounds in 1 ml of ethyl ether. Mark the origin approximately 1 cm from the bottom of each TLC plate and spot each TLC plate at this origin with the solutions of trimyristin and myristic acid. The horizontal distance between the two spots at the origin should be at least 1 cm. *Crystals of the organic compounds to be analyzed must not be transferred from stock solutions and crystals must not be allowed to form during the spotting procedure;* when crystals are applied to a TLC plate, streaking of the spot occurs during plate development.

Develop the TLC plates in hexane and intermediate mixtures of hexane:ethyl ether (75:25 and 50:50) according to the procedure in Figure 7.1. Prior to placing a TLC plate in a developing chamber add two drops of glacial acetic acid to the developing solvent; acetic acid minimizes streaking of the spots for carboxylic acids during development. After the plate has been developed and dried, place the plate in an iodine chamber to visualize the spots on each of these slides. Describe the appearance of the slides, the presence of more than one spot for each solution applied to the TLC plate, the shape of the spots, and their relative color intensity. Determine the R_f values for trimyristin and myristic acid in each of the developing solvents and select the solvent system that provides optimum separation of these compounds.

Postlab Questions

1. Tabulate the results that you obtained for solubility tests and chemical characterization tests (Br_2/CCl_4, and $KMnO_4$) of trimyristin, myristic acid, limonene, and hexane. Use + to designate a positive test and - to designate a negative test.

2. Why do crystals at the spot on the origin of a TLC plate cause streaking in TLC analyses?

3. (a) Describe what you would expect to observe in TLC analysis of trimyristin if the sample that was analyzed contained 10% of a very polar impurity. (b) Describe the limitations of TLC for purity determinations.

4. What would you expect to observe after plate development and visualization as a result of the following errors in the use of TLC: (a) the solvent level in the developing chamber is higher than the spotted sample, (b) too much sample is applied to the TLC plate, and (c) the TLC plate is allowed to remain in the developing chamber after the solvent level has reached the top of the plate.

5. Explain how acetic acid minimizes streaking of the spots for carboxylic acids during plate development in a nonpolar solvent during thin layer chromatographic analysis. Why is streaking observed when acetic acid is not included in the developing solvent?

Experiment Eight

Substitution Reactions of Organic Compounds: Preparation of 3-Chloro-3-methylpentane

Chemical reactions in which a functional group of an organic compound is replaced by a different functional group are known as substitution reactions. Substitution reactions are one of the most common transformations of organic compounds, and a wide variety of substitution reactions are known. (Compare, for example, Experiment 8 with Experiments 17 and 20.) In this experiment we will limit ourselves to consideration of the replacement of one functional group by another in saturated organic compounds:

Substitution reaction

$$-\overset{|}{\underset{|}{C}}-Z + Y: \longrightarrow -\overset{|}{\underset{|}{C}}-Y + Z:$$

These substitution reactions are especially important for preparing new organic compounds.

Many alcohols are naturally occurring and many others are prepared for specific purposes by chemical industry. The use of substitution reactions to replace the hydroxyl functional group of alcohols by a different functional group represents a realistic and economical starting point for the preparation of new organic compounds from available alcohols.

Alcohols undergo substitution reactions when the hydroxyl group is activated by complexation with an acid, and water or a derivative of water (*the leaving group*) is displaced by a nucleophilic reagent (*the nucleophile*):

$$\underset{\textit{Alcohol}}{R{-}OH} + \underset{\textit{Acid}}{H^+} \rightleftharpoons \underset{\textit{Activated alcohol}}{R{-}\overset{H}{\overset{|}{O^+}}{-}H}$$

$$R{-}\overset{H}{\overset{|}{O^+}}{-}H + \underset{\textit{Nucleophile}}{Nuc:^-} \longrightarrow \underset{\textit{Substitution product}}{R{-}Nuc} + H_2O$$

Numerous reagents have been successfully employed to convert alcohols to their corresponding substitution products. The reagents that are most commonly employed are Brønsted acids and the reaction that occurs involves the displacement of water by the conjugate base of the Brønsted acid (Nuc:$^-$ = B:$^-$):

$$\underset{\textit{Alcohol}}{ROH} + \underset{\textit{Brønsted acid}}{HB} \longrightarrow \underset{\textit{Substitution product}}{R{-}B} + \underset{\textit{Water}}{H_2O}$$

Substitution reactions can be used for the efficient preparation of alkyl halides by treating the alcohol with a hydrogen halide (HB = HCl, HBr, or HI). Tertiary alcohols react rapidly with hydrogen halides to yield tertiary alkyl halides. Secondary alcohols are less reactive than tertiary alcohols towards substitution reactions with hydrogen halides. Primary alcohols do not generally react with hydrogen halides to give good yields of alkyl halides at room temperature.

In this experiment you will react a tertiary alcohol, 3-methyl-3-pentanol, with a concentrated aqueous solution of hydrogen chloride (hydrochloric acid):

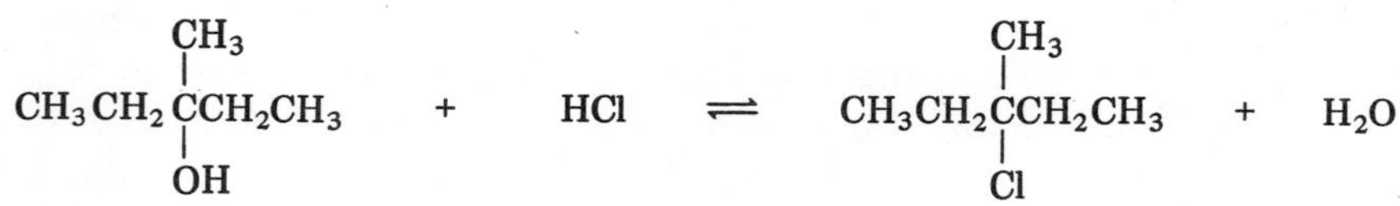

3-Methyl-3-pentanol	*Hydrochloric acid*	*3-Chloro-3-methyl-pentane*
M.w. = 102.2 g/mol	37% HCl	M.w. = 120.6 g/mol
b.p. = 122°C	12*M*	b.p. = 116°C
d_{20} = 0.83 g/ml		d_{20} = 0.89 g/ml

The alcohol is only partially soluble in hydrochloric acid and the two phases must be mixed by shaking to ensure complete reaction. The product, 3-chloro-3-methylpentane, is insoluble in the hydrochloric acid solution and is separated from the acid solution by the use of the separatory funnel. The product is then washed with a solution of sodium bicarbonate to remove any acid that is dissolved in the product. This latter step involves the neutralization of the acid and results in the generation of carbon dioxide gas:

$$NaHCO_3 + HCl(aq) \rightleftharpoons NaCl + H_2O + CO_2(g)$$

After the bicarbonate solution has been separated, small amounts of water dissolved in the product are removed by use of a chemical drying agent. The product is then purified by distillation.

The following flow diagram outlines the separation procedure for this experiment. (See page 97)

Drying Organic Liquids

Organic liquids that are prepared by reactions in aqueous media or are washed with aqueous solutions contain small amounts of water. This water is removed

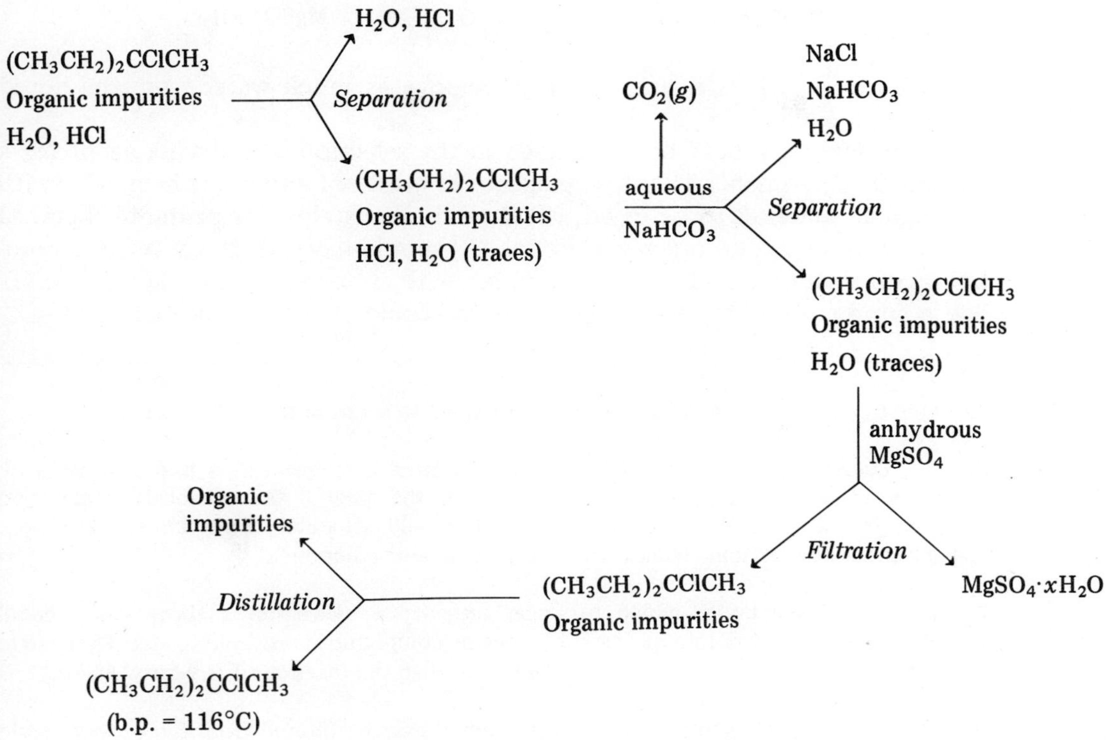

before the liquid is distilled by the use of a drying agent. The drying agent is generally an inexpensive inorganic compound that will efficiently combine with the water. The combination of drying agent and water is then removed from the organic liquid. For example, anhydrous magnesium sulfate combines with water to form a solid hydrated salt ($MgSO_4 \cdot xH_2O$). Up to seven moles of water can combine with one mole of the anhydrous salt, but an excess of

Table 8.1 Drying Agents

Agent	Capacity	Speed	Comments
$CaSO_4$	low	very fast	Sold commercially as Drierite. Applicable to all compounds.
$CaCl_2$	medium	very fast	Not applicable to hydroxyl, amino, and acid compounds. Useful for hydrocarbons, ethers, and alkyl halides.
$MgSO_4$	medium	fast	Applicable to all compounds.
Na_2SO_4	medium	slow	Useful for most compounds. Slow and inefficient.
Molecular sieves	high	medium	Very efficient, but too expensive for general use.
NaOH, KOH	high	fast	Useful for amines and hydrocarbons but not applicable to most other compounds.
P_2O_5	high	fast	Applicable to hydrocarbons, halides, and nitriles, but not to most other compounds.

$$MgSO_4\,(\text{anhydrous}) \quad + \quad x\,H_2O \quad \underset{[x \leqslant 7]}{\rightleftharpoons} \quad MgSO_4 \cdot xH_2O$$

the anhydrous salt is usually used to remove as much water from the liquid phase as possible.

Four factors should be considered in the selection of a drying agent for a particular experiment. The drying agent (1) must not combine chemically with the organic material to be dried, (2) should not catalyze or promote chemical decomposition of the organic material, (3) should have a high capacity to combine with water, and (4) should combine with water at a rapid rate. Table 8.1 lists a number of common drying agents and some of their properties.

Prelab Questions

1. What significant side reactions would you expect to occur in this experiment?

2. The Lucas reagent is a solution of zinc chloride in concentrated hydrochloric acid. If *tert*-butyl alcohol is added to Lucas reagent, the alcohol first dissolves in the acid solution, but an insoluble organic compound rapidly separates from the acid solution. Explain these observations. What is the insoluble organic compound?

3. If *sec*-butyl alcohol is added to Lucas reagent (see Question 2 above) the alcohol first dissolves. After 4–5 min an insoluble organic compound is observed to start to separate from the acid solution. Why is this reaction slower than the reaction of *tert*-butyl alcohol?

4. Why must drying agents such as magnesium sulfate be filtered from the organic liquid before the liquid is distilled? (*Hint:* The anhydrous magnesium sulfate is prepared by heating the hydrated salt in an oven.)

EXPERIMENTAL PROCEDURE

Place 25 ml (20.7g, 0.20 mol) of 3-methyl-3-pentanol and 60 ml of concentrated hydrochloric acid in a 250-ml Erlenmeyer flask. Close the flask with a rubber stopper and thoroughly mix the contents of the flask at regular intervals for 20 min. After each mixing loosen the stopper briefly to release any excess gas pressure. At the end of the 20-min period, add 6 g of solid calcium chloride and shake until the salt has dissolved. Transfer the liquid mixture to the separatory funnel and remove the aqueous layer (lower layer) as completely as possible. Slowly with swirling, add 20 ml of 5% sodium bicarbonate solution to the separatory funnel. When gas evolution has subsided, stopper the separatory funnel and shake cautiously. Release excess gas pressure frequently by inverting the funnel (hold the stopper in place) and

---

CAUTION: Carbon dioxide gas will be evolved when the organic layer is washed with the sodium bicarbonate solution. Do not stopper the separatory funnel until gas evolution has subsided. Swirl the unstoppered separatory funnel to mix the two layers.

opening the stopcock. Remove the aqueous layer and then pour the organic layer into a dry 125-ml Erlenmeyer flask. Add 1 to 2 g of drying agent (anhydrous calcium chloride or magnesium sulfate) and dry the organic layer for 10 min. Set up the distillation apparatus described in Figure 3.4 using the 50-ml flask for the distillation flask. Since 3-chloro-3-methylpentane boils well above the temperature attainable on a steam bath, a burner or heating mantle will be necessary to heat the liquid to boiling. Filter the organic layer through fluted filter paper to remove the drying agent and distill the resulting liquid. Collect and save the fraction boiling at 108 to 118°C. Weigh the amount of product obtained and calculate the percentage yield of the isolated product.

Postlab Questions

1. Would this procedure be useful for the preparation of 1-chlorohexane from 1-hexanol?

2. Suggest a method and general procedure that could be used to prepare 2-iodo-2-methylbutane from 2-methyl-2-butanol.

3. Why should the 3-chloro-3-methylpentane be dried before the distillation?

4. If the organic layer is left in contact with the aqueous bicarbonate solution for several hours, a reduced yield of alkyl halide is obtained. Why?

Experiment Nine

Elimination Reactions of Organic Compounds: Preparation of 3-Methyl-2-pentene

Elimination reactions are commonly used to prepare alkenes. These reactions occur with functionalized organic compounds through the loss of both the functional group (the *leaving group*, —Z) and a hydrogen bonded to a carbon adjacent (β) to the carbon bearing the leaving group. The loss of HZ from adjacent carbon atoms of an organic compound with the resulting formation of a carbon-carbon double bond is called a β- or *1,2-elimination reaction:*

1, 2-Elimination Reaction

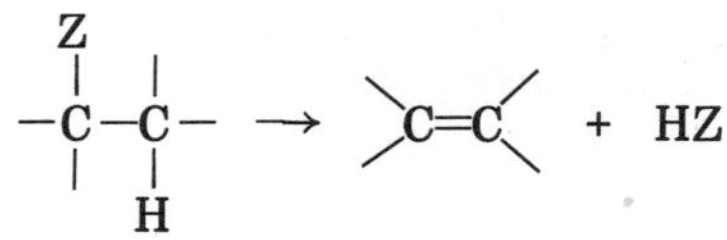

Although a number of different functionalized organic compounds under the appropriate reaction conditions undergo elimination of a leaving group and hydrogen, the two elimination processes most commonly employed for the preparation of alkenes involve the loss of hydrogen halide from an alkyl halide or the loss of water from an alcohol. In this experiment you will prepare 3-methyl-2-pentene from one of two available starting materials, 3-chloro-3-methylpentane (Experiment 8) or 3-methyl-3-pentanol, using reaction conditions that are conducive to elimination.

Dehydrohalogenation

Alkyl halides that are readily available from substitution reactions (Experiment 8) or from direct heat or light induced halogenation of alkanes are starting materials in a preparative method for alkene formation known as dehydrohalogenation. Dehydrohalogenation reactions involve the elimination of hydrogen halide from adjacent carbon positions of the alkyl halide and employ a base ($B:^-$) to aid in removing the hydrogen β to the carbon bearing the halogen. Sodium and potassium hydroxide are both useful bases for dehydrohalogenation

Dehydrohalogenation reaction

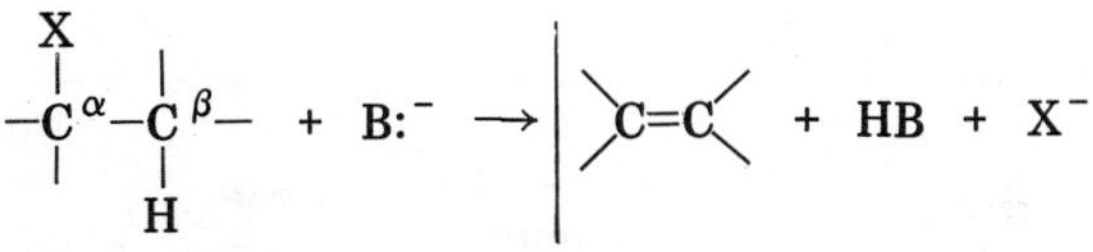

reactions. However, potassium hydroxide is often preferred since it is more soluble in those organic solvents that will dissolve both the base and the alkyl halide. Polar organic solvents such as ethanol, ethylene glycol (1,2-ethanediol), and 2-ethoxyethanol are commonly used as solvents for dehydrohalogenation reactions because they meet these solubility requirements.

Chemists have investigated the factors that affect the course of the dehydrohalogenation reaction. Results of these studies have shown that for a series of alkyl halides that differ only in the halide leaving group, the alkyl iodide is more reactive than the alkyl bromide which, in turn, is more reactive than the alkyl chloride. Alkyl fluorides are generally unreactive in base promoted elimination reactions. In addition, for a given halide leaving group, tertiary alkyl halides are more reactive than secondary alkyl halides, and primary alkyl halides are the least reactive. These general observations suggest that dehydrohalogenation of 3-chloro-3-methylpentane will require more severe reaction conditions (higher concentration of base, stronger base, higher temperature, for example) than those used for elimination from the corresponding alkyl bromide or iodide. However, 3-chloro-3-methylpentane is a tertiary halide and therefore should readily undergo elimination.

Unsymmetrical alkyl halides like 3-chloro-3-methylpentane can form two or more alkene isomers, depending upon which β hydrogen is lost in the reac-

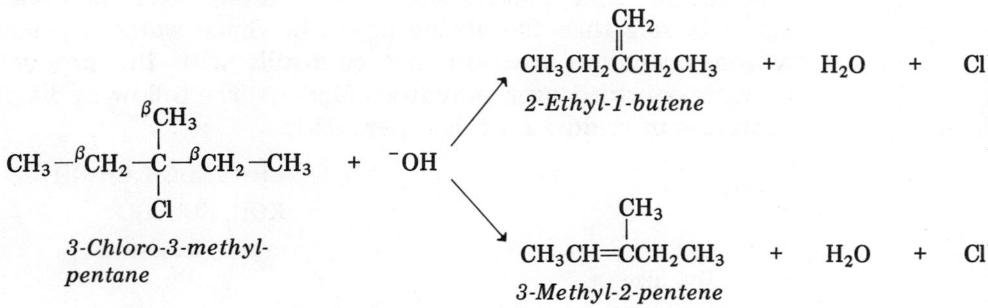

tion. Dehydrohalogenation reactions generally form the most stable alkene in the highest percentage yield. The most stable alkene is usually the most highly substituted alkene. Thus, 3-methyl-2-pentene, which is a trisubstituted alkene, is expected to be the major product, and 2-ethyl-1-butene, a disubstituted alkene, is expected to be the minor product in the dehydrohalogenation of 3-chloro-3-methylpentane.

In this experiment you will prepare 3-methyl-2-pentene by reacting 3-chloro-3-methylpentane with potassium hydroxide in a solution of 2-ethoxyethanol. Actually, a total of three isomeric alkenes are expected products of this reaction, and in Experiment 10 you will analyze the products to determine the relative amounts of each of the isomers in the product. Table 9.1 contains

Table 9.1 Reagents and Products

Name	Structural formula	Molecular weight (g/mol)	Boiling point (°C)	Density at 20°C (g/cm^3)
3-Chloro-3-methylpentane	$CH_3CH_2\overset{CH_3}{\overset{\vert}{C}}ClCH_2CH_3$	120.6	116	0.89
trans-3-Methyl-2-pentene	$(CH_3)(H)C{=}C(CH_3)(CH_2CH_3)$	84.2	70.4	0.70
cis-3-Methyl-2-pentene	$(CH_3)(H)C{=}C(CH_2CH_3)(CH_3)$	84.2	67.7	0.69
2-Ethyl-1-butene	$CH_3CH_2\overset{CH_2}{\overset{\Vert}{C}}CH_2CH_3$	84.2	64.7	0.69
2-Ethoxyethanol	$CH_3CH_2OCH_2CH_2OH$	90.1	135	0.93
Potassium hydroxide	KOH	56.1	—	—

information on the physical properties of the reactants and expected products of this reaction.

After the reaction is complete, the alkene products can be distilled directly from the reaction mixture since the reaction solvent and reactants are much higher boiling than the alkene products. Since water is produced in the dehydrohalogenation reaction and co-distills with the product, the reaction products are dried over anhydrous $MgSO_4$. The following diagram outlines the separation procedure for this experiment:

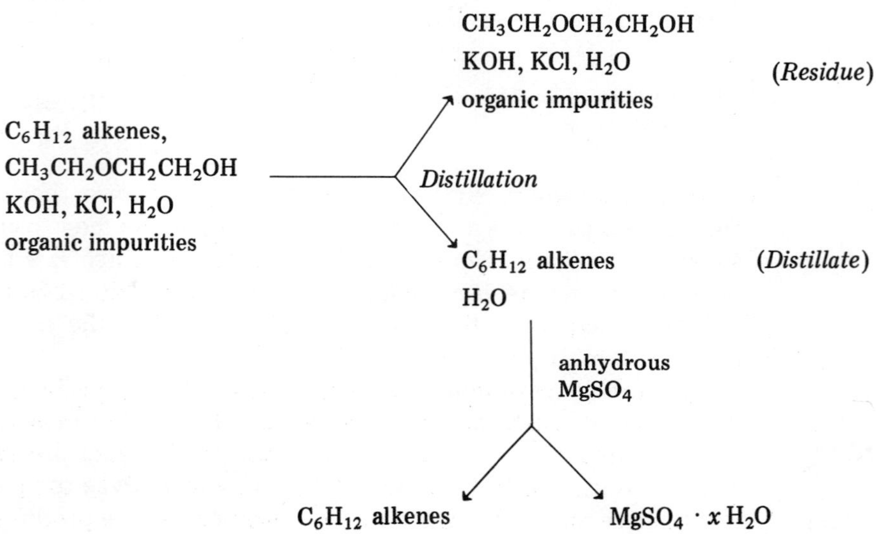

Dehydration

Alcohols are the starting materials in a preparative method for alkene formation known as dehydration that involves the loss of the elements of water from adjacent carbon atoms of the alcohol. Dehydration reactions require the presence of a strong acid catalyst to promote the loss of the $-OH$ group from

Dehydration reaction

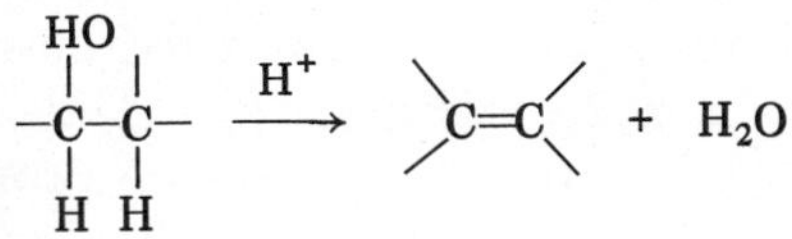

the alcohol. The acid activates the hydroxyl group towards elimination just as it activated the hydroxyl group towards substitution in Experiment 8. Tertiary alcohols are more reactive than secondary alcohols in elimination reactions, and primary alcohols are the least reactive. The dehydration of tertiary alcohols is most useful for alkene preparation. Secondary and primary alcohols often undergo structural rearrangements under the reaction conditions employed for acid-catalyzed dehydration.

3-Methyl-2-pentene can be prepared by the acid-catalyzed dehydration of 3-methyl-3-pentanol:

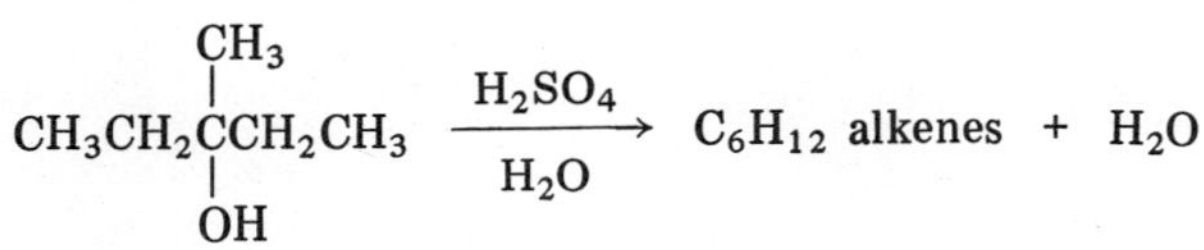

The alkene products produced in this reaction are expected to be the same compounds that are obtained by the dehydrohalogenation of 3-chloro-3-methylpentane, but their relative yields are not expected to be the same. The different relative isomer distributions reflect the different reaction pathways for product formation in the two reactions. The alkenes can be isolated from the aqueous acid mixture by fractional distillation. The product is then purified by washing with dilute sodium bicarbonate solution to remove traces of acid and dried over anhydrous $MgSO_4$.

Elimination Versus Substitution

Elimination reactions and substitution reactions are carried out under similar reaction conditions, and the two reactions are often found to be in competi-

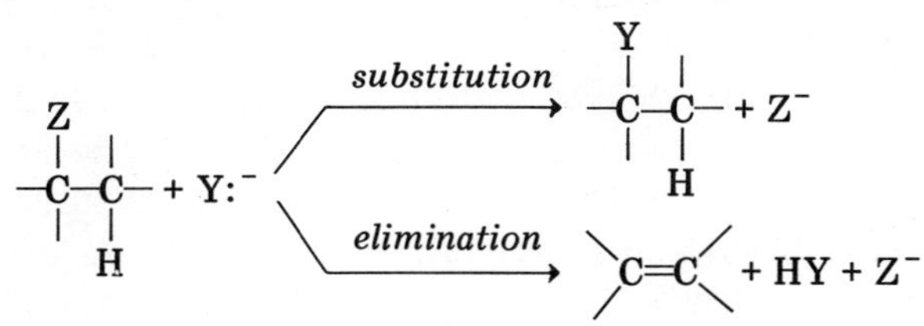

tion. The reagents employed in substitution and elimination reactions (Y:⁻) possess a nonbonding pair of electrons and, therefore, have the properties of both a nucleophile (electron pair donor in the Lewis definition) and a base (proton acceptor in the Brønsted definition). Substitution is favored when Y:⁻ is a good electron pair donor and a weak Brønsted base. Reagents such as halide, X^-, fit these criteria and tend to give products from substitution. Elimination reactions are favored if the reagent Y:⁻ is a strong Brønsted base such as hydroxide. A strong Brønsted base is also a strong Lewis base (a good nucleophile). Simple primary alkyl halides favor substitution over elimination even when the Y:⁻ reagent is a strong Brønsted base. In contrast, tertiary alkyl halides give significant amounts of elimination products even with weakly basic reagents. In these cases the alkyl groups block nucleophilic attack by the basic reagent at the carbon bearing the leaving group, and removal of the β hydrogen is preferred. Secondary alkyl halides are borderline and favor either elimination or substitution depending on the exact reaction conditions. Elimination is the preferred process in the reaction of 3-chloro-3-methylpentane with the strong base potassium hydroxide and, in contrast to the result of Experiment 8, competes effectively with substitution. In fact, careful examination of the product formed in Experiment 8 will show some alkene by-products resulting from elimination.

Solvents for Chemical Reactions

The choice of solvent for a chemical reaction is often critical in determining the outcome of the reaction. The solvent may not only affect the yield of a reaction, but it can also alter the course of a reaction and result in the formation of undesired or new products.

Solvents provide a homogeneous medium in which reactant molecules can effectively undergo collisions that result in the formation of products. An undissolved solid will react with a solid or liquid reagent only at the surface of the solid. Likewise, an insoluble liquid will react with a second liquid reagent only at the interface between the two phases. Two reagents dissolved in a solvent are not prevented from effective mixing and molecular collisions by an impenetrable solid or liquid. Although the solvent should not undergo chemical reactions with the reactants or products, in some reactions it is possible to use one reagent in large excess so that the excess reagent will serve as solvent for the other reagents. For example, in the hydrolysis of trimyristin (Experiment 4) water was used as a reactant and as part of the ethanol-water solvent mixture. Water without added ethanol would not have been an acceptable solvent for this reaction because trimyristin is not soluble in water.

The solvent selected for a chemical reaction (Table 9.2) should not interfere with the isolation and purification of products. For example, if a product is to be isolated by distillation from the reaction mixture, the solvent should be selected with a boiling point that is significantly different from the boiling

Table 9.2 Some Common Solvents for Organic Reactions

Name	Structural formula	Boiling point (°C)	Solubility in water
Acetone	CH_3COCH_3	56	miscible
Acetonitrile	CH_3CN	82	miscible
Dichloromethane	CH_2Cl_2	40	insoluble
N,N-Dimethylformamide	$(CH_3)_2NCHO$	152	miscible
Dimethyl sulfoxide	CH_3SOCH_3	189	miscible
Ethanol	CH_3CH_2OH	78	miscible
Ethylene glycol	$HOCH_2CH_2OH$	198	miscible
Ethyl ether	$(CH_3CH_2)_2O$	35	insoluble
2-Ethoxyethanol	$CH_3CH_2OCH_2CH_2OH$	135	miscible
Methanol	CH_3OH	65	miscible
Pentane	$CH_3(CH_2)_3CH_3$	36	insoluble
Toluene	$C_6H_5CH_3$	111	insoluble

point of the product. The boiling point of the solvent is also an important consideration if the reaction is to be heated at a constant temperature by refluxing the reaction solution (Experiment 4).

The safety and cost of a solvent should be considered when selecting a solvent for a reaction. Safety considerations should include an evaluation of relative health and environmental hazards as well as flammability. Solvents such as benzene and chloroform that recently have been linked to cancer in laboratory test animals should generally be replaced by other solvents, such as toluene and dichloromethane, in reaction procedures that call for their use.

A special class of organic solvents, called *polar aprotic solvents*, has a marked activating effect on many reactions, particularly substitution reactions. Polar aprotic solvents are organic solvents with a relatively high dipole moment but which do not have hydroxyl (—OH) or amino ($-\overset{|}{N}-H$) groups. Examples of these solvents include acetonitrile, CH_3CN; acetone, CH_3COCH_3; dimethylformamide, $HCON(CH_3)_2$; and dimethyl sulfoxide, CH_3SOCH_3. Polar aprotic solvents have the ability to dissolve organic compounds as well as many ionic compounds. However, they do not form hydrogen bonds to anions as do polar protic solvents such as alcohols and water. Because of this lack of hydrogen bonding in solvation, anions of salts dissolved in polar aprotic solvents tend to be very reactive nucleophiles and/or bases.

Prelab Questions

1. Could ethanol be used as the reaction solvent for dehydrohalogenation of 3-chloro-3-methylpentane if the reaction products are to be distilled directly from the reaction solution? Explain.

2. Suggest an alkyl halide that could be used to prepare 2-ethyl-1-butene by dehydrohalogenation.

3. Why is sulfuric acid rather than hydrochloric acid used to catalyze the dehydration of alcohols?

4. Why is water not a useful solvent for the dehydrohalogenation of alkyl halides?

5. List the side reactions that you expect to compete with the dehydrohalogenation of 3-chloro-3-methylpentane.

6. Outline a flow diagram for the separation procedure used to purify the product obtained from the dehydration of 3-methyl-3-pentanol.

EXPERIMENTAL PROCEDURE: DEHYDROHALOGENATION REACTION

Place 50 ml of 2-ethoxyethanol and 11.5 g(0.20 mol) of potassium hydroxide pellets in a 250-ml, round-bottom flask. Swirl the solution

---

CAUTION: Potassium hydroxide and solutions of potassium hydroxide are strongly caustic. Do not handle potassium hydroxide pellets with your fingers. If you should accidently spill hydroxide on your skin, wash the area immediately with a copious amount of water.

---

in the flask to dissolve a portion of the pellets. Add 13.5 ml (12.0 g, 0.10 mol) of 3-chloro-3-methylpentane to the flask. Clamp the flask to a ring stand and attach the condenser for heating under reflux as depicted in Figure 4.2. Add several boiling stones to the solution and heat the solution on a steam bath for 2.5 hours. During the first part of this heating period the solution will not boil, but as product forms the solution will start to boil and to reflux. After completing the reflux period, assemble your apparatus for fractional distillation as shown in Figure 6.4 using the 250-ml flask containing the reaction mixture as the distillation flask. Add several new boiling stones to the distillation flask. Distill the product from the reaction mixture, and collect the product fraction boiling between 50 and 80°C. Dry the product fraction over anhydrous magnesium sulfate (approximately 0.5 g) and then filter the solution to remove the drying agent. This product may be redistilled, but it is sufficiently pure at this point for use in Experiments 10 and 11. Weigh your product and calculate the percentage yield. The product should be stored in a small glass-stoppered flask for use in the next two experiments.

NOTE: The higher boiling material left in the reaction flask is still strongly basic. After cooling, dispose of this material according to the directions of your laboratory instructor.

OPTIONAL EXPERIMENTAL PROCEDURE: DEHYDRATION REACTION

With the aid of a long-stemmed funnel, pour 50 ml of 9*M* sulfuric acid in a 250-ml round-bottom flask set up for reflux as shown in Figure

---

CAUTION: Sulfuric acid will burn the skin upon contact. Wash any acid off the skin at once with large amounts of water.

4.2. Add 24.6 ml (20.4 g, 0.20 mol) of 3-methyl-3-pentanol and several boiling stones to the acid solution. Heat the resulting solution on the steam bath for 15 minutes. Set up the fractional distillation apparatus as depicted in Figure 6.4 using the 250-ml round-bottom flask containing the reaction mixture as the distillation flask. Add several new boiling stones to the distillation flask and distill the product directly from the acidic reaction mixture. Collect the product fraction boiling over the temperature range of 50 to 80°C. Transfer the distillate to a separatory funnel, and slowly with swirling add 10 ml of 5% sodium bicarbonate solution to neutralize any acid impurities. After gas evolution has subsided, remove the lower aqueous layer and pour the organic layer into a 25-ml Erlenmeyer flask. Dry the product over anhydrous magnesium sulfate (approximately 1 g) and then filter the solution to remove the drying agent. The product may be redistilled, but it is of sufficient purity without redistillation for use in Experiments 10 and 11. Weigh the amount of isolated product and calculate the percentage yield.

NOTE: The high boiling residue left in the distillation flask is strongly acidic. This material should be carefully disposed of according to the directions of your laboratory instructor.

Postlab Questions

1. Explain how you could identify the chemical composition of the white solid that precipitates from the dehydrohalogenation reaction mixture in this experiment.

2. Why is the 3-methyl-2-pentene product cloudy when it is distilled from the reaction mixture?

3. Explain how, by the proper choice of solvent, a reflux apparatus can be used to maintain a desired, constant reaction temperature.

4. Why is the boiling range of your distilled product greater than the boiling point range of the expected alkene products?

5. What experimental evidence have you obtained that suggests that the starting material or the product from substitution does not comprise a major portion of your product?

Experiment Ten

Product Analysis: Gas-Liquid Partition Chromatography

Many chemical reactions produce more than one product, and isolated reaction products are often contaminated with side products, unreacted starting materials, or solvent. Thus, it is necessary to analyze the material isolated from a chemical reaction to determine both the *identities* and *quantities* of the compounds that were obtained. In this experiment, the volatile product that was isolated from the dehydrohalogenation of 3-chloro-3-methylpentane in the previous experiment will be analyzed. Although the nature of the dehydrohalogenation process and the boiling point range of the isolated product suggest that this material is a mixture of three C_6H_{12} alkene isomers, the identity and the yield of each component in the isolated product must be determined. Gas-liquid partition chromatography (GLPC)* is an especially useful technique for analyzing this reaction product mixture since it is capable of quantitatively separating small amounts of volatile compounds, even if the compounds have very similar chemical structures and physical properties. The analysis of the product mixture from Experiment 9 would not be practical by fractional distillation because the boiling points of the C_6H_{12} isomers differ by only a few degrees (see Table 9.1).

Gas-Liquid Partition Chromatography

All chromatographic separation techniques are based on the differential migration of the components of a mixture when the mixture is partitioned between a mobile and a stationary phase. In thin layer chromatography (Experiment 7) the components of the mixture are partitioned between a mobile liquid phase, the solvent, and a stationary solid phase, the adsorbent. In gas-liquid partition chromatography, the components of a mixture are partitioned between a mobile *carrier gas* and a *stationary liquid phase.* This partition process occurs inside a tubular column that is packed with the stationary liquid phase coated on a *solid support.* The carrier gas flows through this packed column. Figure

*Gas-liquid partition chromatography is also known as vapor phase chromatography (VPC), gas-liquid chromatography (GLC), and gas chromatography (GC).

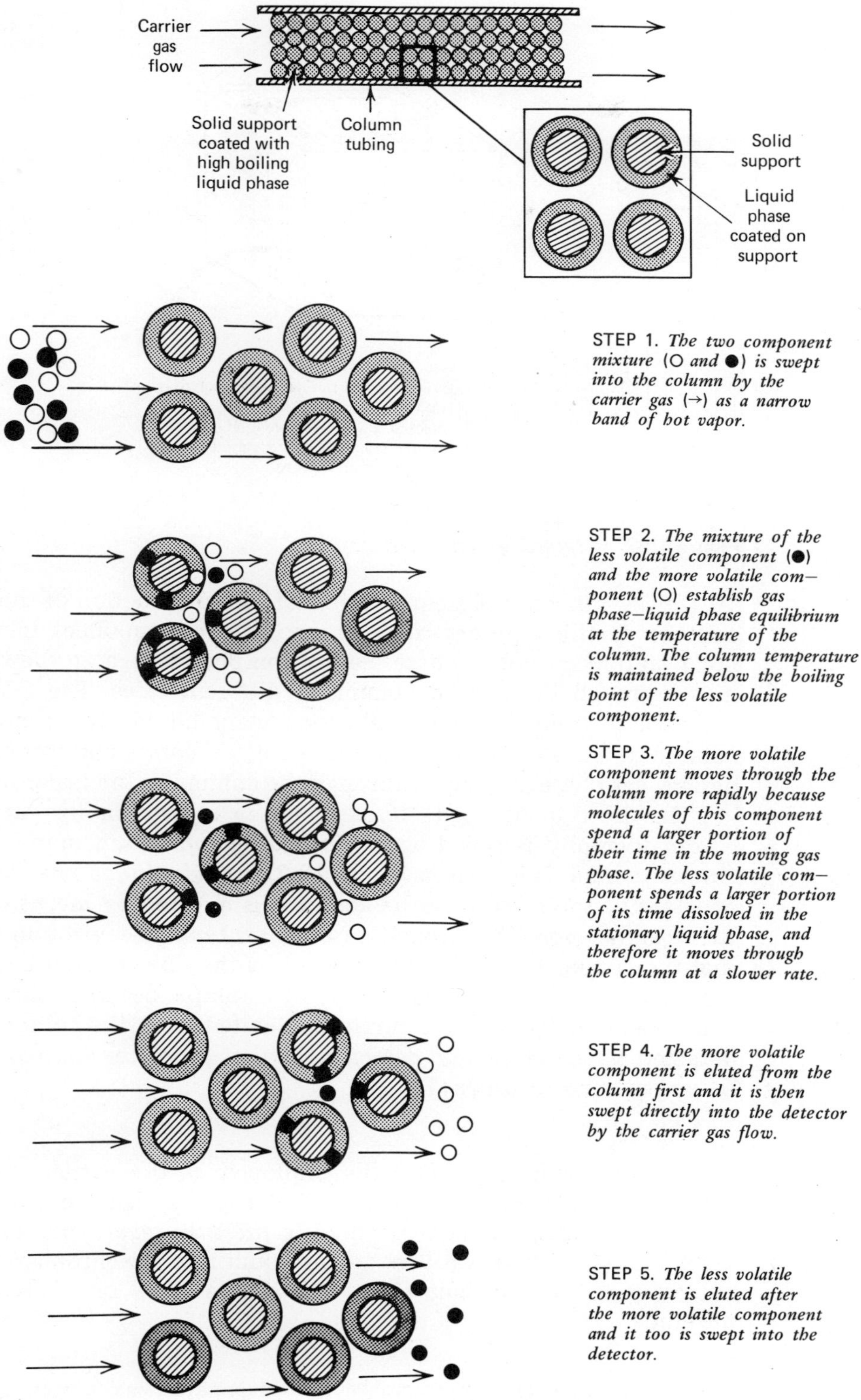

Figure 10.1 *The chromatography column and the vapor-liquid partition process.*

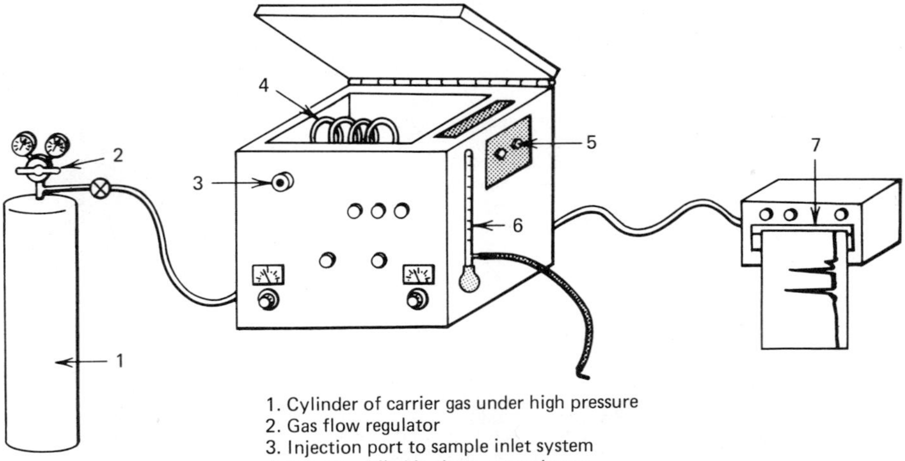

Figure 10.2 *A typical gas chromatograph.*

10.1 gives a diagramatic explanation of the construction of the packed column and the partition process for a nonpolar two-component mixture on a nonpolar stationary liquid phase. Separation occurs because the two components move through the packed column at different rates. The component that is less volatile (more soluble in the stationary liquid phase) moves through the column at the slower rate. The more volatile compound is more soluble in the moving gas phase and moves through the column at the faster rate.

The instrument used to carry out a separation by gas-liquid partition chromatography is called a *gas chromatograph.* A schematic diagram of a gas chromatograph is shown in Figure 10.2. In operation, a small sample (0.010 ml or less)* of the mixture to be analyzed is injected by means of a syringe into the *injection port* (3) through a *rubber septum.* The injection port is heated so that the sample is volatilized and swept by the *carrier gas* into the coiled *column* (4). Separation of the components of the sample occurs in the column. As each component of the original mixture is eluted from the column, its presence is detected by an electronic *detector* (5) that generates a signal that is recorded on a *strip chart recorder* (7).

Carrier Gas

Helium or nitrogen is normally employed as the carrier gas for GLC. The carrier gas is contained in a steel cylinder under high pressure, and its flow to the chromatograph is controlled by a *pressure regulator* attached to the gas cylinder. Gas flow controllers are also built into the chromatograph to give the necessary control to maintain a constant gas flow rate between 30 and 120 ml/min.

*Units of microliters (μl) are commonly used to express this small a volume. 1 μl = 0.001 ml.

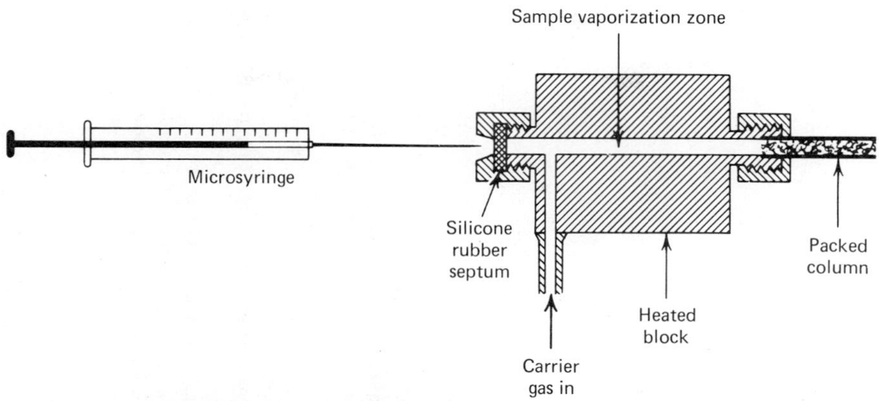

Figure 10.3 *Cross section of sample inlet system.*

Sample Inlet

The sample inlet system (Figure 10.3) is designed to allow the introduction of the sample without disturbing the flow of carrier gas. The inlet system is heated to a temperature sufficient to instantaneously vaporize all of the sample as a narrow band of vapor. The carrier gas sweeps the band of vapor directly from the inlet system into the column. For the analysis of the product from Experiment 9, an inlet temperature above 90°C should be used (>20°C above the highest boiling component in the mixture).

Column

The heart of the gas chromatograph is the packed chromatography column. The standard column is constructed of glass, stainless steel, aluminum, or copper tubing, $\frac{1}{8}''$ or $\frac{1}{4}''$ in diameter and from one to more than 20 feet in length. Glass columns are often preferred for separations of sensitive organic compounds because the glass surfaces do not adsorb organic compounds or catalyze their decomposition as readily as do the metal column surfaces. However, glass columns are very fragile, and for routine work the metal columns are satisfactory. The column is packed with a *solid support material* of small particle size that has been coated with a high-boiling *liquid phase.*

Solid Support

The solid support is a heat-stable, inert material that has a large surface area and a small particle size (diameter). The particle size of the solid support is commonly reported as a *mesh size* range. The 30–60 mesh (0.60–0.25 mm) and 60–80 mesh (0.25–0.15 mm) size supports are generally employed for GLC analysis. Narrow particle-size distribution produces more efficient columns than those with broad distributions because the carrier gas flow is more uniform around particles of similar size. Commonly used solid supports include pink diatomaceous silica (Chromosorb® P), white diatomaceous silica (Chromosorb® W and G), and porous beads that are made from polyaromatic cross-linked polymers (Porapak® and Chromosorb® 101–108). Glass beads and polytetra-

Table 10.1 Common Liquid Phases

Liquid phase	Structure	Max temp. (°C)	Applications
Squalane	Hydrocarbon	150	Hydrocarbons
SE-30	Methyl silicone	250	Generally useful nonpolar phase
QF-1	Trifluoropropyl methyl silicone	250	Halogenated hydrocarbons, hydrocarbons
OV-17	Phenyl methyl silicone	350	Natural products, pesticides, fatty acid esters
OV-225	Cyanopropyl phenyl methyl silicone	250	Generally useful moderately polar phase
Carbowax 20M	Polyethylene glycol	225	Alcohols, ethers, ketones
DEGS	Polyester	190	Esters, acids, ketones, alcohols
β,β'-Oxydipropionitrile	Cyano ether	100	A polar phase. Alcohols, acids

fluoroethylene particles are also used as supports for the separation of very polar compounds.

Stationary Liquid Phase

There are literally thousands of different liquid materials that have been used as the stationary phase in GLC, but most separations can be performed on a relatively few different liquid phases. Table 10.1 describes some common liquid phases, ranging from the nonpolar to the very polar. Nonpolar liquid phases are best for separating nonpolar samples, and conversely, polar liquid phases are more useful for separating polar samples. Nonpolar liquid phases such as squalene and SE-30 separate nonpolar materials on the basis of their relative boiling points. The more volatile materials are eluted from the column first. For samples of equal boiling point, the material that is most soluble in the liquid phase is eluted from the column last. Thus, polar samples elute more rapidly than nonpolar samples of the same boiling point on nonpolar liquid phases, and nonpolar samples elute more rapidly than polar samples of the same boiling point on polar liquid phases. For the analysis of the nonpolar alkene mixture from Experiment 9, a nonpolar liquid phase such as SE-30 gives good separation.

Detector

Several different types of detectors are used in gas chromatographs to sense the vapors of the separated components as they leave the column. The detector provides an electrical signal that is proportional to the amount of vapor eluted from the column. The detector is maintained at a temperature that is sufficiently high to prevent condensation of the sample vapors.

The *thermal conductivity detector* (TCD), widely used for applications where very high sensitivity is not required, is shown in Figure 10.4. The filament is a resistance wire that is heated by the electric current passing through the wire. The temperature of the wire will increase until the rate of heating is

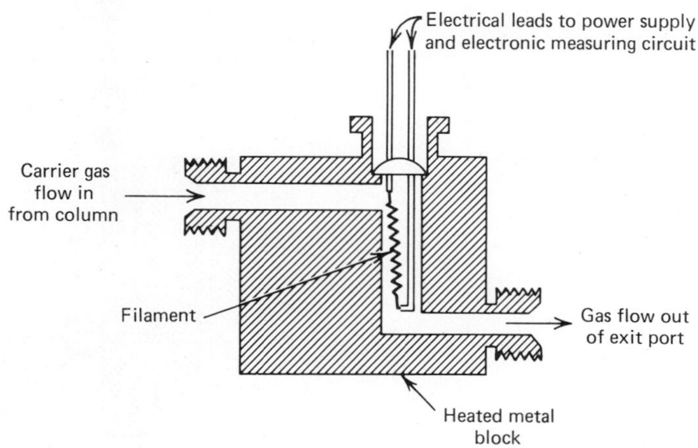

Figure 10.4 *Schematic diagram of thermal conductivity cell.*

exactly matched by the rate of cooling caused by the carrier gas flowing over the wire. For thermal conductivity detectors, a carrier gas with a high thermal conductivity such as helium or nitrogen is normally used. Organic vapors have a lower thermal conductivity than helium or nitrogen. Thus, when an organic vapor elutes from the column, the composition of the gas in the cell changes. The resultant decrease in the thermal conductivity of the gas produces a decrease in the cooling effect on the filament. The filament heats up slightly and the electrical resistance of the filament changes. This resistance change produces an electrical signal. The larger the amount of organic vapor that flows over the filament, the greater the temperature change, and the larger the resulting electrical signal.

The thermal conductivity detector responds to all organic and inorganic compounds. It is also nondestructive so the eluted sample may be collected. The lower detection limit for thermal conductivity detectors is approximately 5×10^{-6} g of sample per ml of carrier gas. Since the filaments of the thermal conductivity detector are oxidized by atmospheric oxygen, carrier gas must be flowing through the detector before turning on the filament current.

The *flame ionization* detector is very sensitive for the analysis of mixtures of oxidizable organic compounds. The lower detection limit for hydrocarbons is less than 5×10^{-9} g of sample per ml of carrier gas. This detector consists of a small hydrogen flame surrounded by an electrostatic field (Figure 10.5). The organic vapor and carrier gas that are eluted from the column are mixed with hydrogen, and the combined vapor is burned. During combustion of the

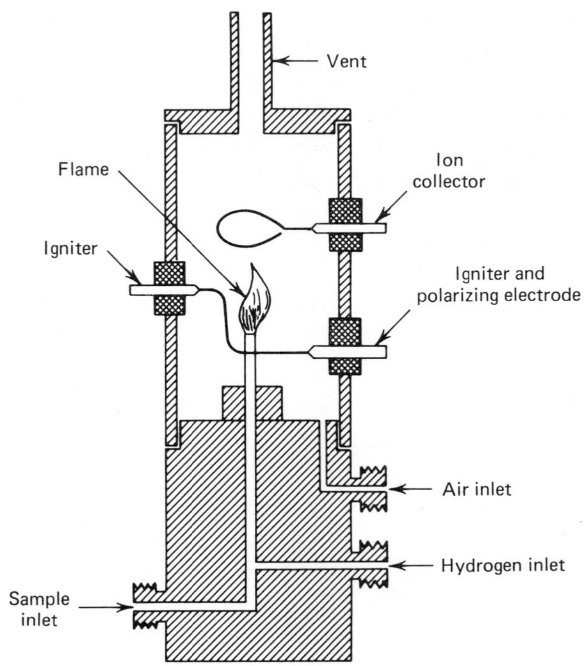

Figure 10.5 *Schematic diagram of a flame ionization detector.*

organic material, free electrons and ionic fragments are formed. These ions, which are collected on the electrode, produce a current proportional to the amount of sample being burned. This detector is insensitive to fully oxidized compounds such as water and CO_2.

The Chromatographic Record

The *strip chart recorder* is used to record the electrical signal from the detector. The recorded plot of signal intensity versus time is referred to as the *chromatogram.* Figure 10.6 illustrates a chromatogram of a three-component mixture. As each of the components is eluted from the column, a *peak* is recorded on the chromatogram. When air is injected with a sample, a peak is recorded as the air elutes from the column. Since the components of air are very volatile, the air flows through most GLC columns at the same rate as the carrier gas.

The time that elapses between the injection of the sample and the maximum signal of a particular peak is the *retention time* of that component. For example, the retention time of component A in Figure 10.6 is represented by T_a and is 3.6 minutes. The retention time of a compound is a constant value under identical chromatographic conditions and does not vary whether the compound is injected as a pure sample or as a component in a mixture. To be compared, retention times must be measured using the same column, with the same temperatures in the chromatograph inlet system, column oven, and with the same flow rate of carrier gas.

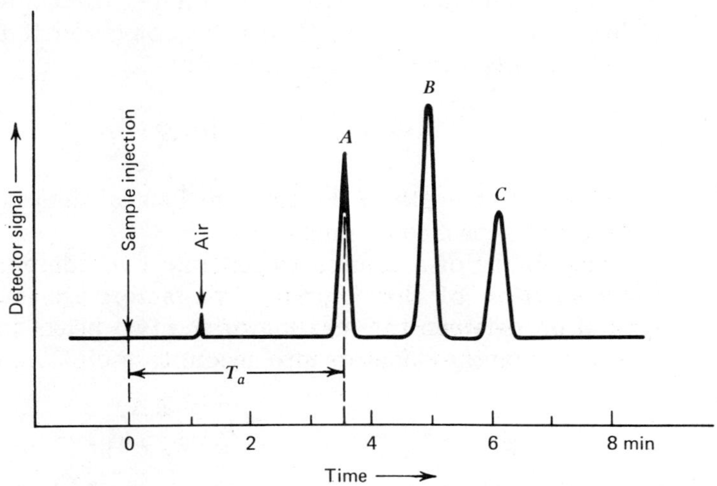

Figure 10.6 *Chromatogram of a three-component mixture.*

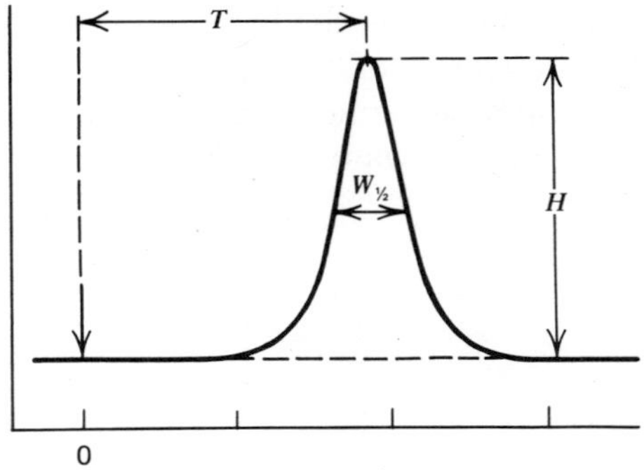

Figure 10.7 *Calculation of column efficiency.*

Factors Affecting the Separation

We have already discussed the importance of the gas chromatography column and indicated that the choice of liquid phase material is an important consideration. When comparing various columns, the chemist usually discusses the column's *efficiency* and, for a particular separation, the column's *resolution.*

The efficiency of a gas chromatography column can be measured empirically by the *theoretical plate number*, n, as

$$n = 5.54\left(\frac{T}{W_{1/2}}\right)^2$$

where T is the retention time of the peak and $W_{1/2}$ is the peak width at one-half the peak height (H) as measured from the baseline (see Figure 10.7). T and $W_{1/2}$ must be measured in the same units, for example, in seconds. As the length of a column increases, the number of theoretical plates also increases. Columns of different length can be compared if the *height equivalent to a theoretical plate* (HETP) is calculated:

$$\text{HETP} = \frac{L}{n}$$

where L is the length of the column. Typical values of n are 500 to 3000 plates per meter for packed columns.

The ability of a column to separate two components in a mixture is called the *resolution* of the column. Two factors are important in resolution: the separation between the maxima of the two peaks and the width of each of the peaks. Taking these factors into account, resolution, R, can be calculated by

$$R = \frac{2\,\Delta T}{W_a + W_b}$$

where W_a is the baseline width of peak A, W_b is the baseline width of peak B, and ΔT is the separation between peak maxima (Figure 10.8). A value of R =

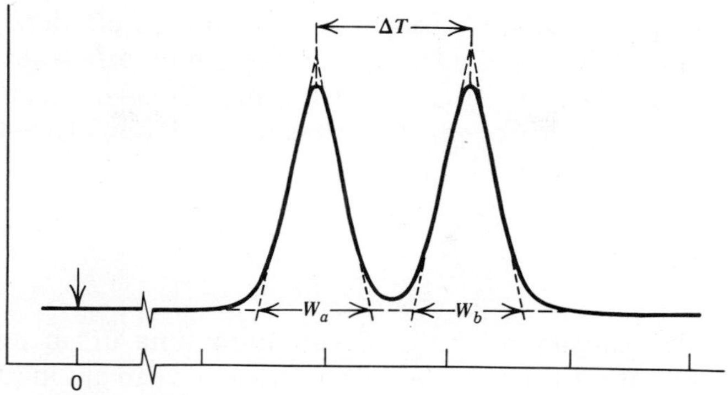

Figure 10.8 *Calculation of column resolution.*

1.5 corresponds to baseline resolution, and a value of $R = 1.0$ implies that the two peaks overlap by about 3% as shown in Figure 10.8.

The amount of liquid phase coated on the support material will affect the efficiency of a column. This amount is recorded as a weight percent; the usual range is 1 to 20%. With a higher liquid phase percentage the retention times of components are longer and more sample can be injected at one time. With low liquid phase loadings analyses are faster, but resolution will suffer unless small sample sizes are used.

The *temperature* of the column affects the ability of the column to separate a mixture. If the column temperature is too high, compounds with low boiling points will be flushed through the column at the same rate as the carrier gas. Lowering the column temperature will cause the vapor pressure of the sample components to decrease. The components will then be less soluble in the gas phase, and separation by vapor-liquid equilibria will be possible. If the temperature is too low, high boiling compounds in the mixture will condense and dissolve in the liquid phase, and these materials may not elute from the column in a reasonable length of time. Increasing the column temperature will decrease the retention time of all components in a sample.

When confronted with the separation of a mixture containing both low and high boiling components, it is useful to start the separation at a low column temperature, where maximum resolution of the low boiling components is possible, and then to increase the column temperature in order to increase the volatility of the high boiling components. This technique is called *temperature programming*, and research gas chromatographs are designed so that the column oven temperature can be automatically increased at a selected rate.

The *carrier gas flow rate* can also be important in determining column efficiency. The longer the time the sample spends in the vapor phase inside the column, the more the sample vapors will undergo gaseous diffusion, and the broader the peak width will be. However, if the carrier gas sweeps the vapors through the column at too high a rate, the vapors will not have sufficient time to establish vapor-liquid equilibrium, and column efficiency will also decrease. The optimum carrier gas flow can be determined experimentally, but the flow range is usually broad for near-optimum efficiency. Therefore, for most $\frac{1}{4}$-in. diameter packed columns, near-optimum efficiency can be achieved with a flow rate between 30 and 120 ml/min. Smaller diameter columns require correspondingly lower gas flow rates for optimum efficiency.

Qualitative Analysis

Comparison of the GLC retention time of an unknown compound with the retention times of known compounds can aid in identifying the unknown compound. If the retention times of the unknown and the known compound are the same under identical conditions, this is considered to be positive, but not absolute, evidence that the two compounds are identical. Just as many different compounds may have the same boiling points, many different compounds may also have the same GLC retention time on a particular column. To confirm the identity of an unknown compound by GLC, the unknown and known samples must be shown to have identical retention times on several columns that differ in the polarity of their liquid phases. If the retention times of a known compound and an unknown compound are not the same (± 3% of retention time) under identical GLC conditions on any one of the several columns, then the two compounds are not identical.

Since the retention times of compounds are subject to a number of instrumental variables, it is essential that the chromatogram of the known sample is run just before the chromatogram of the unknown sample in order to compare retention times under exactly the same conditions. An alternate technique, called *peak enhancement*, requires that pure samples of each of the suspected components of a mixture be added, one by one, to the injected sample of the mixture while looking to see which of the peaks will have its height increased relative to the unmodified mixture. By this method, you can be certain that the pure sample and the mixture have been chromatographed under identical conditions.

It is also possible to *collect samples* of the separate components as the vapors are swept from the exit port of the detector of the gas chromatograph (Figure 10.1) by the use of a cold *collection trap.* Figure 10.9 illustrates two

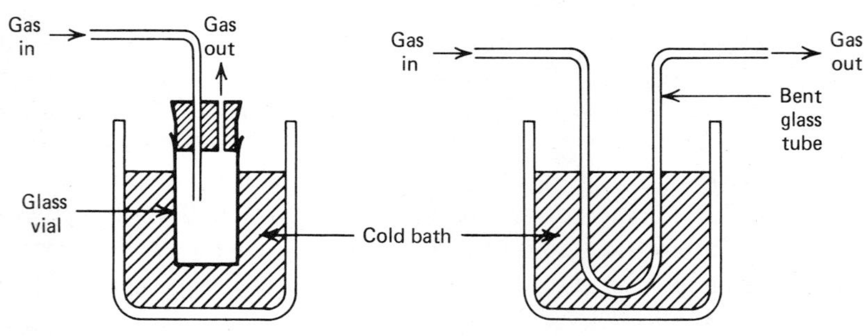

Figure 10.9 *Collection traps.*

easily constructed collection traps. The collection trap is connected to the gas exit port on the gas chromatograph during the time that the recorder indicates that the vapors of the desired component are passing through the detector. Many repetitive collections may be necessary to obtain sufficient pure sample for further analysis.

Quantitative Analysis

Gas-liquid partition chromatography is a very powerful tool for the quantitative analysis of a small sample of a volatile mixture. The area under a chromatographic peak is proportional to the number of moles of compound eluted from the column. Quantitative analysis of the relative amounts of the components in a mixture separated by GLC involves the measurement of the relative peak areas.

The simplest technique for measuring peak area is to approximate the peak area as a triangle and to obtain the area by multiplying the peak height, H, times the width at half-height, $W_{1/2}$ (Figure 10.7). This method is fast and gives good accuracy with symmetrical peaks of measurable width. If peak

$$\text{Area} = H \times W_{1/2}$$

widths are too narrow for accurate measurement, a faster recorder chart speed can be used.

Relative peak areas can also be determined by cutting peaks out with a scissors and weighing the pieces of paper on an analytical balance. Since the weight per unit area of good quality chart paper is uniform, the relative weights of the cut-out peaks are proportional to the relative areas of the peaks. This method can be very accurate if the paper is uniform in weight and if care is taken in cutting out the peaks. It is an especially useful technique if the peaks are not symmetric. A photocopy of the chromatogram can be cut out if the original chromatogram must be saved.

Digital electronic peak integrators are available that automatically measure area and retention time for each peak in the chromatogram, calculate the area percentage, and print this data on a paper tape. Although these units are expensive, they are widely used in research and industrial laboratories because they are very fast and can measure both small and large peak areas with precision.

Once the area of each peak has been determined, the mole percentage of composition for a component in the mixture can be calculated by dividing the peak area of each component in the mixture by the sum of all of the peak areas and multiplying this result by 100%:

$$\%\text{M}_i = \frac{A_i}{A_1 + A_2 + \cdots + A_n} \times 100$$

See Table 10.2 for an example of the calculation of the relative mole percentages of the three components in the chromatogram illustrated in Figure 10.6.

Table 10.2 Calculation of the Relative Mole Percentages of the Three Components in the Chromatogram in Figure 10.6

Peak	Retention time (min)	H (mm)	$W_{1/2}$ (mm)	Area = $H \times W_{1/2}$
A	3.6	93	5.8	540
B	4.9	114	6.0	680
C	6.1	62	8.4	520
			Total Area	1740

$$\% \text{ Component A} = \frac{540}{1740} \times 100 = 31\%$$

$$\% \text{ Component B} = \frac{680}{1740} \times 100 = 39\%$$

$$\% \text{ Component C} = \frac{520}{1740} \times 100 = 30\%$$

This calculation makes the assumption that the detector has equal sensitivity to all of the compounds analyzed, and this is not necessarily true. However, when a thermal conductivity detector is employed, the results obtained by this calculation are reasonably accurate, especially if the compounds in the mixture are isomers. If the compounds in the mixture are of the same molecular weight, the mole percentage will correspond to the weight percentage.

If very accurate results (better than ±2%) are necessary, a standard mixture of known percent composition must be prepared and analyzed. From the relative peak areas of the standard mixture and the known percent composition, the *relative response factor* for each compound can be calculated. These relative response factors are then used when making calculations for unknown samples to correct for the small differences in response that thermal conductivity detectors exhibit for different compounds.

Figure 10.10 shows a chromatogram of the hydrocarbon fraction (b.p. = 50–80°C) collected after distilling and drying the product prepared as described in Experiment 9. One microliter of sample was injected into a gas chromatograph equipped with a thermal conductivity detector. The column in the chromatograph was a $\frac{1}{4}$ in. by 8 ft coiled aluminum tube packed with

---

Figure 10.10 *Chromatogram of the product mixture isolated in Experiment 9. Note: No peak was observed with retention time greater than 9 min.*

GLC Conditions

Varian 920 GC with TC detector.
Column: $\frac{1}{4}$ in X 8 ft (aluminum) packed with 20% SE-30 on 60/80 mesh Chromosorb W.
Column temp: 30°C; Injector temp: 120°C; Detector temp: 200°C.
He gas flow: 120 ml/min.
Sample size: 1 μl; Attenuation: 8; Chart speed: 0.50 in/min.

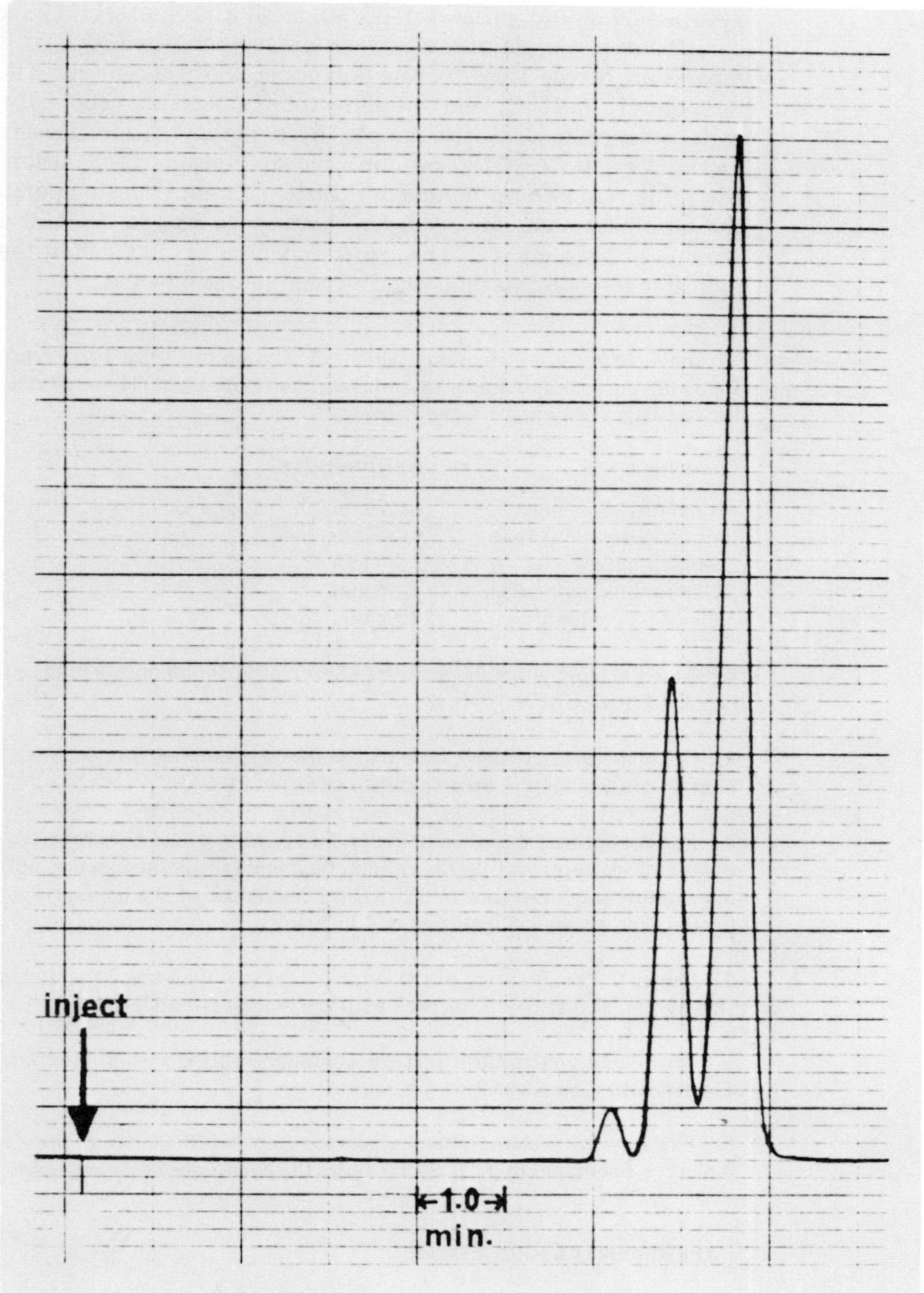

inject
1.0
min.

20% SE-30 liquid phase on 60/80 mesh Chromosorb® W solid phase. The column oven was set at a relatively low temperature (30°C) in order to achieve separation of the very volatile components of the sample. The injector block temperature (120°C) and the detector temperature (200°C) were sufficiently high to vaporize all of the sample. By experimentation it was determined that a carrier gas flow of 120 ml/min gave the optimum separation of components with this sample and column (at 30°C column oven temperature). To record the chromatogram, the recorder chart speed was set at 0.50 in./min, and the attenuation was set at 8. The arrow drawn on the left side of the chromatogram indicates the point in time that the sample was injected.

Prelab Questions

1. Immediately after recording the chromatogram in Figure 10.10, chromatograms of the following pure compounds were obtained under the identical conditions, and the retention time of each compound was measured:

Compound	*Retention Time*
cis-3-Methyl-3-pentene	6.7 min
trans-3-Methyl-2-pentene	7.4 min
2-Ethyl-1-butene	6.0 min
3-Chloro-2-methylpentane	24.8 min
3-Methyl-3-pentanol	19.2 min

What are the most probable identities of each of the components in the product mixture analyzed in Figure 10.10?

2. Calculate the retention time and percentage composition for each component in the mixture analyzed in the chromatogram in Figure 10.10.

3. For the analysis described in Figure 10.10, what would have been the effect on retention times and resolution if: (a) the column temperature was increased to 50°C? (b) the detector temperature was decreased to 170°C? (c) the length of the column was increased to 12 ft? (d) the chart speed was increased to 1 in./min?

4. Calculate the efficiency of the column used to obtain the chromatogram in Figure 10.10. (Base your calculation on the peak with the longest retention time.)

5. How could you confirm that the compound eluted with a retention time of 7.4 min in Figure 10.10 is an alkene?

6. Explain why 3-chloro-3-methylpentane (b.p. = 116°C) has a *longer* retention time than 3-methyl-3-pentanol (b.p. = 122°C) under the conditions described above in Question 1.

EXPERIMENTAL PROCEDURE

Preliminary Chromatograph Set-up. In most cases, the gas chromatograph will be set up and ready for the analysis of the mixture of alkenes from Experiment 9. The following steps will have already been taken by your instructor to prepare the chromatograph for this experiment:

1. The appropriate column has been installed. An 8 to 10 ft column with a 10 to 20% SE-30 liquid phase on Chromosorb® P or W is suitable for this experiment.

2. The inlet (injector) and detector temperatures have been adjusted so that they are at temperatures at least 20°C above the boiling point of the highest boiling component in the mixture to be separated. The detector temperature must have stabilized before the chromatograph can be used; otherwise, the baseline of the chromatogram will drift up or down as the detector temperature changes during the analysis.

3. The carrier gas flow rate has been adjusted to the proper flow rate (30 to 120 ml/min) by the use of the pressure regulator attached to the gas cylinder and the gas flow controllers.

4. The column oven temperature has been adjusted to the proper setting. For this experiment a column oven temperature of 30°C gives acceptable resolution and retention times.

5. The filament current on the thermal conductivity detector has been set to the proper operating value. *NOTE: The filament current must not be turned on unless carrier gas is flowing through the detector.*

6. The balance and recorder zero controls have been adjusted so that the recorder pen is in the proper position for baseline at all instrument sensitivity (attenuator) settings.

7. The instrument sensitivity or attenuator control has been adjusted so that the recorder pen will not go off scale on the largest peak. If the recorder should go off scale when you are recording a chromatogram, you can decrease the sample size or you can decrease the sensitivity of the chromatograph by increasing the attenuator setting. Increasing the attenuator from one setting to the next higher value reduces the sensitivity and recorder scale by a factor of one-half.

8. The recorder speed has been set at the desired speed. One inch per minute is normal, but faster or slower speeds are useful to expand or compress the scale and to conserve paper. You will turn the recorder from *Standby* to *On* in order to activate the chart drive just before injecting your sample.

Recording Instrument Parameters. Before you inject your sample into the chromatograph you should record the following information:

Instrument identification__________
Column: diameter__________ length__________
liquid phase__________ percentage__________
solid phase__________ mesh size__________
Column oven temperature__________ Injector temperature__________
Detector temperature__________
Attenuator setting__________ Chart speed__________
Gas flow rate__________

This information is conveniently recorded directly on the chromatogram, and any changes that are made in these parameters can be recorded on the chromatogram when the changes are made.

Measuring the Gas Flow Rate. A simple soap bubble flow meter (Figure 10.11) is used to measure the gas flow through a chromatograph. The metal tubing attached to the plastic tubing of the flow meter is temporarily connected to the exit port of the gas chromatograph. The rubber bulb is squeezed once so that the soap solution covers the side arm. The carrier gas entering the tube from the side arm causes a bubble to form and then displaces the bubble up the graduated tube. To measure the flow rate, use a sweep second hand or stop watch to determine how long it takes the soap bubble to travel the 10 ml between the markings on the graduated tube. It is necessary to wet the walls of the tube by forming several bubbles before finally measuring the flow rate. Disconnect the flow meter from the exit port after making the flow measurement. The flow rate is calculated in units of ml/min.

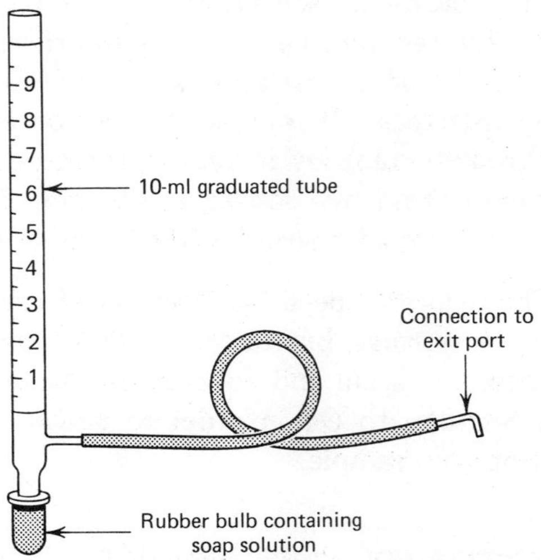

Figure 10.11 *Gas flow meter.*

Sample Injection. A microliter syringe is used to inject the liquid sample into the gas chromatograph. In filling the syringe it is desirable to exclude air from the sample to be injected. This can be accomplished by repeatedly drawing a sample of liquid into the syringe and rapidly squirting the sample back into the liquid. Care must be taken to push squarely on the plunger so that the wire plunger is not bent. In this experiment a two microliter (0.002 ml) sample will be injected. Draw up an amount of sample into the syringe that is slightly larger than the amount to be injected, and then, while holding the syringe vertically with the needle pointing up, adjust the plunger until it reads the desired amount. Wipe the needle with a tissue to remove excess sample.

To inject the sample into the chromatograph the syringe is held in two hands. One hand is used to guide the needle into the center of the rubber septum, and the other hand is used to hold the glass barrel of the syringe. The thumb of the hand holding the barrel is used to hold the plunger in place while inserting the needle as far as possible through the rubber septum. The plunger is smoothly and quickly depressed as soon as the needle has been inserted to the maximum depth. After a second or two, the syringe needle is withdrawn. While inserting and withdrawing the needle from the septum, it is critical to hold the syringe perpendicular to the septum at all times. A bent needle will result if the syringe is held at an angle to the septum.

As soon as the sample is injected make a mark on the chart paper to indicate the injection point. The progress of the chromatogram is followed by watching the recorder. When the chromatogram is complete the recorder is turned from *On* to *Standby* to stop the chart drive. The chromatogram of this particular sample is complete after the three closely spaced peaks have eluted from the chromatograph (see Figure 10.10).

After the microliter syringe has been used, it should be rinsed out with a volatile solvent such as ethyl ether. This can be accomplished by repeatedly pulling the solvent into the syringe and then squirting it back out. The plunger is then removed from the syringe and the solvent remaining in the barrel is allowed to evaporate.

After you have obtained the chromatogram of your product from Experiment 9, determine the retention times for each peak that is apparent in the chromatogram. Also calculate the percentage of each of the three alkene isomers in the product.

Postlab Questions

1. Give the probable identity of each of the peaks that are observed in the chromatogram of your product. Explain how you can assign the probable identity from the information available.

2. Do the relative amounts of each of the three alkene isomers obtained in this experiment correspond with Saytzeff's rule, which states that the most stable alkene will be produced in the largest amount? Explain your answer.

3. The alkene isomer mixture that was analyzed to give the chromatogram in Figure 10.10 was obtained by the dehydration of 3-methyl-3-pentanol. Compare the relative percentage of each alkene obtained by this reaction with the product that you obtained by dehydrohalogenation, and explain why the isomer ratios from the two reactions are different.

4. If the size of the sample that is injected into the gas chromatograph is increased by 20%, what would be the effect on the calculated percentage of each isomer?

5. Calculate the number of theoretical plates obtained in the column that you used for this experiment.

6. Suggest three ways that the resolution obtained in this analysis could be improved.

Experiment Eleven

Addition Reactions of Organic Compounds: Preparation of 3 - Bromo - 3 - methylpentane

Chemical reactions that involve addition of a reagent across a double or triple bond are known as addition reactions. The most common reactions of alkenes are addition reactions in which certain reagents are added across the carbon-carbon double bond:

Addition reactions of alkenes

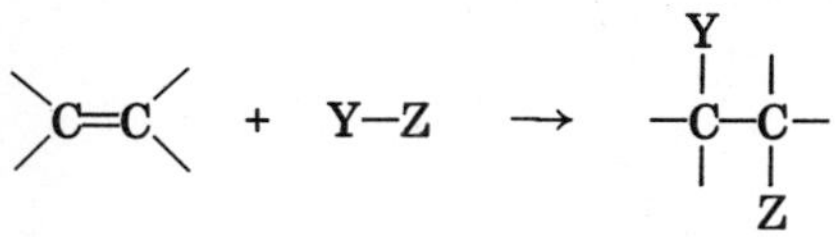

In Experiment 7 bromine (Br_2) addition was shown to be a characteristic reaction of alkenes. Hydrogen halides also undergo addition to carbon-carbon double bonds. In this experiment 3-bromo-3-methylpentane will be prepared by the addition of hydrogen bromide (HBr) to 3-methyl-2-pentene:

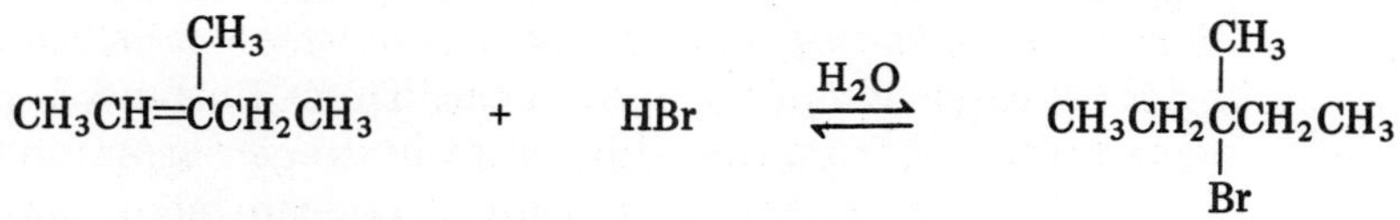

3-Methyl-2-pentene	*Hydrobromic acid*	*3-Bromo-3-methylpentane*
M.w. = 84.2 g/mol	48% HBr (by wt in H_2O)	M.w. = 165.1 g/mol
b.p. = 68–70°C	b.p. = 126°C	b.p. = 130°C
d_{20} = 0.69 g/ml	d_{20} = 1.49 g/ml	d_{20} = 1.18 g/ml

Addition of Hydrogen Halides

Alkenes readily add hydrogen halides (HCl, HBr, or HI) to the carbon-carbon double bond to yield alkyl halides. The reaction is conveniently carried out in

Addition of hydrogen halides

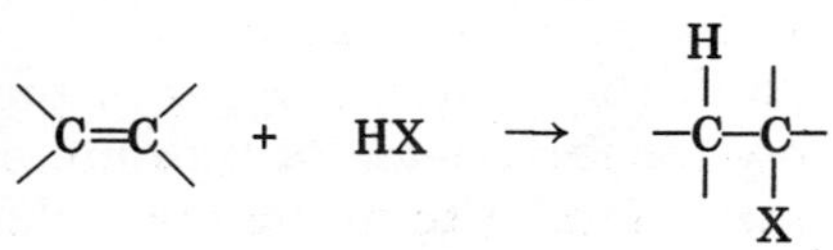

the laboratory by heating an aqueous solution of the hydrogen halide with the alkene. This procedure offers a practical method for the synthesis of a number of alkyl halides from commercially available alkenes.

The factors that affect the course of the addition of hydrogen halides to alkenes have been carefully investigated. Results of these studies have shown that highly substituted alkenes are more reactive than less substituted alkenes. Since 3-methyl-2-pentene has three alkyl groups substituted on the carbons of the carbon-carbon double bond, rapid addition of hydrogen bromide would be expected to occur.

Unsymmetrically substituted alkenes, such as 3-methyl-2-pentene, can potentially yield two different alkyl halides depending on the orientation of the addition of hydrogen halide to the carbon-carbon double bond:

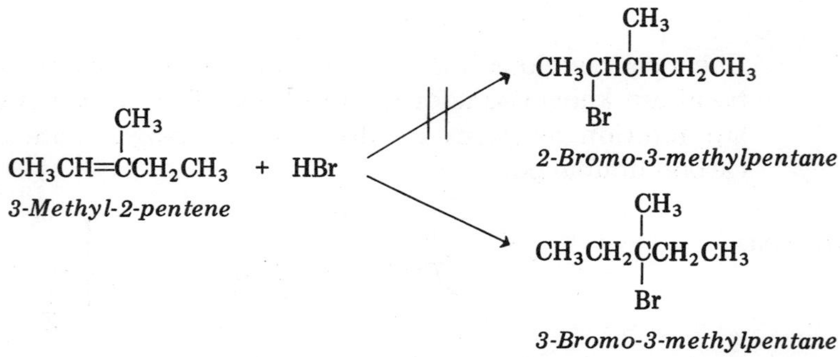

However, studies have shown that in an aqueous solution of hydrogen bromide only the product in which the halogen is bonded to the more substituted carbon atom is formed. These results are the basis of Markovnikov's rule, which states that *in the addition of a polar molecule to a carbon-carbon double bond, the more positive end of the polar molecule becomes attached to the carbon possessing the greater number of hydrogens.* Thus, 3-bromo-3-methylpentane is the expected product from the addition of hydrogen bromide to 3-methyl-2-pentene.

In this experiment, a mixture of concentrated, aqueous hydrobromic acid and the isomeric alkenes prepared in Experiment 9 are heated on the steam bath. The reaction can be monitored by observing that the initial, vigorous refluxing slows down as the lower boiling alkenes are converted into the higher boiling alkyl bromide. After the reaction is complete, the product is separated from the aqueous acid, washed with aqueous sodium bicarbonate, and dried over anhydrous magnesium sulfate. The product is then purified by distillation. The following diagram outlines the purification procedure for this experiment. (See page 129)

The distillation apparatus shown in Figure 11.1 is useful for distillations of small volumes of liquids. The hold-up volume in this apparatus, with a 25- or 50-ml flask, is small. Several side-arm test tubes can be used to collect the fractions of distillate. If regular test tubes are substituted for the side-arm test tubes, the cork (or rubber) stopper in the test tube must have a slit cut in one side to allow for the release of pressure during heating. The test tube is cooled in a beaker of ice water to insure that volatile materials are condensed.

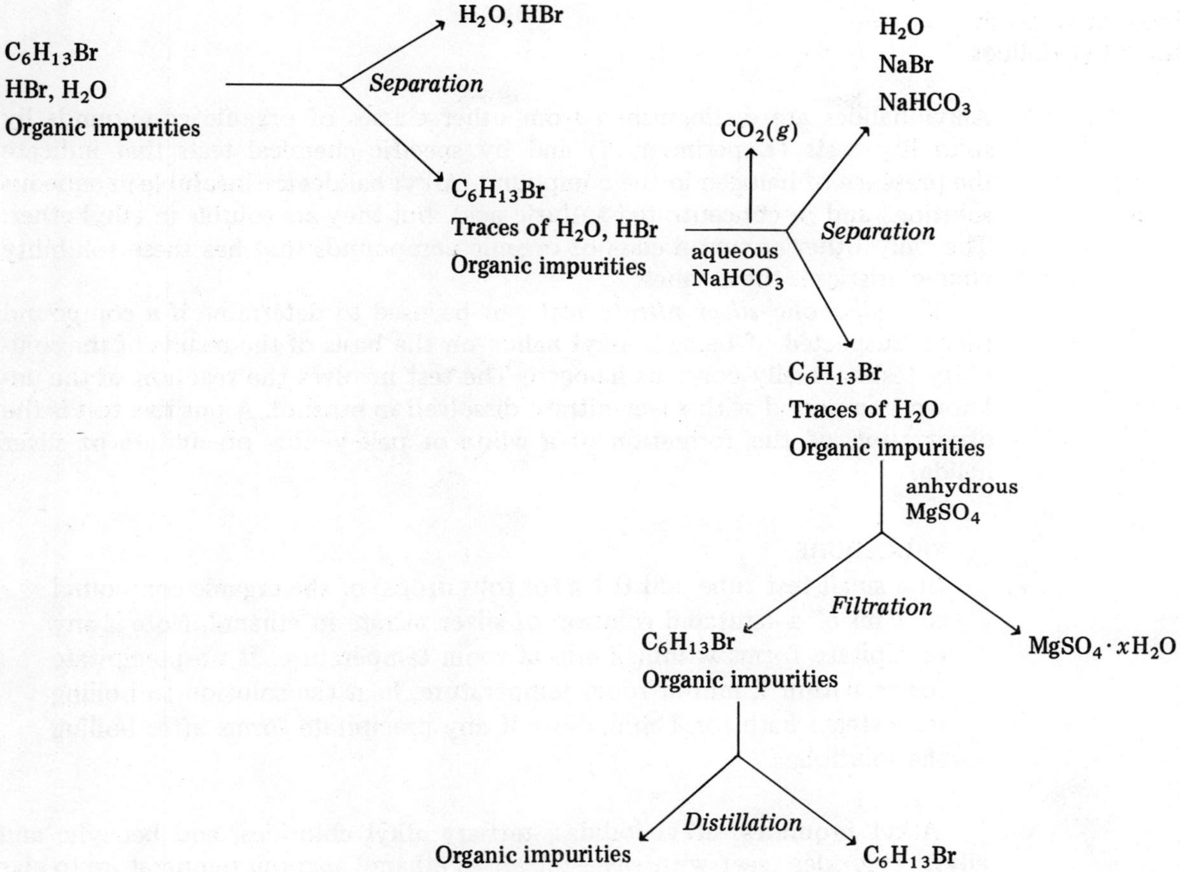

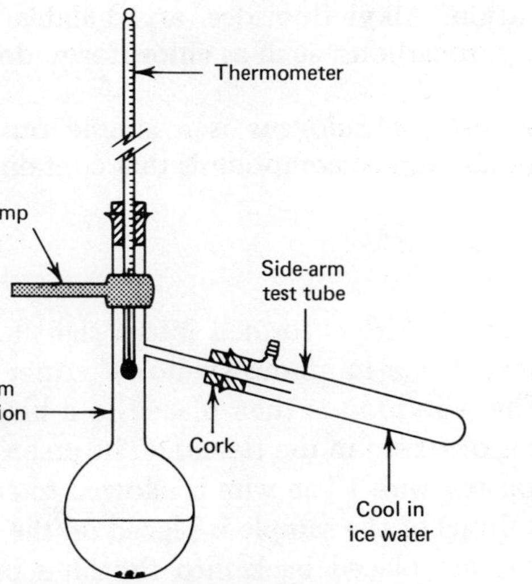

Figure 11.1 *Apparatus for small-scale distillation.*

Product Analysis for Alkyl Halides

Alkyl halides are distinguished from other classes of organic compounds by solubility tests (Experiment 7) and by specific chemical tests that indicate the presence of halogen in the compound. Alkyl halides are insoluble in aqueous solutions and in concentrated sulfuric acid, but they are soluble in ethyl ether. The only other common class of organic compounds that has these solubility characteristics is the alkanes.

The *alcoholic silver nitrate test* can be used to determine if a compound that is suspected of being a alkyl halide on the basis of the results of the solubility tests actually contains halogen. The test involves the reaction of the unknown compound with silver nitrate dissolved in ethanol. A positive test is the observation of the formation of a white or pale-yellow precipitate of silver halide.

PROCEDURE

In a small test tube add 0.1 g (or four drops) of the organic compound to 1 ml of a saturated solution of silver nitrate in ethanol. Note if any precipitate forms within 2 min at room temperature. If no precipitate forms within 2 min at room temperature, heat the solution to boiling on a steam bath for 1 min. Note if any precipitate forms after boiling the solution.

Alkyl bromides, alkyl iodides, tertiary alkyl chlorides, and benzylic and allylic chlorides react with silver nitrate in ethanol at room temperature to give a precipitate of silver halide. Primary and secondary alkyl chlorides do not react with silver nitrate in ethanol at room temperature, but they will react in the boiling ethanol solution. Alkyl fluorides, aryl halides, vinyl halides, and some polyhalogenated hydrocarbons, such as chloroform, do not give a precipitate in this test.

The *Beilstein flame test for halogens* is a simple but sensitive test for halogens and is useful for all organic compounds that contain chlorine, bromine, and iodine.

PROCEDURE

A 10-cm length of copper wire is formed into a tight loop at one end. This loop is then heated in a luminous (yellow) burner flame to form an oxide coating. The wire loop is then placed in a blue burner flame until no green color is observed in the flames. (The green color is due to halogen impurities on the wire.) The wire is allowed to cool, and then a few milligrams (one drop) of the sample is placed on the wire loop. The wire loop and sample are placed back into the blue burner flame. A green flame, which appears quickly and may last for only a few seconds, indicates the presence of halogen in the organic compound (a positive Beilstein test).

Prelab Questions

1. Give the structure of the expected Markovnikov product(s) for the addition of aqueous hydrogen bromide to: *cis*-3-methyl-2-pentene, *trans*-3-methyl-2-pentene, and 2-ethyl-1-butene.

2. Give the structure of the expected Markovnikov product(s) for the addition of aqueous hydrogen bromide to *cis*- and *trans*-2-pentene. Would this reaction be a good method to prepare 2-bromopentane? What alkene will react with hydrogen bromide under these conditions to yield only 2-bromopentane?

3. By the use of chemical characterization tests how can you show that the product obtained in Experiment 8 is 3-chloro-3-methylpentane and not 2-chloro- or 1-chloro-3-methylpentane?

4. Predict the solubility of each of the following compounds in ethyl ether, water, 10% aqueous HCl, 10% aqueous NaOH, and conc. H_2SO_4 (see Experiment 7): 3-methyl-3-pentanol, 3-chloro-3-methylpentane, 3-methyl-2-pentene, and 3-bromo-3-methylpentane.

5. Calculate the number of moles of hydrogen bromide in 20 ml of 48% hydrobromic acid.

6. Two liquid-liquid separations are required in the purification procedure for this experiment. In each separation, predict which phase (organic or aqueous) will be the upper phase and which will be lower phase.

EXPERIMENTAL PROCEDURE

Add 4.0 g of 3-methyl-2-pentene and 20 ml of concentrated hydrobromic acid (48% aqueous solution) to the 50-ml round-bottom flask. Add several boiling stones, and attach the condenser to the flask for reflux. Gently heat the mixture (steam bath) for 1.5 hr. (*Vigorous heating or inefficient condensation will result in the loss of volatile alkene.*) Allow the reaction mixture to cool. Then transfer the mixture into a separatory funnel, add 10 ml of pentane, and thoroughly mix the two phases. Carefully drain the acidic aqueous phase out of the separatory funnel. Slowly, with swirling, add 15 ml of 5% sodium bicarbonate

---

CAUTION: Carbon dioxide gas will be evolved. Do not stopper the separatory funnel until gas evolution has subsided.

---

solution to the separatory funnel. When gas evolution has subsided, thoroughly mix the two layers. Separate the organic layer from the aqueous layer and dry the organic layer with anhydrous magnesium sulfate (~0.5 g). After drying, distill the product and collect the fraction boiling over the range of 120 to 135°C. Record the weight of product obtained and calculate the percentage yield.

Determine the solubility classification of your product by using the procedure described in Experiment 7. Perform the ethanolic silver nitrate and the Beilstein flame tests on your product. Record your observations of these classification tests.

Postlab Questions

1. Write a balanced chemical equation for the reaction that produces carbon dioxide gas when the aqueous bicarbonate is added to the separatory funnel.

2. Some decomposition of 3-bromo-3-methylpentane is observed during the distillation of the product. Suggest how this product can be distilled at a lower temperature in order to avoid thermal decomposition. (Co-distillation with water will not be suitable. Why?)

3. Why is it essential to remove the last traces of water and acid from the organic layer before the organic material is distilled?

4. What experimental evidence do you have that the isolated product is actually 3-bromo-3-methylpentane?

5. One student noted that his reaction mixture did not reflux when heated over a steam bath, so he used a heating mantel to obtain higher temperatures and vigorous refluxing during the addition reaction. The resulting isolated yield was only 12% of the theoretical yield. Explain why vigorous heating of this reaction would result in a poor yield.

Experiment Twelve

Infrared Spectroscopy

In previous experiments solubility tests and chemical characterization tests were used to gain information about the structures of organic compounds. The use of chemical reactions for identifying functional groups present in organic compounds is limited by a number of factors. In contrast, the analysis of organic compounds through their interaction with electromagnetic radiation in the infrared region of the spectrum is applicable to most compounds, is nondestructive, and affords information concerning a variety of functional groups from a single experiment. In this experiment the interaction between infrared radiation and an organic compound will be measured and recorded on an infrared spectrophotometer. A study of these interactions will then provide detailed information about the molecular structure of the compound.

Since its introduction in the 1950s, infrared spectroscopy has revolutionized the procedures employed for structural identification and analysis of organic compounds. The combination of infrared spectral analysis with chemical characterization and solubility tests is now used to rapidly provide structural information for the identification of unknown compounds and for the characterization of new compounds.

Infrared Absorption Spectroscopy

Infrared absorption spectroscopy is the measurement of the amount of infrared radiation absorbed by a compound as a function of the frequency (or wavelength) of the infrared radiation. Thus, the data obtained from an infrared absorption experiment is a graph of infrared radiation frequency (or wavelength) versus the relative amount of infrared radiation of that frequency (or wavelength) transmitted through a sample of the compound (see Figure 12.1). This graph is called the *infrared spectrum* of the compound.

The infrared region of the electromagnetic spectrum most useful to organic chemists is radiation with frequencies between 4000 and 650 cm^{-1} (cm^{-1} = reciprocal centimeter or wave number) and with wavelengths between 2.5 and

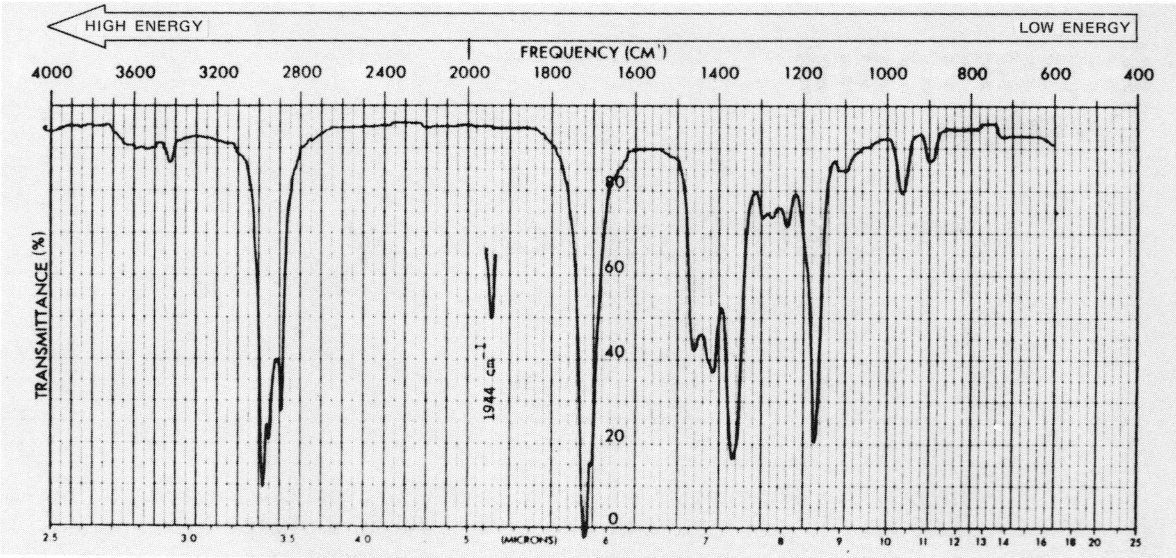

Figure 12.1 *Infrared spectrum of 2-pentanone.*

The horizontal axis is linear in frequency (cm^{-1}) with a scale change at 2000 cm^{-1}. The vertical scale is in units of percent transmittance. When the percent transmittance is near 100%, the sample is not absorbing infrared radiation at that frequency. The absorption of infrared radiation by the sample is indicated by a downward deflection of the graph and a corresponding decrease in percent transmittance. The mark on the spectrum labeled 1944 cm^{-1} is a calibration mark used to correct for any misalignment of the frequency scale. The peak is from the spectrum of a standard polystyrene film and is known to occur at 1944 cm^{-1}

15 μm (1 μm = 1 micrometer = 10^{-6} meters). The wavelength and frequency of radiation are inversely related (see Table 12.1):

$$\text{wavelength } (\mu m) = \frac{10^4}{\text{frequency } (cm^{-1})}$$

The frequency of radiation is directly proportional to the energy of the radiation. The energy of infrared radiation varies from 1.9 kcal/mol (650 cm^{-1}) to 11 kcal/mol (4000 cm^{-1}). Absorption of radiation in this energy range by a molecule results in the increased vibrational energy of the molecule. The high vibrational energy state of the molecule is short-lived, and the absorbed energy is lost as heat.

Molecular Vibrations

A molecule is not a rigid structure but, instead, can be described as a flexible assembly of balls (the atoms) held together by springs (the chemical bonds) of different strengths and lengths. The molecule has two primary ways of flexing: *bond stretching*, in which the bond length increases or decreases,

Table 12.1 The Relationship Between Frequency, Wavelength, and Energy of Radiation

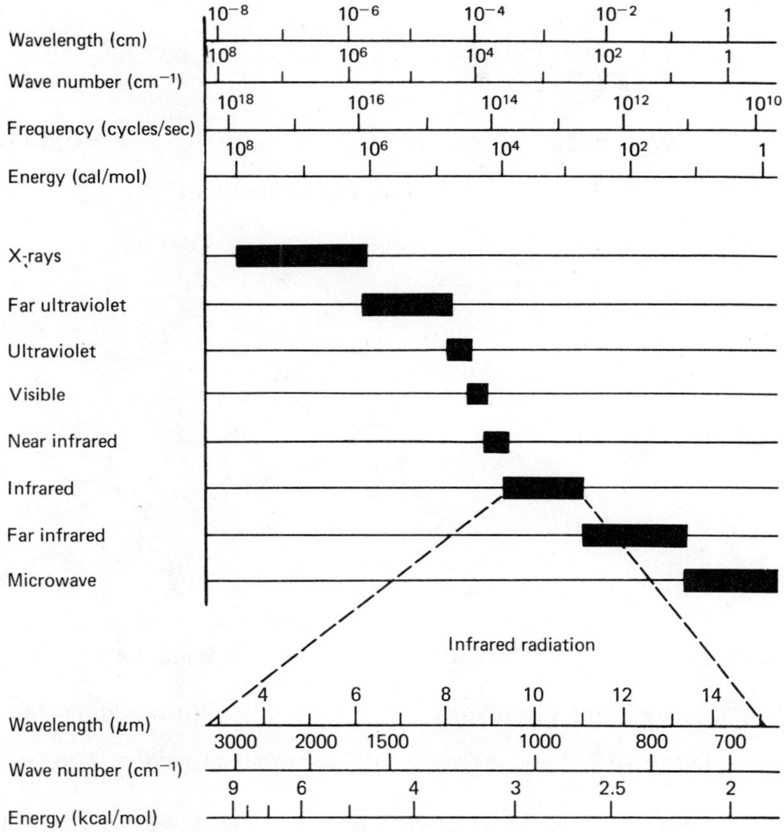

and *bond bending*, in which the angle between bonds is expanded or contracted. Bond stretching and bond bending are called *fundamental modes of vibration.* The various stretching and bending vibrations of a molecule occur only at certain quantized frequencies. The molecule can only absorb infrared radiation if the frequency of the radiation corresponds to the frequency of the vibrational modes in the molecule. The energy that is absorbed causes an increase in the amplitude of the vibrational motion of the bond. Figure 12.2 illustrates the different stretching and bending vibrations that can occur in a molecule.

The frequencies of stretching vibrations are found to occur in the order of bond strengths. For example, the stretching frequency of the carbon-carbon triple bond of an alkyne occurs in the region of 2260 to 2100 cm^{-1}. The stretching frequency of the carbon-carbon double bond of an alkene requires less energy and occurs in the region of 1680 to 1620 cm^{-1}. The stretching frequency of the carbon-carbon single bond of an alkane requires even less energy and occurs in the region of 1200 to 800 cm^{-1}.

The frequency of the stretching vibration of a bond between two atoms is also dependent on the mass of the atoms. Thus, when the bond involves a very

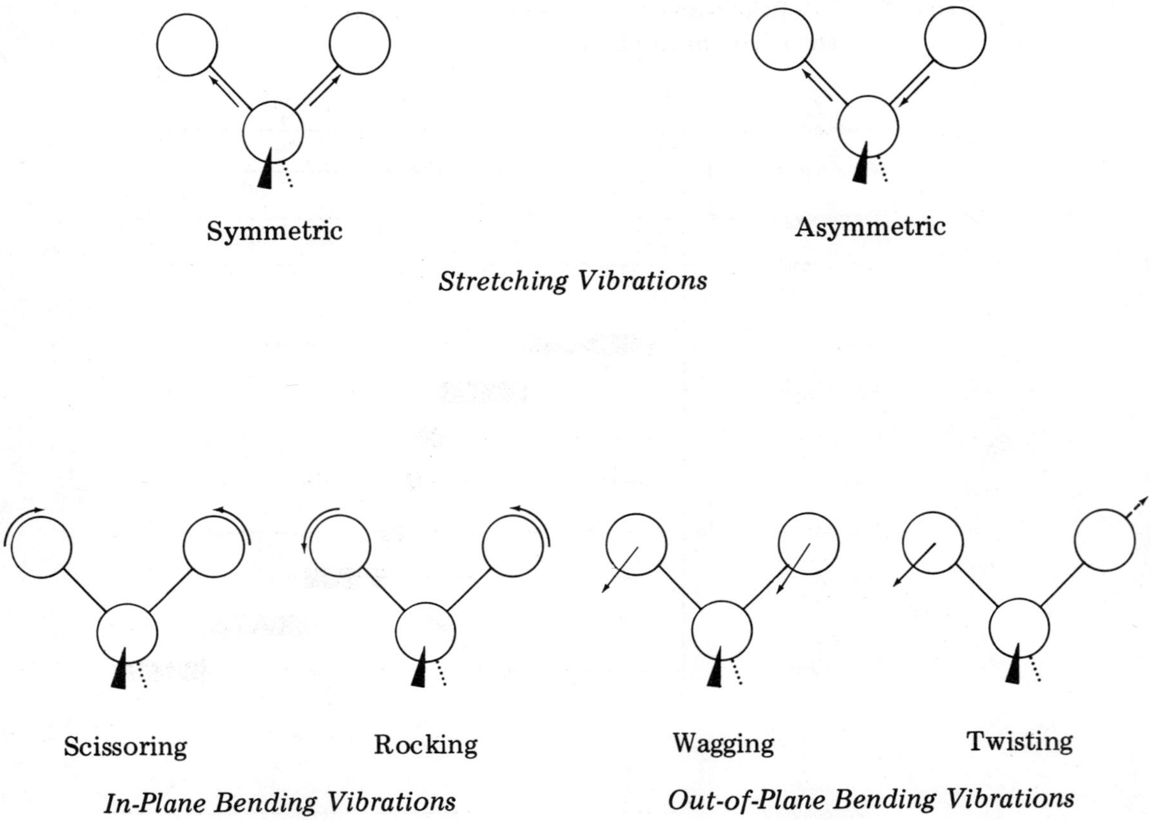

Figure 12.2 *Various fundamental vibrations of a group of atoms.*

light atom, such as a proton in a C—H, O—H, or N—H bond, the stretching vibration is at a much higher frequency (3700 to 2600 cm^{-1}) than when two heavier atoms are involved. For example, the stretching frequency for the O—H bond of water is 3750 cm^{-1}, but the stretching frequency for the O—D bond of D_2O is at 2630 cm^{-1}. In this example, the substitution of a deuterium atom doubles the mass of the hydrogen isotope and results in a large decrease in the bond stretching frequency without weakening the chemical bond.

By assuming that a chemical bond vibrates like a simple spring, the stretching frequency of a bond, $\bar{\nu}$, in cm^{-1}, between two atoms can be approximately related to the mass of the atoms, m_1 and m_2, in atomic mass units, and the force constant of the bond, k, in dyne/cm:

$$\bar{\nu} = \frac{1}{2\pi c}\sqrt{\frac{k}{\frac{m_1 m_2}{(m_1 + m_2)}}}$$

where c is the velocity of light, 3×10^{10} cm sec^{-1}. Single, double, and triple bonds have force constants that are approximately 5, 10, and 15×10^5 dyne/cm, respectively.

The Infrared Spectrophotometer

The infrared absorption spectrum of an organic compound is measured on an infrared spectrophotometer. Figure 12.3 illustrates the components of a typical infrared spectrophotometer.

The source of infrared radiation is commonly either an electrically heated rod, called a Nernst glower, or a simple resistance wire. The infrared radiation from the source is divided into two beams by the use of reflecting mirrors. One beam is focused through a cell that contains the sample. The second beam, which is called the reference beam, is directed along a similar path, except that this reference beam does not pass through the cell that contains the sample. The radiation transmitted along both the reference beam path and the sample beam path is directed to a rotating sector mirror called a chopper. As this mirror rotates, it alternately directs radiation from the sample beam and the reference beam onto the narrow entrance slit of the monochromator. The monochromator employs either a prism or a grating to disperse the radiation so that the detector can measure the relative intensity of radiation that is transmitted through each beam path as a function of frequency (or wavelength). Absorption of radiation at a particular frequency by the sample results in the sample beam being less intense than the reference beam at that frequency. The detector measures this difference in intensity and produces a corresponding electrical signal. The electrical signal is then recorded as an absorption peak on the spectrum. Detectors commonly used in infrared spectrophotometers are thermistors or thermocouples.

Glass and quartz absorb strongly throughout most of the infrared region so they cannot be used as cells to contain samples or to construct prisms to disperse infrared radiation. Materials that do not contain covalent bonds, such as metal halide salts, do not usually absorb infrared radiation. Therefore, sample cells and prisms for infrared spectroscopy are commonly constructed of crystalline salts such as sodium chloride or potassium bromide.

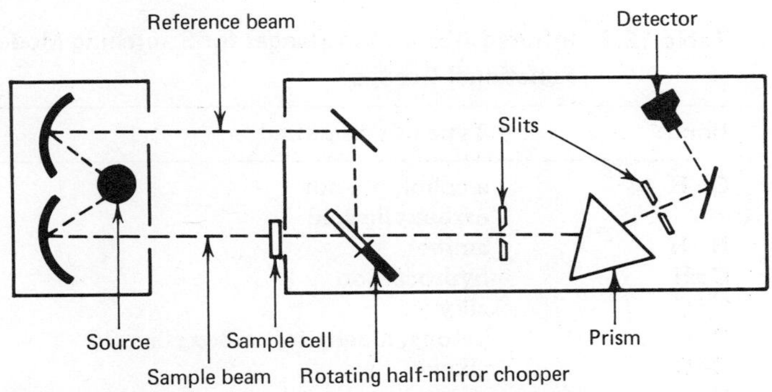

Figure 12.3 *Schematic optical path of an infrared spectrophotometer.*

The Interpretation of Infrared Spectra

The frequency range in which a particular functional group absorbs infrared radiation can generally be predicted with considerable accuracy. This predictability is based on infrared spectral correlations of large numbers of compounds that possess the same functional group. Table 12.2 lists the infrared absorption ranges for stretching modes of common organic functional groups. Thus, the spectrum of limonene is predicted to show a characteristic absorption in the range of 1680 to 1620 cm^{-1} due to the presence of the carbon-carbon double bonds in the molecule and a characteristic absorption in the frequency range of 3300 to 2840 cm^{-1} due to the presence of carbon-hydrogen bonds in the molecule. The infrared spectrum of limonene (Figure 12.4) exhibits an absorption at 1640 cm^{-1} that is attributable to stretching vibrations of the carbon-carbon double bonds and absorptions in the range of 3100 to 2840 cm^{-1} that are due to carbon-hydrogen stretching modes.

Most of the absorption bands that occur in the 4000 to 1400 cm^{-1} region of the spectrum can be assigned to a particular functional group in the molecule. This region of the spectrum is called the *diagnostic region.* The region below 1400 cm^{-1} often contains many absorption bands that cannot be assigned to a particular functional group. This region of the spectrum is referred to as the *fingerprint region* of the spectrum.

A description of the shape and intensity of an infrared absorption band is often given in addition to the frequency of the absorption band. The shape of a band can be characterized as *broad*, such as the absorption band in the region of 3500 to 2500 cm^{-1} in Figure 12.13, or *sharp* such as the band at 1640 cm^{-1} in Figure 12.4. The relative intensity of an absorption band is described as *strong* (*s*), *medium* (*m*), or *weak* (*w*). If the peak is among the most intense (lowest percent transmittance) in a spectrum, it is classified as strong. The carbonyl absorption band at 1715 cm^{-1} in Figure 12.1 is a strong absorption band. Strong absorptions are the most useful for diagnostic purposes. Weak absorption bands in a spectrum are those having a percent transmittance that is less than one-fifth the value of the most intense absorption. The carbon-carbon

Table 12.2 Infrared Absorption Ranges for Stretching Modes of Common Organic Functional Groups

Bond	Type of compound	Frequency range (cm^{-1})
O—H	alcohol, phenol	3650–3100
	carboxylic acid	3200–2500
N—H	amine	3600–3200
C—H	hydrocarbon	3330–2840
C≡C	alkyne	2260–2100
C=O	ketone, aldehyde, carboxylic acid	1800–1650
C=C	alkene	1680–1620
C—O	alcohol, ether, ester	1300–1000

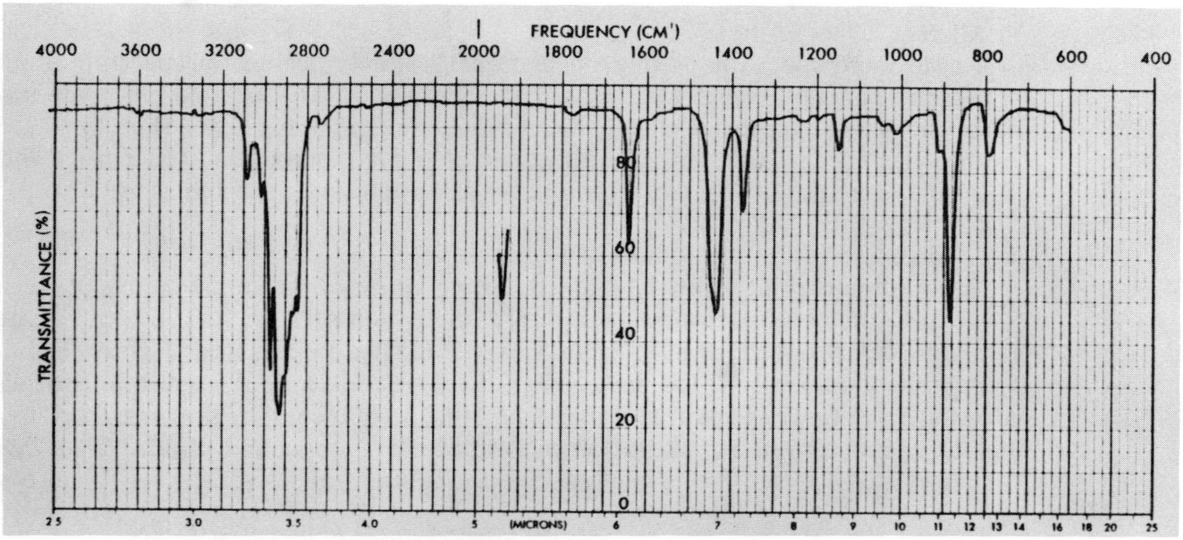

Figure 12.4 *Infrared spectrum of limonene (thin film).*

double bond stretching absorption is usually a weak absorption peak (e.g., the peak at 1635 cm^{-1} in Figure 12.6).

Although compounds of similar structure may have similar infrared spectra in the region of 4000 to 1400 cm^{-1}, there will almost always be a discernible difference in the fingerprint region of the spectrum. Thus, it is usually possible to establish that two samples are identical if the samples have identical infrared spectra when measured under the same conditions. Files of infrared spectra are available for comparison with the spectra of an unidentified organic compound.

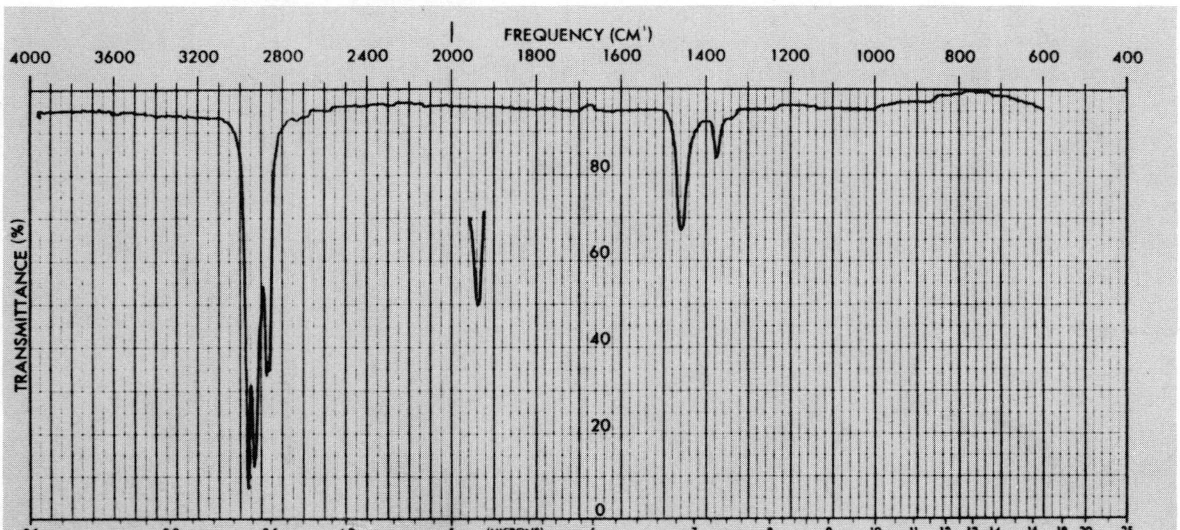

Figure 12.5 *Infrared spectrum of hexane (thin film).*

Alkanes

The infrared spectrum of hexane, Figure 12.5, is typical of the spectrum of an alkane. The C—H stretching absorptions between 3000 to 2840 cm^{-1} are observed in the spectra of all organic compounds containing C_{sp^3}—H groups. The absorptions due to bending vibrations of $-CH_2-$ groups (1458 cm^{-1}) and C—CH_3 groups (1460 to 1380 cm^{-1}) are also typical of most hydrocarbons.

Alkenes

The carbon-carbon double bond stretching vibration of an alkene results in a weak to medium intensity absorption in the region of 1680 to 1620 cm^{-1}. The carbon-hydrogen stretching vibrations of alkenes occur at a slightly higher frequency (3100 to 3000 cm^{-1}) than do the C—H stretching vibrations of alkanes. The C—H bending modes in the 1000 to 650 cm^{-1} region are often useful in predicting the substitution pattern of the double bond (Table 12.3). The infrared spectrum of 1-hexene is shown in Figure 12.6.

Alkynes

The C≡C stretching band is found as a sharp absorption in the region of 2260 to 2100 cm^{-1}. Terminal alkynes also show C_{sp}—H stretching absorption in the region of 3330 to 3270 cm^{-1}.

Aromatic Compounds

Aromatic compounds show characteristic absorptions in several regions of the spectrum. There are two to four absorption bands in the 1670 to 1430 cm^{-1} region that are particularly diagnostic of aromatic structure. These bands, which occur near 1600, 1580, 1500, and 1450 cm^{-1}, are generally sharp and of variable intensity. The bands are due to carbon-carbon bond vibrations of

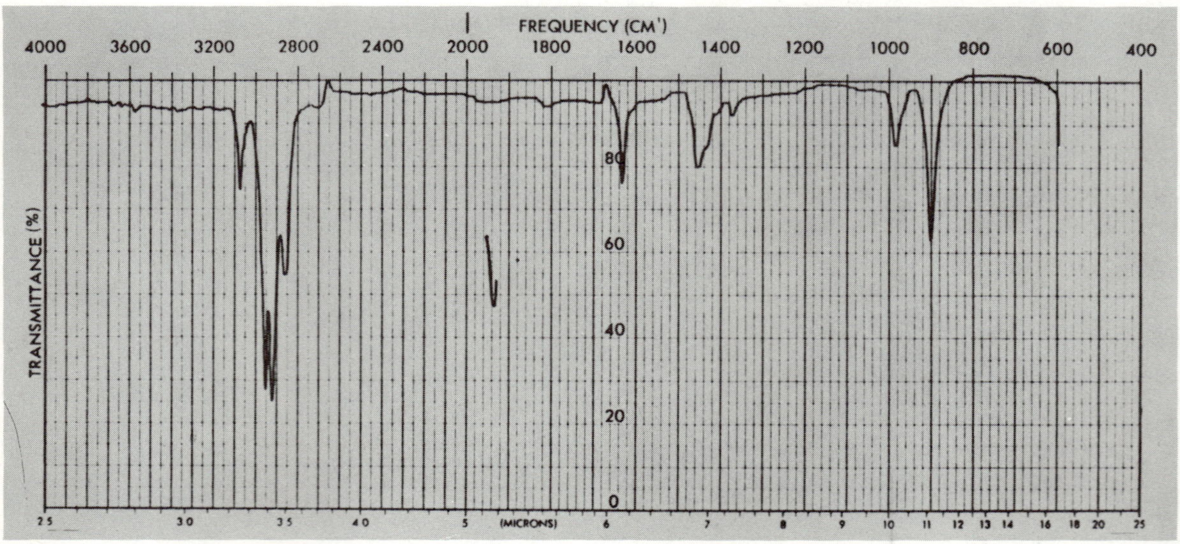

Figure 12.6 *Infrared spectrum of 1-hexene (thin film).*

Table 12.3 Olefinic C—H Bending Vibrations

Substitution pattern	Frequency range (cm^{-1})
$RCH{=}CH_2$	1000–980 and 915–900
$R_2C{=}CH_2$	900–880
trans-RCH=CHR	1000–950
cis-RCH=CHR	840–700
$R_2C{=}CHR$	840–790

Table 12.4 Aromatic C—H Bending Frequencies for Mono- and Disubstituted Aromatic Compounds

Substitution	Frequency range (cm^{-1})
monosubstitution	770–730 and 710–690
o-disubstitution	770–735
m-disubstitution	900–860, 810–760, and 710–670
p-disubstitution	860–800

the aromatic ring. The absence of absorption by a compound in this region indicates that the compound is not aromatic.

Aromatic C—H stretching absorption bands occur in the region of 3100 to 3000 cm^{-1}. In addition, absorption bands due to aromatic C—H bending vibrations in the region of 900 to 650 cm^{-1} are diagnostic of the substitution pattern of the aromatic ring (Table 12.4). The infrared spectrum of toluene (Figure 12.7) clearly shows the diagnostic pattern for monosubstitution in the 900 to 650 cm^{-1} region of the spectrum.

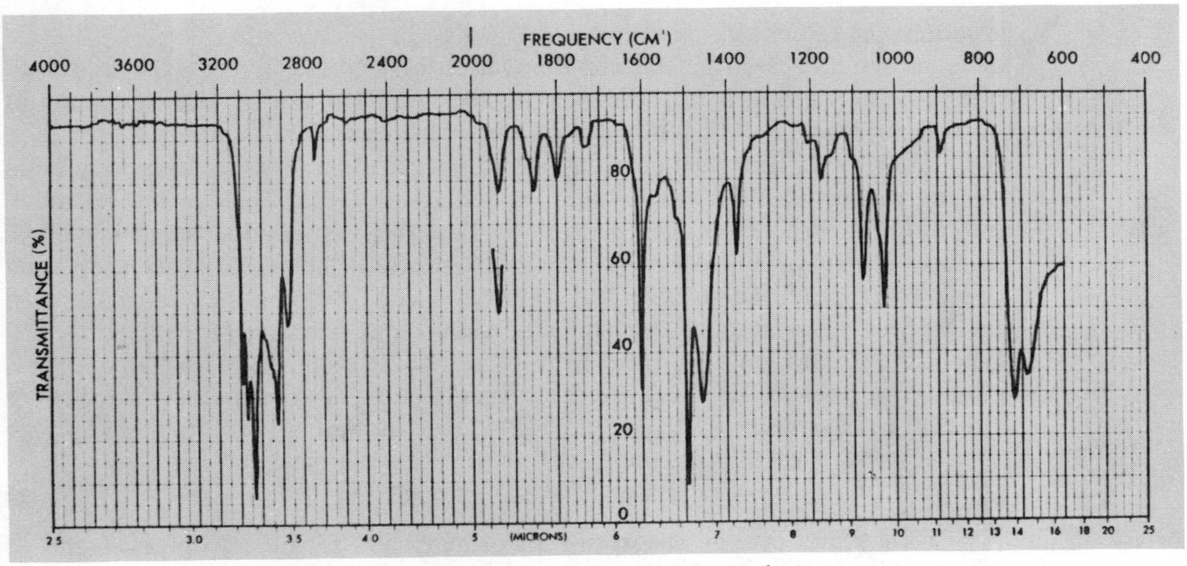

Figure 12.7 *Infrared spectrum of toluene (thin film).*

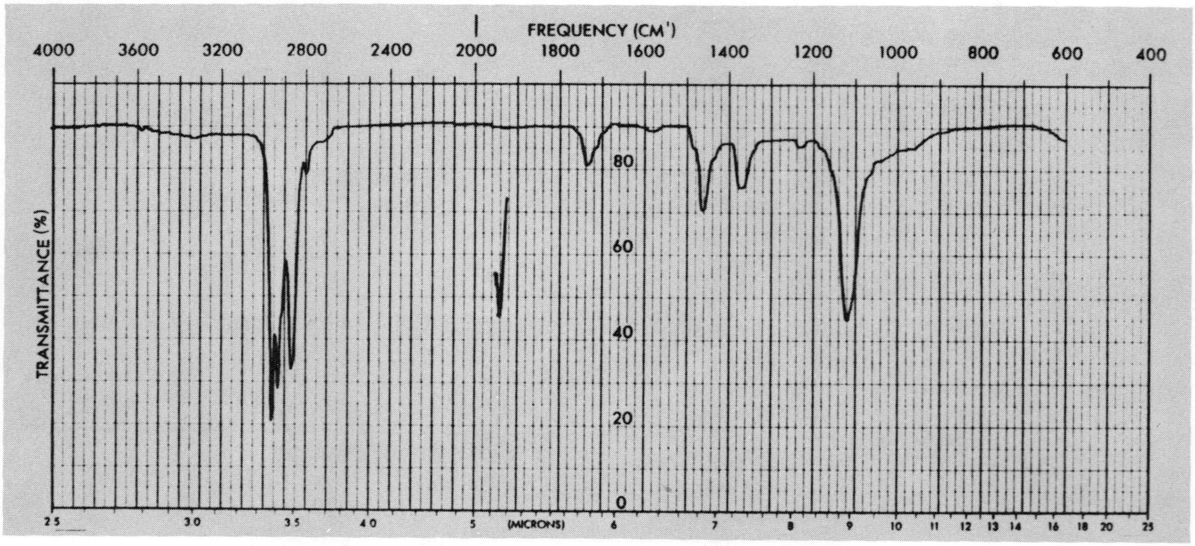

Figure 12.8 *Infrared spectrum of n-butyl ether (thin film).*

Ethers

The characteristic strong absorption band for aliphatic ethers is associated with vibrations of the C—O bond in the region of 1150 to 1090 cm^{-1}. Aryl alkyl ethers generally exhibit two strong absorption bands—one at 1280 to 1200 cm^{-1} and a second at 1080 to 1020 cm^{-1}. These absorptions occur in the fingerprint region of the spectrum. Although other absorption bands, such as C—C stretching bands, also occur in this same region, the absorption bands due to the C—O vibrations can be distinguished since they are among the strongest bands in the spectrum. Figure 12.8 is the spectrum of *n*-butyl ether.

Alcohols and Phenols

The most characteristic infrared absorption of alcohols and phenols is due to O—H bond vibrations in the region of 3650 to 3100 cm^{-1}. The frequency of the absorption band depends on the degree of hydrogen bonding. Absorption by "free" O—H (not participating in hydrogen bonding) occurs as a sharp band in the region of 3650 to 3600 cm^{-1}, but free hydroxyl groups are not commonly observed unless the alcohol is dissolved in nonpolar solvent such as carbon tetrachloride to form a very dilute solution. The O—H absorption band from hydrogen-bonded hydroxyl groups is usually strong and broad, as seen in Figures 12.9 and 12.10. The C—O stretching vibrations in alcohols and phenols produce a strong band in the 1260 to 1000 cm^{-1} region of the spectrum. The position of this absorption band can be used to indicate the class of alcohol as illustrated in Table 12.5. The frequency range for the C—O absorption is lowered by 30 to 60 cm^{-1} if the alcohol is allylic.

Amines

The most characteristic absorption of primary and secondary amines is due to the N—H stretching vibrations in the region of 3600 to 3200 cm^{-1}. Amines like

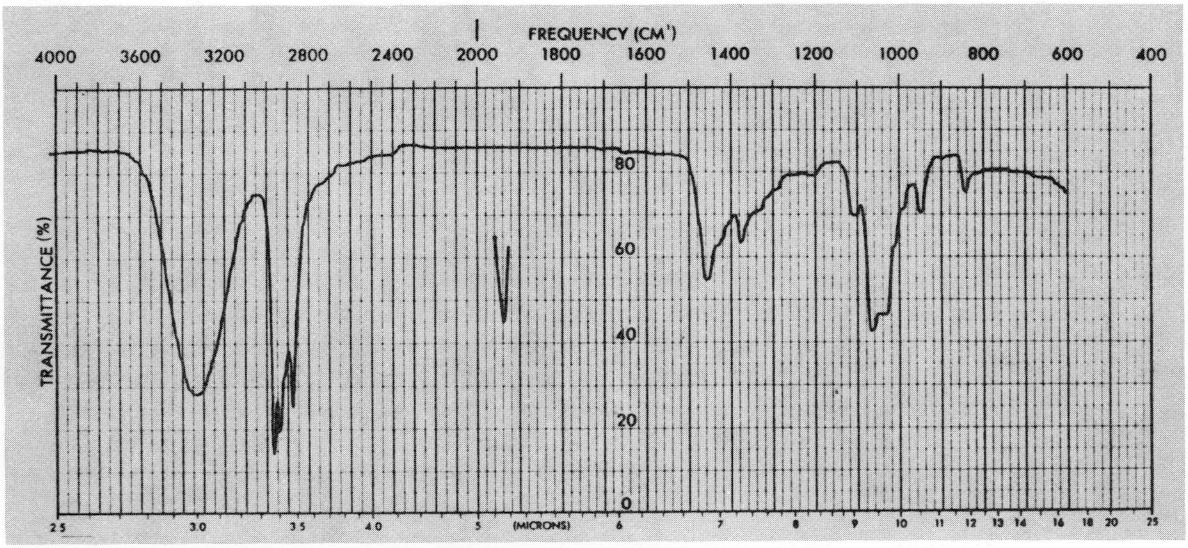

Figure 12.9 *Infrared spectrum of 1-butanol (thin film).*

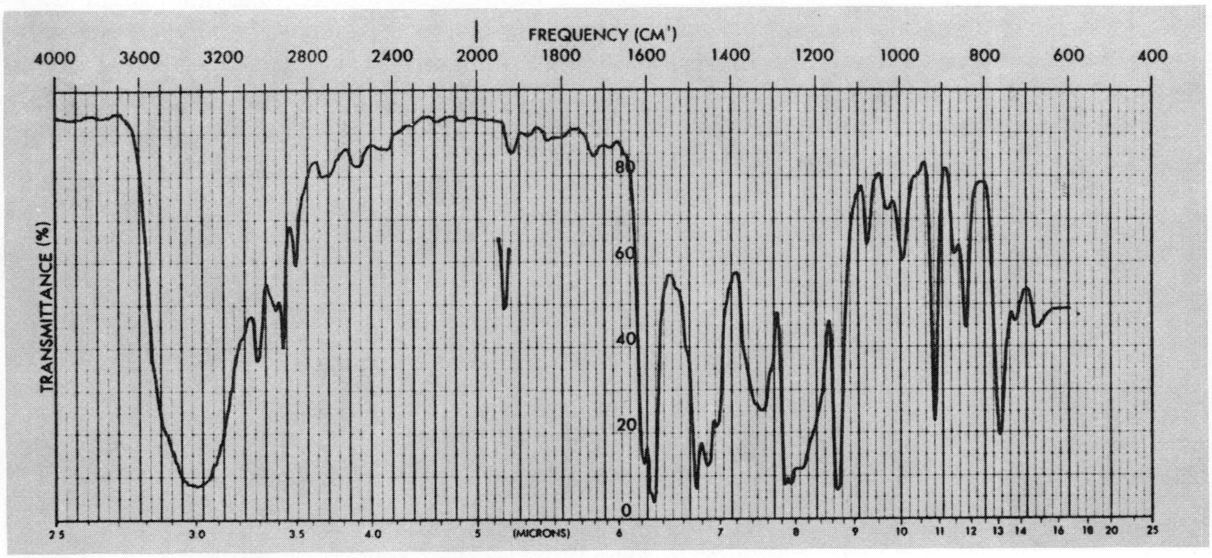

Figure 12.10 *Infrared spectrum of m-cresol (thin film).*

Table 12.5 Alcohol C—O Absorption Bands

Structural type		Frequency range (cm^{-1})
Phenol	ArOH	1260-1180
Saturated tertiary	R_3COH	1200-1125
Saturated secondary	R_2CHOH	1125-1090
Saturated primary	RCH_2OH	1085-1050

alcohols are generally hydrogen bonded and exhibit broad absorption bands, but the absorption due to N—H stretching is much weaker than that due to O—H stretching. The difference is illustrated by the intensity differences in the 3600 to 3200 cm^{-1} regions for 1-butanol (Figure 12.9) and isobutylamine (Figure 12.11). Tertiary amines do not exhibit a N—H stretching absorption.

The absorptions due to C—N vibrations occur in the region of 1360 to 1020 cm^{-1}, but they are not as strong as the corresponding C—O vibrations and, consequently, are not generally useful as a diagnostic indicator of structure.

Ketones

The C=O stretching frequency of ketones occurs in the region of 1750 to 1660 cm^{-1} (Figure 12.1). The carbonyl absorption frequency is sensitive to ring strain, conjugation, and electronegativity effects of substituents, among other factors. The "normal" position for an aliphatic carbonyl absorption in saturated acyclic ketones is near 1715 cm^{-1}. Cyclohexanones and larger cycloalkanones exhibit carbonyl absorption at this frequency, but cyclopentanones have a carbonyl absorption near 1750 cm^{-1}. Conjugation of the carbonyl group with a carbon-carbon double bond lowers the frequency of vibration by about 25 to 50 cm^{-1}. Thus, α,β-unsaturated ketones and phenyl ketones absorb in the region of 1690 to 1665 cm^{-1}.

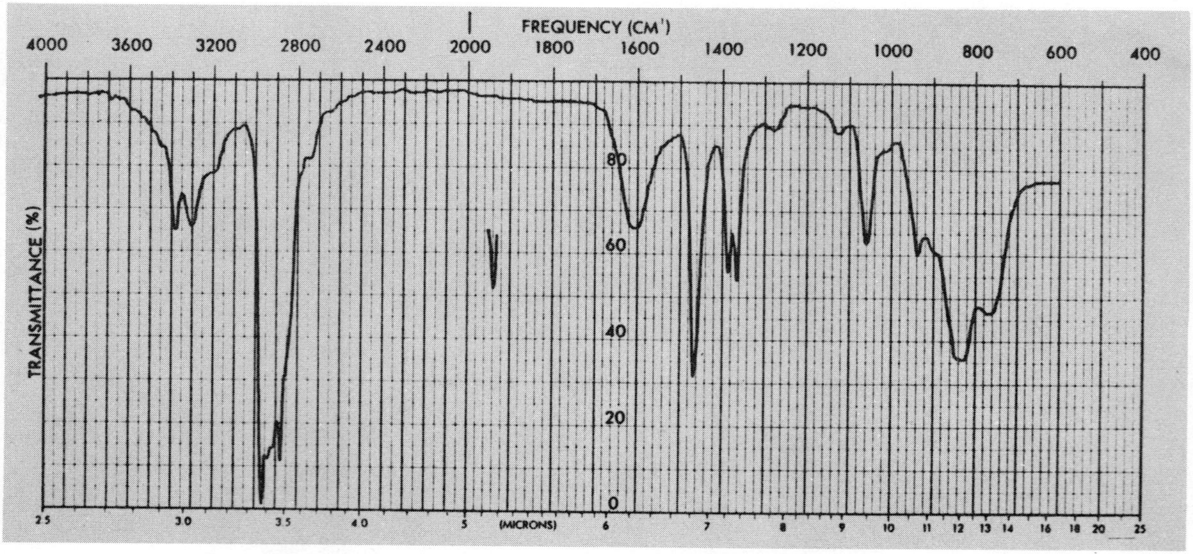

Figure 12.11 *Infrared spectrum of isobutylamine (thin film).*

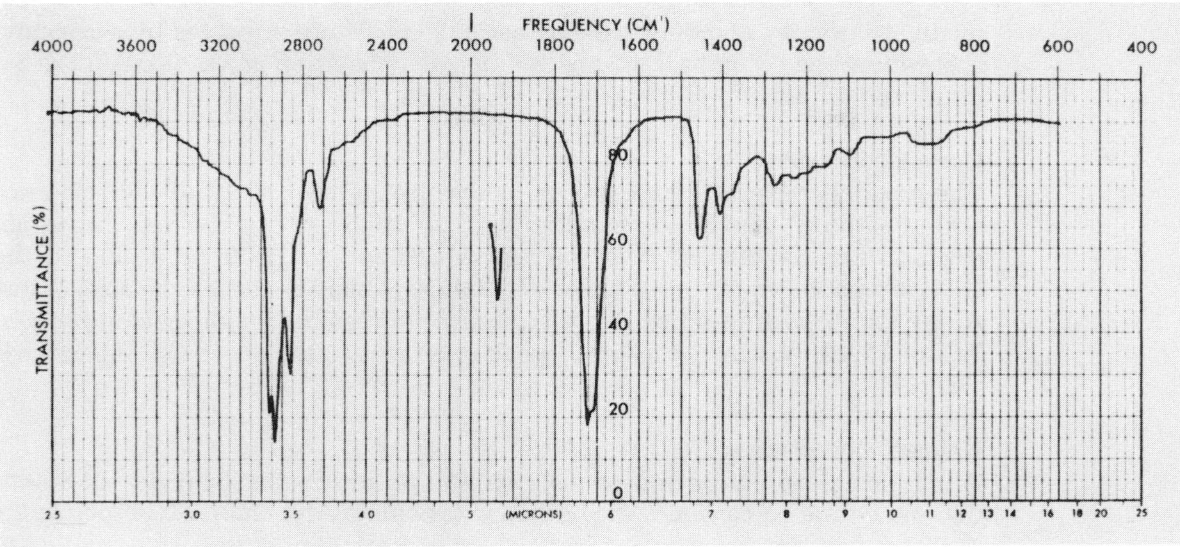

Figure 12.12 *Infrared spectrum of heptaldehyde (thin film).*

Aldehydes

The carbonyl groups of aldehydes and ketones absorb in approximately the same frequency range. The "normal" position of carbonyl group absorption for saturated aliphatic aldehydes is near 1730 cm^{-1}. The frequency of absorption is lowered by conjugation. α,β-Unsaturated aldehydes and aromatic aldehydes absorb in the region of 1710 to 1685 cm^{-1}. A distinguishing feature of the infrared spectrum of an aldehyde is absorption in the 2850 to 2700 cm^{-1} region due to C—H vibrations from the aldehyde functional group. Thus, a

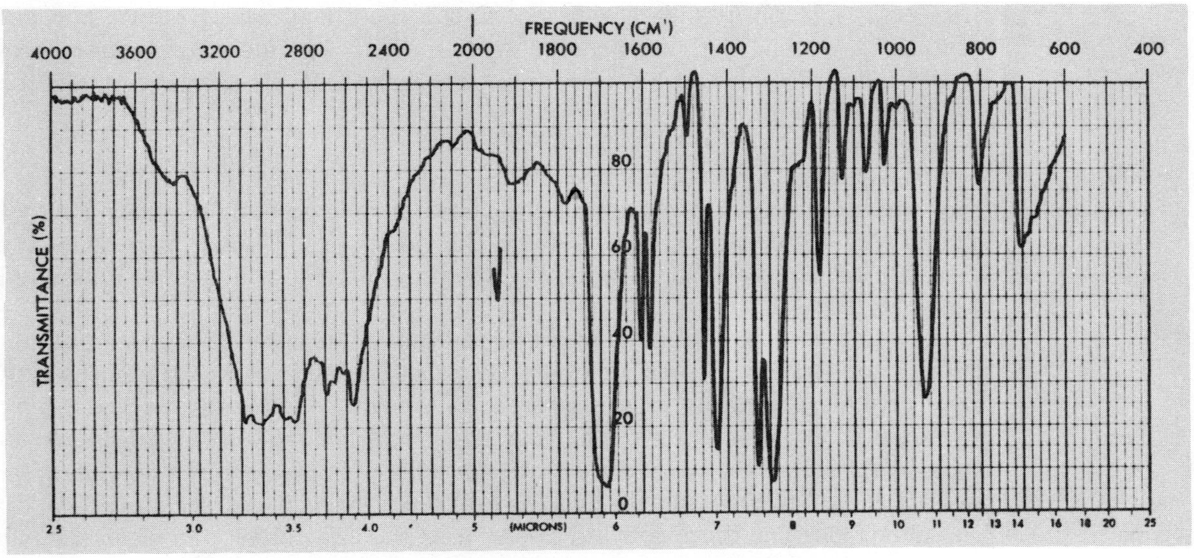

Figure 12.13 *Infrared spectrum of benzoic acid (KBr pellet).*

medium intensity absorption band near 2720 cm^{-1} accompanied by a carbonyl absorption band is good evidence for an aldehyde group in a molecule (Figures 12.12 and 14.4*a*).

Carboxylic Acids

The most characteristic absorptions in the spectrum of a carboxylic acid are a strong carbonyl absorption band at 1720 to 1680 cm^{-1} and a very broad absorption band extending over the region of 3200 to 2500 cm^{-1} due to the O—H stretching vibrations. The very broad O—H band in carboxylic acids is due to the strong hydrogen bonding that occurs between carboxylic acid molecules. Figure 12.13 illustrates the broad peak exhibited in the spectrum of a carboxylic acid.

Halogen Compounds

Aliphatic chloro compounds usually exhibit absorption in the region of 850 to 650 cm^{-1} due to carbon-chlorine bond stretching. Aryl chlorides absorb in the 1100 to 1090 cm^{-1} region.

Aliphatic bromo and iodo compounds absorb in the 690 to 500 cm^{-1} region of infrared spectrum, and these absorption bands are not generally observed on the standard infrared spectrum. The carbon-fluorine bonds of organofluorine compounds absorb strongly in the region of 1400 to 750 cm^{-1} due to C—F bond stretching.

Appendix B contains a convenient list of characteristic infrared absorption frequencies for a number of functional groups.

Sample Preparation

Infrared spectrophotometers will record the spectra of solid, liquid, and gas samples. A special cell constructed with sodium chloride or potassium bromide windows is used to hold or mount the sample in the sample beam of the spectrophotometer.

Liquid Samples

The simplest method for preparing a liquid sample consists of placing a thin film of the liquid between two sodium chloride plates. The salt plates with the thin film of sample held between are then mounted in a holder as illustrated in Figure 12.14.

Large (2.5 cm) polished crystals of sodium chloride are used as the sodium chloride plates. Even small amounts of moisture will damage the surface of these crystals. *Therefore, it is important to remember to handle the crystals by the edges to avoid marring the surface with moisture from your fingers, to always be sure that the sample does not contain water, to clean the crystals after use with an anhydrous solvent, and to store the crystals in a desiccator.*

After recording the spectrum, disassemble the cell holder, and clean the salt plates by rinsing them with an anhydrous volatile solvent.

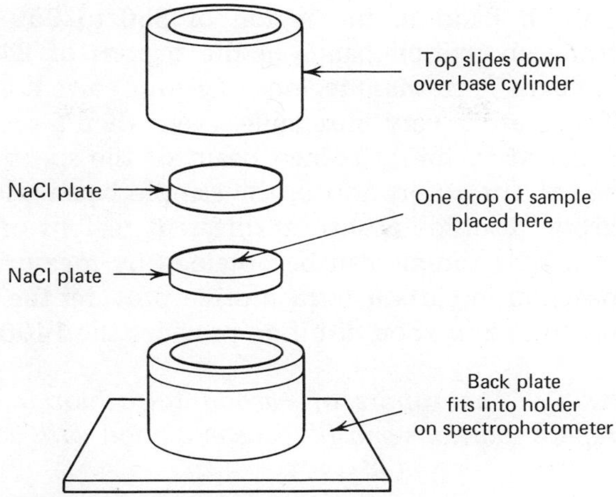

Figure 12.14 *Thin-film technique for preparing samples.*

Solution Cells

Cells with a fixed path length (spacing between the salt plates) are also available to contain neat liquid samples or solutions of solid or liquid samples. A typical solution cell is illustrated in Figure 12.15.

Solutions containing a liquid or solid sample in a solvent for infrared spectroscopy must be made using a solvent that does not have a large number of intense absorption bands in the infrared spectrum or the solvent absorptions will obscure the sample absorptions. The solvent must also dissolve the sample to make a 5 to 10% solution. Few solvents have a limited number of absorption bands in the infrared region. Thus, the choice of solvent is generally limited to either carbon tetrachloride or carbon disulfide even though many compounds have limited solubility in these solvents. Carbon tetrachloride has a strong ab-

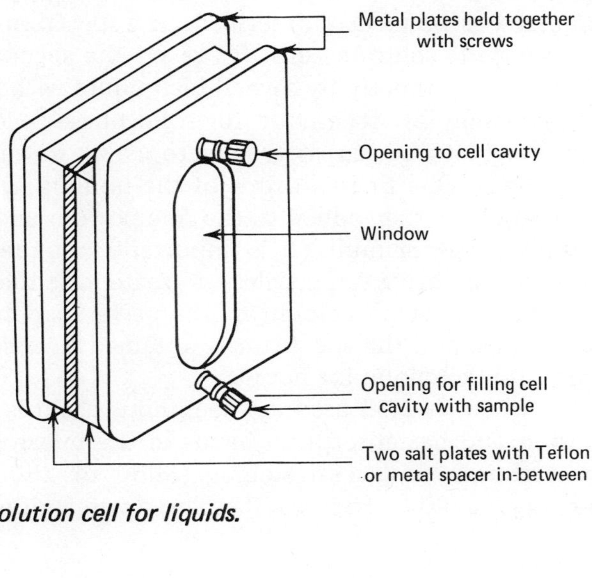

Figure 12.15 *A solution cell for liquids.*

sorption band in the region of 850 to 650 cm^{-1}, and carbon disulfide has strong absorption bands in the regions of 2400 to 2100 cm^{-1} and 1600 to 1400 cm^{-1}. In these regions of strong solvent absorption, the infrared spectrum of a solute is very unreliable, even when a second cell filled with pure solvents is placed in the reference beam of the spectrophotometer to compensate for the solvent absorption in the sample beam. However, carbon tetrachloride and carbon disulfide absorb in different regions of the spectrum. A complete spectrum of a sample can be obtained by measuring two spectra of the sample; a spectrum in carbon tetrachloride provides the 4000 to 1000 cm^{-1} region and a spectrum in carbon disulfide provides the 1400 to 650 cm^{-1} region.

NOTE: The vapors of carbon tetrachloride and carbon disulfide are toxic. Prepare solutions and fill cells in a hood with adequate ventilation.

To prepare a solution sample, first dissolve 0.1 to 0.2 g of sample in 2 ml of solvent. Fill the cell by means of a syringe through the lower opening of the solution cell; this ensures that the air in the cell will escape from the top opening. If a bubble appears in the cell window, gently tap the cell with your finger to move it out of the window area. A second cell that has an identical path length is filled with pure solvent in the same manner. This cell is the *reference cell.* The cell containing a sample is mounted in the sample beam of the spectrophotometer and the reference cell is mounted in the reference beam. The solvent in the reference cell will compensate for the weak or medium intensity absorptions of infrared radiation by the solvent in the sample cell.

After the spectrum has been obtained, the cells are cleaned by flushing with pure solvent and dried by drawing dry air through the cell cavity by means of an aspirator. The cells are stored in a desiccator.

Solid Samples

Solutions of solid samples in carbon tetrachloride or carbon disulfide can be prepared as previously described. The spectrum of the sample is then measured by using the solution cells. However, the spectrum of a solid phase sample can be measured directly by several techniques without dissolving the sample.

The simplest technique for solid phase spectra is the *mull technique.* The sample (10 to 20 mg) is ground to a fine powder by means of a small mortar and pestle. One or two drops of the mulling agent, which is normally mineral oil (Nujol), is then added to the fine powder and the mixture is ground to form a thick paste, or mull. (It is important that the solid be ground to a very small particle size because particles of greater size than 2 μm will scatter the infrared radiation.) A small amount of the paste is applied as a thin film between two salt plates, and the spectrum is obtained as described in the discussion on the thin film technique for liquids.

The mineral oil used as the mulling agent is a high molecular weight hydrocarbon and has absorption bands in the infrared spectrum (Figure 12.16) due to carbon-hydrogen stretching (3000 to 2800 cm^{-1}) and carbon-hydrogen bending (1420 to 1350 cm^{-1}). These two regions of a spectrum of a mineral oil

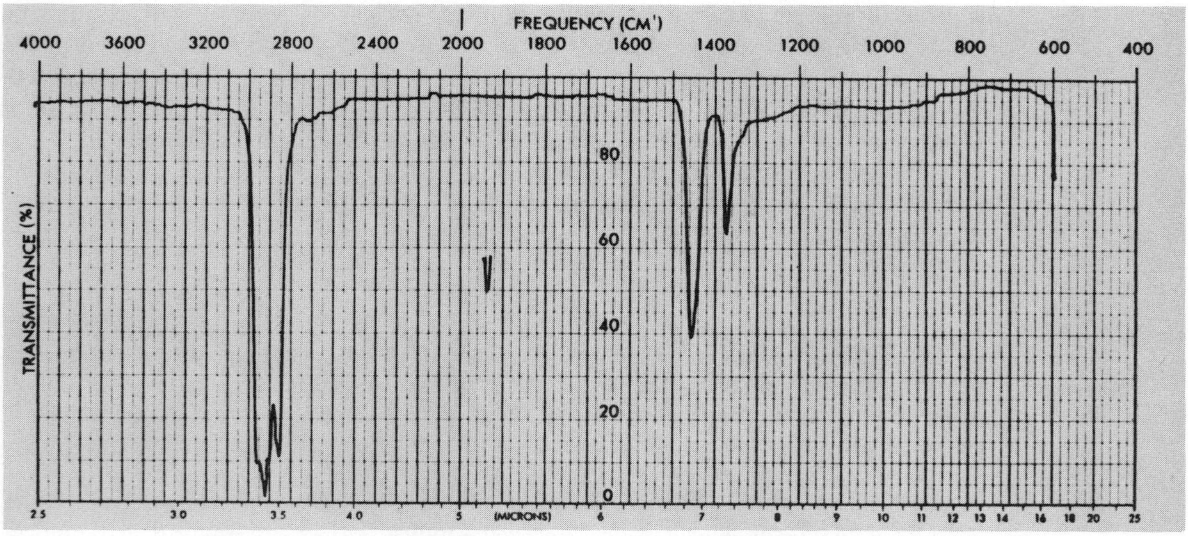

Figure 12.16 *Infrared spectrum of mineral oil (thin film).*

mull are not useful for characterizing the sample, but the remainder of the spectrum is free of interference from the mineral oil absorptions.

The *pressed pellet technique* is usually the method chosen for obtaining the infrared spectrum of a solid since no absorption bands from solvents or mulling agents interfere with the spectrum of the sample. In this technique a homogeneous solid mixture is made from finely ground potassium bromide and the solid sample. By applying high pressure to the powdered mixture in a special press, a thin pellet that can transmit light is formed.

In this technique, 1 to 2 mg of sample is thoroughly ground with 100 mg of anhydrous potassium bromide powder by means of a ball mill (Wig-L-Bug). The press used to form the KBr pellet consists of three parts (Figure 12.17). The press is assembled by placing one bolt in the end of the barrel and tightening the bolt until it is fully inside the barrel. The bolt is then backed out one full turn (counterclockwise). Deposit approximately 50 mg of the prepared KBr powder evenly over the surface of the lower bolt (hold the barrel vertically and gently tap with your finger). Insert the second bolt from the top and advance the bolt until it is finger-tight. Using the $\frac{9}{16}$" wrench and the bench vise attached to the counter top, gradually exert pressure on each bolt, and then use

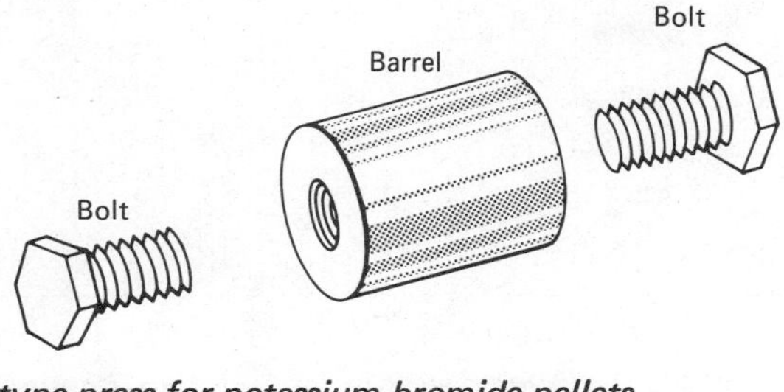

Figure 12.17 *Bolt-type press for potassium bromide pellets.*

as much pressure as can be easily applied with a 6″ wrench for about one minute. The bolts are carefully removed. The pellet should remain fixed in the center of the barrel. The barrel is mounted in the sample beam of the spectrophotometer by means of the plastic holder. A reference beam attenuator is usually necessary to adjust the pen to 90% Transmittance at 2000 cm^{-1}.

The press is cleaned by removing the pellet from the barrel by the use of a pencil. The barrel and bolts are washed in water (to remove KBr), rinsed with acetone, and dried.

Prelab Questions

1. The infrared spectra of 3-methyl-3-pentanol, 3-chloro-3-methylpentane, and *trans*-3-methyl-2-pentene are shown in Figure 12.18. In each spectrum assign as many of the observed infrared absorption bands as possible to a given functional group in the sample. It is generally possible to assign all bands above 1400 cm^{-1} to a particular functional group.

2. The infrared spectra of two different compounds, A and B, are given in Figure 12.19. Each of the two compounds has the molecular formula C_3H_6O. Determine the structure of each of the compounds.

3. Describe how infrared spectroscopy could be used in Experiment 8 to determine if the conversion of 3-methyl-3-pentanol to 3-chloro-3-methylpentane had gone to completion.

4. Figure 12.20 illustrates two infrared spectra of the compound polystyrene that is used as a standard for calibration of the frequency scale on an infrared spectrum. The first spectrum was recorded with a linear frequency scale (except for the scale change at 2000 cm^{-1}); the second was recorded with a linear wavelength scale. Do the major peaks in the two spectra occur at the same frequency? Could the two spectra be easily compared to determine the identity of the samples?

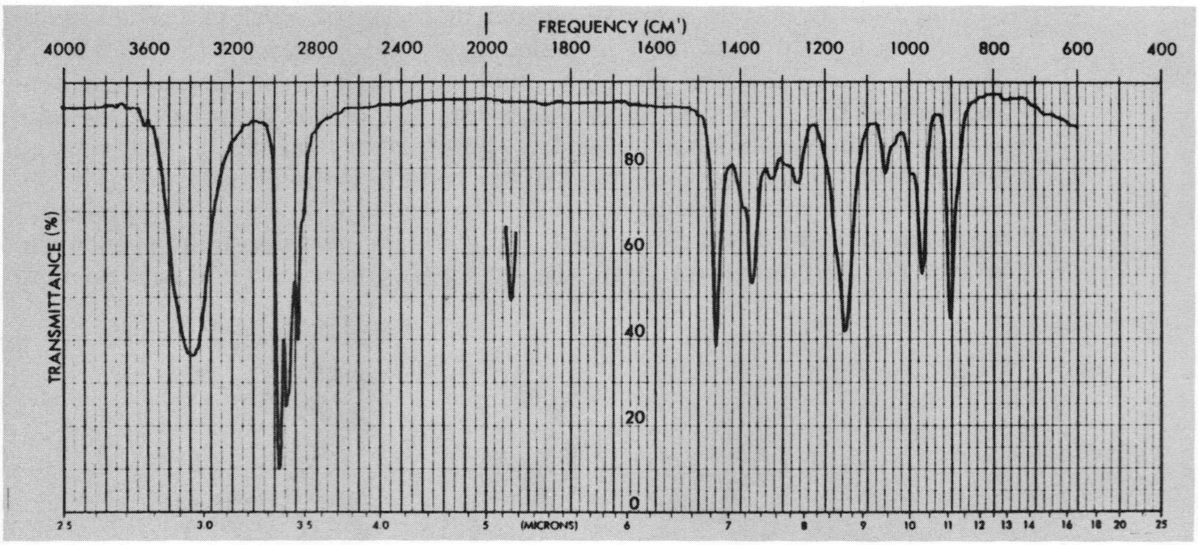

Figure 12.18a *Infrared spectrum of 3-methyl-3-pentanol (thin film).*

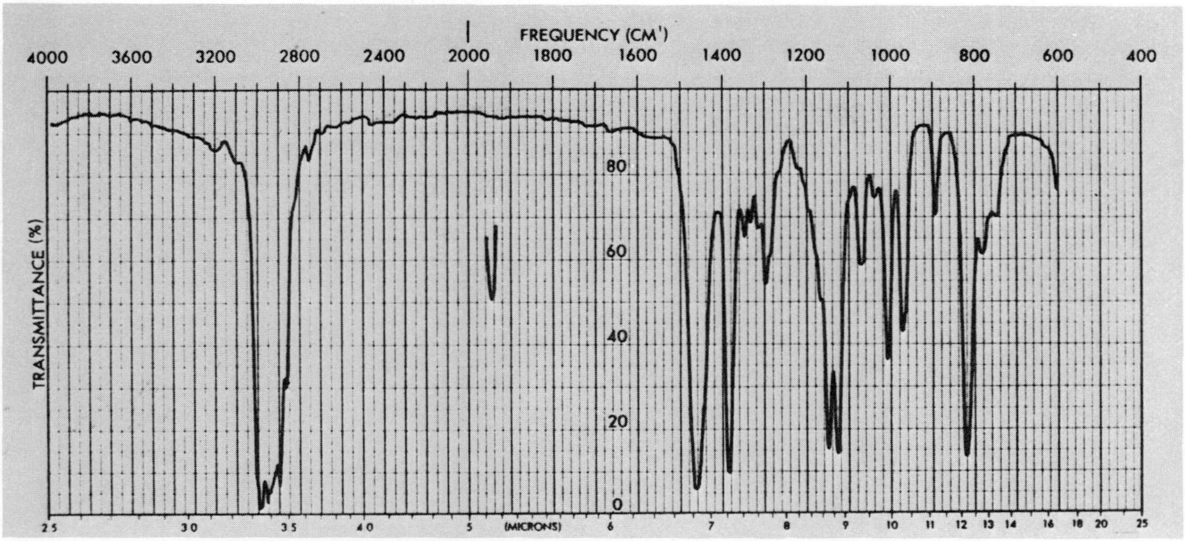

Figure 12.18b *Infrared spectrum of 3-chloro-3-methylpentane (thin film).*

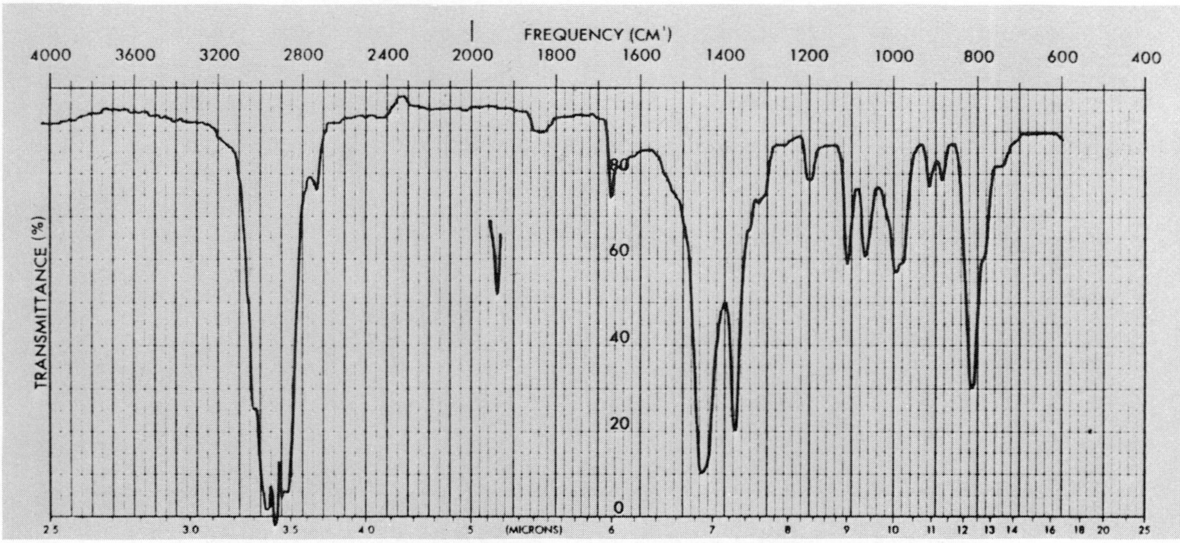

Figure 12.18c *Infrared spectrum of* trans-*3-methyl-2-pentene (thin film).*

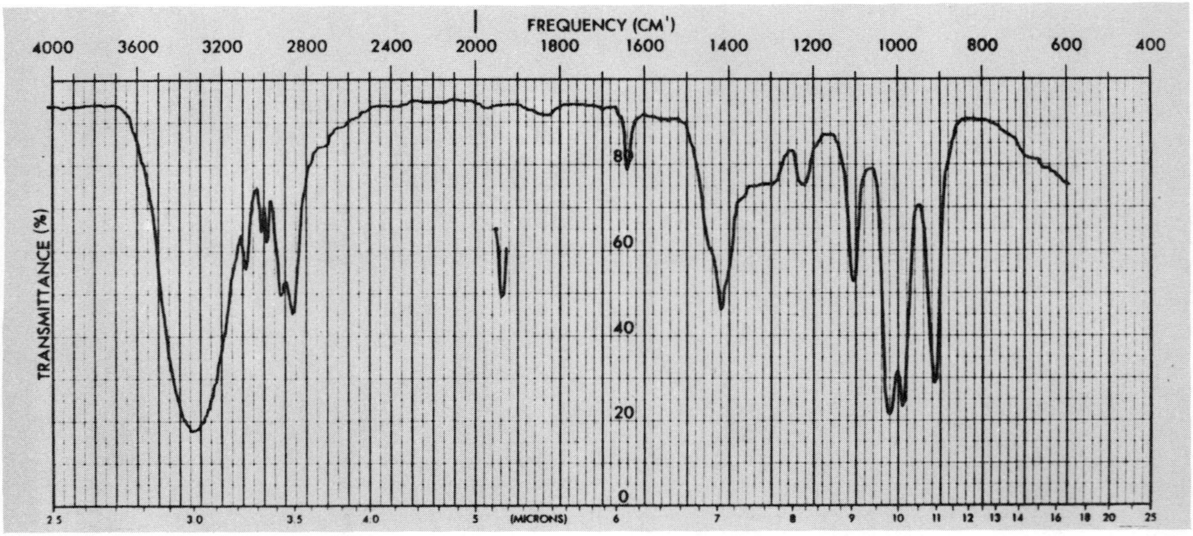

Figure 12.19a *Infrared spectrum of compound A for Question 2 (thin film).*

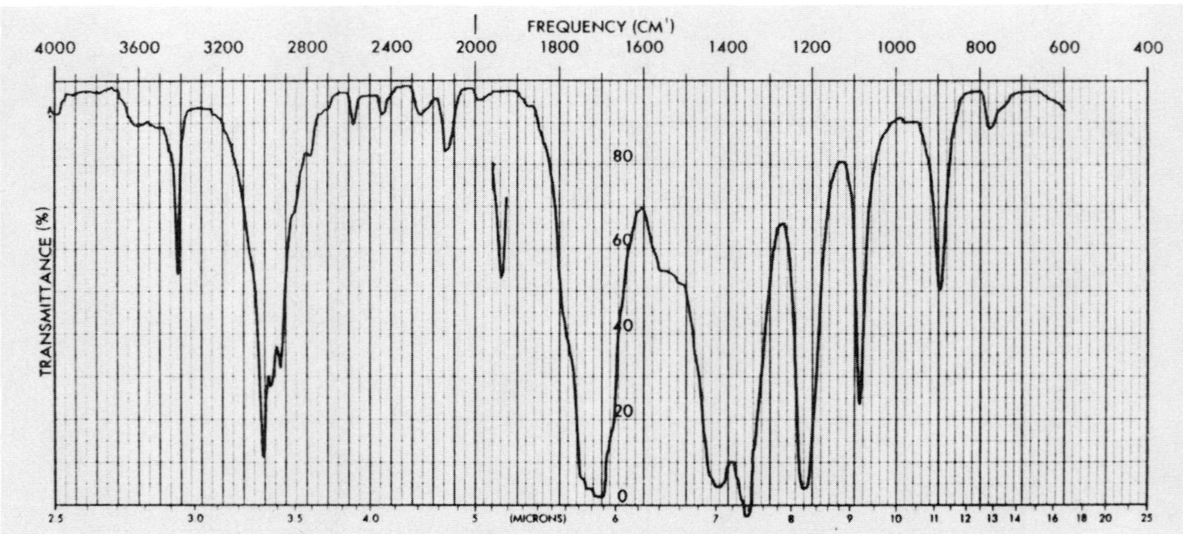

Figure 12.19b *Infrared spectrum of compound B for Question 2 (thin film).*

Note: The absorption band at 3420 cm^{-1} is an overtone band of the 1715 cm^{-1} band. Overtone bands occur at approximately twice the frequency of a very strong absorption band.

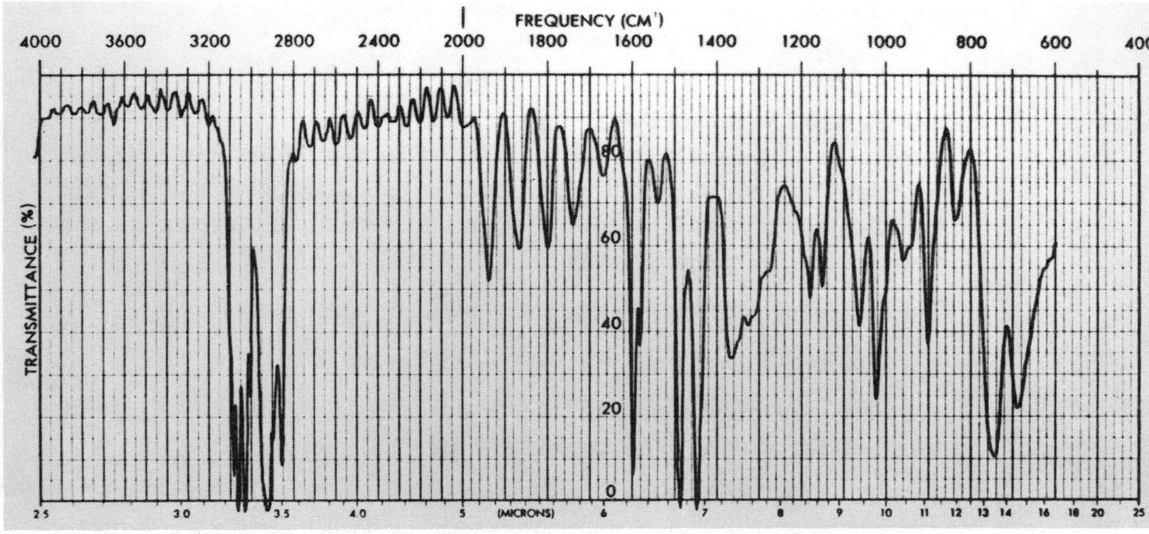

Figure 12.20a *Infrared spectrum of polystyrene film (frequency scale linear).*

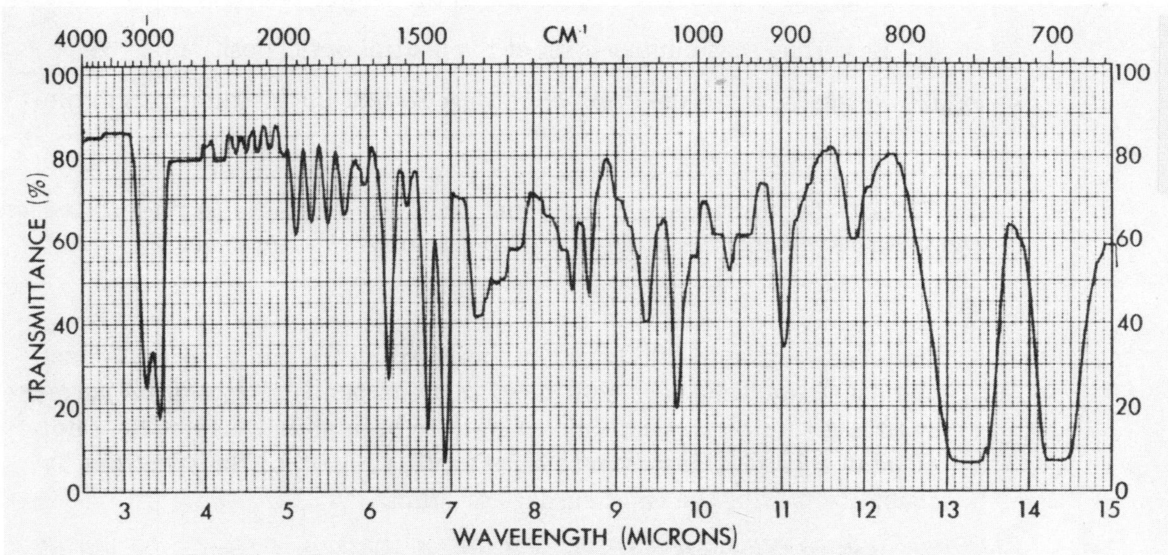

Figure 12.20b *Infrared spectrum of polystyrene film (wavelength scale linear).*

5. Assign the major peaks observed in the spectra shown in Figures 12.5-12.13 to the corresponding functional groups in the molecules.

EXPERIMENTAL PROCEDURE

A number of different models of infrared spectrophotometers are used to record spectra of organic compounds. Your instructor will give you specific instructions for the operation and care of the instrument that you will use for this experiment.

The following procedure is used to record the spectrum:

1. Sign in the instrument logbook (if one is provided).

2. Turn the power switch to ON position. The instrument should warm-up for about 10 minutes before use.

3. Affix the chart paper to the carriage or drum. Be sure that the paper is correctly aligned with the frequency scale or index mark on the instrument.

4. Place the cell holder with sample in the sample beam.

5. Be certain scan control is set at the neutral or off position.

6. Manually move the drum or carriage so that the instrument is in the

---

CAUTION: Do not manually turn the drum or move the carriage unless the chart drive mechanism is disengaged. Do not move the pen manually.

---

frequency region of 2000 cm^{-1} as indicated by the pen position. Adjust the 100% transmittance control so that the pen is at a position corresponding to 90% transmittance. Check to be sure that the pen does not go above 100% transmittance in any other region of the spectrum by manually moving the chart carriage or drum.

7. Return the chart carriage or drum to 4000 cm^{-1} (pen at far left of chart) and set the pen on the chart paper.

8. Turn on the scan control and allow the instrument to automatically scan the complete spectrum.

9. Turn the scan control off.

10. Remove the sample and insert the standard film of polystyrene (in the adapter provided) over the sample-beam window. One or more absorption peaks of the polystyrene standard are plotted on the chart

paper to be used as a calibration guide for the frequency scale. The peaks at 1601 or 1944 cm^{-1} are commonly used for this purpose.

11. Turn the scan control off and remove the chart paper and sample.

12. Clean all equipment used to prepare your sample and return each item to the proper place.

13. If no one is likely to use the instrument in the next few hours, turn off the power switch.

Record the infrared spectra of 3-methyl-3-pentanol and 3-bromo-3-methylpentane. For each spectrum, assign the diagnostic absorption bands to the functional group in the molecule causing the absorption of infrared radiation at that frequency.

Experiment Thirteen

Structural Identification of Organic Compounds

The structural identification of organic compounds is a problem often encountered in organic chemistry. Organic compounds are so numerous and of such wide variety that the problem of identification is formidable unless approached in a systematic and logical manner. Both spectroscopic and chemical techniques are useful for solving these problems, and these techniques often complement each other.

In this experiment you will be given a sample of an organic compound and asked to identify it. The unknown will be either an alkane, alkene, alcohol, or an alkyl halide and will contain no more than one functional group. These limitations (monofunctional compounds and four chemical classes) greatly simplify the problem of identifying the unknown.

The general method employed to identify an unknown compound in this experiment is applicable, when expanded in Experiments 21–25, to the identification of most organic compounds that contain carbon, hydrogen, oxygen, nitrogen, and halogen atoms. The method is designed for the identification of compounds that have been previously characterized and reported in the chemical literature. An unknown compound is identified by comparing its physical and chemical properties with those of known compounds until a known compound that has identical properties is found. If the properties of the unknown are identical to those of one and only one known compound, the two compounds are identical and the unknown has been identified. Thus, there are two parts to the solution of this problem: the physical and chemical properties of the unknown must be accurately determined, and the literature must be searched for a description of a compound that has the properties observed for the unknown.

Identification of an unknown can often be accomplished by determining just a few of the physical and chemical characteristics of the unknown. Interpretation of the unknown compound's infrared spectrum also provides significant information about the identity of the compound. If the infrared spectra of the unknown compound and a known compound are exactly the same, there is very strong evidence that the two compounds are identical.

The general procedure for the identification of an unknown organic compound consists of the following steps:

1. *Determination of the physical constants of the unknown compound*, e.g., melting point and boiling point.
2. *Qualitative elemental analysis*—for elements present in the unknown other than carbon, hydrogen, and oxygen.
3. *Determination of the solubility properties of the unknown.*
4. *Interpretation of the infrared spectrum of the unknown*—to determine the chemical class of the unknown.
5. *Chemical characterization tests*—to determine the chemical class of the unknown.
6. *Literature search*—for compounds that belong to the same chemical class as the unknown and that have physical constants similar to the unknown.
7. *Further experimental comparisons*—between the unknown compound and known compounds to provide additional data that are necessary to complete and confirm the identification. This step generally involves additional chemical tests, comparison of spectra, and the preparation of one or more solid derivatives (known compounds) from the unknown compound.

In the following sections we will discuss steps 1–6 of this general procedure. The preparation of derivatives will be discussed in Experiments 22–24.

Determination of Physical Properties

Observable physical properties of the unknown such as physical state (solid or liquid), size and shape of crystals (see Figure 2.3), and color should be carefully noted. Generally the most important physical constants that can be obtained for an organic compound are its melting point and boiling point. This is true not only because some structural inferences can be made on the basis of these data, but also because many literature sources that list physical and chemical properties of organic compounds are organized such that compounds within a chemical class are listed in order of increasing boiling points and increasing melting points.

The boiling point range of a liquid unknown is conveniently determined by distilling the sample in the simple distillation apparatus illustrated in Figure 11.1. Purification of the sample can also be accomplished by this procedure. As the liquid is distilled, the first portion, approximately one-fourth of the sample, should be collected as a first fraction. The second fraction of distillate should consist of about one-half of the sample. The boiling point range of the second fraction should be recorded as the boiling point of the sample. Generally, the first fraction of distillate collected will contain water or more volatile organic impurities, and the second fraction will consist of the pure substance.

The boiling point range and the melting point range are useful indications of the purity of the unknown compound. If the melting point range of a solid unknown is greater than 3°C, the compound is probably impure and must be recrystallized and dried before an accurate melting point range can be obtained. If the boiling point range of the second fraction from the distillation of a liquid sample is greater than 3°C, the sample should be purified by redistillation using a fractionating column (Figure 6.4).

Other physical constants that are useful for comparison include the freezing point, the refractive index (discussed in Experiment 21), and the density of the purified sample.

Qualitative Analysis of Elements

The number of elements that may be found in organic compounds is quite large, but in this text the analyses will be limited to the detection of the most common elements; nitrogen, chlorine, bromine, iodine, and sulfur. Nitrogen, the halogens, and sulfur are detected as ionic compounds that are formed by reaction of the unknown compound with sodium at high temperatures. This process is referred to as *sodium fusion.* In the sodium fusion process the organic compound is decomposed, organic nitrogen is converted to sodium cyanide, the halogens are converted to sodium halide, and sulfur is converted to sodium sulfide.

Sodium fusion

$$\text{Organohalogen compound} \xrightarrow[\text{heat}]{\text{Na}} \text{NaX}$$

$$\text{Organonitrogen compound} \xrightarrow[\text{heat}]{\text{Na}} \text{NaCN}$$

$$\text{Organosulfur compound} \xrightarrow[\text{heat}]{\text{Na}} \text{Na}_2\text{S}$$

The cyanide ion and the sulfide ion are detected by color reactions with *p*-nitrobenzaldehyde in dimethyl sulfoxide [$CH_3S(O)CH_3$]. The presence of halide ions is detected by formation of the silver halide precipitate. The individual halide ions are identified on the basis of the ability of dilute nitric acid to selectively oxidize iodide ions to iodine and of excess concentrated nitric acid to oxidize bromide ions to bromine. Chloride ions are not oxidized by concentrated nitric acid.

$$\text{I}^- \xrightarrow{\text{dilute HNO}_3} \text{I}_2 \text{ (purple in CCl}_4\text{)}$$

$$\text{Br}^- \xrightarrow{\text{conc. HNO}_3} \text{Br}_2 \text{ (red-brown in CCl}_4\text{)}$$

$$\text{Cl}^- \xrightarrow{\text{conc. HNO}_3} \text{(No Reaction)} \xrightarrow{\text{AgNO}_3} \text{AgCl (white ppt)}$$

PROCEDURE FOR SODIUM FUSION

CAUTION: Sodium metal reacts violently with water. It is essential that samples containing more than traces of water be dried before fusion with sodium. Sodium must not be added to water, but excess sodium may be carefully destroyed by additions to anhydrous n-butyl alcohol. When working with sodium metal, all skin contact must be avoided. A safety shield is recommended for this procedure.

Using forceps, remove a wafer of sodium metal dispersion in paraffin from the bottle and place it directly into a clean, dry, 4-in. test tube. Clamp the tube in a vertical position with an all-metal clamp and gradually heat the bottom of the tube with a Bunsen burner until the wafer melts. While being careful not to hit the sides of the warm tube with the sample, drop a few milligrams (two drops of liquid) of the sample directly onto the melted sodium-paraffin. Heat the bottom of the tube until the lower portion (2 cm) of the glass is red-hot. Continue heating for at least two minutes or until it is evident that vapors are no longer being evolved from the mouth of the test tube. (These vapors are organic decomposition products and may ignite during the heating process.) Allow the test tube to cool to room temperature and then add five drops of methanol. With a clean glass stirring rod, thoroughly mix the solid residue and methanol to destroy excess sodium. (If gas bubbles are produced, wait for the reaction to subside.) Add 2 ml of distilled water and a boiling stone to the test tube, and then gently heat the mixture to boiling. Filter the solution and rinse the filter paper and residue with 4 ml of distilled water. Collect the filtrate and rinse in a small test tube.

If at this point the filtrate is dark colored, it is probable that the mixture was not heated to a sufficiently high temperature during fusion or the amount of sample was too large. The fusion procedure should be repeated if the filtrate is so dark as to obscure future tests.

*Detection of Nitrogen (and Sulfur).** Pour about 0.5 ml of the sodium fusion filtrate into a small test tube and saturate the solution with solid sodium bicarbonate (undissolved solid should remain). Then add one or two drops of this saturated solution to a test tube containing 1 ml of a 1% *p*-nitrobenzaldehyde solution in dimethyl sulfoxide. The presence of nitrogen is indicated if the solution changes from the original yellow

*This test was adapted from: J.A. Vinson and W.T. Grabowski, *J. Chem. Ed.* **54**, 187 (1977).

color to purple. A color change to green indicates that sulfur is present. If both nitrogen and sulfur are present, only a purple color will be observed.

Detection of Halogens. Pour 0.5 ml of the sodium fusion filtrate into a small test tube, and acidify the solution by adding, dropwise, 3*M* nitric acid. (Test with *p*H or litmus paper to be sure that the solution is acidic.) If either nitrogen or sulfur is present in the filtrate, add 1 ml of distilled water and a boiling stone, and then (in a hood) boil the solution for two minutes to expel the H_2S or HCN as gases. Cool the solution. Add three drops of 5% silver nitrate solution. A white to yellow precipitate indicates the presence of halogen (chloride, bromide, or iodide).

Detection of the Individual Halogens. Pour 0.5 ml of the sodium fusion filtrate into a small test tube. Add 0.5 ml of carbon tetrachloride and three drops of concentrated nitric acid. Shake the solution and

---

CAUTION: Use only nitric acid that is colorless for this test. Nitric acid that is yellow or brown may react exothermically with the carbon tetrachloride.

---

then allow the layers to separate. The presence of iodine is indicated by a purple coloration of the carbon tetrachloride layer. If iodine is present, remove the carbon tetrachloride layer by pipet. Add 0.5 ml of fresh carbon tetrachloride and one drop of concentrated nitric acid to the original test solution. Shake the mixture, and then allow the layers to separate. If the carbon tetrachloride layer is still purple, remove it by pipet and repeat the procedure until the carbon tetrachloride layer remains colorless.

Add 2 ml of concentrated nitric acid to the test tube containing the test solution and the colorless carbon tetrachloride layer. Carefully shake the mixture, and then allow the layers to separate. The presence of bromine is indicated by a light brown or brown-red coloration of the carbon tetrachloride layer. If bromine is present, remove the carbon tetrachloride layer by pipet. Extract the test solution with 1 ml portions of carbon tetrachloride until the carbon tetrachloride layer remains colorless.

Add three drops of a 5% silver nitrate solution to the test solution. The immediate formation of a white precipitate indicates the presence of chloride ions.

In this experiment, the unknown will not contain either nitrogen or sulfur since the possible functional groups are limited. However, the tests for the detection and identification of halogen in the unknown are essential since other methods of analysis, such as infrared spectroscopy, do not generally give clear evidence for the presence or absence of halogen atoms. The sodium fusion test for halogen is generally more reliable than the Beilstein test described in Experiment 11. The sodium fusion test is also applicable to all types of organohalogen compounds; whereas, the alcoholic silver nitrate test (Experiment 11) will give a positive test for halogen with only limited types of organic compounds containing halogen.

Classification by Solubility

The use of solubility tests to classify organic compounds has been introduced (Experiment 7). In this experiment solubility tests in water and in concentrated sulfuric acid can be used to aid in the identification of the unknown. Since the unknown is limited to neutral compounds (alkanes, alkenes, alkyl halides, and alcohols), tests in aqueous acid or base and in ether will not distinguish between possible classes.

Alkanes and alkyl halides are insoluble in all solubility test solutions except ether (Table 7.1). Alkenes are insoluble in water, but they are soluble in ether and concentrated sulfuric acid (with reaction and discoloration).

Alcohols have a polar functional group that is capable of hydrogen bonding. Therefore, alcohols containing four carbons or less are soluble in water. Some branched-chain and secondary alcohols of five carbons are also soluble in water, but other five-carbon and larger alcohols are not soluble in water under our test conditions (0.1 g in 3 ml). Alcohols are soluble in ether and in concentrated sulfuric acid (with reaction and discoloration).

A general scheme for the classification of the unknown in this experiment on the basis of solubility is given in Figure 13.1.

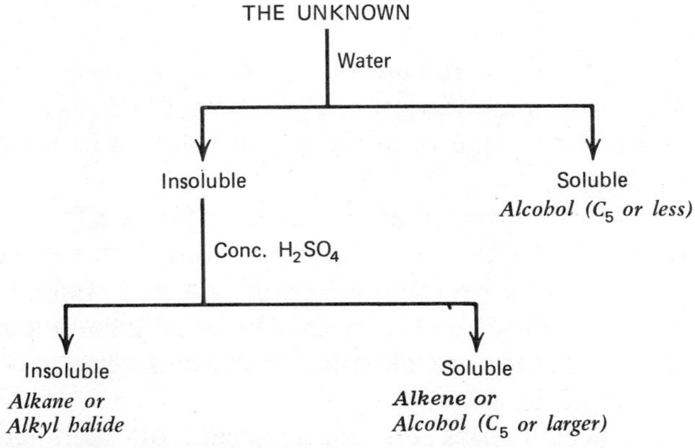

Figure 13.1 *Classification of alkanes, alkenes, alkyl halides, and alcohols on the basis of solubility.*

Interpretation of the Infrared Spectrum

The interpretation of the infrared spectrum of the unknown compound will provide considerable information about the structure of the compound. All common organic compounds, except perhalogenated materials such as carbon tetrachloride or tetrachloroethene, will have absorption bands in the region of 3330 to 2840 cm^{-1} due to C—H stretching vibrations.

If the unknown is an alcohol, a characteristic, strong absorption band is observed in the region of 3650 to 3100 cm^{-1} due to O—H vibrations. The position of the C—O stretching vibration will give an indication of the class of alcohol (Table 12.5). The spectrum of an unknown sample that contains water as an impurity will also show an absorption band in the 3650 to 3100 cm^{-1} region. If the absorption is due to water, it will decrease in intensity after the sample is dried over anhydrous magnesium sulfate.

If the spectrum of the unknown has a weak to medium intensity absorption band in the region of 1680 to 1620 cm^{-1}, the unknown may be an alkene. Alkenes, except tetrasubstituted alkenes, also have absorption bands in the region of 3100 to 3000 cm^{-1} due to olefinic C—H stretching and in the region of 1000 to 650 cm^{-1} due to olefinic C—H bending (Table 12.3).

Infrared spectra of alkanes and alkyl halides will appear similar in the 4000 to 1400 cm^{-1} region. Absorption bands due to C—H stretching and C—H bending will be the only apparent bands in this region for these compounds. Alkyl chlorides usually exhibit an absorption in the region of 850 to 650 cm^{-1}, but if the absorption occurs below 650 cm^{-1}, it will not be observed with standard infrared spectrophotometers. Appendix B contains detailed spectra-structure correlation information.

Chemical Characterization Tests

The use of chemical characterization tests was introduced in Experiment 7. The bromine addition test and the permanganate oxidation test for alkenes is described in that experiment and should be referred to for the characterization of alkenes.

An unknown that is classified as an alkane on the basis of solubility tests, infrared spectrum, and lack of halogen in the elemental analysis should be subjected to the bromine addition test and the permanganate oxidation text since it is possible to confuse an alkene with an alkane without the results of these tests. Alkanes are chemically characterized by their lack of reactivity in the described tests.

Alkyl halides are characterized by their lack of any functional group except halogen. The alcoholic silver nitrate test (Experiment 11) can be used to indicate the subclass of alkyl halide to which the compound belongs.

Ceric Ammonium Nitrate Test

The presence of the hydroxyl function in an organic molecule can be detected by the use of the *ceric ammonium nitrate reagent.* The reagent, which is yellow in aqueous solution, forms a complex with alcohols that is red in color. A positive test for an alcohol is indicated by a change in color of the reagent from yellow to red when a few drops of the unknown are added. Tetrahydrofuran is used as a co-solvent for the test if the compound to be tested is insoluble in water. Alcohols of more than 10 carbons and sterically hindered alcohols may not complex and give a color change.

PROCEDURE

(*a*) *For Water Soluble Compounds.* Dilute five drops of the ceric ammonium nitrate reagent* with 3 ml of distilled water in a test tube and mix thoroughly. Add six drops of the compound and shake the mixture. Note any color change. The color change should be immediate.

(*b*) *For Water Insoluble Compounds.* Add 3 ml of tetrahydrofuran to five drops of the ceric ammonium nitrate reagent in a test tube and thoroughly mix the solution. If a precipitate forms, add three or four drops of water and redissolve the reagent with shaking. The solution should be a yellow color at this point. If the solution is orange, red, or colorless, new tetrahydrofuran should be used to prepare more of the test solution. Add six drops of the compound to be tested, shake to mix, and note any color change. If the compound to be tested is a solid, dissolve 0.20 g of the solid in 1 to 2 ml of tetrahydrofuran, and then add the tetrahydrofuran solution to the test solution.

Alcohols can be divided into three subclasses on the basis of their structure. Several characterization tests are available that will indicate the subclass of an unknown alcohol. The Lucas test and the chromic acid oxidation test are the most useful of these tests.

Structural type:	$R^1{-}CH_2OH$	$R^2{-}\overset{R^1}{\overset{\vert}{C}}HOH$	$R^2{-}\overset{R^1}{\overset{\vert}{\underset{R^3}{\underset{\vert}{C}}}}{-}OH$
Classification:	*primary alcohol*	*secondary alcohol*	*tertiary alcohol*

Chromic Acid Oxidation Test

Chromic acid reacts rapidly with primary and secondary alcohols to yield carboxylic acids and ketones, respectively (Experiment 14). These reactions occur within 15 seconds to give a distinct color change from the orange chromic

*The ceric ammonium nitrate reagent is prepared by dissolving 90 g of $(NH_4)_2Ce(NO_3)_6$ in 100 ml of distilled water. Warming will accelerate the solution process.

acid solution to a blue-green solution (or precipitate) of chromium(III) compounds. Tertiary alcohols give no reaction. This reagent will also react with other easily oxidized organic compounds such as aldehydes and phenols.

Chromic acid test

$$\underset{\textit{orange}}{RCH_2OH + H_2CrO_4} \xrightarrow{<15\ \text{sec}} \underset{\textit{blue-green}}{RC(=O)OH + \text{Cr(III) compounds}}$$

$$\underset{\textit{orange}}{R_2CHOH + H_2CrO_4} \xrightarrow{<15\ \text{sec}} \underset{\textit{blue-green}}{RC(=O)R + \text{Cr(III) compounds}}$$

$$\underset{\textit{orange}}{R_3COH + H_2CrO_4} \xrightarrow{>1\ \text{min}} \underset{\textit{orange}}{\text{No Reaction}}$$

PROCEDURE

Dissolve 0.02 g (one drop) of the unknown alcohol in 1 ml of reagent grade acetone and add one drop of the chromic acid reagent.* The orange color of the reagent disappears completely and a blue-green solution or precipitate is formed if the alcohol is primary or secondary.

Lucas Test

A distinction may be made among the three subclasses of alcohols by using the Lucas reagent (hydrochloric acid-zinc chloride). The test is limited to alcohols which are soluble in the reagent (generally C_6 or less). Tertiary alcohols, benzylic alcohols, and allylic alcohols react immediately with the reagent to form an insoluble alkyl chloride. Secondary alcohols also react to form an insoluble alkyl chloride, but the reaction generally takes 2 to 8 minutes. Primary alcohols require more than 10 minutes for reaction with the reagent. A reaction is indicated by the clouding of the solution due to the formation of the in-

Lucas Test

$$R_3COH \xrightarrow[HCl/ZnCl_2]{<1\ \text{min}} R_3CCl$$

$$R_2CHOH \xrightarrow[HCl/ZnCl_2]{2\text{–}8\ \text{min}} R_2CHCl$$

$$RCH_2OH \xrightarrow[HCl/ZnCl_2]{>10\ \text{min}} RCH_2Cl$$

Soluble in Lucas reagent — *Insoluble in Lucas reagent*

**CAUTION: Chromic acid solutions and chromium trioxide are corrosive and strong oxidizing agents.* Dissolve 5 g chromium trioxide in 15 ml of distilled water in a 250-ml Erlenmeyer flask. Carefully, with swirling, add 5 ml of conc. sulfuric acid. Mix to obtain a homogeneous solution.

soluble alkyl chloride. Often a second layer of alkyl chloride will slowly form. Since this reagent will absorb water on exposure to the atmosphere, the reagent may vary in activity. It is essential to run this test on a known secondary and tertiary alcohol to compare rates of reaction.

PROCEDURE

Add five drops of the alcohol to 2 ml of the Lucas reagent* in a small test tube. Thoroughly mix the reagents and allow the solution to stand at room temperature. Note the length of time required for the solution to turn cloudy. Compare this time to the time observed for the known compounds.

Chemical characterization tests must be performed carefully. Instructions for tests should be followed closely, and measurements of amounts of reagents should be approximated with care. It is critical to record exact observations of test results; do not report just the conclusion that a given test was positive or negative.

To correctly interpret the results from chemical characterization tests with an unknown compound, control tests should be run with "known" compounds. One control test should be done with a known compound that should give a positive result, and one test should be done on a compound that should give a negative result. Running control reactions not only checks your technique and illustrates test results, but it also checks the quality of reagents. Many test reagents decompose with time and may then give false or misleading test results.

Searching the Chemical Literature

Once the functional group class and physical properties of the unknown have been determined, the next step is to search the literature to find all compounds that have similar chemical and physical properties. Appendix A contains a list of common organic compounds and their physical and chemical properties. More extensive compilations of data can be found in the texts and handbooks listed in the References at the end of this book.

In comparing physical constants with values reported in the literature, it is important to remember that the literature values are generally measured on carefully purified materials. Therefore, the experimental values and the literature values cannot be expected to agree exactly unless the unknown has also been carefully purified. Usually compounds with melting points or boiling points higher and lower than the experimental value must be considered for

**CAUTION: This reagent is very corrosive.* The Lucas reagent is prepared by dissolving 16 g of anhydrous zinc chloride in 10 ml of concentrated hydrochloric acid. The reagent is hygroscopic and will lose activity if left open to the atmosphere.

comparison of other properties. In general, if the observed melting point or boiling point range is narrow, it should be safe to first consider only compounds with listed melting points or boiling points that are within 10°C above the upper limit of the observed range and within 5°C below the lower limit of the observed range. Obviously compounds with physical constants closest to the experimental values should be considered first. If the compound does not melt or boil over a narrow range, then the limits of the search must be expanded.

If the compound is a high boiling liquid, it may also be necessary to compare the physical properties of similar compounds that are listed as low-melting solids. Many compounds with melting points within 20°C above room temperature will remain as liquids at room temperature unless they are carefully purified and cooled.

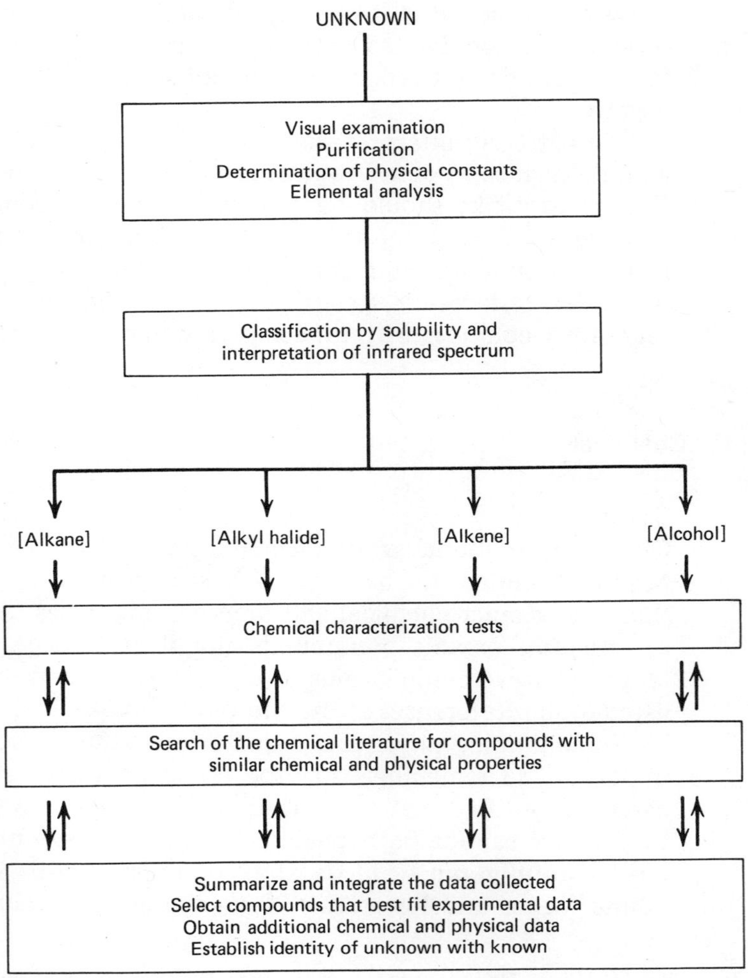

Figure 13.2 *Scheme for the identification of an unknown alkane, alkene, alkyl halide, or alcohol.*

After comparing the literature data on known compounds with the chemical and physical properties of the unknown compound, there will often be several compounds with properties similar to the data obtained for the unknown. It is then necessary to determine what additional data on the unknown would be useful in distinguishing between the possible choices. Further chemical tests and physical measurements on the unknown will be necessary to provide the additional data. A detailed comparison of the infrared spectrum of the unknown with the infrared spectra of possible compounds is also useful in establishing the identity of the unknown compound. The scheme for the identification of an unknown alkane, alkene, alkyl halide, or alcohol is outlined in Figure 13.2.

Prelab Questions

1. Unknown compound A is soluble in ether and concentrated sulfuric acid but not in water. The infrared spectrum of compound A has no absorption band in the region of 3650 to 3200 cm^{-1}. To what chemical class does compound A belong (alkane, alkene, alkyl halide, or alcohol)? How would you confirm this conclusion?

2. Unknown B is only soluble in ether and boils over the range of 46 to 48°C. Upon addition of a few drops of silver nitrate solution to the acidified sodium fusion filtrate, a heavy white precipitate formed. After mixing 0.5 ml of the sodium fusion filtrate, 0.5 ml of carbon tetrachloride, and 2 ml of concentrated nitric acid, no color was observed in the organic layer. To what class of compounds does compound B belong? Compound B reacts rapidly at room temperature with alcoholic silver nitrate to form a white precipitate. What is the identity of compound B?

3. Compound C (b.p. = 65 to 67°C) is only soluble in ether. When the acidified sodium fusion filtrate from compound C was treated with a few drops of silver nitrate solution, no precipitate formed. What is the identity of compound C?

4. Compound D (b.p. = 94 to 96°C) was soluble in ether, water, and concentrated sulfuric acid. Sodium fusion and elemental analysis results indicated no halogen was present. Compound D did rapidly discolor bromine in carbon tetrachloride and gave a dark precipitate upon treatment with potassium permanganate solution. Upon addition of D to ceric ammonium nitrate solution, the solution turned red. Upon addition of D to zinc chloride in concentrated hydrochloric acid, the solution rapidly (<1 min) turned cloudy. What functional groups are present in D? What is the identity of D?

EXPERIMENTAL PROCEDURE

You will be expected to identify an organic compound by making use of the procedures discussed in this experiment. The unknown compound will be either an alkane, alkene, alkyl halide, or alcohol. You should carefully plan your experiments so that you do not waste materials. The compound that you are given may require purification before accurate physical data can be determined. Characterization tests should be performed with samples of known compounds as well as with the unknown.

Although the exact type of notebook to be kept and the form of the summary report will differ with each laboratory, it is suggested that the following list be used as a guide for what should be covered in the notebook for each unknown studied:

1. Date the sample was received.

2. Results of visual examination, determination of purity, purification methods, determination of physical constants.

3. A determination of elements present.

4. A determination of solubility properties.

5. An interpretation of the infrared spectrum.

6. A list of characterization tests that should be performed based on preliminary evidence of functional group classification.

7. Observations and conclusions for all chemical characterization tests.

8. A list of all compounds from the literature that have similar chemical and physical properties.

9. Observations and results from additional tests used to eliminate possible compounds from the list.

10. A summary report giving the chemical and physical properties of the unknown and, based on this data, a logical presentation of your evidence and conclusions regarding the probable identity of your unknown. A comparison of your experimental data with the data from the literature is essential in this presentation.

Experiment Fourteen

Oxidation Reactions of Organic Compounds: Oxidations of Alcohols

---

Chemical reactions in which there is a complete loss of one or more electrons from a compound are known as oxidation reactions. These reactions are among the most important synthetic transformations of alcohols because they provide a direct route from available alcohols to aldehydes, ketones, and carboxylic acids. Primary alcohols are oxidized to aldehydes by the removal of two hydrogens: one bonded to oxygen and the second bonded to the adjacent carbon. However, aldehydes are themselves susceptible to further oxidation to carboxylic acids by a process that involves the replacement of the hydrogen bound to the carbonyl group with a hydroxyl group. Oxidations of secondary alcohols usually form ketones. Tertiary alcohols, because they do not possess a hydrogen bound to the carbon bearing the hydroxyl group. are usually unreactive in oxidation reactions.

Oxidation of Alcohols

$$\underset{\textit{Primary alcohol}}{RCH_2OH} \xrightarrow{[O]} \underset{\textit{Aldehyde}}{RCH{=}O} \xrightarrow{[O]} \underset{\textit{Carboxylic acid}}{RCOOH}$$

$$\underset{\textit{Secondary alcohol}}{R_2CHOH} \xrightarrow{[O]} \underset{\textit{Ketone}}{R_2C{=}O}$$

The half-reaction that describes the oxidation of an alcohol to an aldehyde or a ketone shows that this transformation involves a formal loss of two electrons:

$$\textit{Half-reaction} \quad R_2CHOH \longrightarrow R_2C{=}O + 2\,H^+ + 2\,e^-$$

Oxidation will occur when the alcohol is treated with a reagent that can accept these two electrons. The reagent that is the electron acceptor in an oxidation reaction is the *oxidizing agent.* The oxidizing agent is reduced by the gain in electrons.

Numerous oxidizing agents, such as potassium permanganate or chromic acid, are suitable for the conversion of secondary alcohols to ketones or of primary alcohols to carboxylic acids. However, few oxidizing agents will oxidize primary alcohols to aldehydes without effecting further oxidation of aldehydes to carboxylic acids.

Ceric Ion Oxidation of Benzyl Alcohol

The ceric ion, Ce^{4+}, is a facile oxidant of organic compounds and, particularly, of alcohols. In these oxidation reactions the ceric ion is reduced to the cerous ion, Ce^{3+}—a net one-electron change in cerium:

$$\textit{Half-reaction} \quad Ce^{4+} + e^- \longrightarrow Ce^{3+}$$

Thus for oxidations of organic compounds that require the transfer of two electrons, two moles of ceric ion per mole of organic compound are necessary to complete the transformation.

Ceric salts are orange, and concentrated aqueous solutions of these salts are deep orange-red. Cerous salts, the reduced cerium products in ceric ion oxidations, are colorless. As a consequence of the color difference between ceric and cerous ions, the visual change of a reaction solution that contains ceric ion from orange-red to colorless signals the reduction of the ceric ion to the cerous ion.

Ceric ammonium nitrate (CAN), $Ce(NO_3)_6(NH_4)_2$, is the cerium reagent most commonly employed for oxidations of organic compounds. Although primary alcohols usually yield complex product mixtures when oxidized by this reagent, benzyl alcohols are cleanly converted to benzaldehydes. In this experiment you will synthesize benzaldehyde by the ceric ammonium nitrate oxidation of benzyl alcohol:

$$C_6H_5CH_2OH \quad + \quad 2\,Ce(NO_3)_6(NH_4)_2 \xrightarrow{H_2O}$$

Benzyl alcohol	*Ceric ammonium nitrate*
M.w. = 108 g/mol	M.w. = 548.3 g/mol
b.p. = 205°C (105°C @ 20 Torr)	(orange-red)
(colorless)	

$$C_6H_5CHO \quad + \quad 2\,Ce(NO_3)_5(NH_4)_2 \quad + \quad 2\,HNO_3$$

Benzaldehyde	*Cerous ammonium nitrate*
M.w. = 106 g/mol	(colorless)
b.p. = 179°C (75°C @ 20 Torr)	
(colorless)	

Two moles of ceric ammonium nitrate are required to stoichiometrically convert one mole of benzyl alcohol to benzaldehyde. Since the only colored reactant or product is ceric ammonium nitrate, oxidation is viewed as complete when the reaction solution is colorless. The detailed mechanism of this oxidation reaction is investigated in Experiments 33 and 34.

Chromic Acid Oxidation of 1-Phenylethanol

Chromic acid, H_2CrO_4, is the most widely used oxidant for alcohols. In these oxidation reactions chromic acid, in which chromium is formally in the +6 oxidation state, is reduced to the chromic ion, Cr^{3+}— a net three-electron change in the oxidation state of chromium:

Half-reaction
$$6\,H^+ + HO{-}\overset{\overset{\displaystyle O}{\|}}{\underset{\underset{\displaystyle O}{\|}}{Cr}}{-}OH + 3\,e^- \longrightarrow Cr^{3+} + 4\,H_2O$$

Thus for oxidations of organic compounds that require the transfer of two electrons, two moles of chromic acid are required to oxidize three moles of organic compound.

Chromic acid is commonly prepared by dissolving anhydrous chromium trioxide (CrO_3), a strong Lewis acid, in water. Hydrated chromium trioxide is a strong Bronsted acid that is comparable in acid strength to sulfuric acid. Chromic acid dimerizes to the dichromate ion, and at concentrations of chromic acid that are greater than $10^{-2}\,M$, the dichromate ion is the dominant chromium species in solution. The dichromate ion has the same oxidation capabilities as does chromic acid.

$$\underset{\substack{\textit{Chromium}\\ \textit{trioxide}\\ \text{(chromic}\\ \text{anhydride)}}}{CrO_3} + H_2O \rightleftharpoons \underset{\substack{\textit{Chromic}\\ \textit{acid}}}{H_2CrO_4}$$

$$H_2CrO_4 \rightleftharpoons HCrO_4^- + H^+$$

$$2\,HCrO_4^- \rightleftharpoons \underset{\substack{\textit{Dichromate}\\ \textit{ion}}}{Cr_2O_7^{2-}} + H_2O$$

Concentrated solutions of either chromic acid or the dichromate ion are deep orange-red. In aqueous solutions the chromic ion that is formed by reduction of chromium(VI)* exists as the blue hexahydrate, $Cr(H_2O)_6^{3+}$. As a consequence of the color difference between the chromium(VI) and chromium(III) species, the visual change of a reaction solution that contains either chromic acid or dichromate from orange-red to green (combination of orange and blue) to blue signals the reduction of chromium(VI) to the chromic ion.

In the oxidation procedure for this experiment, 1-phenylethanol is oxidized by chromic acid (prepared from chromium trioxide) using the Jones procedure (also known as the *Jone's oxidation*):

*The Roman numeral notation specifies the oxidation state of the metal ion without reference to the structure of the ion. For example, chromium(VI) denotes either chromic acid or the dichromate ion in this discussion.

$$CrO_3 + H_2O \rightleftharpoons H_2CrO_4$$

Chromic anhydride
M.w. = 100 g/mol
m.p. = 196°C
(red-violet)

Chromic acid
(orange-red)

$$3\ C_6H_5CH(OH)CH_3 + 2\ H_2CrO_4 + 3\ H_2SO_4 + 4\ H_2O \xrightarrow[H_2O]{CH_3COCH_3} 3\ C_6H_5C(=O)CH_3 + 2\ Cr(H_2O)_6^{3+} + 3\ SO_4^{2-}$$

1-Phenylethanol
M.w. = 122 g/mol
b.p. = 204°C
(88°C @ 10 Torr)

Chromic acid
(orange-red)

Acetophenone
M.w. = 120 g/mol
b.p. = 202°C
(78°C @ 10 Torr)

Chromic sulfate (blue)

In the Jones procedure an aqueous solution of chromic acid is added to an acetone solution of the alcohol that is to be oxidized. In this experiment oxidation of acetophenone by chromic acid competes with oxidation of acetone; acetone, which is present in large excess, will be preferentially oxidized in this competing side reaction. After oxidation is complete the acetone, which is infinitely soluble in water, is conveniently separated from the alcohol oxidation product by extraction with water.

Distillation at Reduced Pressure

Organic liquids that have a high boiling point at atmospheric pressure or that decompose during distillation at atmospheric pressure often can be distilled at pressures that are lower than atmospheric pressure (*reduced pressures*). As defined in Experiment 3, the boiling point of a liquid is the temperature at which $P_{atm} = P_{vp}$. By reducing P_{atm}, the vapor pressure at which $P_{atm} = P_{vp}$ is also reduced, and this results in a lowering of the boiling point. Although the relationship between the boiling point and pressure is complex, as can be seen from the pressure-boiling point correlations in Figure 14.1, lowering the pressure for distillation from atmospheric pressure (760 Torr) to 25 Torr generally lowers the boiling point by approximately 100 to 125°. For each subsequent halving of the pressure below 25 Torr, the boiling point is lowered by nearly 10°.

Two practical problems are associated with distillations at reduced pressures. The first is *bumping*—irregular eruptive boiling—caused by the explosive escape of vapor from the liquid. Boiling stones, which are normally effective at atmospheric pressure for controlling regular boiling, are generally ineffective

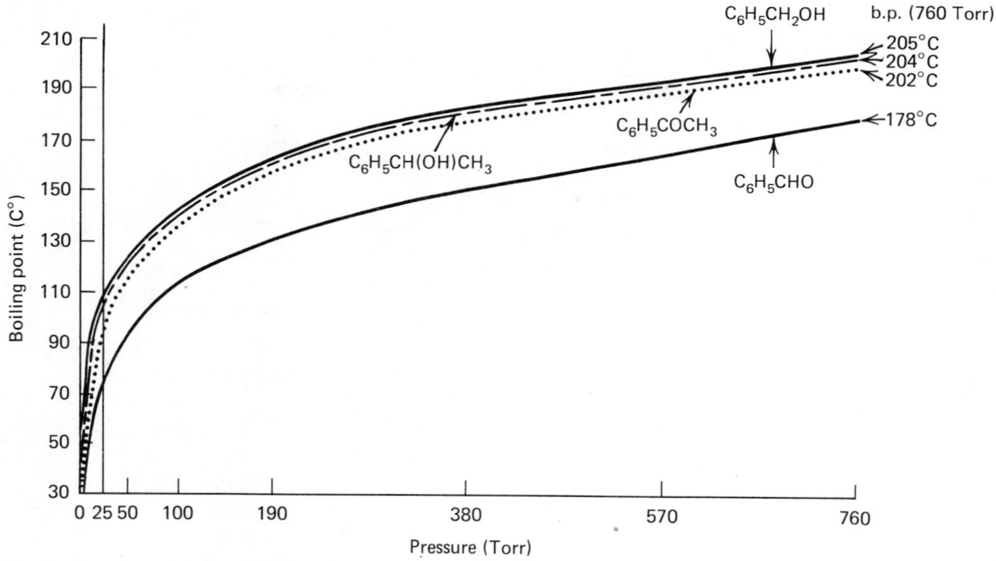

Figure 14.1 *Pressure–boiling point relationships for benzyl alcohol, benzaldehyde, 1-phenylethanol, and acetophenone.*

for distillations at reduced pressures. Wood applicator sticks, broken-off to a length sufficient for one end to touch the bottom of the distilling flask and the other to remain above the liquid level, are usually more efficient in promoting regular boiling. A capillary air inlet tube is another apparatus often used to prevent bumping; however, it cannot be employed in distillations of air-sensitive compounds such as benzaldehyde. A magnetic stirring bar that is rotated in the distilling flask by a magnetic stirrer is generally most effective in controlling the boiling behavior of a liquid at reduced pressures.

The second problem normally encountered in distillations at reduced pressures is *control of the rate of distillation* to maintain vapor-liquid equilibrium. At reduced pressures vaporization is more difficult to control than at atmospheric pressure since the volume of vapor is expanded at reduced pressure ($PV = nRT$ for an ideal gas) and the volume of the distillation apparatus is limited. There is less hold-up of volatile material in a distillation at reduced pressure. Generally, the rate of distillation is effectively controlled by maintaining a slow, steady condensation of liquid by heating the distilling flask at a temperature that is less than 30° higher than the head temperature during distillation.

Reduced pressures are obtained by connecting the distillation apparatus to either a water aspirator or a mechanical oil (vacuum) pump. By employing a water aspirator pressures that are as low as 20 Torr can be obtained; with a vacuum pump pressures as low as 0.01 Torr can be realized. Both distillation procedures are commonly referred to as *vacuum distillations.*

Figure 14.2 describes the apparatus used for vacuum distillations that employ a water aspirator. The distillation apparatus shown in this figure is interchangeable with the apparatus in Figures 3.4, 5.3 (without steam inlet but with

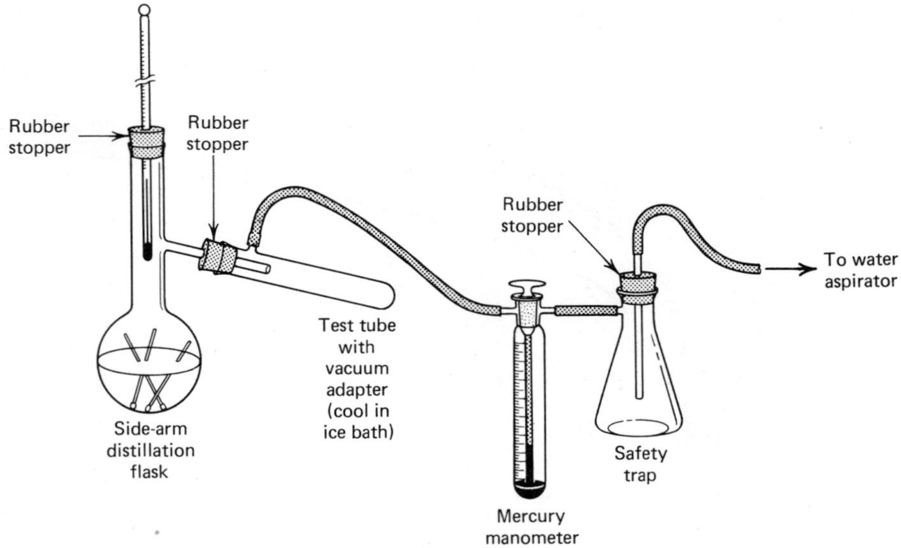

Figure 14.2 *Apparatus for vacuum distillation under aspirator pressure.*

thermometer in distillation head), or 6.4. The safety trap is essential for distillations under aspirator pressures since pressure changes caused by a decrease in the rate of heating or by a change in the water flow will cause water to back-up. The manometer that is attached to the distillation apparatus at the trap measures the reduced pressure. The following procedure should be followed in vacuum distillations that employ a water aspirator:

1. Assemble the appropriate apparatus. Use only thick-walled glassware and examine your glassware for cracks. The stress on evacuated glassware is substantial; weak points in glassware may give way to an implosion.
2. Lubricate all glass joints in the system to be evacuated to minimize air leaks and prevent the glass joints from "freezing." Be certain that all joints are fitted tightly together before evacuating the system.
3. Use only thick-walled rubber tubing for the connections from the aspirator to the vacuum adapter and manometer. The rubber tubing must not "collapse" during the distillation.
4. The size of the distilling flask must be sufficiently large so that the liquid fills the flask less than half-full. Volatile components in a mixture, such as ether, are distilled at atmospheric pressure prior to vacuum distillation.
5. Place boiling aids in the distilling flask prior to evacuation of the system. If the vacuum distillation is interrupted, new boiling aids must be added to the filter flask.
6. Check that a reduced pressure can be obtained from your aspirator by detecting suction at the aspirator and at the end of your vacuum tubing that is to be attached to the vacuum adapter.
7. Turn on the water at the aspirator prior to heating the distilling flask. The

aspirator should be turned on slowly at first to allow trapped air and volatile liquids to escape. After the initial boiling surge use full water pressure for the distillation.

8. After distillation is complete, terminate heating and slowly release the vacuum by opening the system to the atmosphere between the vacuum adapter and the water trap. Turn off the water after the system is at atmospheric pressure.

Figure 14.3 describes the apparatus used for vacuum distillations that employ a vacuum pump. (An alternate distillation apparatus may be substituted for the one shown in this figure.) The vacuum trap is employed to condense vapor that escapes from the distillation apparatus (volatile liquids) and thereby prevent these vapors from entering the mechanical oil pump. The trap is usually cooled to - 78° by a slurry of dry ice (solid carbon dioxide) in isopropyl alcohol

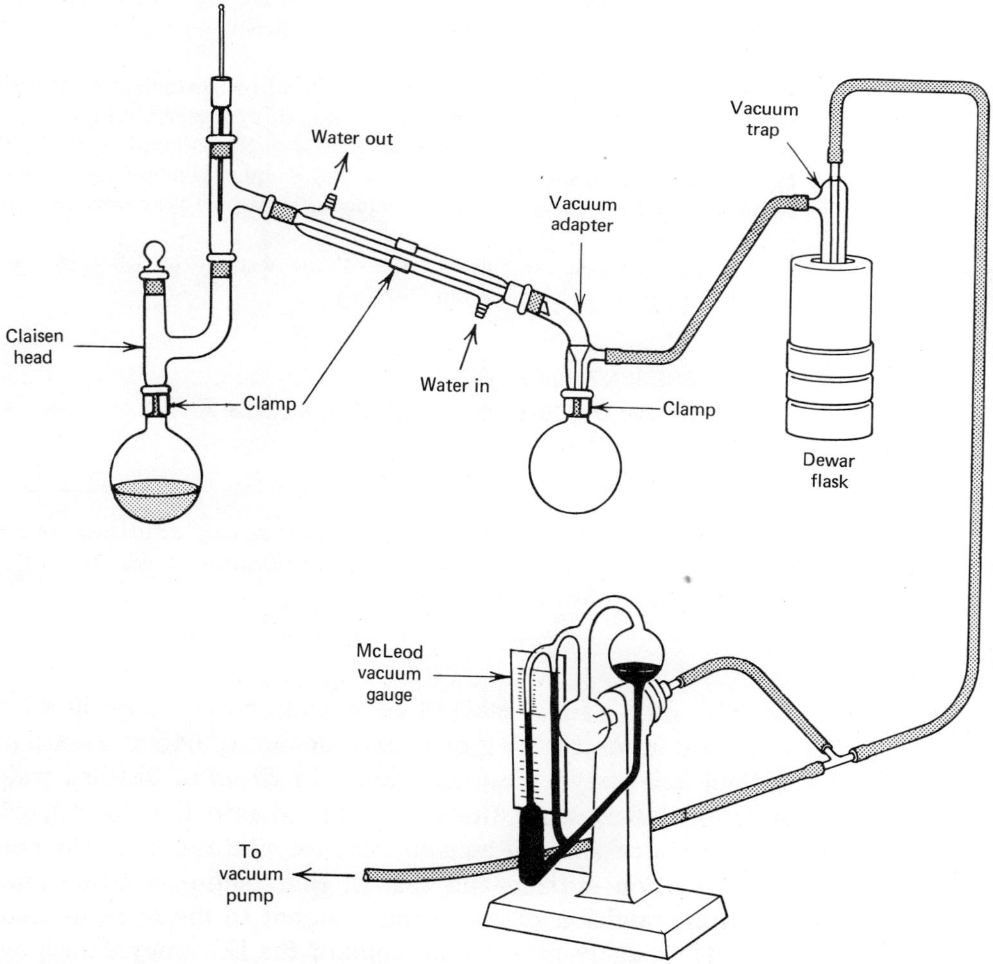

Figure 14.3 *Apparatus for vacuum distillation that employs a vacuum pump.*

that is contained in a Dewar flask. The same procedure is followed for this vacuum distillation except that Steps 6 and 7 are modified to accommodate the use of a vacuum pump.

Prelab Questions

1. If an excess of ceric ammonium nitrate (CAN) is used to oxidize benzyl alcohol to benzaldehyde, how could you visually determine when oxidation is complete?

2. Benzaldehyde reacts with CAN to form benzoic acid but this oxidation is slow compared to the oxidation of benzyl alcohol to benzaldehyde. By comparison, benzyl alcohol is also oxidized by chromic acid, but benzaldehyde reacts with this oxidizing agent at a rate comparable to the rate for the oxidation of benzyl alcohol. For the chromic acid oxidation of benzyl alcohol, which is complicated by subsequent reactions of the primary product (*secondary reactions*), which procedure would be more advantageous for obtaining a higher yield of the primary product: slowly adding the oxidizing agent to the alcohol or slowly adding the alcohol to the oxidizing agent? Why? Is there any advantage in following the opposite addition procedure for the oxidation of benzyl alcohol by CAN?

3. (a) If one mole of benzyl alcohol is oxidized to benzaldehyde by two moles of CAN in 1.0 l of water, what is the molar concentration of hydrogen ion in the aqueous solution after oxidation is complete? (b) If three moles of 1-phenyl-ethanol is oxidized to acetophenone by two moles of chromic acid in 1.0 l of 3.0*M* aqueous sulfuric acid, what is the molar concentration of the hydrogen ion in the aqueous solution after oxidation is complete?

4. Using the ideal gas law calculate the volume occupied by 1.0×10^{-3} mol of benzaldehyde at its boiling point at both 10 and 760 Torr.

EXPERIMENTAL PROCEDURE: CERIC ION OXIDATION OF BENZYL ALCOHOL

CAUTION: Ceric ammonium nitrate is a strong oxidizing agent. If this reagent contacts your skin or clothing, immediately wash the affected area with copious quantities of water.

Dissolve 54.8 g (0.10 mol) of ceric ammonium nitrate in 80 ml of distilled water. Weigh 5.0 g of benzyl alcohol (0.046 mol) into a separate 250-ml Erlenmeyer flask and then add 20 ml of distilled water to the alcohol. Slowly pour the CAN solution into the 250-ml Erlenmeyer flask that contains the aqueous alcohol mixture. Note the color of the CAN solution initially and that of the reaction solution immediately following addition of the cerium reagent to the aqueous alcohol mixture. Thoroughly mix the contents of the Erlenmeyer flask by swirling and *gently* heat the reaction mixture on a steam bath. Continue agitation of the reaction mixture until no further color change is observed. When no further color change is observed or when the reaction mixture is colorless, cool the contents of the Erlenmeyer flask to room temperature, and then transfer the cooled reaction mixture to a separatory funnel. Add 50 ml of ether to the separatory funnel and thoroughly mix the ether phase with the aqueous phase by vigorous shaking. Separate the aqueous phase from the ether phase, wash the ether phase with two 25-ml portions of water, and then extract the ether solution with 25 ml of a 5% aqueous sodium bicarbonate solution. Dry the ether solution over anhydrous magnesium sulfate, filter the magnesium sulfate from the ether, and distill the ether using the simple distillation apparatus. Dispose of the collected ether in the container provided for waste ether. Pour the residue remaining after distillation of the ether into a small distilling flask and distill the benzaldehyde under aspirator pressure or at atmospheric pressure. Record the boiling point range for your distilled product and the pressure at which distillation occurred. Weigh your product and calculate the percentage yield of the isolated benzaldehyde.

EXPERIMENTAL PROCEDURE: CHROMIC ACID OXIDATION OF 1-PHENYLETHANOL

Weigh 5.0 g of chromium trioxide (0.050 mol) and carefully add this reagent to 15 ml of distilled water in a 250-ml Erlenmeyer flask. Swirl the mixture, then add 5 ml of concentrated sulfuric acid, and again agitate the mixture until a homogeneous solution is obtained.

CAUTION: Chromium trioxide and chromic acid are strong acids and strong oxidizing agents. If either of these reagents contact your skin or clothing, immediately wash the affected area with copious quantities of water.

Weigh 6.1 g of 1-phenylethanol (0.050 mol) into an Erlenmeyer flask and then add 100 ml of reagent-grade acetone. Pour this solution into a 500-ml three-necked round-bottom flask that is fitted with an addition funnel (your separatory funnel will also function as an addition funnel), thermometer adapter, and glass stopper. Insert the thermometer through the adapter so that the bulb of mercury is below the surface of the acetone solution. The three-necked flask, which is clamped from the central joint to a ringstand, is placed in an ice-water bath. The chromic acid solution is then poured into the addition funnel by the use of a long-stem funnel.

When the temperature of the acetone solution is below 20°C, add the chromic acid solution dropwise to the alcohol-acetone solution while agitating the three-necked flask to obtain thorough mixing of the reagents. Control the rate of chromic acid addition so that the reaction temperature remains between 20 and 30°C. Note the color of the reaction solution during the addition and continue the addition until the light orange color of chromic acid persists. Continue agitation of the reaction solution for 15 min after completion of the addition.

After the reaction is complete slowly add solid sodium carbonate until the solution reaches neutral *p*H. Filter the resulting mixture by gravity and wash the precipitate twice with 30-ml portions of ether. Pour the filtrate into a separatory funnel, add 100 ml of a saturated aqueous solution of sodium chloride, and after thorough shaking, separate the aqueous layer from the ether layer. Pour the ether layer into an Erlenmeyer flask that contains approximately 1 g of anhydrous magnesium sulfate. Wash the aqueous layer twice with 20-ml portions of ether and combine the ether washings with the prior ether extract. Filter the magnesium sulfate from the ether, and distill the ether using the simple distillation apparatus. Dispose of the collected ether in the container provided for waste ether. Pour the residue remaining after distillation of the ether into a small distilling flask and distill the acetophenone under reduced pressure or at atmospheric pressure. Record the boiling point range for your distilled product and the pressure at which distillation occurred. Weigh your product and calculate the percentage yield of the isolated acetophenone.

Postlab Questions

1. Suggest an explanation for the change in color of the ceric ion upon addition of the CAN solution to benzyl alcohol (*Hint:* See Experiment 33).

2. If benzoic acid is produced from the oxidation of benzyl alcohol by ceric ammonium nitrate, from which solution—the aqueous phase, the sodium bicarbonate washing, or the ether solution—could you expect to isolate this product? Describe how benzoic acid could be isolated.

3. What evidence do you have from your experimental procedure for the oxidation of benzyl alcohol by CAN that indicates the presence or absence of benzyl alcohol? Describe how you could identify benzyl alcohol in your isolated reaction product and how you could quantitatively determine the percentage yield of benzaldehyde. (If the Jones oxidation is performed, answer the same questions for 1-phenylethanol.)

4. Chromic acid oxidation of acetone competes with chromic acid oxidation of acetophenone. If the rate for oxidation of acetophenone is 100-times faster than that for oxidation of acetone, what molar excess of acetone over acetophenone is required so that less than one percent of acetophenone is oxidized during the oxidation procedure? (Assume that the reaction solution contains only chromic acid, acetone, and acetophenone.)

5. Compare the boiling point that you obtained for your product with the data in Figure 14.1. Is the pressure that you recorded the same as that expected from the boiling point-pressure curve?

Product Analysis for Aldehydes and Ketones

Solubility Tests. Like other classes of neutral, polar organic compounds, aldehydes and ketones are soluble in ethyl ether and in concentrated sulfuric acid; low molecular weight aldehydes and ketones, such as acetone and acetaldehyde, are also soluble in water. Aldehydes and ketones that have more than four carbons are not appreciably soluble in water. However, carbonyl compounds can generally be distinguished from other classes of organic compounds by their solubility in aqueous sodium bisulfite. Aldehydes and ketones from *bisulfite addition products* with sodium bisulfite:

Bisulfite addition test

$$\underset{\text{Aldehyde or ketone (water insoluble)}}{R_2C{=}O} + \underset{\text{Sodium bisulfite}}{NaHSO_3} \xrightarrow{H_2O} \underset{\text{Bisulfite addition product (water soluble)}}{R_2C(OH)SO_3^-Na^+}$$

Not all carbonyl compounds produce the water-soluble addition product. Diaryl ketones, aryl alkyl ketones, such as acetophenone, and ketones with bulky alkyl groups react slowly or not at all with sodium bisulfite. For many carbonyl compounds, however, the bisulfite addition reaction provides a convenient solubility characterization test and is the basis of a useful procedure to separate carbonyl compounds from mixtures through extraction.

SOLUBILITY TEST PROCEDURE

3.0 ml of 40% aqueous sodium bisulfite is added in 1.0-ml portions to 0.1 g (0.1 ml) of the water-insoluble organic compound contained in a small test tube. The mixture is shaken vigorously after the addition of each portion of the bisulfite solution to allow thorough mixing. The organic compound has formed the addition product if the organic compound dissolves completely in water. If a solid precipitate is obtained, add 3 ml of water to determine if the precipitate is water soluble.

2,4-Dinitrophenylhydrazine Test. Aldehydes and ketones are easily distinguished from other classes of organic compounds by chemical characterization tests for the carbonyl group. 2,4-Dinitrophenylhydrazine is particularly valuable as a chemical characterization reagent since it reacts with most aldehydes and ketones to form a relatively insoluble *2,4-dinitrophenylhydrazone product:*

2,4-Dinitro phenylhydrazone formation

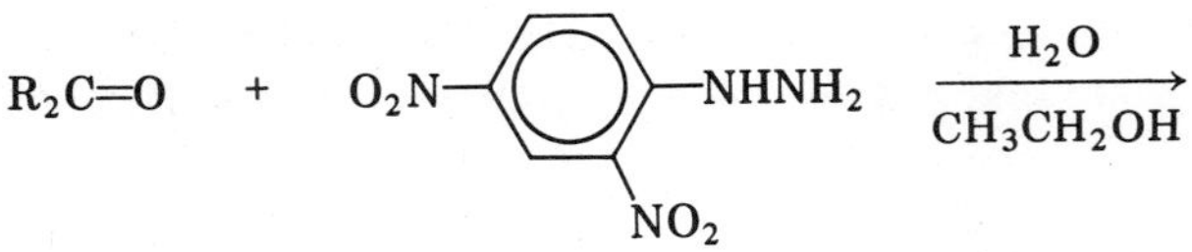

Aldehyde or ketone

2,4-Dinitro-phenylhydrazine
m.p. = 194°C
(red solid, orange solution)

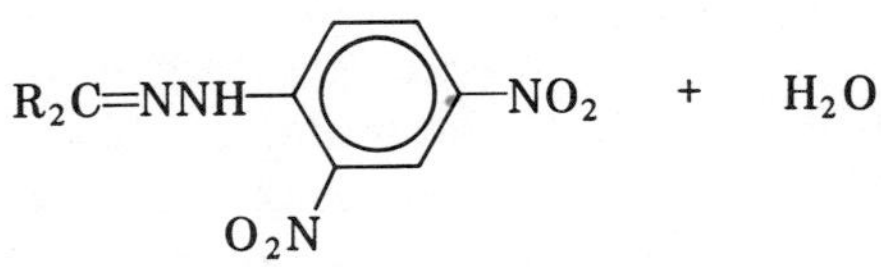

A 2,4-dinitro-phenylhydrazone
(yellow to orange-red solid precipitate)

Care must be taken in interpreting the results from the 2,4-dinitrophenylhydrazone characterization test since a small amount of a solid 2,4-dinitrophenylhydrazone can be obtained in analyses of alcohols that contain aldehyde or ketone impurities. A positive test is the formation of the 2,4-dinitrophenylhydrazone in an amount that corresponds to the number of moles of the limiting reagent; the formation of a small amount of the hydrazone may be due to carbonyl compounds that are impurities or that are formed by oxidation during the characterization test.

PROCEDURE

In a small test tube add 0.05 g (or two drops) of the organic compound to 2 ml of reagent grade 95% ethanol. To this solution add 1 ml of the 2,4-dinitrophenylhydrazine reagent* and shake the test tube vigorously. If no precipitate forms immediately, heat the solution on a steam bath for 1 min and then add five drops of water. The formation of a yellow to orange-red precipitate is a positive test.

The color of the 2,4-dinitrophenylhydrazone often gives an indication of the structure of the aldehyde or ketone from which the hydrazone is formed. Aldehydes and ketones that are not conjugated, such as acetone and acetalde-

*The 2,4-dinitrophenylhydrazine reagent is prepared by dissolving 1.0 g of 2,4-dinitrophenylhydrazine in 5 ml of concentrated sulfuric acid and then adding 7 ml of water dropwise while stirring the acid solution. Add 20 ml of reagent grade 95% ethanol to the warm solution and, after thorough mixing, filter if a solid has precipitated. *CAUTION: 2,4-dinitrophenylhydrazine is a potentially explosive material that should not be subjected to unusual shock or stress.* The reagent grade 95% ethanol used to prepare this reagent and employed as a solvent in this test must be free of aldehyde or ketone impurities. The 2,4-dinitrophenylhydrazine reagent can be used to test for these carbonyl contaminants.

hyde, yield yellow 2,4-dinitrophenylhydrazones. Conjugation with a carbon-carbon double bond or an aromatic ring, as is the case with benzaldehye and acetophenone, renders the 2,4-dinitrophenylhydrazone an orange-red color.

Chromic Acid Test. Aldehydes are readily oxidized to carboxylic acids whereas ketones are less reactive or are unreactive towards the same oxidants. This chemical difference provides the basis for chemical characterization tests that distinguish aldehydes from ketones. Two such tests—the *chromic acid test* and *Tollens' test*—are described in this experimental section.

Chromic acid reacts rapidly with aldehydes to form carboxylic acids. In general, aliphatic aldehydes form a distinct blue-green precipitate of chromic sulfate within 15 sec; aromatic aldehydes require a longer time but usually less than 1 min. Ketones are more resistant to oxidation and require reaction times that are longer than 2 min for the appearance of the distinctive precipitate:

Chromic acid test

$$RCHO + H_2CrO_4 \xrightarrow{15\ sec} [Cr(H_2O)_6]_2(SO_4)_3$$

$$ArCHO + H_2CrO_4 \xrightarrow{<1\ min} [Cr(H_2O)_6]_2(SO_4)_3$$

$$R_2CO + H_2CrO_4 \xrightarrow{>2\ min} [Cr(H_2O)_6]_2(SO_4)_3$$

orange-red solution — *blue-green precipitate*

Although this test distinguishes aldehydes from ketones, it does not differentiate aldehydes from alcohols, olefins, or other easily oxidized organic compounds (Experiment 13).

PROCEDURE

In a small test tube add 0.02 g (one drop) of the organic compound to 2 ml of reagent-grade acetone and then add one drop of the chromic acid reagent (preparation described in Jones oxidation, Experiment 14). The rate of formation of the blue-green precipitate determines the nature of the carbonyl compound.

Tollens' Test. Unlike the chromic acid test, which is relatively unselective, the Tollens' test is specific for aldehydes and distinguishes them from ketones, alcohols, olefins, and other monofunctional organic compounds. Aldehydes are oxidized to carboxylate salts by the silver ion complexed with ammonia (*Tollens' reagent*):

Tollens' test

$$\underset{\text{Aldehyde}}{RCHO} + 2\ \underset{\text{Diammine-silver complex}}{Ag(NH_3)_2^+} + 2\ OH^- \longrightarrow 2\ \underset{\text{Silver mirror}}{Ag^\circ} + RCOO^- + H_2O + NH_4^+ + 3\ NH_3$$

Metallic silver is usually formed as a mirror on the walls of the reaction vessel in this oxidation.

PROCEDURE

The Tollens' reagent is prepared immediately prior to use and never stored. In a clean small test tube add one drop of 10% aqueous sodium hydroxide to 1 ml of 5% aqueous silver nitrate. Shake the tube and than add a dilute solution of ammonium hydroxide dropwise and with shaking until the precipitate of silver hydroxide just dissolves. (Avoid adding a large excess of the ammonia solution.) Add 0.05 g (or two drops) of the organic compound to the Tollens' reagent, shake the tube, and allow the solution to stand for 10 min. If no reaction has occurred after 10 min, warm the test tube in a beaker of warm water for 5 min. A silver mirror on the walls of the test tube or a metallic precipitate in the mixture constitutes a positive Tollens' test. After the test is complete wash the contents of the tube into a container of dilute nitric acid that is specifically designated for this purpose. Rinse the test tube with dilute nitric acid.

Iodoform-Formation Test. Treatment of carbonyl compounds that possess α-hydrogens with halogen in basic media results in the substitution of halogen for hydrogen at the α-carbon position. Methyl ketones undergo exhaustive halogen substitution followed by cleavage of the carbonyl carbon-methyl carbon bond. When the halogen is iodine, methyl ketones produce the easily-characterized, yellow solid iodoform (m.p. = 119°C):

Iodoform-formation test

$$\underset{\text{A methyl ketone}}{RCOCH_3} + 3\,I_2 + 4\,OH^- \longrightarrow RCOO^- + \underset{\substack{\text{Iodoform} \\ \text{yellow solid} \\ (\text{m.p.} = 119^\circ\text{C})}}{CHI_3} + 4\,H_2O + 3\,I^-$$

A positive *iodoform-formation test* is the formation of a yellow precipitate that has a melting point of 119°C. Methyl ketones such as acetophenone give a positive iodoform test. However, a positive test is also obtained from compounds that are oxidized to methyl ketones. Thus methylcarbinols—2° alcohols having the $RCH(OH)CH_3$ structure—also produce iodoform:

$$RCH(OH)CH_3 \xrightarrow[OH^-]{I_2} RC(=O)CH_3 \xrightarrow[OH^-]{I_2} CHI_3$$

PROCEDURE

To 0.2 g (or four drops) of the organic compound in a small test tube add the minimum volume of tetrahydrofuran necessary to completely dissolve that compound (<2 ml). Add 2 ml of 10% aqueous sodium hydroxide followed by dropwise addition with vigorous shaking of the iodine-potassium iodide reagent.* Continue the addition of the iodine reagent until an excess of iodine is observed in the reaction solution. Heat the tube in a beaker that contains water at 55 to 65°C and continue the addition of the iodine reagent. Addition is complete when the dark color persists after two minutes at the elevated temperature. Excess iodine is consumed by adding 10% aqueous sodium hydroxide dropwise and with shaking until the dark color just disappears. Pour the contents of the tube into an Erlenmeyer flask that contains 10 ml of cold water. Iodoform, if produced, will precipitate as a yellow solid. Filter the yellow solid and take its melting point to confirm that the solid is iodoform.

Infrared Spectroscopy is a powerful tool for the characterization of aldehydes and ketones (Experiment 12). Aldehydes and ketones exhibit a characteristic intense carbonyl stretching vibration in the infrared between 1760 and 1660 cm^{-1}. Aliphatic aldehydes have a distinguishing carbonyl absorption at 1740 to 1720 cm^{-1}, and conjugated aldehydes such as benzaldehyde exhibit the carbonyl group absorption at 1710 to 1685 cm^{-1}. Aldehydes also have characteristic aldehydic C—H absorptions at 2775 to 2700 cm^{-1}. Aliphatic ketones have a distinguishing carbonyl group absorption at 1725 to 1705 cm^{-1}, and conjugated ketones such as acetophenone exhibit the carbonyl group absorption at 1690 to 1665 cm^{-1}. The infrared spectra of benzaldehyde and acetophenone are given in Figure 14.4.

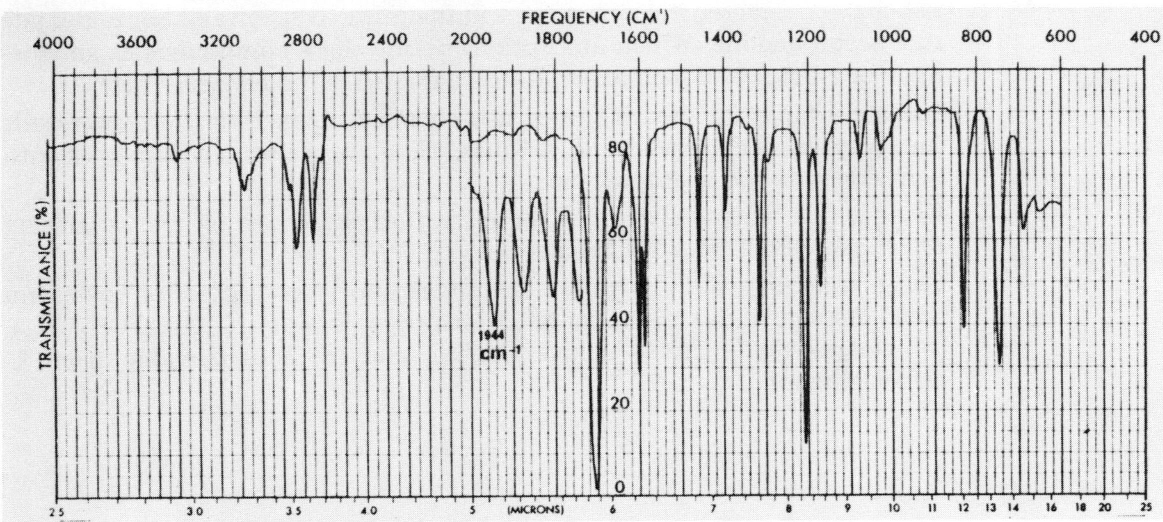

Figure 14.4a *Infrared spectrum of benzaldehyde (thin film).*

*This reagent is prepared by adding 2.0 g of potassium iodide and 1.0 g of iodine to 8 ml of water. The resulting dark solution is aqueous potassium triiodide. Iodine is only partially soluble in water.

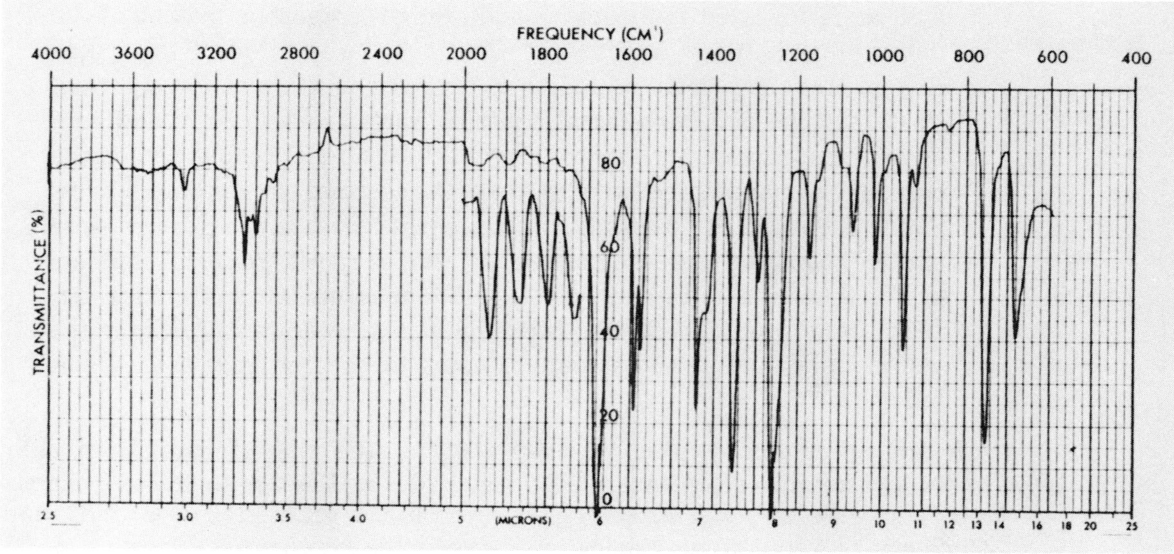

Figure 14.4b *Infrared spectrum of acetophenone (thin film).*

Characterization of Benzaldehyde and Acetophenone

1. Label the characteristic absorptions in the infrared spectra of benzaldehyde and acetophenone. Which absorptions define these compounds as an aldehyde or a ketone? Which absorptions characterize the benzene ring?
2. Prepare a table for solubility test and chemical characterization test results for analyses of aldehydes and ketones. Tabulate the results from tests with benzaldehyde and acetophenone.
3. Using a flow diagram, describe an extraction procedure for the separation of benzaldehyde from benzyl alcohol.
4. Using solubility and chemical characterization tests, describe how you could distinguish (a) acetophenone from propiophenone ($C_6H_5COCH_2CH_3$), (b) benzaldehyde from benzyl alcohol, and (c) acetophenone from 1-phenylethanol.

Experiment Fifteen

The Grignard Synthesis: Preparation of 1,2-Diphenylethanol

Organometallic compounds are among the most versatile reagents for organic synthesis. By definition these reagents possess a carbon-metal bond and, as a

Organometallic compound

$$-\overset{|}{\underset{|}{C}}{}^{\delta-}\!-\!M^{\delta+}$$

consequence of the electronegativity difference between carbon and the metal, there is a substantial partial negative charge at the metal-substituted carbon atom. The usefulness of organometallic reagents in organic syntheses is derived from their carbanionic character which empowers these reagents with the ability to form carbon-carbon bonds in nucleophilic addition reactions and nucleophilic substitution reactions. In this experiment you will prepare benzylmagnesium chloride, an organomagnesium halide, and synthesize 1,2-diphenylethanol by nucleophilic addition of benzylmagnesium chloride to the carbonyl group of benzaldehyde:

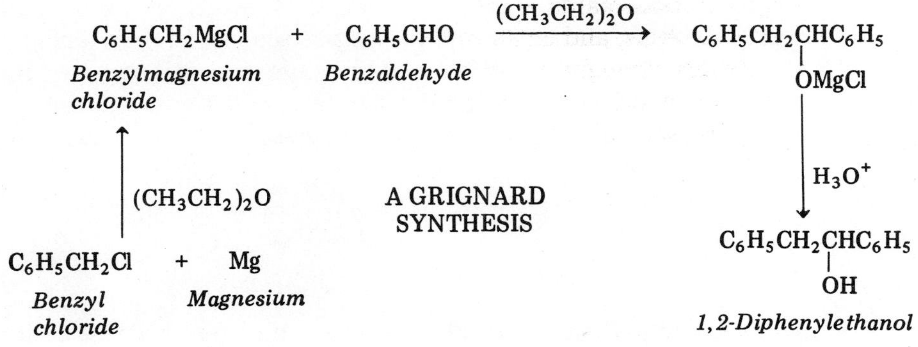

Preparation of Organomagnesium Halides (Grignard Reagents)

Organomagnesium halides—discovered by Victor Grignard in 1900 and commonly known as *Grignard reagents*—are prepared by reacting an organic halide with magnesium in an ether solvent, usually ethyl ether:

$$\underset{\textit{Organic halide}}{RX} + \underset{\textit{Magnesium}}{Mg} \xrightarrow{(CH_3CH_2)_2O} \underset{\textit{Organomagnesium halide}}{RMgX}$$

To prepare the Grignard reagent, thin turnings of metallic magnesium are covered with dry ether and the organic halide is added to this mixture to initiate reaction. Magnesium atoms at the surface of the magnesium turnings insert between carbon and halogen of the organic halide. As the reaction progresses, layer after layer of magnesium in the turnings react to form the ether-soluble organomagnesium halide.

Initiation of the reaction between magnesium and organic halide is often the most difficult step in the Grignard synthesis. Surface oxides on the metal turnings inhibit contact of magnesium atoms with the organic halide. Exposing new surfaces of the magnesium turnings by crushing the turnings under the ether solvent against the surface of the reaction flask, however, enables the exposed metal to react with the organic halide. Etching the magnesium surface by adding a small crystal of iodine often has a similar reaction initiating effect.

The quality of the ether employed for the preparation of Grignard reagents is critical. If water is present in the reaction solvent (or in the reaction flask before ether is added), reaction between the organic halide and magnesium is severely inhibited. Even if formed in a wet environment, organomagnesium halides are consumed in a rapid acid-base reaction with water to form the corresponding hydrocarbons:

Reaction with water

$$\underset{\textit{Organomagnesium halide}}{RMgX} + H_2O \longrightarrow RH + (HO)MgX$$

Organic compounds that possess an O—H or N—H bond, such as carboxylic acids, alcohols, and amines, also react as acids with organomagnesium halides.

Ethers strongly coordinate magnesium in organomagnesium halides and are therefore good solvents for the formation of these reagents. Two ether molecules are strongly bound to each magnesium. Anhydrous ethyl ether is most

The Grignard reagent in ether

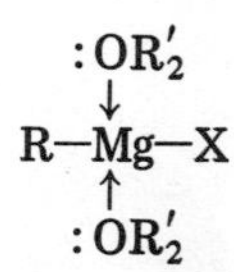

often employed as the reaction solvent, not only because of its ability to coordinate with magnesium in organomagnesium halides, but also because the high vapor pressure of ethyl ether at the temperatures used for the preparation of Grignard reagents excludes oxygen. Oxygen reacts with organomagnesium halides to form magnesium halide salts of alcohols:

Reaction with oxygen

$$2\ RMgX + O_2 \longrightarrow 2\ ROMgX$$

Organomagnesium halide → *Alkoxymagnesium halide*

The reactivity of organic halides with magnesium follows the order RI > RBr > RCl. Saturated organic halides (alkyl halides) are more reactive than unsaturated organic halides (vinyl halides and aryl halides). Unsaturated organic chlorides generally do not react with magnesium. The preparation of benzylmagnesium chloride in this experiment will require an induction period of approximately 15 min for initiation of this exothermic reaction. Longer induction periods or no reaction of benzyl chloride with magnesium will result from acid impurities in the benzyl chloride or from the presence of water in the reaction medium. As soon as the reaction begins, oxygen is excluded from

Preparation of benzylmagnesium chloride

$$C_6H_5CH_2Cl + Mg \xrightarrow{(CH_3CH_2)_2O} C_6H_5CH_2MgCl$$

Benzyl chloride
M.w. = 126.6 g/mol
b.p. = 179°C
d_{25} = 1.100

Magnesium
M.w. = 24.31

Benzylmagnesium chloride

the reaction solution by adding benzyl chloride to the ether-covered magnesium at a rate sufficient to maintain ether at reflux. The rate of addition of benzyl chloride to the reaction medium is also important in controlling the coupling reaction between the Grignard reagent and the organic halide (a nucleophilic substitution reaction):

Coupling reaction

$$RMgX + RX \longrightarrow R{-}R + MgX_2$$

Organomagnesium halide + *Organic halide*

At lower concentrations of the organic halide the reaction of the organic halide with the insoluble magnesium turnings is favored over the coupling reaction with the ether-soluble organomagnesium halide. Use of dilute solutions for the preparation of the Grignard reagent also minimizes this competing side reaction.

Addition Reactions of Organomagnesium Halides

Grignard reagents undergo rapid nucleophilic addition to the carbonyl group of aldehydes, ketones, carbon dioxide, and carboxylic acid derivatives. These reactions involve the formation of a carbon-carbon bond between the carbonyl carbon and the metal-substituted carbon of the Grignard reagent. With benzaldehyde, the carbonyl substrate employed in this experiment, benzylmagnesium chloride forms the magnesium chloride salt of 1,2-diphenylethanol. This salt is only partially soluble in ether and precipitates from solution. Subsequent addition of aqueous mineral acid to the reaction mixture produces 1,2-diphenylethanol:

Preparation of 1,2-Diphenylethanol

$$\underset{\substack{\textit{Benzaldehyde}\\ \text{M.w.} = 106\text{ g/mol}\\ \text{b.p.} = 178^\circ\text{C}}}{C_6H_5CHO} + \underset{\substack{\textit{Benzylmagnesium}\\ \textit{chloride}}}{C_6H_5CH_2MgCl} \xrightarrow{(CH_3CH_2)_2O} \underset{\substack{|\\ OMgCl}}{C_6H_5CHCH_2C_6H_5}$$

$$\underset{\substack{|\\ OMgCl}}{C_6H_5CHCH_2C_6H_5} \xrightarrow{H_3O^+} \underset{\substack{|\\ OH\\ \textit{1,2-Diphenylethanol}\\ \text{M.w.} = 198\text{ g/mol}\\ \text{m.p.} = 67^\circ\text{C}}}{C_6H_5CHCH_2C_6H_5} + Mg^{2+} + Cl^-$$

Prelab Questions

1. Air contains carbon dioxide as well as oxygen. Write the chemical reaction that describes the fate of benzylmagnesium chloride in the presence of carbon dioxide. How can this reaction be minimized in your experimental procedure?

2. If you stop your synthesis after the preparation of the Grignard reagent and wait to perform the addition of benzaldehyde until the next laboratory period, which chemical reactions will cause a depletion of the Grignard reagent? Can you store the reaction solution with the assurance that these reactions can be prevented?

3. Benzyl chloride is hydrolyzed by water to benzyl alcohol and hydrogen chloride. Benzyl alcohol reacts with benzyl chloride to form benzyl ether and hydrogen chloride. What reactions can be expected between these hydrolysis products and either benzylmagnesium chloride or magnesium? How could you determine if your benzyl chloride reagent contained these impurities?

4. Describe the flow diagram for the purification procedure in this experiment. Explain how 1,2-diphenylethanol can be separated from excess reagents and from the products of competing reactions in this diagram.

EXPERIMENTAL PROCEDURE

In this experiment you will employ a 500-ml three-neck round-bottom flask, an addition funnel (separatory funnel), and a reflux condenser. This equipment can be conveniently dried in an oven at temperatures above 100°C and immediately stoppered with cotton after removal from the oven. In an alternate drying procedure, the assembled glassware and magnesium turnings can be flame dried with a Bunsen burner. Pieces of cotton (or drying tubes containing anhydrous calcium chloride or calcium sulfate) at the openings of the equipment to the air are usually effective in removing atmospheric moisture as air enters the glassware upon cooling. *Do not begin this experiment until your glassware is at room temperature.*

Anhydrous ethyl ether must be used for the Grignard synthesis. If only technical grade ether is available, it must be dried over a suitable drying agent (calcium chloride or magnesium sulfate, Table 8.1) prior to use. Commercial anhydrous ether that is dried over molecular sieves results in shorter induction periods for initiation of the reaction between organic halides and magnesium.

CAUTION: Ethyl ether is highly volatile and extremely flammable. Be sure that there are no flames in your area before you handle this solvent.

Add 1.2 g of magnesium turnings (0.050 mol) to a 500-ml three-neck round-bottom flask that is fitted with an addition funnel, reflux condenser, and a glass stopper. When the reaction flask is at room temperature add 10 ml of anhydrous ether to cover the magnesium. Loosely stopper the condenser and addition funnel openings with pieces of cotton. Turn on the water to the condenser. Prepare an ice bath for use in cooling the reaction mixture in the event that ether refluxes too vigorously during reaction.

Dissolve 6.3 g (5.7 ml) of benzyl chloride (0.050 mol) in 40 ml of anhydrous ethyl ether and pour this solution into the addition funnel.

CAUTION: Benzyl chloride is a lachrymator—a substance that irritates the eyes and produces tears. This reagent should be rapidly measured into a container that is easily stoppered and used in a well-ventilated area.

Add 10 ml of this solution to the reaction flask and gently agitate the reaction solution to ensure thorough mixing. Place a small crystal of iodine into the reaction solution, again gently agitate the reaction solution, and commence crushing the magnesium turnings firmly against the bottom of the flask with a dry spatula or with the blunt end of a stirring rod. (*Be careful so that you do not punch a hole in the flask.*) A change in the color of the solution from yellow (due to dissolved iodine) to colorless, the formation of a cloudy solution, and observation of bubbles forming with increasing rapidity at the surface of the magnesium are evidence that the reaction has commenced. As the reaction proceeds, ether will be observed to reflux vigorously. At this point begin dropwise addition of the benzyl chloride solution at a rate sufficient to maintain moderate reflux of the ether. If the refluxing action of the ether is slow, heat the reaction mixture on a steam bath. If refluxing is too rapid, cool the reaction mixture in an ice bath. After the addition is complete, gently heat the reaction mixture on a steam bath for 15 min. Only residual pieces of magnesium remain when the reaction is complete.

Cool the flask that contains the Grignard reagent in an ice-water bath and add 25 ml of anhydrous ether to the reaction solution through the addition funnel. Dissolve 4.2 g of benzaldehyde (0.040 mol) in 25 ml of anhydrous ether and pour this solution into the addition funnel used in the preparation of benzylmagnesium chloride. Add the benzaldehyde solution dropwise to the reaction flask with continual agitation of the mixture. Moderate refluxing of the ether should be maintained during the dropwise addition. After complete addition of the benzaldehyde solution, reflux the reaction mixture for 15 min and then cool the reaction flask in an ice bath.

Pour 10 ml of concentrated sulfuric acid into an Erlenmeyer flask that contains 25 g of ice in 25 ml of water. Slowly add this acidic solution to the ice-bath-cooled reaction flask with frequent agitation of the mixture. The white precipitate that was formed during addition of benzaldehyde will decompose completely. Transfer the resulting mixture to a separatory funnel and separate the water layer from the ether layer. Pour the ether layer into an Erlenmeyer flask. Wash the water layer once with a 20-ml portion of technical grade ether and combine this ether washing with the previous ether extract. Wash the combined ether solution with 30 ml of 5% aqueous sodium bicarbonate solution and then dry the ether solution over anhydrous magnesium sulfate. Filter the ether solution into a 250-ml round-bottom flask and distill the ether on a steam bath. Dispose of the collected ether in the container provided for waste ether. Add 15 ml of hexane to the residue remaining after distillation of ether and continue heating the flask to remove the last traces of ether. While still hot, transfer the hexane solution to an Erlenmeyer flask and then allow the solution to cool. Collect the solid product and obtain its melting point. Recrystallize the product if the melting point range is below 62°C. Weigh the collected 1,2-diphenylethanol and calculate its percentage yield. Save this compound for use in Experiment 16.

Postlab Questions

1. Benzaldehyde is readily oxidized in air to benzoic acid. Using chemical reactions describe the fate of benzylmagnesium chloride in the presence of benzoic acid. What product(s) are expected from these reactions after the acid workup of the Grignard reaction mixture?

2. Which reactant is the limiting reagent in this experiment? In your experimental observations what evidence do you have that the preparation of the Grignard reagent did not occur in 100% yield? If the Grignard reagent is formed in only 50% yield, which reactant is the limiting reagent?

3. Grignard reagents react with carbon dioxide to form magnesium halide salts of carboxylic acids. In the experimental procedure for these reactions, the Grignard reagent is added to crushed solid carbon dioxide rather than the reverse (normal) addition used in this experiment. Why?

4. Predict the results of solubility tests (Experiment 7) and alcohol chemical characterization tests (Experiment 13) for 1,2-diphenylethanol.

5. Describe how you could detect the presence of each of the following compounds in the hexane solution from this experiment prior to crystallization of 1,2-diphenylethanol, using infrared spectroscopic analysis or chemical characterization tests: (a) benzaldehyde, (b) benzyl chloride, (c) stilbene (1,2-diphenylethene), and (d) benzoic acid. Each of these compounds are possible impurities in the desired 1,2-diphenylethanol solution.

Experiment Sixteen

Dehydration of 1,2-Diphenylethanol

In Experiment 16, 1,2-diphenylethene (*stilbene*) is prepared by dehydration of 1,2-diphenylethanol. *trans*-Stilbene, the thermodynamically more stable alkene isomer, is the exclusive product of this acid-catalyzed reaction. *cis*-Stilbene, if initially formed, is rapidly isomerized to *trans*-stilbene under the acidic reaction conditions. The procedure employed for this synthesis is similar to that described for the dehydration of 3-methyl-3-pentanol in Experiment 9.

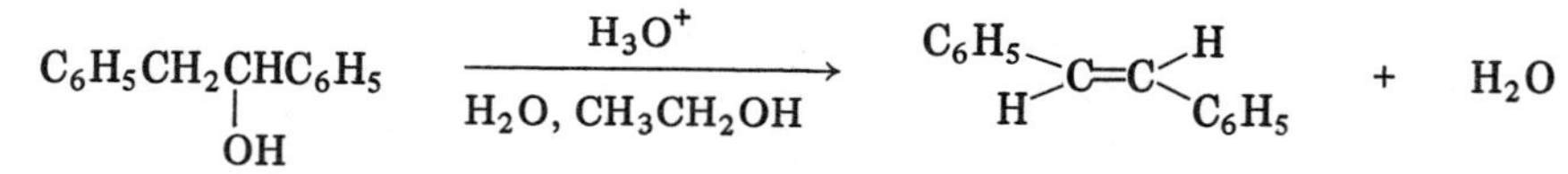

1,2-Diphenylethanol
M.w. = 198 g/mol
m.p. = 67°C

trans-Stilbene
M.w. = 180 g/mol
m.p. = 124°C

Experiment 16 completes a reaction sequence through which benzyl alcohol can be converted to *trans*-stilbene. (Conversion of benzyl alcohol to benzyl chloride is possible through a procedure similar to that described in Experiment 8 for the preparation of 3-chloro-3-methylpentane.) The experiments included in this reaction sequence involve separate laboratory operations in which the product of one reaction is the reactant for the next reaction. The amount of product recovered from a prior reaction in the sequence determines the amount of reactant that can be used in subsequent reactions (the limiting reagent). Subsequent experiments will continue to combine laboratory operations in sequence for the synthesis of desired organic products.

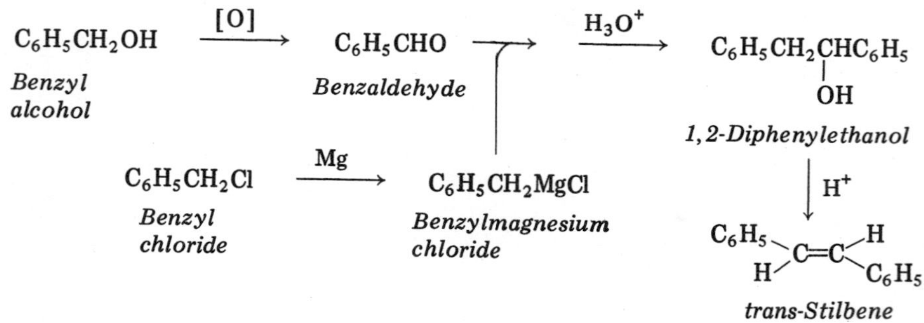

Decolorizing Compounds by the Use of Activated Charcoal

Products obtained from organic reactions often contain impurities that have a yellow to brown color. Even in small amounts, these colored impurities cause what should be a white or colorless product to appear colored, and it is desirable to remove these impurities. Impurities that are highly colored have extended conjugation (Experiment 30) and usually have a high molecular weight. Normally, recrystallization is not effective in removing all of these impurities.

The molecular characteristics that give rise to color are also those that often cause preferential adsorption on solid surfaces like those found on activated charcoal. Consequently, the use of *activated charcoal* (also called *decolorizing carbon* or *Norit*) in a recrystallization procedure can often lead to a dramatic visual improvement in the purity of a sample. Activated charcoal is a carbon preparation that possesses a large, active surface for the adsorption of high molecular weight organic impurities. Those colored impurities not adsorbed on activated charcoal are usually sufficiently soluble that they either remain in the filtrate following filtration of the recrystallized sample or can be washed from the surface of the filtered crystals.

The amount of activated charcoal that should be employed for purification is dependent on the amount of impurity that can be adsorbed on the charcoal surface. If a solution is only lightly colored, no more than 20 mg of activated charcoal per gram of solid material should be used. If the solution is dark, appreciably more activated charcoal may be used—but never more than 50 mg per gram of solid. The use of more charcoal than required for the removal of an impurity causes larger than necessary losses of the desired product due to adsorption of the product on the activated charcoal.

The use of activated charcoal follows Step 1 of the General Procedure for Recrystallization (Figure 2.5). The decolorizing carbon is added to the solution prior to filtration (Step 3), and the solution is agitated by swirling to wet the carbon. While maintaining the temperature of the solution near the boiling point of the solvent to prevent crystallization of the solid, the decolorizing carbon is filtered from the solution. Adsorption is rapid and is most effective at low temperatures; the boiling temperature is maintained solely to prevent crystallization of the solid from a saturated solution.

Prelab Questions

1. Outline a flow diagram for the separation procedure used to purify *trans*-stilbene in this experiment.

2. List the side reactions that might occur with dehydration of 1,2-diphenylethanol.

3. The extent of reaction in the conversion of 1,2-diphenylethanol to *trans*-stilbene can be followed by thin layer chromatography. Briefly describe procedures for application of the reaction solution to the TLC plate, for development of the TLC plate, and for spot visualization.

EXPERIMENTAL PROCEDURE

With the aid of long-stem funnel, pour 20 ml of 9*M* sulfuric acid into a 100-ml round-bottom flask that has been set up for reflux as shown in

---

CAUTION: Sulfuric acid will burn the skin upon contact. Wash acid from your skin at once with copious amounts of water.

Figure 4.2. Dissolve 1.0 g of 1,2-diphenylethanol (5.0 mmol*) in 5.0 ml of absolute ethanol and then add this solution and several boiling stones to the acid solution. Heat the resulting solution with swirling on the steam bath for 30 min and then allow this solution to cool to room temperature. (If thin layer chromatographic analysis is to be performed, remove samples from the reaction mixture immediately following the mixing of the acid and ethanol solutions and at the end of the 30-min reaction period.) After cooling to room temperature, extract the reaction solution with 50 ml of hexane, wash the hexane solution with 50 ml of water, and dry the resulting hexane extract over anhydrous magnesium sulfate. Filter the hexane solution into a clean, dry 100-ml round-bottom flask and then remove hexane by simple distillation until 15 ml of solution remains in the distillation flask. Transfer the hot hexane solution to a 50-ml Erlenmeyer flask, cool the flask and, if the solution is colored, employ activated charcoal to decolorize the solution. Collect the white crystalline solid by suction filtration and obtain its melting point. If the melting point of the isolated solid is less than 121°C, recrystallize your product from hexane or purify the solid by sublimation. Weigh your product and calculate the percentage yield.

Sublimation

Solids that have a relatively high vapor pressure, as is the case for *trans*-stilbene, can be purified by the process of sublimation. When the impurities in a sample have vapor pressures substantially lower than that of the desired material, sublimation is an effective alternative to recrystallization for the purification of a solid. The sublimation process depends on the solid-vapor equilibrium: solid and vapor phases are interconverted without passing through an intermediate liquid phase. As shown in Figure 16.1, a typical phase diagram relating solid, liquid, and vapor phases with pressure and temperature, the liquid phase is thermodynamically unstable and does not exist at temperatures and pressures

*Mmol = millimol = 10^{-3} mol.

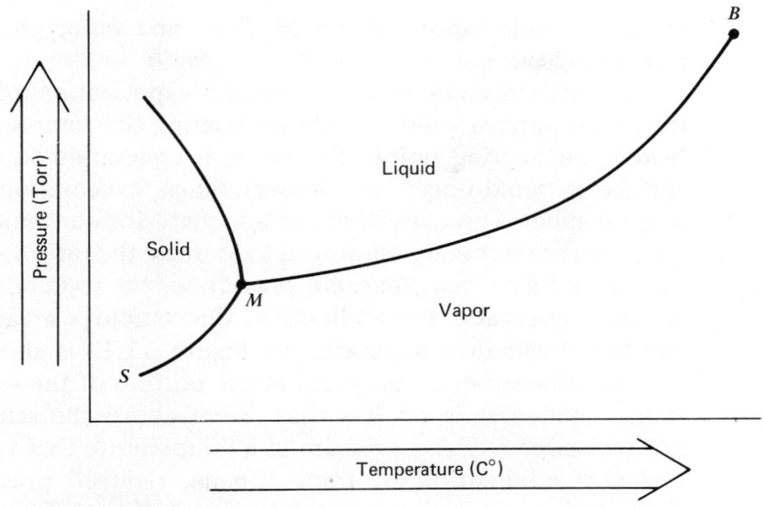

Figure 16.1 *Single-component phase diagram.*

Point M is the melting point of the pure compound and point B is the boiling point (v.p. = 760 Torr at 1 atm external pressure for most organic compounds). Note that the liquid-vapor curve M-B is similar to that previously given in Figure 5.1.

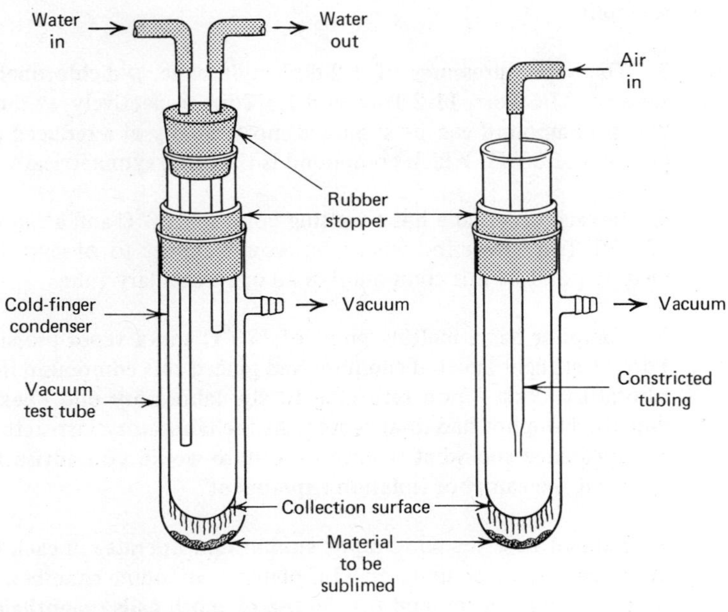

Figure 16.2 *Apparatus for sublimation.*

along the solid-vapor curve S–M. Solid and vapor phases are in direct equilibrium along this curve.

The sublimation method involves vaporization of the desired compound from the impure solid sample by heating the sample at a temperature that is below the melting point. The vapor is condensed to the solid state on a cold surface (a "cold-finger" condenser). Since few compounds have vapor pressures at atmospheric pressure that are adequate for convenient sublimation, reduced pressures are normally employed to increase the rate of sublimation. *Sublimation occurs rapidly when the vapor pressure of the solid equals the applied pressure.* A useful apparatus for sublimation that employs a vacuum test tube, like that for the distillation apparatus in Figure 11.1, is shown in Figure 16.2. The sample to be sublimed is placed at the bottom of the vacuum tube and the cold-finger condenser is set less than 2 cm above the sample. The sample is then heated under reduced pressure at a temperature that is below its melting point. For the sublimation of *trans*-stilbene, reduced pressure (approximately 20 Torr) and steam bath temperatures are sufficient for purification. The purified sample is deposited on the inner "cold-finger" condenser; it can then be removed, after careful separation of apparatus, by scraping the crystals from the surface of the condenser.

Postlab Questions

1. Describe the method or methods that would allow you to detect the presence or absence of *cis*-stilbene in your isolated product. Assume that you have authentic samples of *cis*- and *trans*-stilbene for your analysis.

2. A solid forms after cooling your reaction solution. What is the probable structure of this solid? Describe an alternate procedure to that used in this experiment for the isolation of this solid.

3. The vapor pressures of 1,2-diphenylethane, *p*-dichlorobenzene, and 1,3,5-trichlorobenzene are 0.06 Torr, 11.2 Torr, and 1.4 Torr, respectively, at their melting points (52 to 54°C). Which compound can be sublimed most rapidly at a reduced pressure of 20 Torr and a temperature of 50°C? Which compound is the most symmetrical?

4. Hexachloroethane has a melting point of 186°C and a vapor pressure at that temperature of 780 Torr. Describe what you would expect to observe if you attempted to obtain a melting point of this compound in an open capillary tube.

5. Camphor has a melting point of 179°C and a vapor pressure at that temperature of 370 Torr. A student isolated camphor and placed this compound in an open beaker in the locked laboratory desk. Upon returning to the laboratory one week later, the student discovered that the camphor had disappeared. As the laboratory instructor, how would you explain this disappearance and what storage procedure would you advise for the student who has begun to repeat the camphor isolation experiment?

6. Explain how the solid-vapor equilibrium operates in each of the following examples: (a) the visualization of spots on TLC plates in an iodine chamber, (b) the evaporation of dry ice at room temperature, and (c) the use of moth balls (naphthalene and *p*-dichlorobenzene) in a clothes closet to repel moths.

Experiment Seventeen

Friedel-Crafts Reactions: Preparations of 1,3-Dimethyl-5-tert-butylbenzene and 2,4-Dimethylacetophenone

Aromatic compounds react with alkyl halides in the presence of catalytic amounts of Lewis acids such as anhydrous aluminum chloride to produce substitution products. The alkyl group of the alkyl halide replaces a hydrogen on the aromatic ring and an alkyl-substituted aromatic compound and hydrogen halide are produced. These reactions are known as Friedel-Crafts alkylation reactions. Aromatic compounds also react with acid halides in the presence of Lewis acids to produce substitution products. The acyl group (RCO—) of the acid halide replaces a hydrogen on the aromatic ring and an aryl ketone (RCOAr) and hydrogen halide are formed. These reactions are known as Friedel-Crafts acylation reactions.

Friedel-Crafts Alkylation Reactions

The alkylation of aromatic compounds by Friedel-Crafts reactions is an impor-

Friedel-Crafts Alkylation

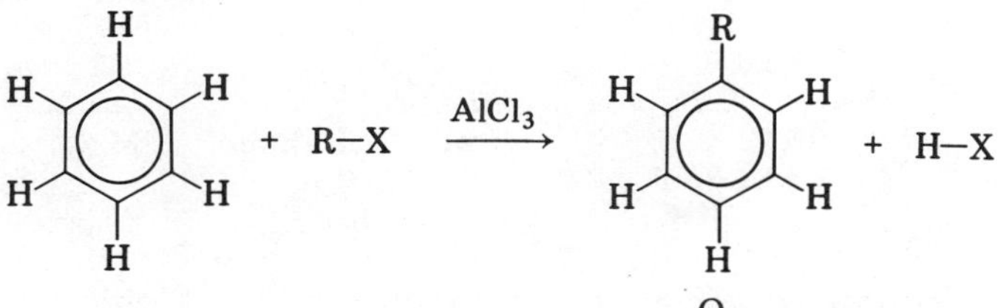

Friedel-Crafts Acylation

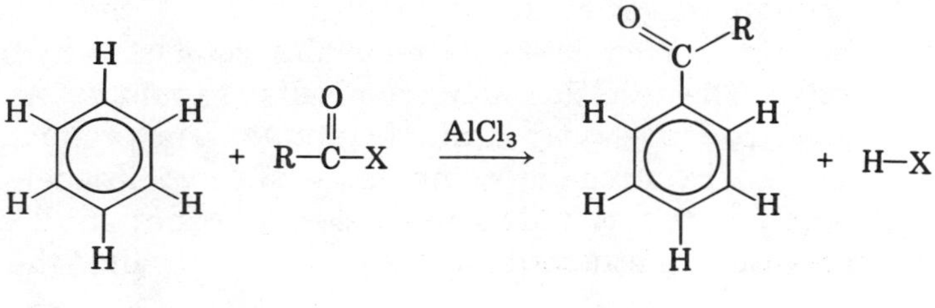

tant process for preparing alkyl substituted aromatic compounds in the laboratory, and it is used on an industrial scale to produce ethyl benzene and isopropylbenzene in multimillion pound per year amounts. The usual laboratory procedure for this reaction employs anhydrous aluminum chloride to catalyze the reaction of an alkyl halide with an aromatic compound to form an alkyl substituted aromatic compound and hydrogen halide.

The Friedel-Crafts reaction is an electrophilic aromatic substitution reaction. The aluminum chloride is a strong Lewis acid that reacts with the alkyl halide to form an *alkyl cation–aluminum tetrahalide complex:*

$$R{-}X \quad + \quad AlCl_3 \quad \rightleftharpoons \quad R^+AlCl_3X^-$$

Alkyl halide	*Aluminum chloride*	*Alkyl cation–aluminum tetrahalide complex*

Subsequent reaction of the electrophilic alkyl cation–aluminum tetrachloride complex with an electron-rich benzene ring results in the formation of the alkylbenzene, hydrogen halide, and regeneration of the aluminum chloride catalyst:

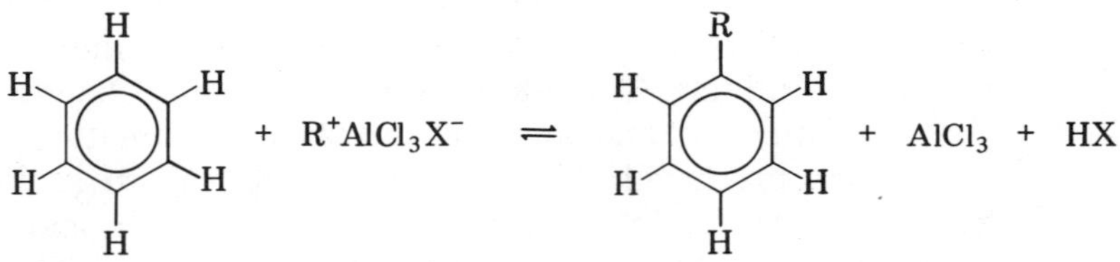

Benzene *Alkyl-benzene*

In this experiment you will prepare 1,3-dimethyl-5-*tert*-butylbenzene by the reaction of *m*-xylene and *tert*-butyl chloride in the presence of anhydrous aluminum chloride catalyst. Since aluminum chloride is not consumed in the

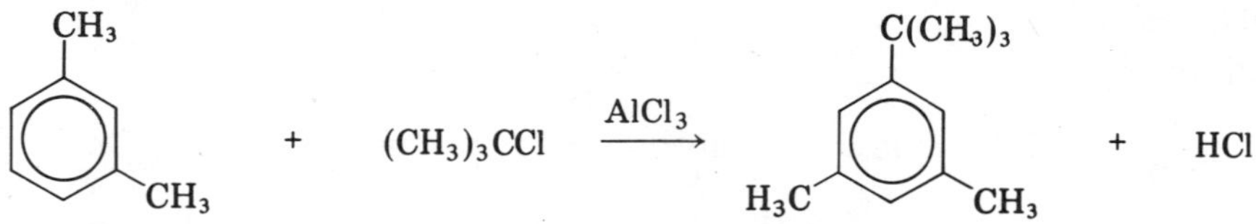

m-*Xylene*	tert-*Butyl chloride*	*1,3-Dimethyl-5-*tert-*butylbenzene*
M.w. = 106.2 g/mol	M.w. = 92.57 g/mol	M.w. = 162.3 g/mol
b.p. = 139°C	b.p. = 52°C	b.p. = 206°C
d_{20} = 0.86 g/ml	d_{20} = 0.85 g/ml	d_{20} = 0.86 g/ml

reaction, less than an equivalent amount of this catalyst is necessary. After the reaction is complete, the remaining catalyst is hydrolyzed by the addition of water. The resulting aluminum salts are soluble in aqueous acid and are separated from the water insoluble product. The product is purified by distillation.

Certain aromatic compounds undergo Friedel-Crafts alkylation reactions rapidly, whereas others react slowly or not at all under the same conditions. Substituents attached to the benzene ring alter the reactivity of the aromatic

Table 17.1 The Effect of Benzene Substituents in Aromatic Substitution Reactions

Activating groups: o,p-directors		*Deactivating groups: m-directors*	
—R	(alkyl, aryl)	—COR	(aldehydes, ketones)
—NRCOR	(amides)	$—CONR_2$	(amides)
$—NR_2$	(amines)	$—NR_3^+$	(ammonium salts)
—OR	(ethers)	$—CO_2H$	(carboxylic acids)
—OH	(phenols)	$—CO_2R$	(esters)
		—CN	(nitriles)
Weakly deactivating: o,p-directors		$—NO_2$	(nitro compounds)
—F, —Cl, —Br, —I	(arylhalides)	$—SO_3H$	(sulfonic acids)

ring towards alkylation. For example, ethylbenzene is more reactive than benzene, chlorobenzene is less reactive than benzene, and nitrobenzene is unreactive in Friedel-Crafts alkylation reactions. Those substituents that result in increased reactivity of the benzene ring relative to the reactivity of benzene itself are called *activating groups*, and those substituents that result in decreased activity are called *deactivating groups*. Table 17.1 contains a summary of the effect of common substituents on the reactivity of the benzene ring in electrophilic substitution reactions. Friedel-Crafts alkylation reactions do not take place with aromatic compounds possessing substituent groups that are more deactivating than the halogens.

Generally, monosubstituted benzenes with an activating substituent or a halogen substituent yield predominantly ortho and para disubstituted products in electrophilic substitution reactions. In contrast, the presence of a substituent that is more deactivating than halogen generally results in the predominant formation of the meta disubstituted product in electrophilic substitution reactions. (These strongly deactivated aromatic compounds do not react at all in Friedel-Crafts reactions.) Thus, the alkylation of toluene would be predicted to yield ortho and para disubstituted products.

In this experiment the reaction of *m*-xylene with *tert*-butyl chloride-aluminum chloride produces the symmetrical 1,3-dimethyl-5-*tert*-butylbenzene and not the unsymmetrical 1,2,4-substituted isomer that would be predicted on the basis of the *o,p*-directing influence of the two methyl groups on the *m*-xylene ring. This result is due to the steric crowding that results from the introduction of a bulky *tert*-butyl group into a position ortho to one of the methyl substituents. The 1,3,5-trialkylbenzene is thermodynamically more stable than either the 1,2,3- or the 1,2,4-trialkylbenzene isomers, and under the reaction conditions employed for this experiment the 1,2,3- and the 1,2,4-trisubstituted isomers rearrange to the thermodynamically more stable 1,3,5-trisubstituted isomer. In reactions where the relative yield of isomeric products is determined by the relative thermodynamic stability of the isomers and not by the relative rate of formation of the isomers, the product distribution is said to be under *thermodynamic control*.

In addition to the limitations of the Friedel-Crafts reaction due to the unreactivity of aromatic compounds substituted with strongly deactivating substituents and to rearrangement processes, the alkylation reaction presents additional experimental problems. Since Friedel-Crafts alkylation of an aromatic compound introduces an activating substituent, the product is usually more reactive than the starting material towards further alkylation. A large excess of the starting aromatic compound relative to alkyl halide is usually employed to reduce the amount of polyalkylated compounds that are formed. (In this experiment the product is not subject to further alkylation since all of the remaining unsubstituted positions are sterically crowded.) Finally, since the Friedel-Crafts alkylation reaction involves the formation of an intermediate alkyl cation–aluminum tetrahalide complex rearrangement of the alkyl groups occurs if a more stable alkyl cation can be formed. For example, alkylation of benzene with isobutyl chloride will yield *tert*-butylbenzene and not isobutylbenzene. The intermediate isobutyl cation rapidly rearranges to the more stable *tert*-butyl cation.

The structure of the product in this experiment, 1,3-dimethyl-5-*tert*-butylbenzene, may be confirmed by infrared spectroscopy (Experiment 12). The aromatic C—H bending frequencies in the 900 to 650 cm^{-1} region of the infrared spectrum are diagnostic of the substitution pattern of the trisubstituted benzene ring (see Figure 17.4a). As shown in Table 17.2, each trisubstituted benzene isomer shows two major absorption bands in this region of the spectrum. The relative intensity of the absorption bands will vary with different substituents.

Table 17.2 Aromatic C—H Bending Frequencies for Trisubstituted Benzene Rings

Substitution	Frequency range (cm^{-1})
1,2,3-trisubstituted	780–760 and 745–705
1,2,4-trisubstituted	885–870 and 825–805
1,3,5-trisubstituted	875–835 and 700–680

Friedel-Crafts Acylation Reactions

The reaction of an acid chloride (or anhydride) with an aromatic compound in the presence of a Lewis acid is an important method for the preparation of aryl ketones. The mechanism of this reaction is quite similar to the mechanism of the alkylation reaction. The Lewis acid, which is generally anhydrous aluminum chloride, reacts with the acid chloride to form an *acylium ion–aluminum tetrachloride complex:*

$$\underset{\textit{Acid chloride}}{R\overset{O}{\overset{\|}{C}}\text{—}Cl} + \underset{\textit{Aluminum chloride}}{AlCl_3} \longrightarrow \underset{\textit{Acylium ion–aluminum tetrachloride complex}}{R\overset{O}{\overset{\|}{C}}{}^{+}AlCl_4^-}$$

The electrophilic acylium ion-aluminum tetrachloride complex then reacts with an electron-rich aromatic ring, and this reaction results in the formation of the aryl ketone, hydrogen chloride, and aluminum chloride. Unlike the alkylation

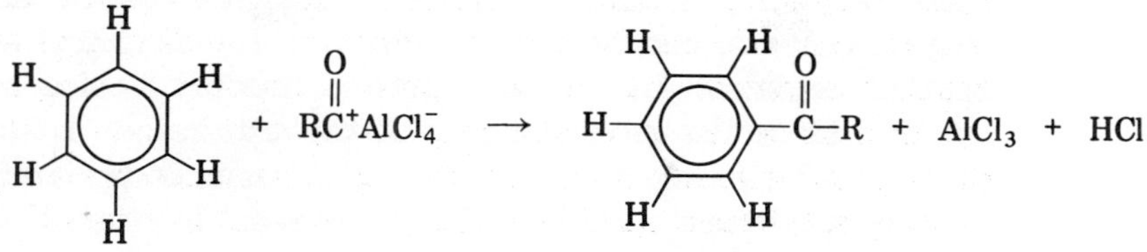

Aryl ketone

reactions where the Lewis acid is regenerated, the aluminum chloride forms a complex with the ketone product. Therefore, at least one equivalent of aluminum chloride is required for acylation reactions.

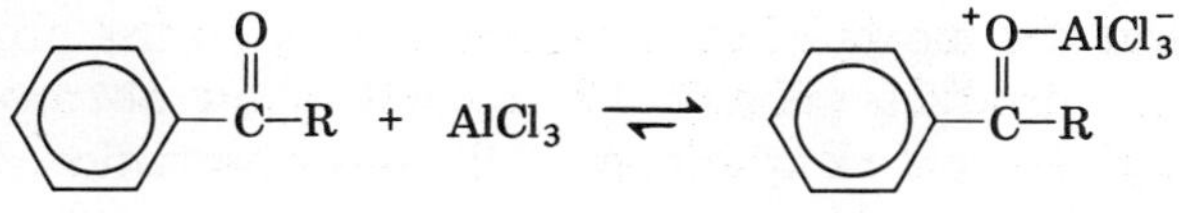

Ketone-aluminum chloride complex

In this experiment you will prepare 2,4-dimethylacetophenone by the reaction of *m*-xylene with acetyl chloride and aluminum chloride. A solvent, dichloromethane, is used in this reaction to ensure adequate mixing of the reactants.

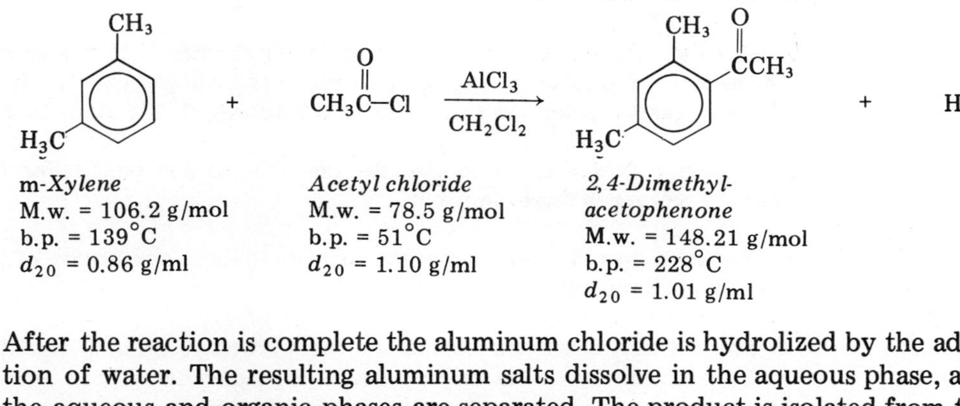

m-*Xylene*	*Acetyl chloride*	*2,4-Dimethyl-acetophenone*
M.w. = 106.2 g/mol	M.w. = 78.5 g/mol	M.w. = 148.21 g/mol
b.p. = 139°C	b.p. = 51°C	b.p. = 228°C
d_{20} = 0.86 g/ml	d_{20} = 1.10 g/ml	d_{20} = 1.01 g/ml

After the reaction is complete the aluminum chloride is hydrolized by the addition of water. The resulting aluminum salts dissolve in the aqueous phase, and the aqueous and organic phases are separated. The product is isolated from the organic phase by fractional distillation.

The product formed in the acylation of *m*-xylene is 2,4-dimethylacetophenone. Formation of this isomer and not the less sterically crowded 3,5-dimethylacetophenone is predicted on the basis of the *o,p*-directing effect of the methyl substituents. Rearrangement of the initially formed 2,4-dimethylacetophenone to the thermodynamically more stable, 3,5-substituted isomer is not observed under conditions of the Friedel-Crafts acylation reaction. The isomer formed at the fastest initial rate is the major product that can be isolated. In chemical reactions where the relative yield of isomeric products is determined by the relative rate of formation of the isomers, the product distribution is said to be under *kinetic control.*

Unlike the carbonium ion intermediates in Friedel-Crafts alkylation reactions, rearrangements of intermediate acylium ions do not occur in Friedel-Crafts acylation reactions. Straight chain carboxylic acid chlorides such as butyryl chloride can be used to synthesize *n*-alkyl aryl ketones in good yields. Further acylation of the aryl ketone product is also not an important side reaction since the aromatic ring of the aryl ketone is strongly deactivated by the acyl substituent. Monosubstituted benzene compounds with a strongly deactivating substituent (Table 17.1) do not react in either Friedel-Crafts alkylation or acylation reactions.

The structure of the product in this experiment, 2,4-dimethylacetophenone, can be confirmed by chemical characterization tests and by infrared spectroscopic analysis. The 2,4-dinitrophenylhydrazone test and the iodoform-formation test (Experiment 14) can be used to confirm the carbonyl group and the methyl ketone functionality in the product. The infrared spectrum of the product (Figure 17.4b) will show absorption bands characteristic of an aryl ketone (Experiment 12) and absorption bands characteristic of a 1,2,4-trisubstituted benzene ring (Table 17.2).

Prelab Questions

1. Compare the relative molar ratios of anhydrous aluminum chloride used in the Friedel-Crafts alkylation and acylation reactions. Explain why less aluminum chloride is necessary in the alkylation reaction.

2. Calculate the number of moles of hydrogen chloride that will be evolved during the Friedel-Crafts alkylation reactions. Assuming that hydrogen chloride is an ideal gas, what volume of gas is evolved under the reaction conditions (21°C, 1 atmosphere pressure)?

3. Prepare a flow diagram to describe the isolation and purification of 1,3-dimethyl-5-*tert*-butylbenzene in this experiment.

4. Prepare a flow diagram to describe the isolation and purification of 2,4-dimethylacetophenone in this experiment.

5. Friedel-Crafts alkylation of toluene with *tert*-butyl chloride will yield two different *tert*-butyltoluene isomers as major products depending on reaction conditions. If the reaction is carried out rapidly with a weak Lewis acid catalyst such as ferric chloride, the major product is *p-tert*-butyltoluene. If the same catalyst is used, but the reaction mixture is allowed to remain in contact with the catalyst for several hours, a mixture of 30% para and 70% meta *tert*-butyltoluene is formed. Explain these experimental results.

EXPERIMENTAL PROCEDURE:

SYNTHESIS OF 1,3-DIMETHYL-5-*tert*-BUTYLBENZENE

Set up the apparatus as illustrated in Figure 17.1 using a thoroughly dry 125-ml filter flask for the reaction flask. The filter flask is connected with rubber tubing to the stem of a funnel that is inverted over water contained in a beaker. The funnel is clamped in position so that the rim of the funnel is slightly (1 mm) below the surface of the water. Hydro-

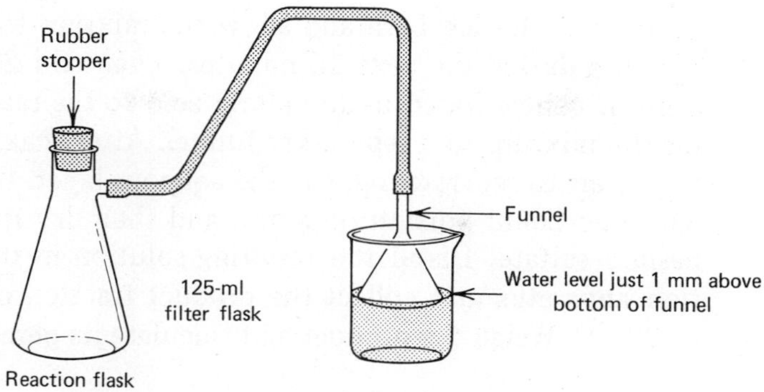

Figure 17.1 *Reaction apparatus for Friedel-Crafts reaction with hydrogen chloride trap.*

gen chloride gas evolved during the reaction is dissolved in the water. The large volume inside the inverted funnel prevents water from backing up into the reaction flask.

CAUTION: Do not allow aluminum chloride to contact your skin or clothes. Aluminum chloride reacts rapidly with water and with moisture in the air to form hydrogen chloride. Do not keep aluminum chloride in an open container.

Place 12.2 ml (10.6 g, 0.10 mol) of *m*-xylene and 12.5 ml (10.6 g, 0.11 mol) of *tert*-butyl chloride in the reaction flask. Stopper the flask and cool the flask in an ice bath. Weigh out approximately 1 g of anhydrous aluminum chloride, and quickly place the powder in a dry test tube. Stopper the test tube. (Also make certain that the top is tightly closed on the reagent bottle.) After the solution of *m*-xylene and *tert*-butyl chloride has cooled to a temperature below 5°C, remove the stopper from the filter flask and add about one-fifth of the aluminum chloride

---

CAUTION: Vigorous evolution of HCl gas may occur on addition of the aluminum chloride.

---

to the solution. Quickly replace the stopper in the flask and swirl the flask to mix the reagents. Keep the flask in the ice bath and add the remaining aluminum chloride in three equal portions during the next 15 minutes. Swirl the contents of the flask after each addition.

When the last of the aluminum chloride has been added and the vigorous reaction (evolution of gas) appears to have subsided, remove the flask from the ice bath and allow the mixture to warm to room temperature during the next 15 minutes. Then add 20 ml of ice water and 5 ml of concentrated hydrochloric acid to the reaction mixture. Transfer the mixture to a separatory funnel. After shaking and then allowing the layers to separate, remove the aqueous layer. Wash the organic layer with one 20-ml portion of water, and then dry it over anhydrous magnesium sulfate. Distill the resulting solution in the small scale distillation apparatus and collect the product fraction over the range of 195 to 206°C. Weigh the product and calculate its percentage yield.

EXPERIMENTAL PROCEDURE:
SYNTHESIS OF 2,4-DIMETHYLACETOPHENONE

Place 14 g of anhydrous aluminum chloride and 25 ml of dichloro-

---

CAUTION: Do not allow aluminum chloride to contact your skin or clothes. Aluminum chloride reacts rapidly with water and with moisture in the air to form hydrogen chloride. Keep aluminum chloride in a closed container.

---

methane in a 250-ml round-bottom flask equipped with an addition funnel, reflux condenser and hydrogen chloride trap as illustrated in Figure 17.2. Mix 12.2 ml (10.6 g, 0.10 mol) of *m*-xylene and 8.0 ml (8.8 g, 0.11 mol) of acetyl chloride in the dropping funnel. Add the mixture in the addition funnel dropwise with shaking or swirling to the aluminum chloride-dichloromethane mixture over a period of about 15 minutes. The reaction mixture will boil during this addition and evolve hydrogen chloride. After the addition is complete, heat the mixture at reflux on the steam bath for 45 minutes. The reaction mixture will be dark at this point. Pour the cooled reaction mixture into 30 g of ice in a 250-ml beaker. Add 20 ml of concentrated hydrochloric acid to the mixture and stir the mixture until all solids have dissolved. Pour this mixture into a separatory funnel and separate the organic layer. If necessary, additional dichloromethane (~15 ml) may be added to aid in differentiating the two layers. Wash the organic layer with 20 ml of an aqueous 5% sodium bicarbonate solution and then dry it over anhydrous magnesium sulfate. Separate dichloromethane from the higher boiling impurities and product by simple distillation using the steam bath as the heat source. Fractionally distill the resulting residue in the simple (side-arm) distillation apparatus and collect the product boiling over the range of 220 to 230°C. If excessive foaming occurs during distillation, a small wad of *loosely* rolled glass wool inserted into the neck of the distilling flask and positioned between the top of the flask

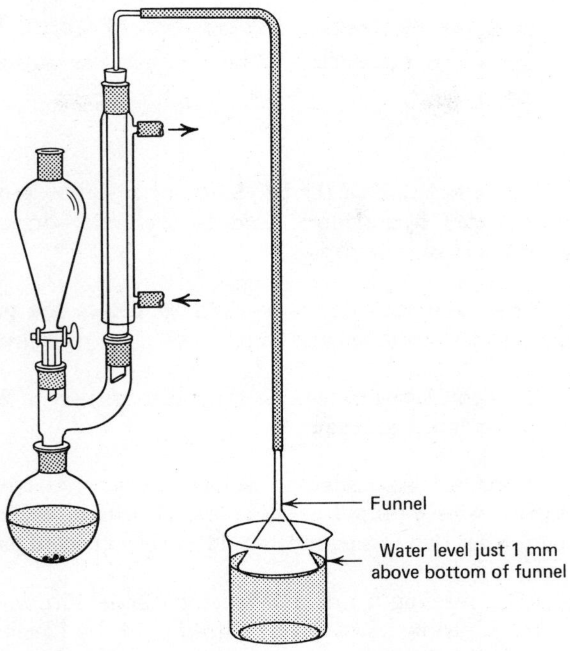

Figure 17.2 *Reaction apparatus for Friedel-Crafts reaction with hydrogen chloride trap.*

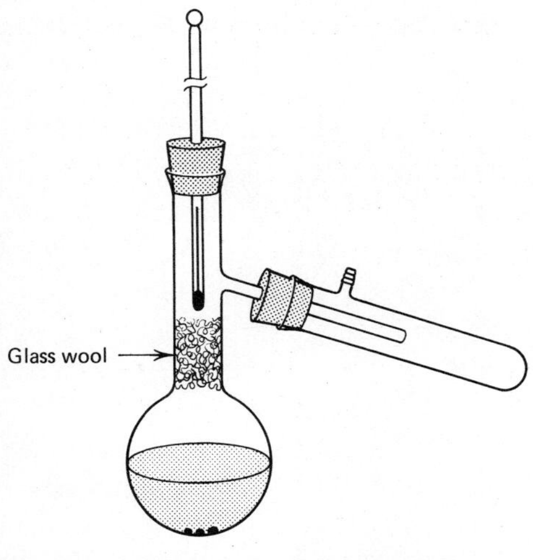

Figure 17.3 *Apparatus for fractional distillation with glass wool in neck to reduce foaming over into the collection tube.*

and the bottom of the side-arm (Figure 17.3) will prevent foaming over into the collection tube. Weigh the product and calculate its percentage yield.

Postlab Questions

1. In the isolation of the alkylation or acylation product, why was it necessary to add the concentrated hydrochloric acid to the water-organic mixture before separation of the aqueous and organic phases?

2. What technique could be used to detect the presence of small amounts of the 2,4-dimethyl-*tert*-butylbenzene (b.p. = 212°C) in your product from the alkylation reaction?

3. How could you determine the molar amount of HCl that was evolved during the alkylation or acylation procedure?

4. A student attempted to prepare *n*-propylbenzene by the Friedel-Crafts alkylation of benzene with *n*-propyl chloride and aluminum chloride and isolated two C_9H_{12} alkylbenzenes. What two compounds would be the expected products in this reaction?

5. After reacting 2,4-dimethylacetophenone with the iodoform reagent, a student isolated the solid, yellow iodoform by filtration of the aqueous solution. The student then acidified the aqueous filtrate and a new white solid crystallized from solution (m.p. = 127°C). What is the identity of the white solid?

6. Outline the sequence of reactions that would be necessary for the preparation of *p-tert*-butylacetophenone from benzene, *tert*-butyl chloride, acetyl chloride, and aluminum chloride.

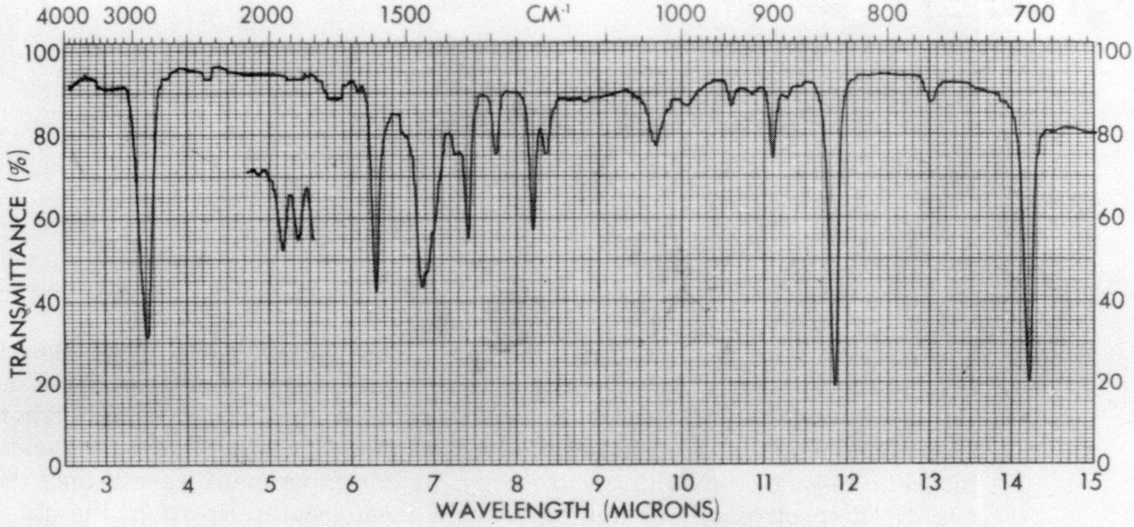

Figure 17.4a *Infrared spectrum of 1,3-dimethyl-5-tert-butylbenzene* (*thin film*). (*Note that the scale is linear in wavelength.*)

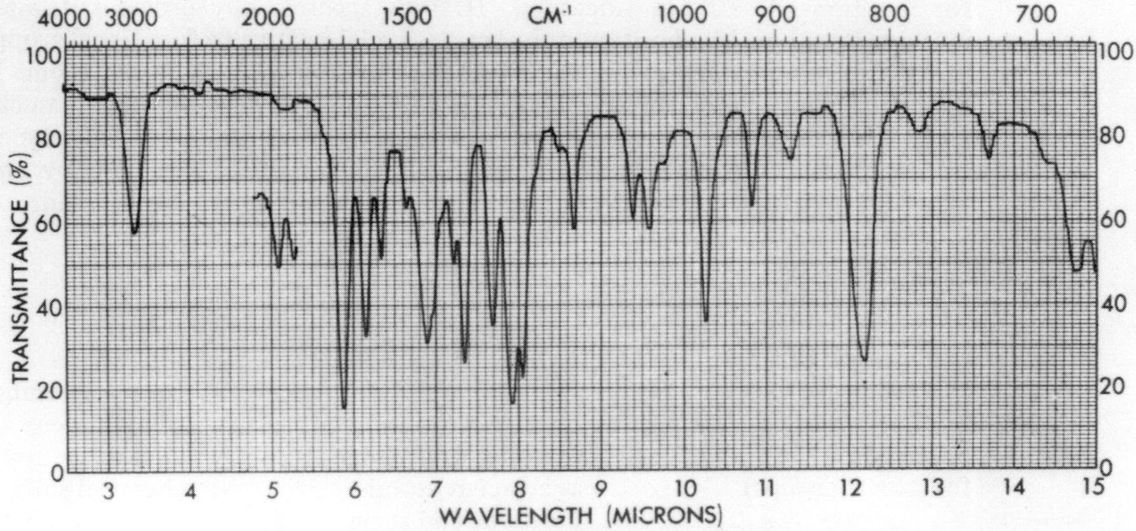

Figure 17.4b *Infrared spectrum of 2,4-dimethylacetophenone* (*thin film*). (*Note that the scale is linear in wavelength.*)

Experiment Eighteen

Nuclear Magnetic Resonance Spectroscopy

The interaction of electromagnetic radiation with certain nuclei in a magnetic field produces signals that characterize the chemical environment of these nuclei in molecules and furnishes the relative number of nuclei at each position. This phenomenon is the basis of the spectroscopic technique known as nuclear magnetic resonance (NMR) spectroscopy. In this experiment we will only discuss NMR spectroscopy of hydrogen nuclei in molecules. However, the use of NMR spectroscopy to probe other nuclei in molecules, particularly ^{13}C, is rapidly becoming accessible and adds new dimensions to the versatility of this spectroscopic technique for structural analyses.

Since its introduction as a structural probe in the late 1950s, proton nuclear magnetic resonance spectroscopy (^{1}H NMR spectroscopy) has proven to be an invaluable method for the structural identification of organic compounds. Unlike infrared spectroscopy (Experiment 12), which is primarily used for functional group identification, ^{1}H NMR spectroscopy provides detailed information about the position of hydrogen nuclei relative to functional groups in molecules (*chemical shift*), the number of hydrogens on adjacent atoms in molecules and their spacial relationship to the absorbing hydrogen nuclei (*spin-spin splitting*), and the relative number of hydrogen nuclei at each chemical shift position (*integration*). Nuclear magnetic resonance spectroscopy provides the structural details of hydrogen arrangements in organic compounds.

^{1}H NMR SPECTROSCOPY

- *Chemical Shift* (position of hydrogen nuclei relative to functional groups)
- *Spin-Spin Splitting* (number of hydrogens at adjacent atomic positions and their spacial relationship to the absorbing hydrogen nuclei)
- *Integration* (relative number of hydrogens at each chemical shift position)

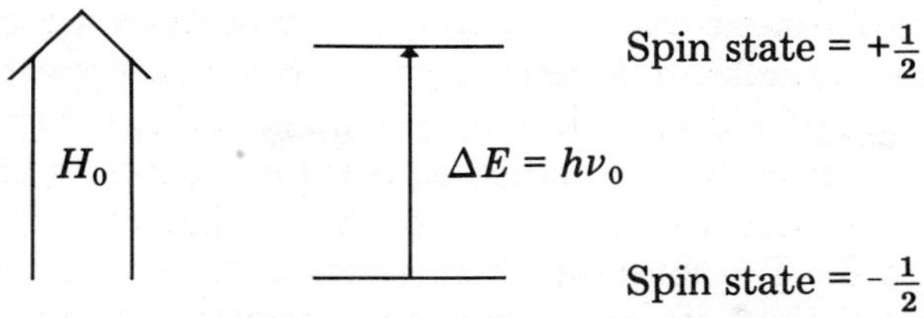

Figure 18.1 *Energy difference between spin states for the hydrogen nucleus.*

Nuclear Magnetic Resonance Spectroscopy

The hydrogen nucleus possesses a nuclear spin of $\frac{1}{2}$. In a uniform magnetic field (H_0) hydrogen nuclei are distributed between two distinguishable spin states, $+\frac{1}{2}$ and $-\frac{1}{2}$ (Figure 18.1). The small energy difference $\Delta E = h\nu_0$ separates these two spin states so that at ordinary temperatures in magnetic fields of 10,000 Gauss (G) the population difference between the two spin states is only a few parts per million. Interaction of a hydrogen nucleus in the lower energy spin state with electromagnetic radiation, whose frequency ν is equal to the energy difference between the two spin states ($\nu = \nu_0$), causes energy absorption by the nucleus with a concurrent transition to the higher energy state. When absorption of energy occurs, the transitional energy of the nucleus (ΔE) and the applied energy are said to be "in resonance."

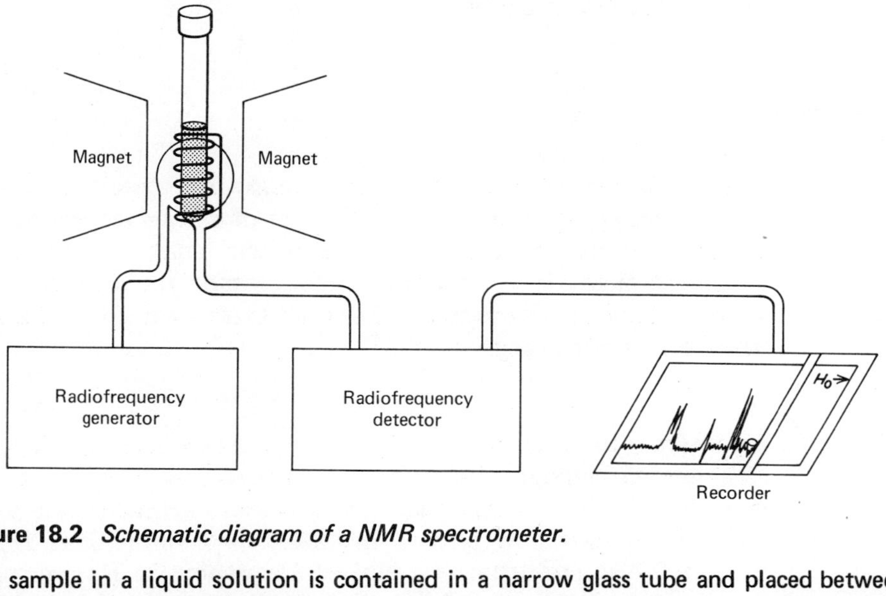

Figure 18.2 *Schematic diagram of a NMR spectrometer.*

The sample in a liquid solution is contained in a narrow glass tube and placed between the poles of the magnet. The radiofrequency generator transmits energy to the sample from a coil that surrounds the sample tube. The resonance signal is detected by the radiofrequency detector and is recorded on a chart paper.

The instrument employed to detect resonance absorptions from the nuclei of a sample is called a *nuclear magnetic resonance spectrometer.* The spectrometer consists of a magnet assembly that is capable of producing a strong homogeneous magnetic field, a radiofrequency generator to induce the nuclear transition, and a radiofrequency detector to measure the resonance absorption (Figure 18.2). The magnetic field strength is proportional to the frequency of radiation required for resonance. In a magnetic field of 14,092 G, for example, resonance absorptions by hydrogen nuclei can be achieved with radiofrequencies of 60 MHz (megahertz, 60×10^6 Hz). Proportionally greater or lesser frequencies are required with magnets whose field strengths are greater or less than 14,092 G. Nuclear magnetic resonance spectrometers with radiofrequencies of 30, 60, 80, 90, 100, and even greater than 300 MHz are now available. However, the vast majority of current information from ^1H NMR spectrometric analyses of organic compounds has been obtained with 60 MHz instruments.

NMR spectral determinations are made from samples in liquid solutions in order to obtain sharp signals. Rapid molecular motions of molecules in nonviscous solutions places individual nuclei within molecules in the same magnetic environment during the lifetime of the NMR observation (10^{-3} sec). The absorptions of nuclei in solids, in which random movement of molecules is negligible, are usually very broad and, consequently, difficult or impossible to detect. In viscous liquids random molecular motions are also restricted, and broadened signals result from the nonidentical environments of individual nuclei in different molecules. To obtain high resolution proton spectra, solid and liquid samples are usually dissolved in solvents that do not themselves produce resonance absorptions; carbon tetrachloride and deuteriochloroform ($CDCl_3$) are most commonly employed.

Chemical Shift

The usefulness of ^1H NMR spectroscopy for structure determinations results from the dependence of the frequency of resonance absorption by a hydrogen nucleus on the magnetic environment that surrounds the nucleus. In a strong magnetic field (H_0), electrons in the immediate vicinity of the absorbing nucleus induce small magnetic fields of their own (H_1) that screen the nucleus from the applied magnetic field:

$$H_{\text{observed}} = H_0 - H_1$$

Therefore, the magnetic field at the absorbing nucleus (H_{observed}) differs from the applied magnetic field (H_0), and the degree to which the electron-induced magnetic field (H_1) influences the observed magnetic field determines the position of the resonance signal in the NMR spectrum (Figure 18.3).

An NMR spectrum is a plot of frequency (in Hz) versus intensity relative to a standard proton reference absorption. The proton absorption for tetramethylsilane (TMS) is universally accepted as the reference standard for NMR spectra of organic compounds. Proton absorptions for organic compounds are

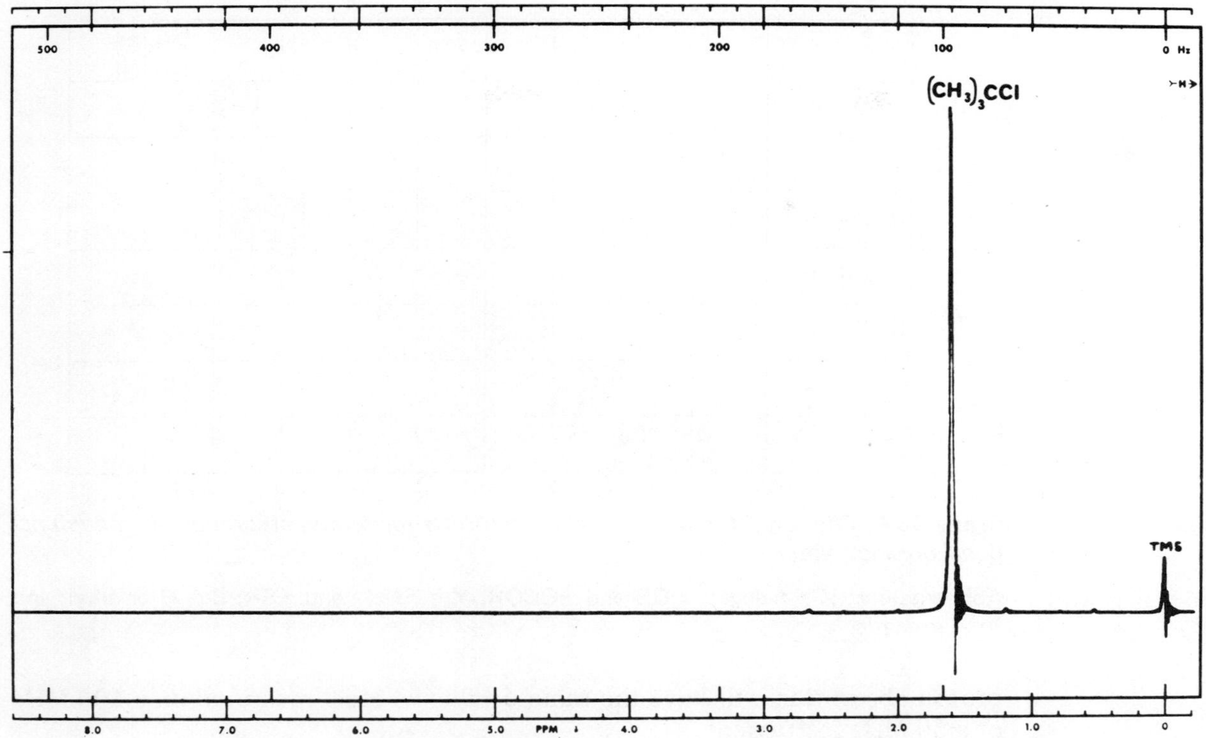

Figure 18.3 *The NMR spectrum of tert-butyl chloride with tetramethylsilane (TMS) as the reference standard using a 60 MHz NMR spectrometer.*

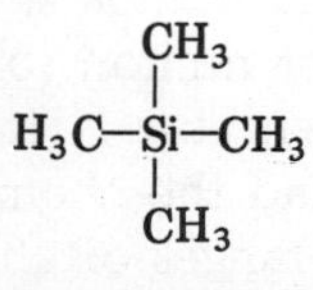

Tetramethylsilane (TMS) (reference standard for NMR spectra of organic compounds)

recorded relative to the TMS absorption. The difference between the absorption positions of the hydrogen nuclei of the sample and the reference standard is known as the *chemical shift.*

Chemical shift is reported in dimensionless units δ (delta) that are obtained by dividing the frequency of absorption in Hz relative to TMS ($\times 10^6$) by the frequency of the spectrometer.* Thus for a 60 MHz spectometer

$$\delta = \frac{\text{chemical shift from TMS in Hz} \times 10^6}{60 \times 10^6 \text{ Hz}}$$

The resonance frequency of TMS is set at 0 Hz which corresponds to 0.00 δ. Delta (δ) units are independent of the frequency of the spectrometer; a sample

*Until recently, an alternate unit, τ (tau), was routinely employed in an alternative NMR chemical shift scale. The tau scale has TMS set at 10.00τ and is related to the δ scale by $\tau = 10.00 - \delta$:

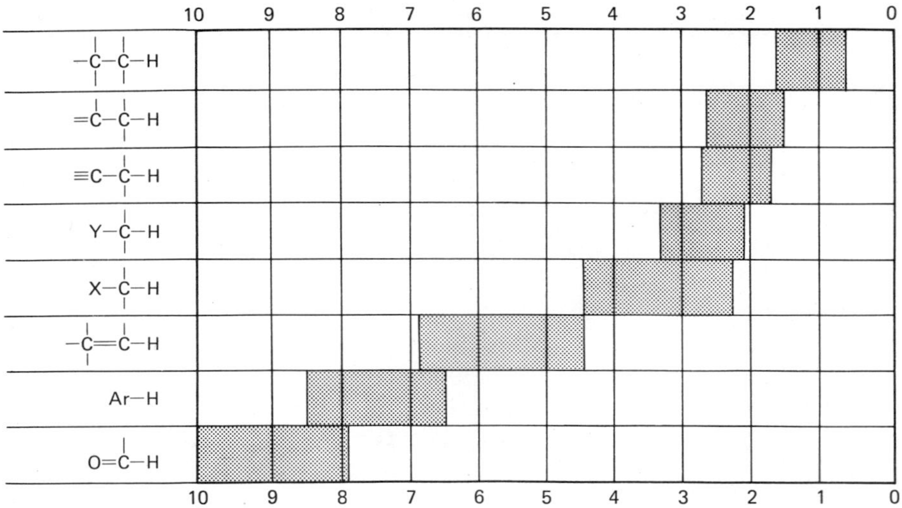

Figure 18.4 *Characteristic ranges of chemical shifts for proton absorptions in monofunctional organic compounds.*

Abbreviations: X = halogen, —OR and —OCOR; Y = —NR_2 and —SR; R = H or alkyl; and Ar = aromatic.

proton absorption will have the same δ-value whether measured on a 100 MHz or a 60 MHz instrument.

The spread of resonance frequencies for proton absorptions in organic compounds is relatively small. At a field strength of 14,092 G, for which the resonance frequency of the hydrogen nucleus is 60×10^6 Hz, the range of proton frequencies in organic compounds is only 600 Hz. Although this range is equivalent to only 10 parts per million (or 10 δ units), proton absorptions can be accurately determined to within ±1 Hz and correlations between the structures of organic compounds and the chemical shifts of proton absorptions can be determined. Characteristic ranges of chemical shifts for proton absorptions in selected monofunctional organic compounds are described in Figure 18.4 and Appendix C.

Spin-Spin Splitting

The magnetic field experienced by an absorbing hydrogen nucleus is influenced by the presence of other hydrogen nuclei. The interaction of the absorbing nuclear spin with the spin states of hydrogen nuclei on adjacent atoms produces a multiplicity of absorption signals that is dependent on the number of interacting spin states and on the effectiveness of the spin-spin interaction. Spin-spin splitting of proton resonance absorptions adds fine structure to the NMR spectrum.

The NMR spectrum of ethyl ether (Figure 18.5) exemplifies the spectral result of spin-spin splitting of the proton absorptions. Two sets of resonance

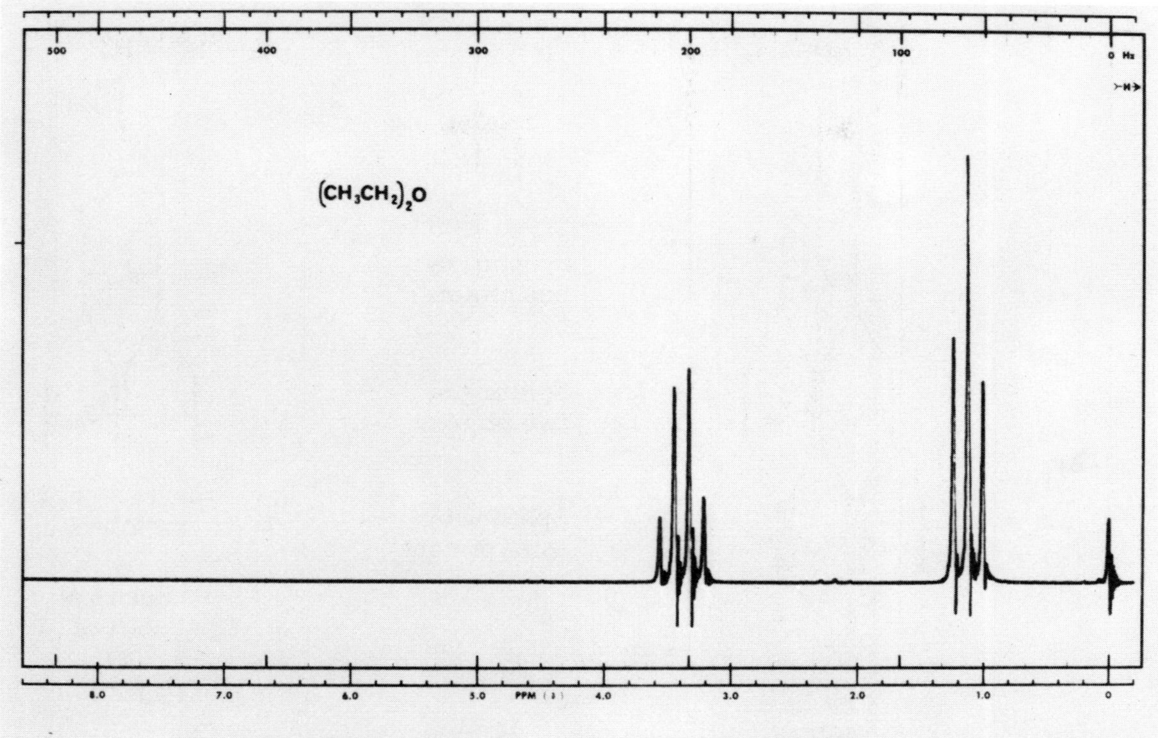

Figure 18.5 *NMR spectrum of ethyl ether in carbon tetrachloride. (Note that carbon tetrachloride does not possess a hydrogen nucleus and, therefore, does not produce an NMR signal.)*

absorptions are observed. The absorption centered at 3.4 δ is in the chemical shift range for absorptions by hydrogen nuclei on carbon adjacent to oxygen (Figure 18.4). The absorption centered at 1.2 δ is in the chemical shift range for absorptions by hydrogen nuclei on carbon adjacent to an sp^3-hybridized carbon. Thus, the chemical shift information describes the proton arrangement in ethyl ether relative to the oxygen functional group:

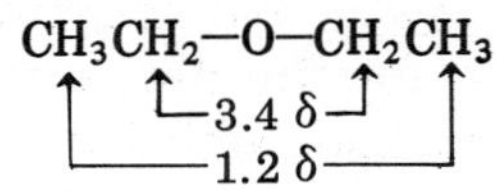

However, the set of absorptions centered at 3.4 δ consists of four peaks in an intensity ratio of 1 : 3 : 3 : 1, whereas the set of absorptions centered at 1.2 δ consists of three peaks in an intensity ratio of 1 : 2 : 1. The four-peak absorption is referred to as a *quartet* and the three-peak absorption is a *triplet*. The triplet absorption results from the interaction of the absorbing methyl protons with the two adjacent hydrogen nuclei of the methylene group (Figure 18.6). Each hydrogen of the methylene group splits the methyl proton absorption. The effectiveness of this spin-spin interaction is the frequency difference between the lines of the triplet. Similarly, the quartet absorption results from the interaction of the absorbing methylene protons with the three adjacent hydrogen

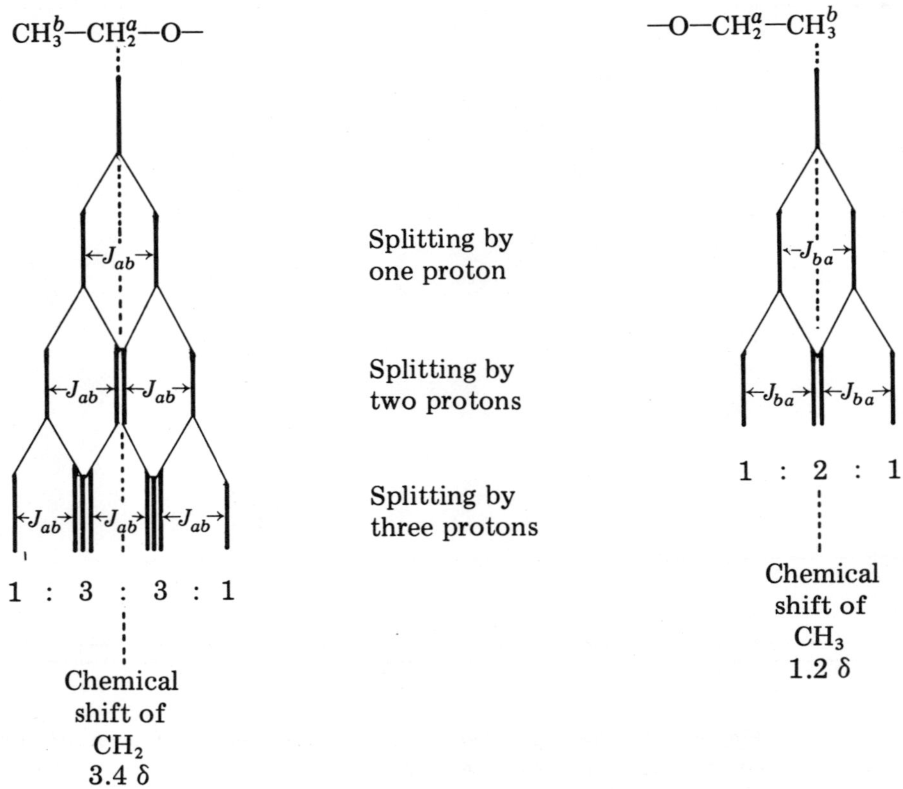

Figure 18.6 *Spin-spin splitting of methyl and methylene hydrogen nuclei in ethyl ether.*

nuclei of the methyl group. Each hydrogen of the methyl group splits the methylene proton absorption. The effectiveness of this spin-spin interaction is the frequency difference between the lines of the quartet.

The frequency difference (in Hz) between adjacent spectral lines caused by nuclear coupling is defined as the *coupling constant*, *J*. For example, the coupling constant for the methyl triplet in Figure 18.6 is J_{ba} and that for the methyl quartet is J_{ab}. For coupled nuclei, $J_{ab} = J_{ba}$, and this identity is a necessary condition for the determination of coupled nuclei in complex spectra. The coupling constant, which is a measure of the effectiveness of the spin-spin interaction, is independent of the strength of the applied magnetic field. For the coupled methyl and methylene splittings of ethyl ether, $J_{ab} = J_{ba} = 7$ Hz.

Spin-spin splitting between nuclei occurs through covalent bonds. For aliphatic compounds such as ethyl ether, the value of the coupling constant falls off rapidly with distance through bonds. Coupling is generally observed through only three bonds and the value of the coupling constant in acyclic aliphatic systems is generally 7 Hz.

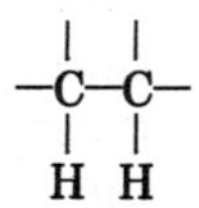

In aliphatic compounds, coupling is observed through only three bonds.

Table 18.1 Coupling Constants Between Hydrogen Nuclei in Representative Organic Systems

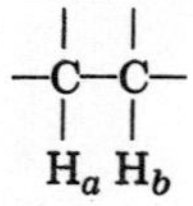	J_{ab} = 7 Hz	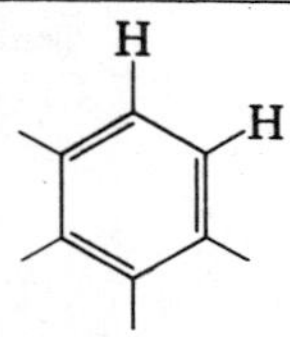	J_{ortho} = 6–9 Hz
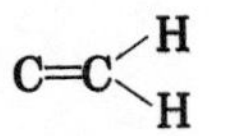	J_{gem} = 0–5 Hz		
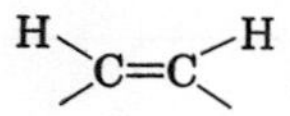	J_{cis} = 3–12 Hz	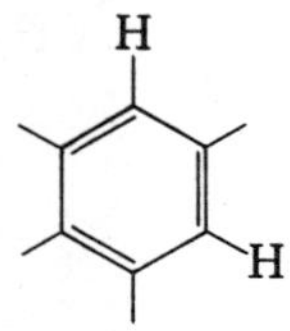	J_{meta} = 1–3 Hz
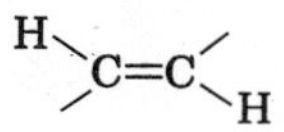	J_{trans} = 11–18 Hz		
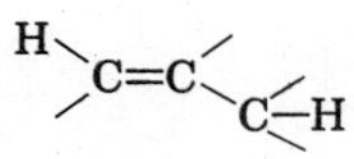	$J_{allylic}$ = 1–2 Hz	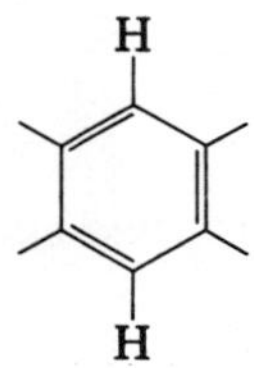	J_{para} < 1 Hz
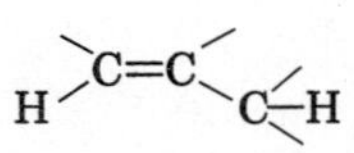	$J_{allylic}$ = 1 Hz		

In unsaturated organic compounds coupling through as many as five bonds can be observed, although the couplng constant decreases rapidly with distance. Coupling constants between hydrogen nuclei in representative organic systems are presented in Table 18.1.

Magnetic Equivalence

The ethyl ether spectral example illustrates an essential rule required for the interpretation of NMR spectra: *when two or more hydrogens are in identical magnetic environments, these hydrogens are magnetically equivalent and resonate at the same frequency* (*same chemical shift*). For example, all nine hydrogens of *tert*-butyl chloride are magnetically equivalent and, consequently, produce a single NMR signal (Figure 18.3). Similarly, ethyl ether possesses two equivalent methyl groups and two equivalent methylene groups. Each of the hydrogens of one methyl group is magnetically equivalent to the remaining methyl-group hydrogens, and the hydrogens of one methyl group are magnetically equivalent to the hydrogens on the opposite methyl group. Magnetic equivalence is also evident for the methylene groups of ethyl ether. Since ethyl ether possesses only two nonequivalent kinds of hydrogens, there are only two sets of absorptions (Figure 18.7). Also, since the two methyl groups are equiva-

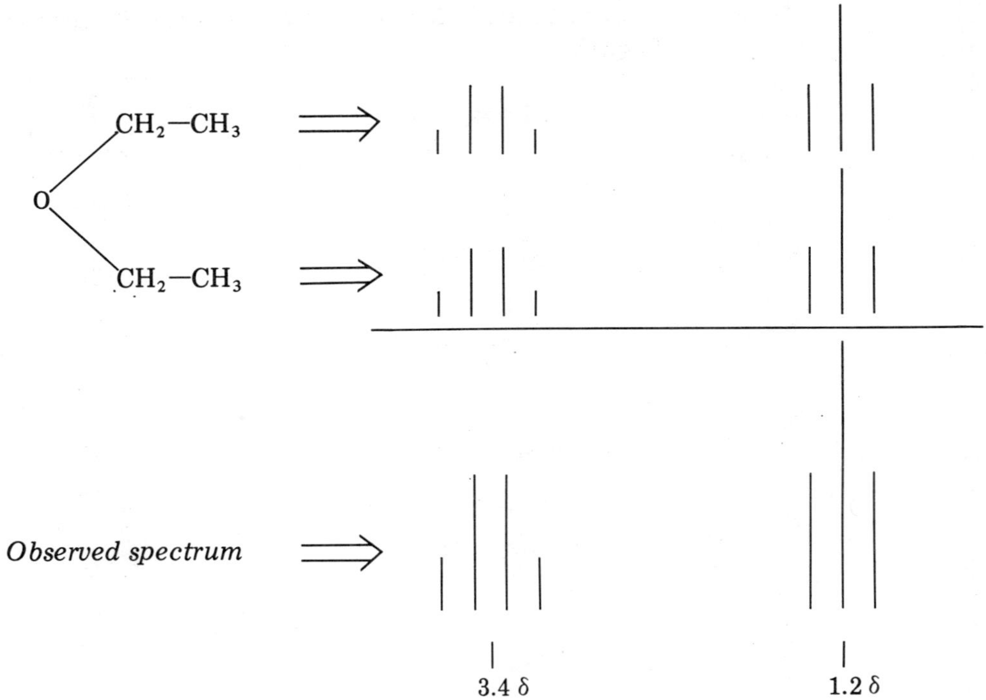

Figure 18.7 *Magnetic equivalence in the NMR spectrum of ethyl ether. The observed spectrum is the sum of the resonance absorptions for each set of magnetically equivalent hydrogens.*

lent and each methyl group is adjacent to a methylene group, the absorption of each methyl group is split by only two adjacent hydrogens. Similarly, the absorption of each methylene group is split by the three adjacent hydrogens of the methyl group.

Coupling is observed only when the interacting nuclei are magnetically nonequivalent. In other words, coupling is observed between hydrogen nuclei that have different chemical shifts. When adjacent hydrogen nuclei are magnetically equivalent, a singlet absorption is observed; for example,

$BrCH_2CH_2Br$

1,2-Dibromoethane
singlet, 3.5 δ

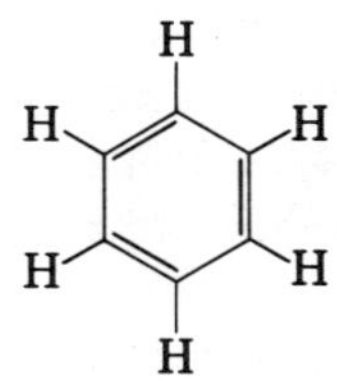

Benzene
singlet, 7.27 δ

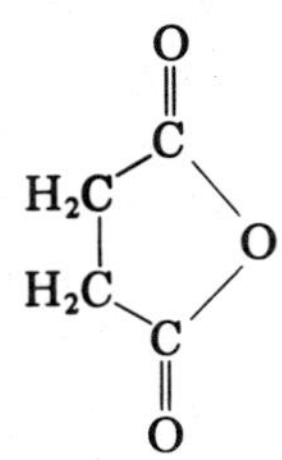

Succinic anhydride
singlet, 3.0 δ

In contrast, the NMR spectrum of limonene (Experiments 5 and 6) shows a

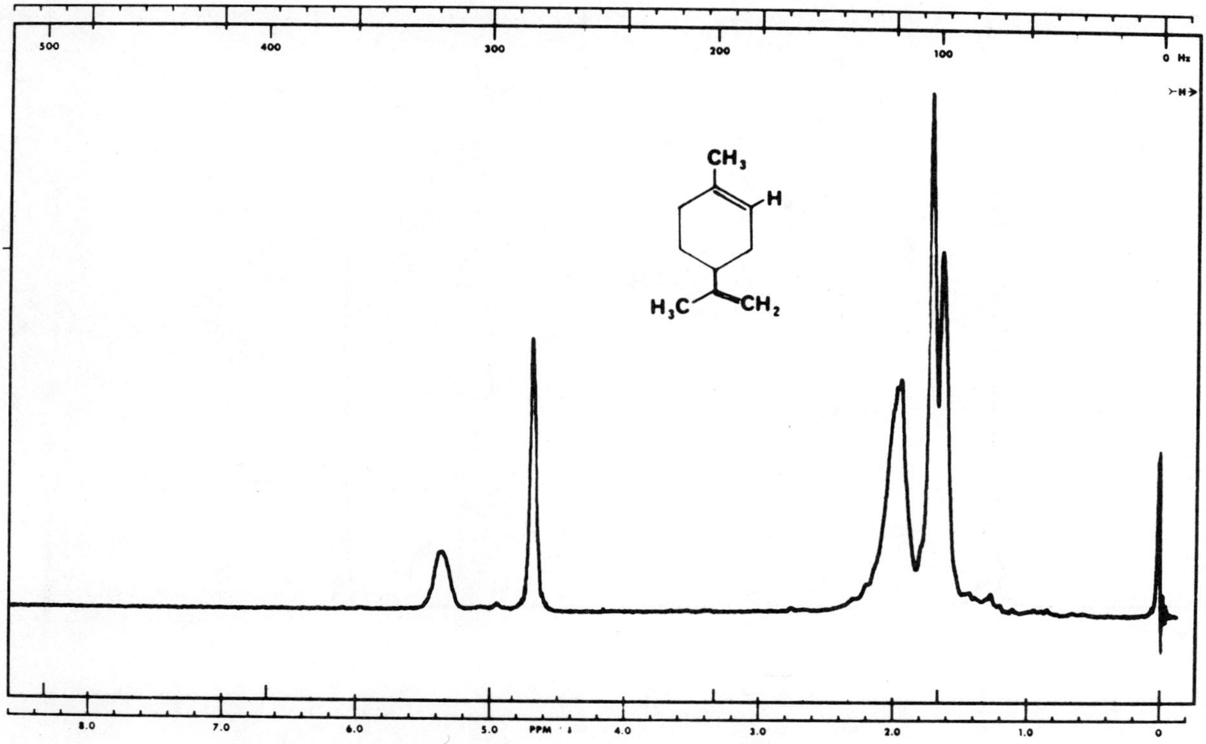

Figure 18.8 *NMR spectrum of limonene.*

complexity of absorptions (Figure 18.8) because of the magnetic nonequivalence of the constitutent hydrogens.

Hydrogen nuclei that are not chemically equivalent may be magnetically equivalent. For example, the aromatic ring hydrogens of toluene are not chemically equivalent. However, the chemical shift difference between ortho, meta, and para hydrogens in toluene is negligibly small and, consequently, these hydrogens do not exhibit spin-spin splitting at 60 MHz. The aromatic ring hydrogens are observed as a singlet in the NMR spectrum of toluene (Figure 18.9).

When the chemical shift difference between coupled hydrogens is less than 2 δ, the multiplet absorptions are perturbed. The inner lines of each multiplet increase in intensity and the outer lines decrease in intensity as the chemical shift difference decreases to 0 δ. The NMR spectra of *p*-nitrobenzoic acid, ethyl *p*-nitrobenzoate, and benzocaine that are given in Experiment 30 exemplify this perturbation.

Integration

The intensity of a NMR absorption is proportional to the area under the absorption curve and to the number of nuclei that produce the signal. Electronic integrators that are available with all commercial spectrometers integrate the

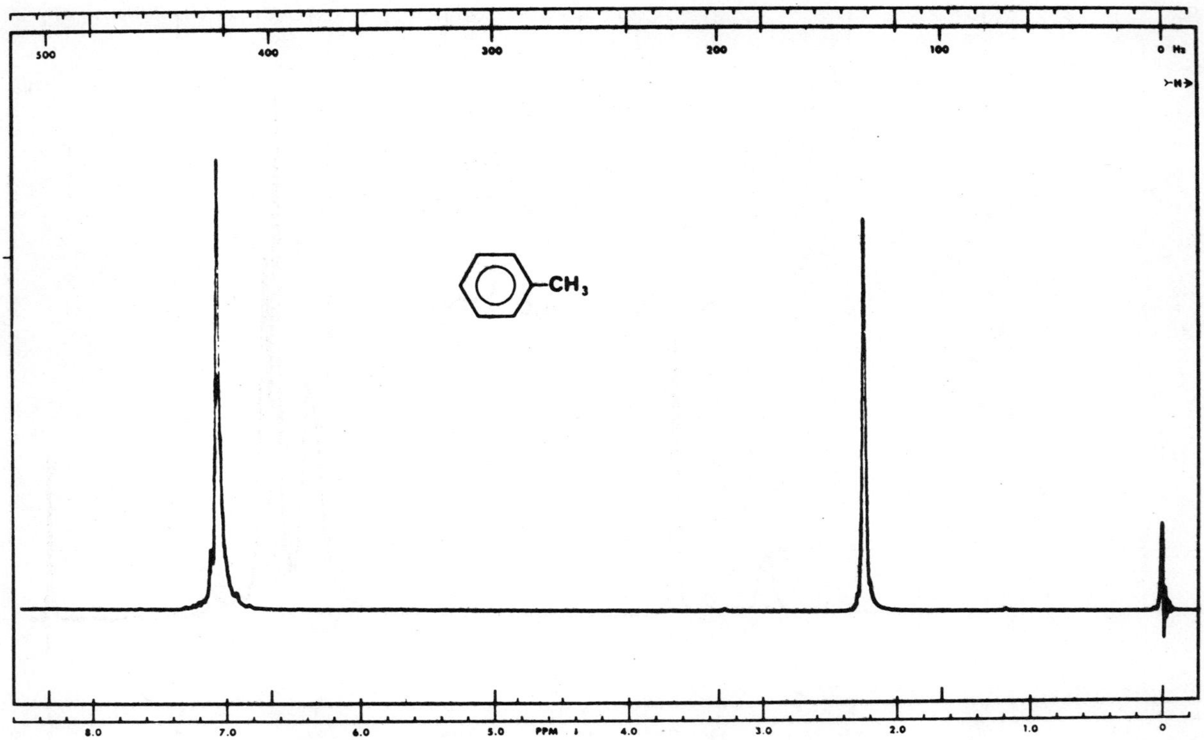

Figure 18.9 *NMR spectrum of toluene.*

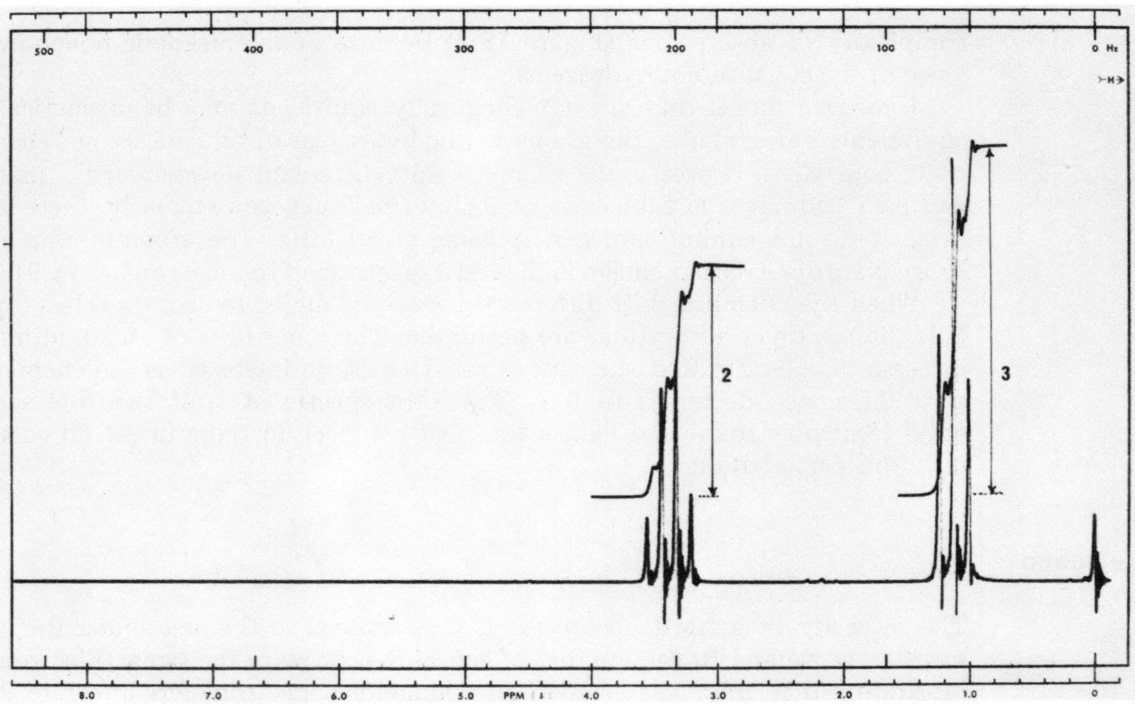

Figure 18.10 *Integrated NMR spectrum of ethyl ether.*

absorption signals and provide a direct measure of the relative number of nuclei for each absorption. The integral is recorded over the absorption spectrum, as shown in Figure 18.10, in an operation that is separate from that used to obtain the NMR spectrum. Comparison of the relative heights of the integrals for each set of absorptions provides the relative number of hydrogen nuclei that produce each absorption. The actual number of hydrogen nuclei is an integral multiple of the relative number obtained by integration. Thus, integration of the NMR spectrum for ethyl ether (Figure 18.10) shows that the ratio of the relative number of hydrogen nuclei which absorb at 3.4 δ and 1.2 δ is 2 : 3; the actual number of hydrogens is a multiple of 2 times the relative ratio.

The entire set of absorptions from a multiplet is counted in order to obtain the relative number of hydrogens. Separate integration of a multiplet measures the relative intensities of the multiplet absorptions. For example, in the spectrum for ethyl ether the absorption centered at 3.4 δ is recorded as a quartet in the predictable integral ratio of 1 : 3 : 3 : 1. Similarly, the absorption centered at 1.2 δ is a triplet whose individual absorptions are in an integral ratio of 1 : 2 : 1. The absorptions at 3.4 and 1.2 δ are in a ratio of 2 : 3, and it is this ratio that is a measure of the relative number of hydrogens in the sample.

NMR Spectroscopy in Structural Determinations

The wealth of information available from the NMR spectrum of a structurally unknown compound is invaluable for structural determinations. As you begin to use NMR spectroscopy for the identification of organic compounds you should be prepared to separately identify individual sets of absorptions and record (a) the chemical shift of the absorption (see Appendix C), (b) the multiplicity of the spin-spin splitting, and (c) the relative integration of the absorption. The unified data is then interpreted with the aid of other structural information obtained for the unknown compound.

Functional Derivatives of Alkanes

Table 18.2 lists the chemical shifts and spin-spin multiplicities for absorptions from selected organic compounds. The chemical shifts for these compounds are consistent with those predicted in Figure 18.4 and indicate that the chemical shift of proton absorptions is dependent on the electron density of the C—H bond. Bonding to electronegative elements causes deshielding of proton absorptions. In addition, an increase in the number of alkyl groups bonded to the carbon to which the absorbing hydrogen nucleus is attached causes greater deshielding by approximately 0.2 to 0.5 δ unit per alkyl group.

The spin-spin multiplicity for absorptions from aliphatic organic compounds describes the number of hydrogens on adjacent carbon atoms; for example,

singlet $\Longrightarrow$ $-\overset{|}{\underset{|}{C}}-C-H$

Table 18.2 Chemical Shifts and Spin-Spin Multiplicities[a] for Selected Organic Compounds, RX (R = Methyl, Ethyl, *n*-Propyl, Isopropyl, and *tert*-Butyl)

	R = CH_3	R = CH_2CH_3		R = $CH_2CH_2CH_3$ [b]			R = $CH(CH_3)_2$		R = $C(CH_3)_3$
X =	CH_3	CH_2	CH_3	αCH_2	βCH_2	CH_3	CH	CH_3	CH_3
—H	0.23(s)	0.86(s)	0.86(s)	0.91(t)	1.33(hep)	0.91(t)	1.33(hep)	0.91(t)	0.89(d)
—CH=CH_2	1.71(d)	2.00(qui)	1.00(t)	—	—	—	1.73(oct)	—	1.02(s)
—C≡N	1.98(s)	2.35(q)	1.31(t)	2.29(t)	1.71(sex)	1.11(t)	2.67(hep)	1.35(d)	1.37(s)
—$COOCH_3$	2.01(s)	2.28(q)	1.12(t)	2.22(t)	1.65(sex)	0.98(t)	2.48(hep)	1.15(d)	1.16(s)
—COOH	2.08(s)	2.36(q)	1.16(t)	2.31(t)	1.68(sex)	1.00(t)	2.56(hep)	1.21(d)	1.23(s)
—$COCH_3$	2.09(s)	2.47(q)	1.05(t)	2.32(t)	1.56(sex)	0.93(t)	2.54(hep)	1.08(d)	1.12(s)
—I	2.16(s)	3.16(q)	1.88(t)	3.16(t)	1.88(sex)	1.03(t)	4.24(hep)	1.89(d)	
—C_6H_5	2.35(s)	2.63(q)	1.21(t)	2.59(t)	1.65(sex)	0.95(t)	2.89(hep)	1.25(d)	1.32(s)
—NH_2	2.47(s)	2.74(q)	1.10(t)	2.61(t)	1.43(sex)	0.93(t)	3.07(hep)	1.03(d)	1.15(s)
—Br	2.69(s)	3.37(q)	1.66(t)	3.35(t)	1.89(sex)	1.06(t)	4.21(hep)	1.73(d)	1.76(s)
—Cl	3.06(s)	3.47(q)	1.33(t)	3.47(t)	1.81(sex)	1.06(t)	4.14(hep)	1.55(d)	1.60(s)
—O—	3.24(s)	3.37(q)	1.15(t)	3.27(t)	1.55(sex)	0.93(t)	3.55(hep)	1.08(d)	1.24(s)
—OH	3.39(s)	3.59(q)	1.18(t)	3.49(t)	1.53(sex)	0.93(t)	3.94(hep)	1.16(d)	1.22(s)
—$OCOCH_3$	3.67(s)	4.05(q)	1.21(t)	3.98(t)	1.56(sex)	0.97(t)	4.94(hep)	1.22(d)	1.45(s)

[a] s = singlet, d = doublet, t = triplet, q = quartet, qui = quintet, sex = sextet, hep = heptet, oct = octet.
[b] Coupling constants for splittings between hydrogens on adjacent carbons are nearly identical.

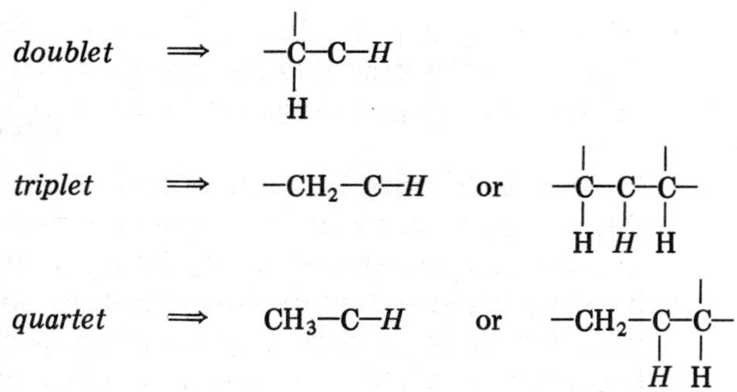

The number of lines in a multiplet of lines is $n + 1$ for n magnetically equivalent nuclei. The intensity ratios for these multiplets are

singlet	1
doublet	1:1
triplet	1:2:1
quartet	1:3:3:1
quintet	1:4:6:4:1
sextet	1:5:10:10:5:1
heptet	1:6:15:20:15:6:1

The outside lines of a sextet and a heptet are often missed in the interpretation of NMR spectra because of their low intensities relative to the center lines.

Functional Derivatives of Aromatic Compounds

Substituents on benzene affect the chemical shift of the ring hydrogens. The magnitudes of the substituent effects on the chemical shifts of ortho, meta, and para hydrogens are described by the data in Table 18.3. When the chemical shift differences for the ring hydrogens are less than or equal to 0.1 δ, as is the case for toluene (Figure 18.9), the resonance absorptions for the aromatic hydrogens appear as a broadened singlet. With greater chemical shift differences, resonance absorptions for the individual ring hydrogens are separated and appear with the spin-spin splitting patterns from coupling between the magneti-

Table 18.3 Chemical Shifts of Aromatic Hydrogens for Selected Substituted Benzenes

	Chemical shift of ring hydrogens (δ)		
Substituent	ortho	meta	para
$-NH_2$	6.5	7.0	6.6
$-OH$	6.8	7.1	6.9
$-Br$	7.0	7.4	7.3
$-Cl$	7.2	7.3	7.3
$-CH_3$	7.1	7.2	7.1
$-CH_2X$	7.3	7.3	7.3
$-CHO$	7.8	7.5	7.6
$-COCH_3$	7.9	7.4	7.6
$-COOH$	8.1	7.4	7.5
$-NO_2$	8.2	7.4	7.6

cally nonequivalent ring hydrogens. The NMR spectrum of acetophenone (Figure 18.11) describes the complex but characteristic absorption pattern for the aromatic ring hydrogens of carbonyl-substituted benzene derivatives.

Hydrogens Bonded to Heteroatoms

The chemical shifts of hydrogens bonded to oxygen and nitrogen (Table 18.4) are strongly influenced by hydrogen bonding. Since the degree of hydrogen bonding is dependent on concentration, solvent, and temperature, the chemical shift for these protons is particularly sensitive to concentration, solvent, and temperature. Unlike hydrogens bonded to carbon, hydrogens bonded to oxygen and nitrogen have variable chemical shifts. The —OH and —NH_2 absorptions in the NMR spectra reported in Experiments 20 and 30 exemplify this variability for chemical shifts of hydrogens bonded to oxygen and to nitrogen.

Hydrogens bonded to nitrogen and oxygen undergo *chemical exchange*—the exchange of protons between molecules; for example,

$$\mathrm{ROH}^a + R\mathrm{OH}^b \rightleftharpoons \mathrm{ROH}^b + R\mathrm{OH}^a$$

These protons become scrambled between molecules and, in a given time period, can exist in the environment of more than one molecule.

Chemical exchange of hydrogens bonded to oxygen or nitrogen (except in

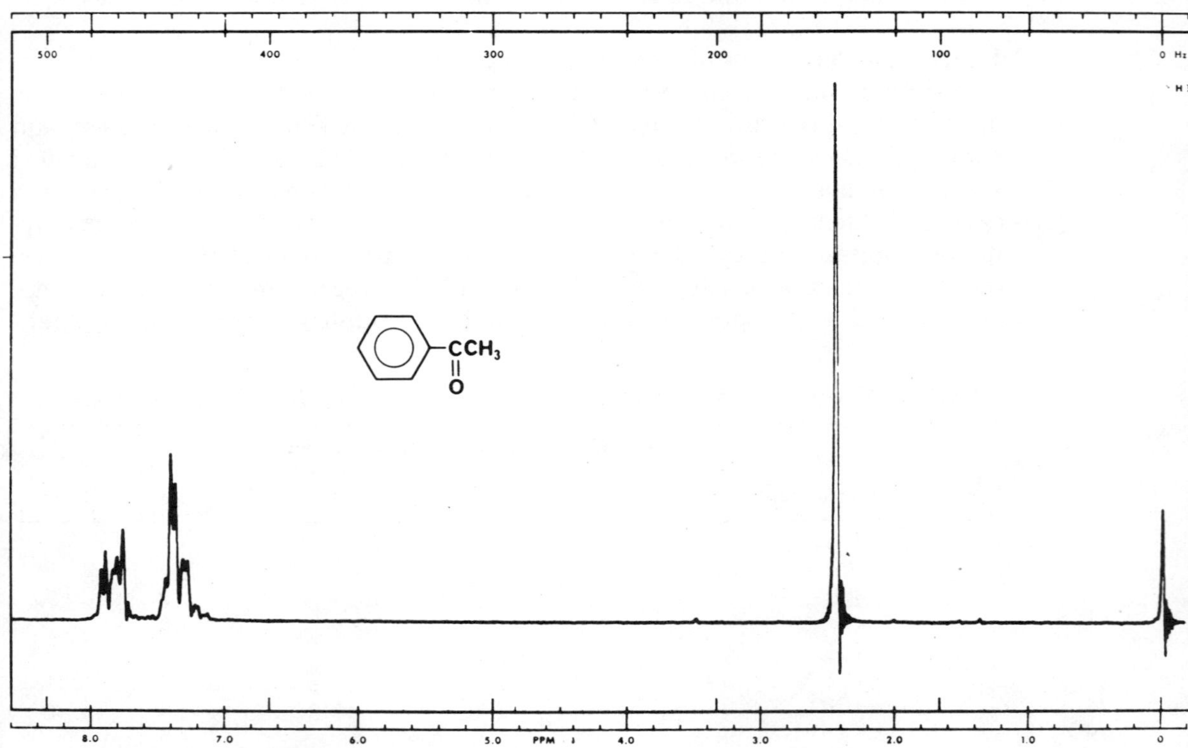

Figure 18.11 *NMR spectrum of acetophenone. The ortho protons are deshielded relative to the meta and para protons.*

Table 18.4 Chemical Shifts for Hydrogens Bonded to Oxygen and Nitrogen

Compound	Structural Type	Chemical shift range (δ)
Alcohol	ROH	1-5
Phenol	ArOH	4-8
Carboxylic acid	RCOOH	10-13
Amine	RNH_2 and R_2NH	0.5-3.5
Arylamine	ArNH_2 and Ar_2NH	3-5
Amide	RCONH_2 and RCONHR′	6-9

amides) is usually rapid on the NMR time scale and, therefore, these hydrogens do not cause spin-spin splitting of hydrogen nuclei on adjacent carbon atoms nor are their absorptions split by adjacent hydrogen nuclei.

Chemical exchange of hydrogens bonded to oxygen or nitrogen with D_2O (deuterium oxide) causes the disappearance of the absorption due to the exchangeable proton and the appearance of a signal from HDO; for example,

$$ROH + D_2O \rightleftharpoons ROD + HOD$$

Since deuterium does not absorb at the frequencies used to observe proton

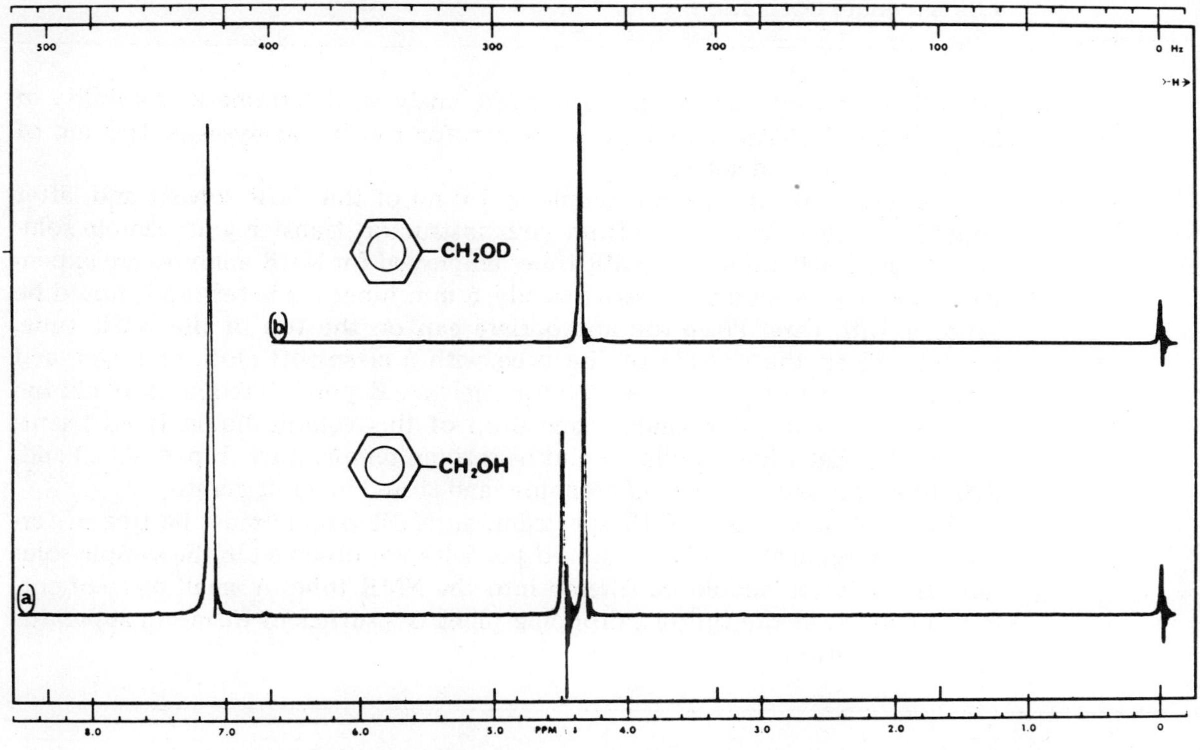

Figure 18.12 *(a) NMR spectrum of benzyl alcohol in CCl_4. (b) Same spectrum as (a) but with added two drops of D_2O (region of 6.8 to 8.6 δ is not shown in this spectrum but absorption at 7.1 δ is unchanged).*

spectra, chemical exchange with D_2O is a useful technique for the detection of —OH or —NH— protons in a compound by NMR spectroscopy. The procedure normally employed for exchange with D_2O involves the following operation: After the integrated spectrum of a compound has been obtained, 1 to 2 drops of D_2O are added to the NMR tube, the contents of the NMR tube are shaken vigorously for 1 to 2 min, and a second spectrum is obtained. This technique is exemplified in Figure 18.12.

Sample Preparation for NMR Spectrometric Analysis

Routine NMR spectra can be obtained with 100 mg of a compound. As mentioned earlier in this experiment, NMR spectra are taken of compounds dissolved in nonprotonic solvents such as carbon tetrachloride or deuteriochloroform.

CAUTION: Skin contact with chlorocarbon solvents and the breathing of their vapors should be avoided. Sample preparation should be performed in a hood with adequate ventilation.

Prior to preparing your sample for NMR analysis, determine its solubility in CCl_4 or $CHCl_3$. The minimum solubility for routine analyses is 100 mg of sample per 1.0 ml of solvent.

Dissolve 100 mg of your sample in 1.0 ml of the NMR solvent and, after obtaining a clean NMR tube from your instructor, transfer your sample solution to the NMR tube. The NMR tubes employed for NMR analyses are expensive, precisely ground glassware (usually 5 mm inner diameter) and should be handled with care. Place the appropriate cap on the top of the NMR tube, carefully clean the outside of the tube with a clean soft cloth or towel, and submit the tube to your instructor for analysis. If you are required to add the TMS reference to your sample, one drop of this volatile liquid is sufficient; tetramethylsilane has a boiling point near room temperature (b.p. = 26°C) and, therefore, it is kept in a *closed* container and stored in a refrigerator.

In order to obtain a NMR spectrum, an NMR sample must be free of ferromagnetic impurities. If undissolved particles are observed in the sample solution, the solution should be filtered into the NMR tube. A small piece of cotton imbedded in the tip of a dropping pipet is a sufficient filtration apparatus for this operation.

Prelab Questions

1. The chemical shift of the methylene hydrogens of ethyl ether is 3.4 δ on a 60 MHz NMR spectrometer. (a) What is the frequency in Hz relative to TMS for this absorption? (b) At what frequency and chemical shift would this absorption occur on a 100 MHz instrument?

2. The coupling constant for the proton absorptions of ethyl ether is 7 Hz on a 60 MHz instrument. What is the coupling constant for these absorptions on a 100 MHz instrument?

3. Separately list the sets of magnetically equivalent nuclei for each of the following compounds: (a) toluene, (b) benzyl alcohol, (c) acetophenone, and (d) limonene. Check your answers against the NMR spectra of these compounds given in this experiment.

4. Assign the proton absorptions for the spectra of (a) toluene, (b) benzyl alcohol, (c) acetophenone, and (d) limonene.

EXPERIMENTAL PROCEDURE

The NMR spectra of selected compounds that were isolated or prepared in Experiments 1-17 are given in Figures 18.13*a*–18.13*f*. Identify each of these compounds.

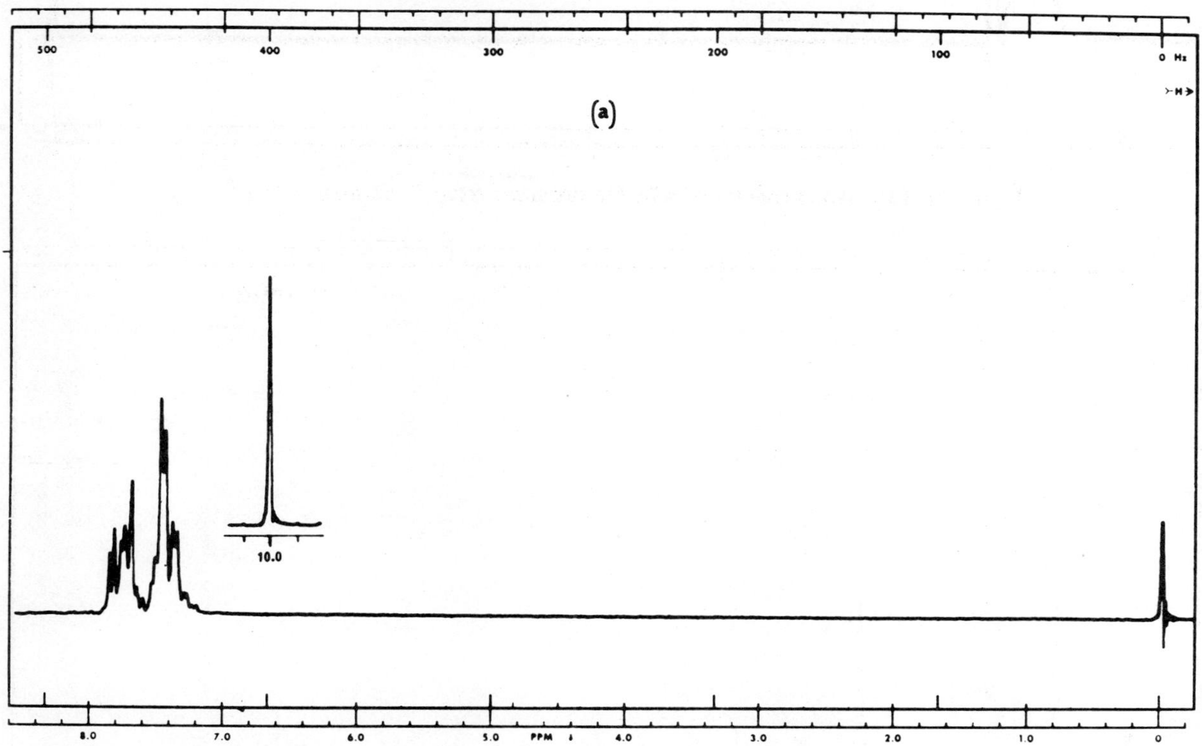

Figure 18.13a *NMR spectra of selected products from Experiments 1–17.*

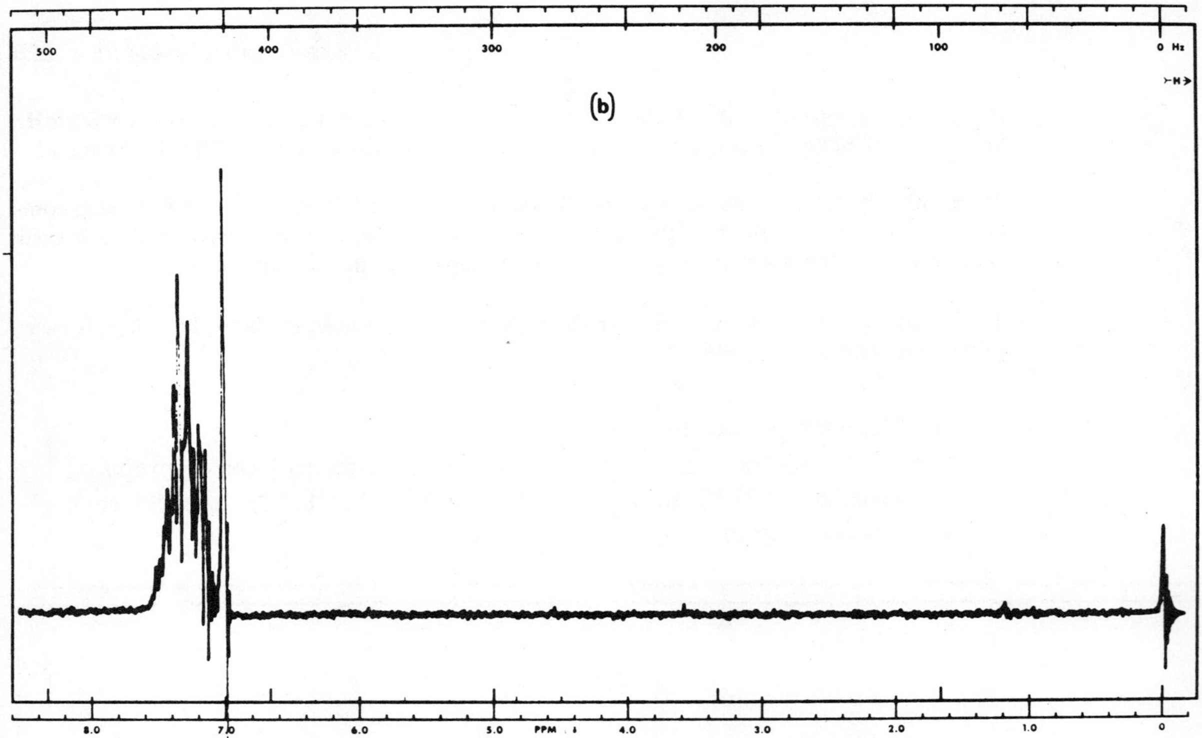

Figure 18.13b *NMR spectra of selected products from Experiments 1–17.*

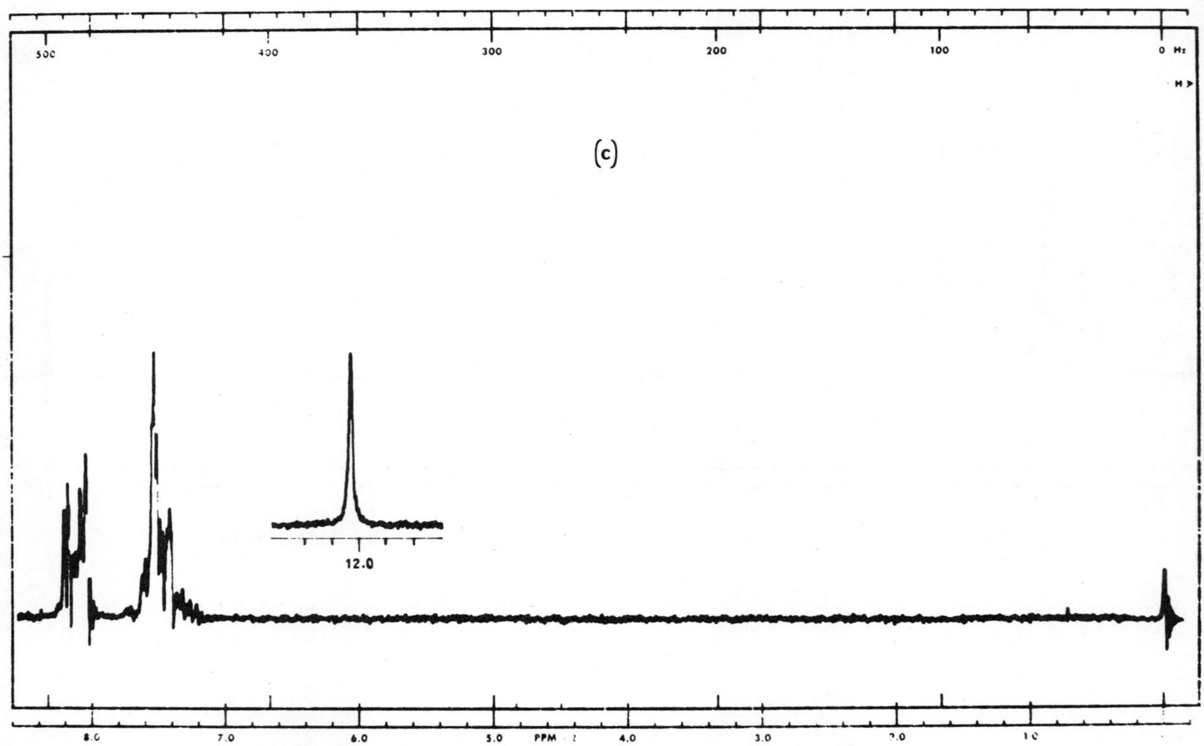

Figure 18.13c *NMR spectra of selected products from Experiments 1–17.*

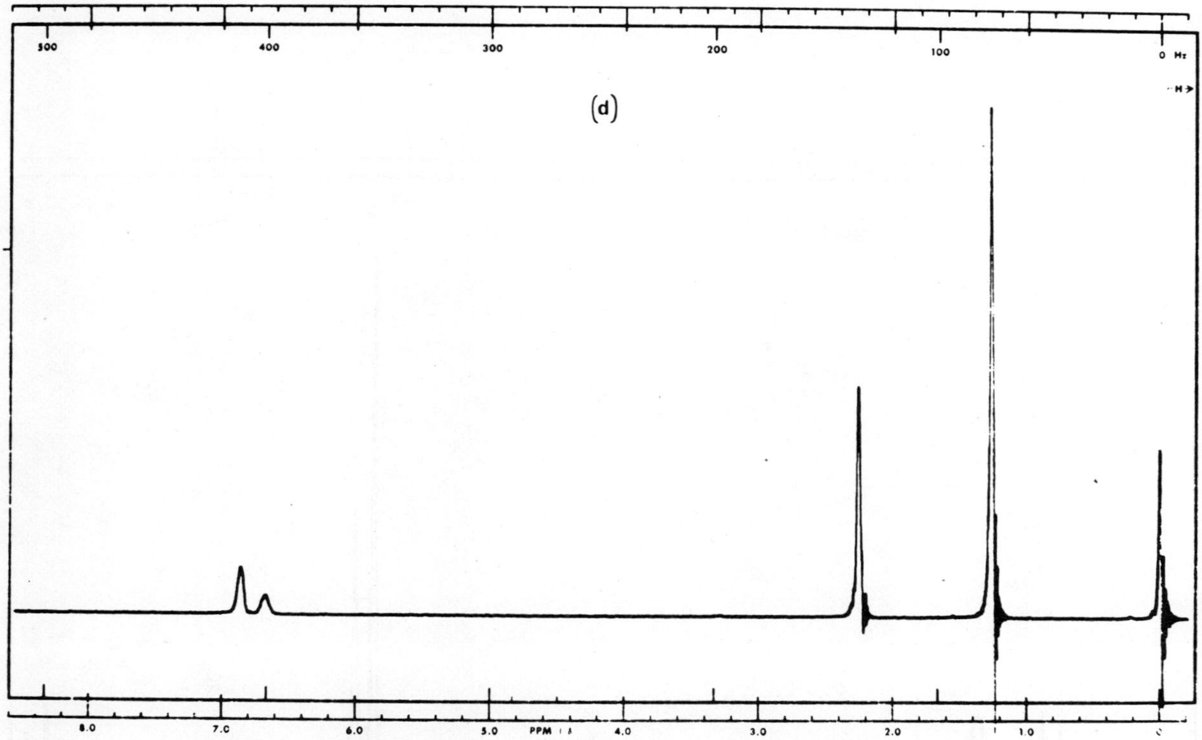

Figure 18.13d *NMR spectra of selected products from Experiments 1-17.*

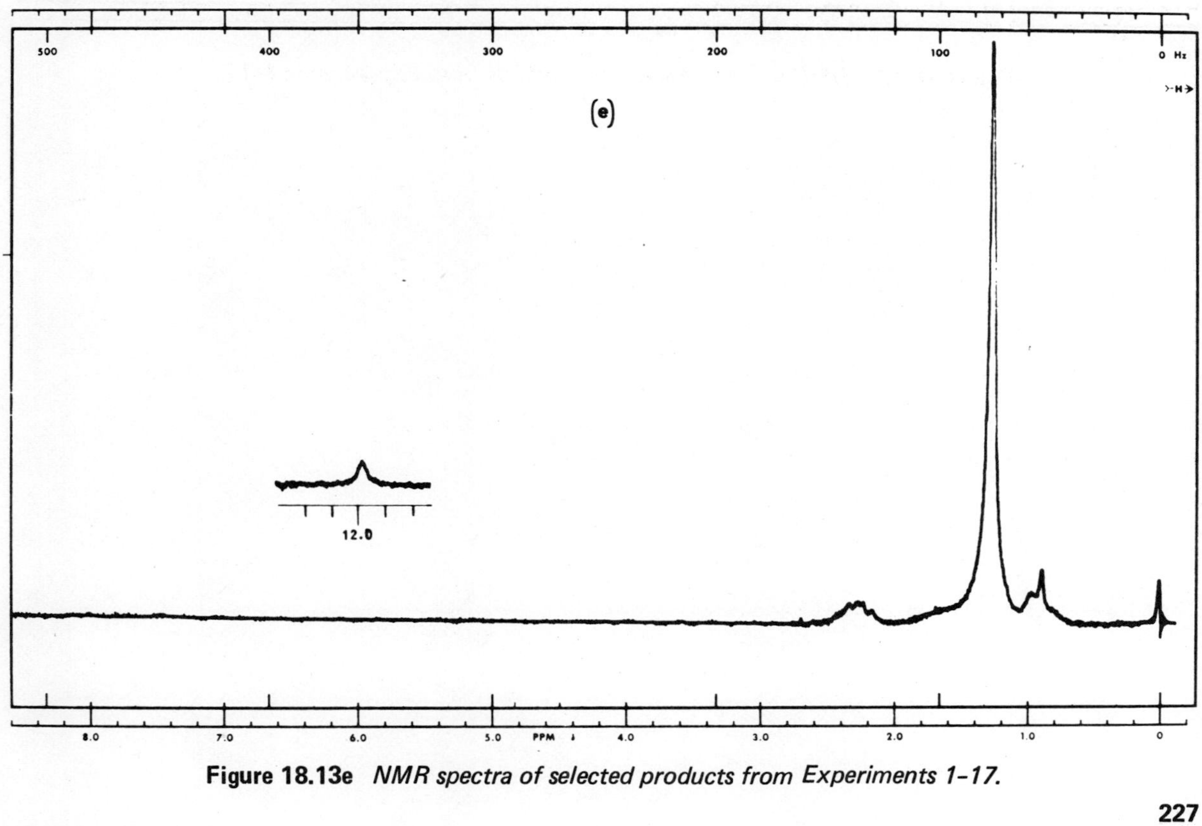

Figure 18.13e *NMR spectra of selected products from Experiments 1-17.*

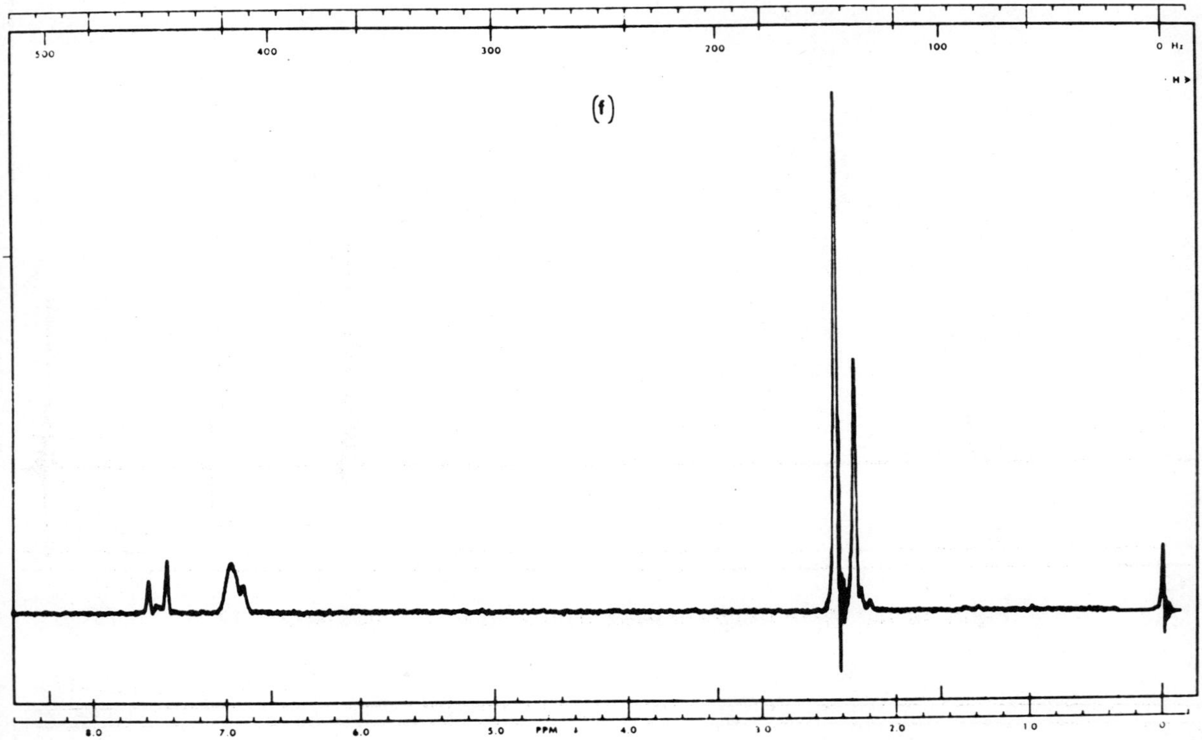

Figure 18.13f *NMR spectra of selected products from Experiments 1-17.*

Experiment Nineteen

Column Chromatography: Synthesis and Purification of o-Nitrophenol

Aromatic compounds react with nitric acid to form nitro substituted derivatives:

Nitration reaction

ArH	+	HNO_3	$\longrightarrow$	Ar—NO_2	+	H_2O
Aromatic compound		*Nitric acid*		*Nitroaromatic compound*		

This reaction, like the Friedel-Crafts reactions in Experiment 17, is an electrophilic aromatic substitution reaction. Nitric acid is the source of the positively charged nitronium ion, NO_2^+, that is the electrophile in aromatic nitration reactions. In strongly acidic media, nitric acid dissociates to form nitronium ion and water:

HNO_3	+	H^+	$\rightleftharpoons$	NO_2^+	+	H_2O
Nitric acid				*Nitronium ion*		

The nitronium ion, which is a strong Lewis acid, replaces hydrogen on the aromatic ring by addition to an unsaturated carbon center followed by loss of a proton:

$$Ar\text{—}H + NO_2^+ \longrightarrow Ar\text{—}NO_2 + H^+$$

In this experiment you will prepare *o*-nitrophenol by nitration of phenol.

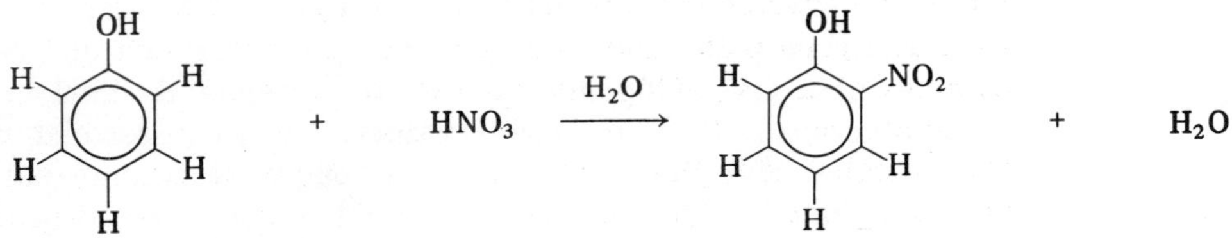

Phenol	*Nitric Acid*	*o-Nitrophenol*
M.w. = 94.1 g/mol	M.w. = 63.0 g/mol	M.w. = 139.11 g/mol
m.p. = 43°C	M = 16 mol/l	m.p. = 45°C

This reaction actually produces a mixture of both the *ortho*- and *para*-nitrophenols in which the relative ratio of ortho to para product is approximately 2:1. Only traces of *m*-nitrophenol are formed since the hydroxyl group is a strong *o*,*p*-directing substituent.

The reaction conditions necessary for the nitration of substituted aromatic compounds depend on the activating or deactivating effects of the substituents on the benzene ring. Since the hydroxyl substituent of phenol strongly activates the benzene ring towards electrophilic substitution reactions, mild reaction conditions—dilute nitric acid and low temperature—are sufficient for the mono nitration of phenol. However, even with these mild conditions, a major side reaction in the nitration of phenol is the formation of dinitrophenol by the nitration of nitrophenol. Compounds with deactivating substituents on the benzene ring generally require more vigorous reaction conditions (concentrated nitric acid, sulfuric acid catalysis, and higher reaction temperatures).

Since nitric acid is a strong oxidizing agent and phenols are generally easily oxidized, the nitration of phenol results in the formation of side products by oxidation of the phenol. These side products are very dark colored and form thick tars.

The desired product in this experiment, *o*-nitrophenol, must be separated from *p*-nitrophenol, dinitrophenol, and the other side products produced in the nitration reaction. Column chromatography will be used in this experiment for the separation of the product.

Column Chromatography

Like thin layer chromatography (Experiment 7), column chromatography is a form of adsorption chromatography. Column chromatography is also called *elution chromatography* since the separated compounds are eluted from the column. Column chromatography and thin layer chromatography (TLC) are similar in principle; compounds in the mixture to be separated are partitioned between a solid adsorbent (*the stationary phase*) and a solvent (*the mobile phase*) that flows past the solid adsorbent. The more strongly a compound is adsorbed onto the stationary phase and the less the compound is dissolved in the moving liquid phase, the slower the compound will migrate along the stationary phase in the direction of the solvent flow. In thin layer chromatography the adsorbent is coated in a thin layer (~0.25 mm) on an inert support such as a glass plate, and solvent flows by capillary action past the absorbent. Samples of a few milligrams or less are conveniently analyzed by thin layer chromatography. In column chromatography the adsorbent is packed into a glass column and the solvent flows through the column past the adsorbent particles. Since a glass column can hold much more adsorbent than a TLC plate and the column size can be selected according to the amount of material to be chromatographed, column chromatography can be used to separate much larger quantities of material than can be analyzed by thin layer chromatog-

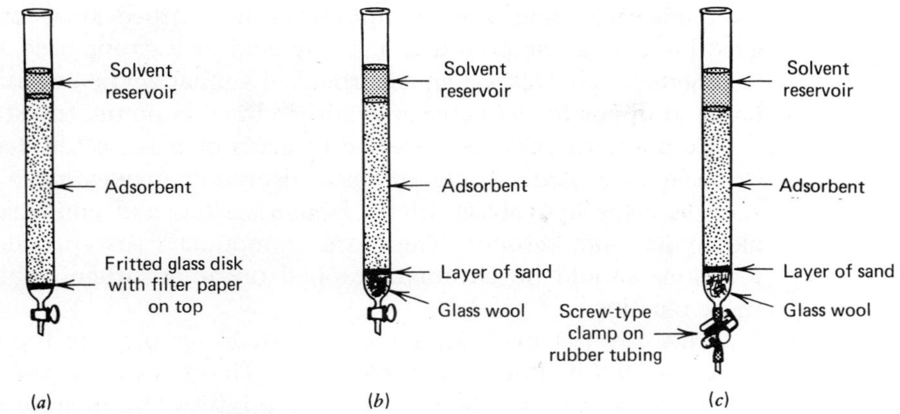

Figure 19.1 *Chromatography colums.*

raphy. Gram quantities of a mixture can be separated on common laboratory columns (1 to 5 cm in diameter and 10 to 100 cm in length).

Several types of laboratory columns are shown in Figure 19.1. Some columns have porous, fritted glass plates in the bottom of the column to hold the adsorbent in the column and a stopcock to control the flow of the liquid phase through the column. The ratio of the column length to the column diameter is at least ten to one.

The two adsorbents most commonly used for column chromatography are alumina (Al_2O_3) and silica gel (SiO_2). Alumina adsorbs organic compounds more strongly than silica gel and is considered to be more active than silica gel. Alumina is used for the separation of most nonpolar and moderately polar organic compounds. Silica gel is generally the adsorbent of choice for the separation of polar organic compounds. The activity of both alumina and silica gel can vary greatly depending on the amount of water present in the adsorbent. The most active adsorbents contain the least amount of water. In this experiment you will use alumina as the adsorbent to chromatograph the nitrophenol mixture.

Alumina is classified into five activity grades. The most active, Grade I, is easily deactivated by the addition of water to prepare the other grades for use in chromatography. The water and alumina are added to a jar. The jar is tightly

	Decreasing Activity →			
Grade of alumina to be prepared	II	III	IV	V
Percent by weight of water added to alumina	3	6	10	15

sealed and then rolled or rotated for several hours to ensure an even distribution of the water on the alumina. The higher activity grades of alumina are most useful for chromatography of nonpolar compounds and the less active grades are used for the chromatography of more polar compounds.

Commercial alumina is available in acidic, neutral, or basic forms, depending on the type of surface washing that was used in its manufacture. Silica gel is normally slightly acidic. Acidic compounds are adsorbed most strongly on a basic adsorbent and basic compounds are adsorbed most strongly on an acidic adsorbent. If a compound is a strong acid or a strong base, the best separation is generally achieved on an adsorbent of similar acidic or basic character; adsorbents of opposite character may adsorb the compound too strongly.

Compounds that are sensitive to acids or bases often decompose on acidic or basic adsorbents. Acidic or basic adsorbents may catalyze chemical reactions such as ester hydrolysis, olefin isomerization, and condensation reactions of aldehydes and ketones. Therefore, compounds susceptible to these types of reactions should be chromatographed on an adsorbent that will not promote these reactions.

The solvents used for column chromatography are the same solvents that were useful for TLC (see Table 19.1). The more polar the solvent employed, the more rapidly all components in a mixture will migrate through a column. Little or no separation of nonpolar components in a mixture will be achieved if a polar solvent is used for the chromatographic separation. On the other hand, if a nonpolar solvent is used to achieve optimum separation of the nonpolar components, then polar components in the mixture will not be eluted from the adsorbent by the solvent. To solve this problem and achieve optimum separation of both polar and nonpolar compounds in a mixture, the solvent composition passing through the column can be gradually changed to a more polar solvent by slowly increasing the percentage of a more polar solvent in a two-solvent mixture (*gradient elution*). Alternatively, stepwise solvent composition changes can be made to obtain increased solvent polarity.

The choice of a solvent system for the separation of a mixture by column chromatography is based upon results published in the literature for similar separations or upon a systematic study of solvents for the analytical separation of the mixture by TLC. However, results obtained on separations by TLC may not be directly transferable to column chromatography since column chro-

Table 19.1 Pure Solvents and Solvent Mixtures in Order of Increasing Eluting Power

Hexane	INCREASING ELUTING POWER ↓
Cyclohexane	
Toluene	
Benzene[a]	
Dichloromethane	
Chloroform[a]	
Cyclohexane-ethyl acetate (80:20)	
Dichloromethane-ethyl ether (80:20)	
Dichloromethane-ethyl ether (60:40)	
Cyclohexane-ethyl acetate (20:80)	
Ethyl ether	
Ethyl ether-methanol (99:1)	
Ethyl acetate	
Tetrahydrofuran	
1-Propanol	
Ethanol	
Methanol	

[a]These solvents should not be used for elution since their vapors are a serious health hazard.

matography generally gives less separation between components than does TLC. A stepwise or gradient elution is most useful if the thin layer chromatogram of the mixture developed in a low polarity solvent shows good separation of nonpolar components and the presence of other polar components at or near the origin.

In this experiment you will use dichloromethane as the eluting solvent to separate the mixture of nitrophenols. Since the *o*-nitrophenol is the least polar of the compounds to be separated, a single low-polarity solvent can be used to achieve separation of the product from the other more polar components of the reaction mixture.

The Chromatography Column

The adsorbent must be packed uniformly into the glass column without air pockets or cracks. The lack of uniformity in the adsorbent disrupts even elution and prevents optimum separation. A generally successful method of packing most adsorbents in a column is the *slurry packing technique.* The column is clamped in a stable, vertical position with the stopcock closed and filled to one-fourth its volume with solvent. If the column is equipped with a fritted glass disk at the bottom, a piece of filter paper cut to the exact diameter of the disk is placed down the column on top of it. If the column is not equipped with a fritted glass disk, a small wad of glass wool is forced down the column with a glass rod to form a filter pad at the bottom. Sand is next poured into the column to form a 1-cm layer on top of the filter paper or glass wool. Solvent and adsorbent are then mixed together in a beaker to form a slurry. The stopcock on the column is opened to allow solvent to drain dropwise into an Erlenmeyer flask. Then the slurry of adsorbent and solvent is poured into the column at a rapid, continuous rate until the desired column height is achieved. The adsorbent should settle gently down through the solvent to produce a uniform packing. *Once the column has been packed, the solvent level should never be allowed to fall below the top surface of the adsorbent or cracks and air pockets will result.* The top one-fourth of the column is usually not filled with adsorbent in order to provide a reservoir for solvent. Once the column has been filled with the adsorbent to the proper height, the solvent is drained from the column until the liquid level just reaches the top surface of the adsorbent. The packed column is now ready to be used.

In the procedure employed for column chromatography the mixture to be separated is dissolved in the smallest volume possible of the solvent selected to start the chromatography.* If the total mixture is not soluble in this solvent, then the least polar solvent that will readily dissolve the mixture should be used. With the aid of a pipet, the solution of the mixture is transferred onto the

*The maximum volume of solvent used to dissolve the sample should not exceed $\frac{1}{20}$ of the volume of the packed column.

The sample is dissolved in the minimum amount of an organic solvent and applied onto the top of the adsorbent in the chromatography column.

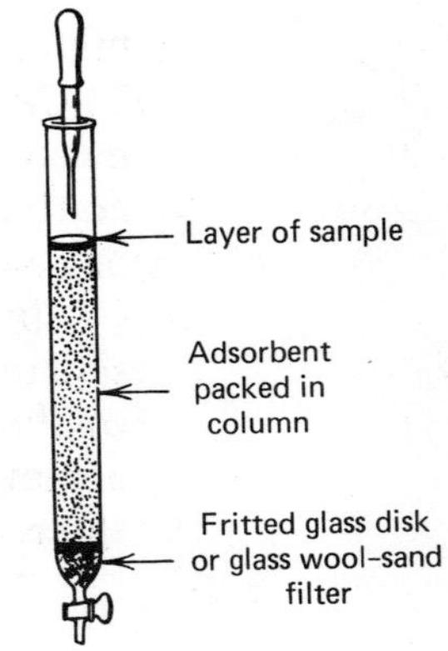

After the sample has been applied and drained down into the adsorbent layer, fresh solvent is added to the top of the column. The stopcock is opened and the solvent flowing from the column is collected in separate fractions. (Do not allow the solvent level in the column to fall below the top of the adsorbent.)

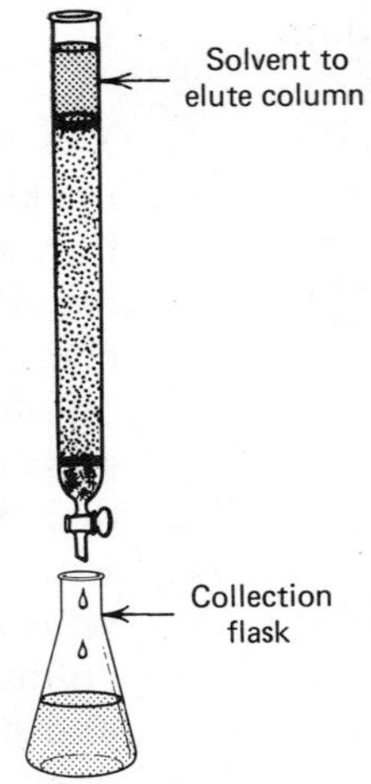

A separatory funnel can be used to provide continuous addition of solvent. The funnel is filled with solvent, tightly stoppered, and placed above the column so that the stem of the funnel is down inside the column. The stopcock is opened and solvent flows from the funnel until the level of solvent is just above the tip of the separatory funnel stem. Solvent will flow from the funnel when the solvent level falls below the tip of the separatory funnel.

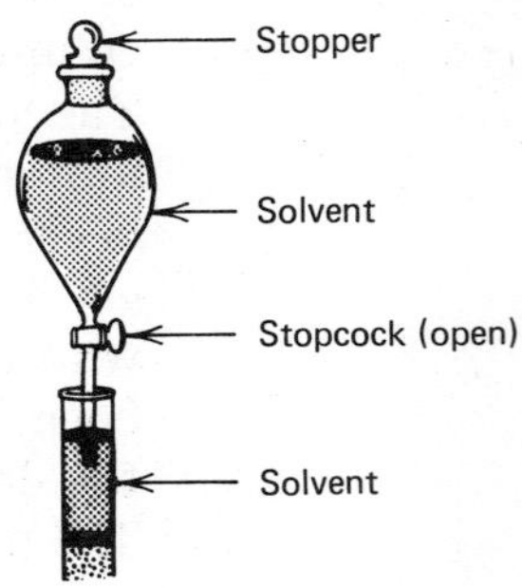

Figure 19.2 *Procedure for column chromatography.*

A nonpolar component in the sample is retarded least by adsorption on the polar surface of the adsorbent and will be eluted from the column in the first few fractions. More polar components are more strongly adsorbed and will not be eluted from the column until later fractions are collected.

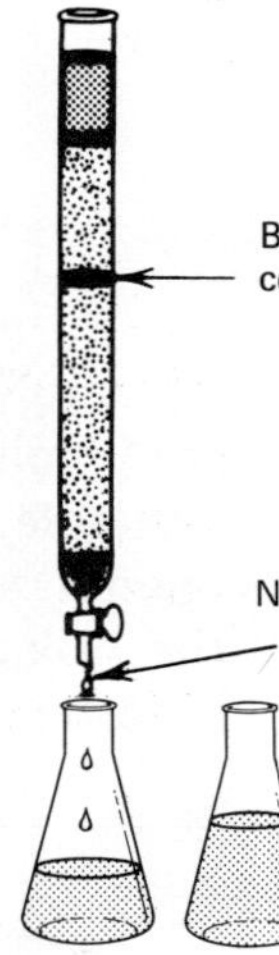

Figure 19.2 *(continued)*

top of the adsorbent in the column (Figure 19.2). The stopcock at the bottom of the column is opened and solvent is slowly drained out the bottom of the column and collected in an Erlenmeyer flask until the sample solution level in the column just reaches the top of the adsorbent. While taking care not to stir up the adsorbent, fresh solvent is added in small portions to the top of the column in order to wash a residual sample into the column. A 1-cm layer of sand may be added on top of the adsorbent to help prevent disturbing the adsorbent in the solvent addition process. After the sample has been washed into the column, the empty upper portion of the column is filled with solvent. The solvent flowing from the column is collected in separate fractions. The flow rate and the size of the fractions depend on the diameter of the column. Generally, 10-ml fractions are collected at a rate of 1 to 2 ml/min from a column that is 1 cm in diameter. Fractions of 50 ml are collected at a rate of 4 to 5 ml/min from a column 2 cm in diameter, and even larger fractions are collected at a faster rate from larger diameter columns. Since components of different polarity migrate through the column at different rates, the components of the mixture are separated and collected in different fractions of the eluting solvent. In this experiment the nitrophenol products are yellow or orange, and it is possible to observe separate, colored bands of material moving down the column. Fractions containing nitrophenol are also colored.

Each fraction collected is concentrated to a small volume by distillation or evaporation of the solvent. The resulting concentrated solution from each fraction is spotted separately on a thin layer chromatography slide (Experiment 7) and analyzed to determine the composition of the material in that fraction. Sequential fractions that contain a pure component of the original mixture are combined and the component is isolated.

In this experiment *o*-nitrophenol is adsorbed least by the alumina, and therefore it is eluted from the column before *p*-nitrophenol or the dinitrophenol. The relatively high mobility of the ortho isomer is a result of strong

Intramolecular hydrogen bonding in o-nitrophenol

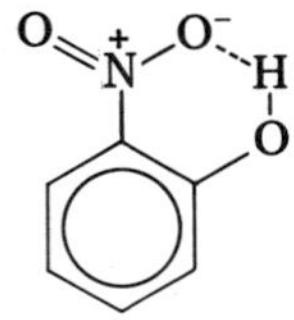

intramolecular hydrogen bonding. Because the polar nitro and hydroxyl groups are involved in intramolecular hydrogen bonding, these groups are no longer as available for hydrogen bonding to the surface of the adsorbent. The *p*-nitrophenol isomer cannot form intramolecular hydrogen bonds, and, therefore, the polar groups are available to form hydrogen bonds to the surface of the adsorbent.

Prelab Questions

1. Suggest a reason why the addition of concentrated sulfuric acid should increase the rate of nitration of an aromatic compound.

2. Which dinitrophenol isomer is expected to be the major product from the dinitration of phenol?

3. A student has obtained a product mixture from the reaction of acetanilide ($C_6H_5NHCOCH_3$) with concentrated nitric acid. Which chromatographic technique (gas-liquid partition chromatography, thin layer chromatography, or column chromatography) would be most useful to determine the number of components in the mixture? Which chromatographic technique would be most useful to separate 3 g of the mixture into its components?

4. Calculate the molar ratio of nitric acid to phenol used in this experiment. Neglecting other side reactions, what is the expected yield of *o*-nitrophenol if the product distribution of ortho to para isomers is in the ratio of 2:1?

5. Explain why *p*-nitrophenol is much more soluble in water than *o*-nitrophenol (para, 1.6 g/100 ml; ortho, 0.21 g/100 ml).

EXPERIMENTAL PROCEDURE

Add 3 ml of concentrated nitric acid to 7 ml of water in a 50-ml Erlenmeyer flask and cool the mixture in an ice bath to less than 5°C. In a test tube, mix 3 g of phenol* with eight drops of water to obtain

---

CAUTION: Avoid contact of phenol with your skin. Phenol can cause chemical burns.

---

*If liquified phenol (88% phenol–12% water) is used, omit the addition of water.

a homogeneous solution. Add the phenol solution dropwise to the aqueous nitric acid with rapid swirling and maintain the temperature below 20° by cooling in an ice bath. After the phenol has been added maintain the temperature between 30 and 40°C by periodic cooling for 15 minutes. Then add 10 ml of ice-cold water to the dark colored reaction mixture and extract with two 15-ml portions of dichloromethane. Wash the combined organic extracts with two 15-ml portions of water, and then dry the organic layer over anhydrous magnesium sulfate. Concentrate the organic solution to a volume of approximately 3 ml by distillation of the solvent on a steam bath.

The reaction mixture should be analyzed by thin layer chromatography (Experiment 7). Silica gel TLC plates developed in either dichloromethane or toluene are suitable for this analysis. The spots on the slides should be detected under ultraviolet light or in an iodine chamber as well as by their color in visible light. Samples of *o*-nitrophenol, *p*-nitrophenol, and phenol should be run for comparison of R_f values.

Pack a glass column* with 25 g of alumina slurried in dichloromethane according to the procedure described in this experiment. Add the concentrated dichloromethane extract from the nitration reaction to the top of the column and then elute the column with dichloromethane. Collect fractions of 10 ml each. A yellow band is clearly visible as the *o*-nitrophenol migrates down the column. Collect a total of ten fractions *after* the yellow product band begins to elute from the column.

Concentrate each of the fractions obtained to about 2 ml, and then analyze each fraction by thin layer chromatography using the same system that was used to analyze the original reaction mixture. Fractions containing pure *o*-nitrophenol are quantitatively transferred to a tared flask and the solvent is evaporated. Determine the weight of pure *o*-nitrophenol and calculate the percentage yield. The product may be recrystallized from ethanol.

Postlab Questions

1. Why is slow addition of phenol with cooling to the nitric acid solution important in this experiment?

2. Why must the thin layer chromatograms be viewed under ultraviolet light or stained with iodine in addition to observing the colored spots?

3. Summarize in tabular form the results of the thin layer chromatographic analysis of each

*A 50-ml buret or similarly sized column is used. A suitable column can be made from a 40 cm length of 1.5 cm O.D. glass tubing with one end drawn-down to fit into a short piece of rubber tubing. A screw-type clamp is used to control the solvent flow through the tubing (see Figure 19.1*c*).

of the fractions from the column chromatography. Describe the color and relative intensity of each spot.

4. What would have been the experimental result if fractions of 3 ml instead of 10 ml had been collected? What would have happened to the observed separation if fractions of 30 ml had been collected?

5. Describe how you could determine if any unreacted phenol remains in the nitration reaction mixture before the mixture is chromatographed on the column.

6. *o*-Nitrophenol and *p*-nitrophenol can be separated by steam distillation. The ortho isomer is volatile and distills with steam, but the para isomer does not. Explain this difference in chemical behavior.

Experiment Twenty

Diazotization of 2,5-Dichloroaniline: Preparation of 2,5-Dichlorophenol

---

Primary amines, RNH_2, react with nitrous acid in acidic solutions to form diazonium salts:

Diazotization reaction

$$\underset{\substack{\textit{1°}\\ \textit{Amine}}}{RNH_2} + \underset{\substack{\textit{Nitrous}\\ \textit{acid}}}{HONO} + \underset{\substack{\textit{Mineral}\\ \textit{acid}}}{HZ} \xrightarrow{H_2O} \underset{\substack{\textit{Diazonium}\\ \textit{salt}}}{RN_2^+Z^-} + \underset{\textit{Water}}{H_2O}$$

($HZ = HCl$ or H_2SO_4 ordinarily)

Aqueous mineral acid solutions of sodium nitrite are generally employed for these conversions, known as *diazotization reactions.* The mineral acid promotes diazotization of primary amines by converting the relatively weak Brønsted acid nitrous acid ($pK_a = 3.29$) into the reactive nitrosonium ion:

$$\underset{\substack{\textit{Nitrous}\\ \textit{acid}}}{HONO} + H^+ \rightleftharpoons H_2\overset{+}{O}{-}NO \rightleftharpoons H_2O + \underset{\substack{\textit{Nitrosonium}\\ \textit{ion}}}{NO^+}$$

Reaction of the nitrosonium ion with the amine (*nitrosation*) results in the formation of the diazonium ion. The anion associated with the diazonium ion is the conjugate base of the mineral acid used for diazotization.

Diazonium ions are formally described by Lewis structures in which molecular nitrogen is bonded to carbon:

$$RN_2^+ \equiv R{-}\overset{+}{N}{\equiv}N\!: \quad \textit{Diazonium ion}$$

Aliphatic diazonium salts are unstable and rapidly lose molecular nitrogen; the resulting alkyl carbonium ion then undergoes its characteristic reactions. In contrast, aromatic diazonium salts are relatively stable, and many of these salts can be isolated for use in a variety of chemical transformations. Generally, however, the isolation of aromatic diazonium salts is avoided because of their sensitivity to heat and shock, and reactions of these compounds are performed in the solutions in which they are prepared.

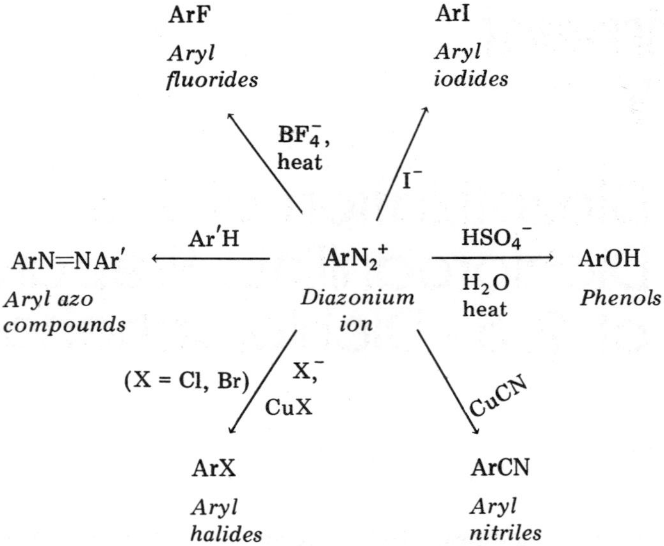

Figure 20.1 *Reaction of aromatic diazonium salts.*

Aromatic diazonium salts are useful synthetic intermediates. Substitution reactions of these diazonium salts and diazo coupling reactions, particularly in the formation of azo dyes, are widely used for the preparation of substituted aromatic compounds (Figure 20.1). In this experiment you will prepare 2,5-dichlorobenzenediazonium hydrogen sulfate by diazotization of 3,4-dichloroaniline with sodium nitrite in aqueous sulfuric acid and employ this diazonium salt for the synthesis of 2,5-dichlorophenol. Since phenols react with diazonium salts to form azo compounds under the same reaction conditions used to prepare phenols, 2,5-dichlorophenol will be removed from the reaction solution as it is formed. The method employed for phenol removal, steam distillation, selectively separates the phenol from the reactant diazonium salt and the expected azo by-products.

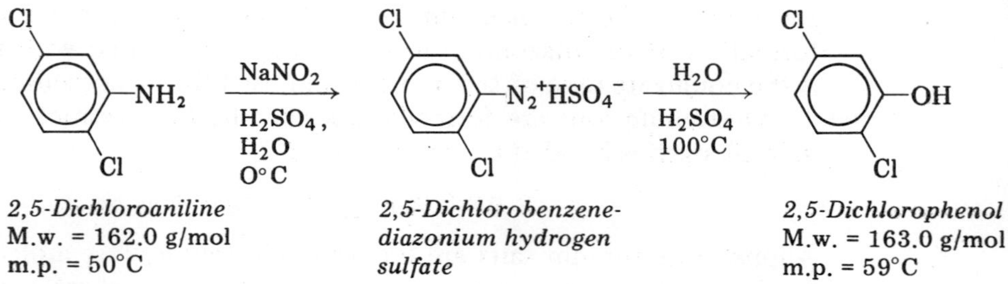

Prelab Questions

1. Write balanced chemical reactions for the conversion of 2,5-dichloroaniline to 2,5-dichlorobenzenediazonium hydrogen sulfate and for the preparation of 2,5-dichlorophenol from 2,5-dichlorobenzenediazonium hydrogen sulfate.

2. Aromatic diazonium ions are known to form 1,3-diaryltriazenes:

$$ArN_2^+ + ArNH_2 \rightleftharpoons ArN{=}N{-}\overset{+}{N}H_2Ar \rightleftharpoons ArN{=}N{-}NHAr + H^+$$

A 1,3-diaryl-triazene

In mineral acid solutions, however, diazotization of aromatic amines is quantitative and 1,3-diaryltriazenes are not observed reaction products. Explain why these compounds are not formed if the diazotization reactions are performed in mineral acid solutions.

3. Phenols are susceptible to electrophilic substitution by diazonium salts that results in the formation of azo compounds. Describe the reaction conditions that will favor phenol formation from diazonium salts and will minimize the production of azo compounds.

4. If upon completion of the procedure for phenol formation your reaction mixture contains unreacted amine (an organic base), diazonium salt, and phenol (a weak organic acid), what method of extraction will allow you to selectively isolate the phenol? Describe the isolation procedure in a flow diagram.

5. Write balanced equations for the side-reactions expected in the preparation of 2,5-dichlorophenol.

EXPERIMENTAL PROCEDURE

Preparation of 2,5-Dichlorobenzenediazonium Hydrogen Sulfate. Prepare 50 ml of a 60% aqueous sulfuric acid solution by slowly adding 30 ml of concentrated sulfuric acid to 20 ml of water. Dissolve 3.30 g of 2,5-dichloroaniline (0.020 mol) in this acid solution* and then carefully pour the resulting solution into a 500-ml three-necked round-bottom flask that is fitted with a thermometer, an addition funnel, and a reflux condensor. Place the round-bottom flask in an ice bath and note the formation of a precipitate upon cooling. Prepare an aqueous sodium nitrite solution by dissolving 1.70 g of sodium nitrite (0.025 mol) in 20 ml of water and pour this solution into the addition funnel. Add approximately 20 g of ice to the reaction mixture, and then commence addition of the sodium nitrite solution at a sufficiently slow rate (about 1 drop/sec) so that the reaction temperature remains below 10°C. Swirl the reaction solution at regular intervals to ensure thorough mixing. Observe and record changes in the color of the reaction solution and changes in the amount of precipitate present during the addition. After all of the sodium nitrite has been added, remove the ice bath, and allow the reaction mixture to warm to room temperature. *Do not store this solution; continue with the synthesis of 2,5-dichlorophenol.*

*Heat the mixture on a steam bath, if necessary, to completely dissolve the amine.

Synthesis of 2,5-Dichlorophenol. Rinse the addition funnel thoroughly with water. Transfer the reaction solution prepared in the previous procedure from the 500-ml three-necked round-bottom flask to the clean additional funnel. Add 100 ml of water followed by 65 ml of concentrated sulfuric acid to the empty 500-ml three-necked flask, thoroughly mix the acid and water, and construct the glassware for

---

CAUTION: By simply pouring concentrated sulfuric acid into water, sulfuric acid and water form distinct liquid layers. Failure to thoroughly mix these two layers may result in flash evaporation of the upper aqueous acid layer during heating of the acid solution to boiling.

---

steam distillation as shown in Figure 20.2. After adding several boiling stones, heat the acid solution to boiling. When the first drop of distillate enters the collection flask, begin adding the diazonium salt solution from the addition funnel at approximately the same rate as that of the distillation (2 to 3 ml/min). Continue the distillation until the con-

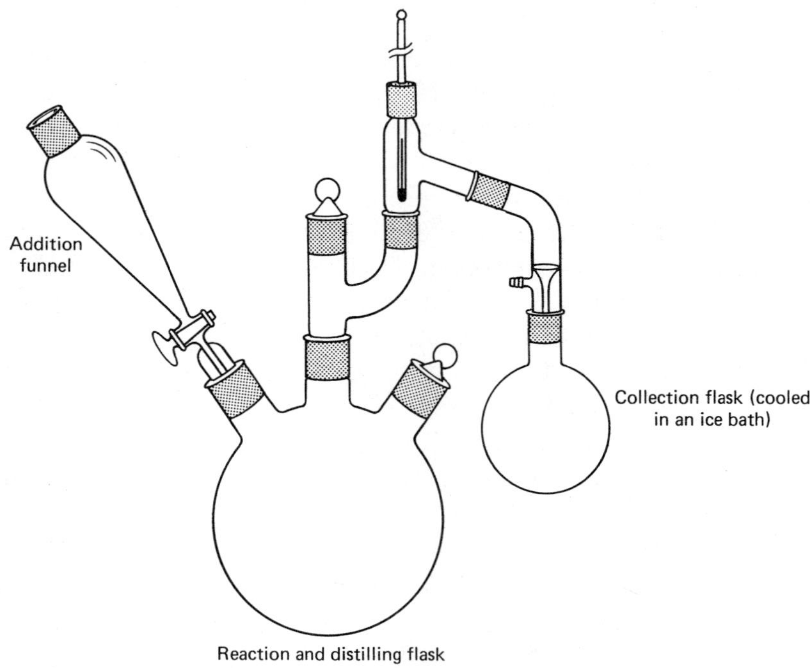

Figure 20.2 *Apparatus for synthesis and steam distillation of 2,5-dichlorophenol.*

A condenser is not employed in this construction because of the relatively high melting point of 2,5-dichlorophenol.

densing liquid is clear (at least 100 ml of water). Collect the solid product from the collection flask by suction filtration. Weigh the solid that is obtained, calculate its percentage yield, and obtain its melting point. Recrystallize the product from ether/pentane if the melting point obtained is less than 57°C.

Postlab Questions

1. Propose the structure of the solid formed upon cooling the aqueous acid solution of 2,5-dichloroaniline.

2. Upon adding sodium nitrite to the acid solution of 1,5-dichloroaniline, the solid re-enters the solution and the solution becomes colored. Explain these observations.

3. In the procedure for the synthesis of 2,5-dichlorophenol, steam distillation is employed to remove the phenol as it is formed. Why? If the solution had been refluxed rather than steam distilled, would you expect to obtain the same yield of 2,5-dichlorophenol?

Product Analysis for Aromatic Amines and Phenols

Solubility Tests

Aromatic amines, like aliphatic amines (Experiment 24), are basic organic compounds that are usually soluble in 10% aqueous hydrochloric acid. Since few classes of organic compounds exhibit basic characteristics, amines can generally be distinguished from other classes of organic compounds by their solubility in aqueous acid. However, aromatic amines are less basic than aliphatic amines and, particularly when high molecular weight anilines are mixed with 10% aqueous hydrochloric acid, often require more time and more vigorous shaking than their aliphatic counterparts for complete dissolution. In addition, anilines having more than one electron-withdrawing substituent at ortho and para positions may be insoluble in 10% aqueous hydrochloric acid ($pH = 0.1$) because of their low basicity.

Unlike alcohols (Experiment 13), phenols are acidic compounds that are usually soluble in 10% aqueous sodium hydroxide. Carboxylic acids are also soluble in 10% aqueous sodium hydroxide ($pH = 14$) but are similarly soluble in 5% sodium bicarbonate solutions ($pH = 8$–9). Most phenols, however, are insoluble in 5% sodium bicarbonate. Thus phenols can generally be distinguished from other common classes of organic compounds by their solubility in 10% aqueous sodium hydroxide and insolubility in saturated sodium bicarbonate solution.

Bromine Water Test

Amino and hydroxyl groups substituted on an aromatic ring activate the aromatic ring in electrophilic substitution reactions. The high reactivity of anilines and phenols towards weak electrophiles such as bromine is the basis for a distinguishing chemical characterization test for these compounds. Anilines that are soluble in water decolorize bromine water to form *ortho-* or *para*-bromoanilinium bromides:

Bromine water test for anilines

$$ArNH_2 + Br_2 \xrightarrow{H_2O} o\text{- or } p\text{-}BrArNH_3^+Br^-$$

An aniline	*Bromine* (orange)	o- *or* p-*Bromoanilinium bromide* (usually colorless)

Further substitution of bromine is prevented by formation of the conjugate acid of the aniline in the initial substitution reaction. In contrast, phenols react with bromine water at all available ortho and para positions:

Bromine water test for phenols

$$ArOH + n\,Br_2 \xrightarrow{H_2O} o,p\text{-}Br_nArOH + n\,HBr$$

A phenol	*Bromine* (n = 1–3) (orange)	o, p-*Bromo-substituted phenol* (usually colorless)

Although the bromine water test is not suitable for anilines that are insoluble in water (<0.1 g/10 ml), it is suitable for most phenols since phenols that are insoluble in water can be dissolved in dilute sodium hydroxide solution.

PROCEDURE

Dissolve 0.02 g of the organic compound in 2 ml of water (or dilute sodium hydroxide, if necessary) and pour this solution into a small test tube. Add a saturated solution of bromine in water dropwise to the dissolved organic compound, and with vigorous shaking of the test tube continue the addition until the bromine color persists. Record the amount of bromine water that is decolorized.

Ferric Chloride Test

Phenols are often characterized by the colored complexes that they form with ferric chloride, $FeCl_3$. The color of these complexes varies with the nature of the phenol. Other classes of organic compounds, including enols and oximes ($R_2C{=}N{-}OH$), also give colored ferric chloride complexes. However, some phenols do not form colored complexes with ferric chloride. Thus the absence of a positive test with ferric chloride does not necessarily signify that the organic compound is not a phenol. Alcohols do not form colored complexes with ferric chloride.

PROCEDURE

Dissolve 0.02 g of the organic compound in 1 ml of water (or in a solution of ethanol and water if the compound is not water soluble) and pour the resulting solution into a small test tube. Add one drop of a 5%

aqueous solution of ferric chloride to the dissolved organic compound and observe the color of the resulting solution. Most phenols form ferric chloride complexes that are colored red, green, blue, or purple.

Ceric Ammonium Nitrate Test

Like alcohols, phenols form colored complexes with ceric ammonium nitrite. In aqueous solution phenols associate with the cerium reagent to form a brown to greenish-brown precipitate; in aqueous tetrahydrofuran a red to brown coloration is usually observed. The test procedure is described in Experiment 13. Care must be taken in interpreting the results from this test since aromatic amines are generally oxidized by the ceric ion reagent to compounds that exhibit the same color as a phenol complex.

Diazo Coupling Test

Primary aromatic amines torm stable diazonium salts that are readily identified through the formation of orange to red azo dyes from coupling reactions with β-naphthol:

Diazo coupling

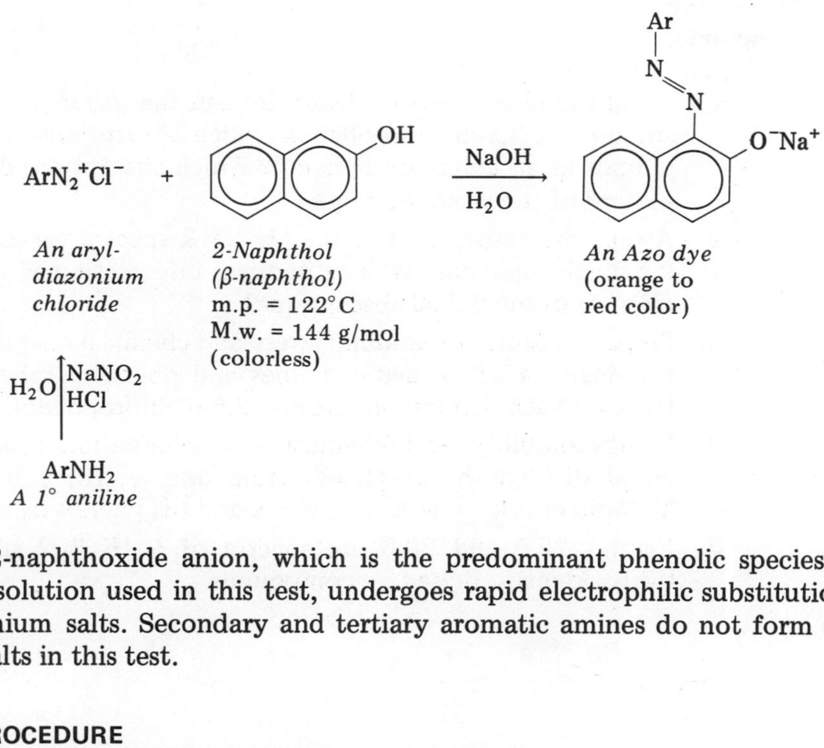

The 2-naphthoxide anion, which is the predominant phenolic species in the basic solution used in this test, undergoes rapid electrophilic substitution with diazonium salts. Secondary and tertiary aromatic amines do not form diazonium salts in this test.

PROCEDURE

In a small test tube dissolve 0.02 g of the organic compound in 1 ml of 10% aqueous hydrochloric acid. Cool this solution to less than 5°C in an ice bath and then add four drops of a 10% aqueous solution of sodium nitrite. In a separate test tube dissolve 0.02 g of β-naphthol in 1 ml of 10% aqueous sodium hydroxide. Using an eyedropper (dropping pipet) transfer five drops of the acidic diazonium salt solution to the basic β-naphthol solution and observe the color of the resulting solution.

Infrared and ^1H NMR spectroscopy provide definitive structural information for aromatic amines and phenols. Characteristic absorptions for the N—H, O—H, C—N, and C—O bonds can be observed in the infrared spectra of these compounds along with characteristic absorptions for ring substitution. Absorptions due to aromatic ring hydrogens that are ortho to the amino or hydroxyl groups of anilines and phenols are usually observed by NMR spectroscopy in the high field aromatic region between 7.1 and 6.6 δ. The N—H proton of anilines is observed as a broad absorption and is usually found in the 5–3 δ spectral region. The chemical shift of the absorption from the hydroxyl proton of phenols is also variable, usually between 7 and 4 δ. The infrared and ^1H NMR spectra of 2,5-dichloroaniline and 2,5-dichlorophenol are given in Figure 20.3 and 20.4, respectively.

Characterization of 2, 5-Dichloroaniline and 2, 5-Dichlorophenol

1. Label the characteristic absorptions in the infrared spectra of 2,5-dichloroaniline and 2,5-dichlorophenol. Which absorptions characterize these compounds as an amine or a phenol? Which absorptions define the substitution pattern of the aromatic ring?
2. Assign the absorptions in the ^1H NMR spectra for 2,5-dichloroaniline and 2,5-dichlorophenol. What structural information is provided by the integrations of individual absorptions?
3. Prepare a table for solubility test and chemical characterization test results for analyses of aromatic amines and phenols. Tabulate the results from tests with 2,5-dichloroaniline and 2,5-dichlorophenol.
4. Using solubility and chemical characterization tests, describe how you could distinguish (a) phenol from aniline, (b) anisole from phenol, (c) *N,N*-dimethylaniline from aniline, and (d) phenol from benzyl alcohol.
5. Figures 20.5 and 20.6 are spectra of $C_6H_5Cl_2N$ and $C_6H_4Cl_2O$, respectively. Identify these two compounds.

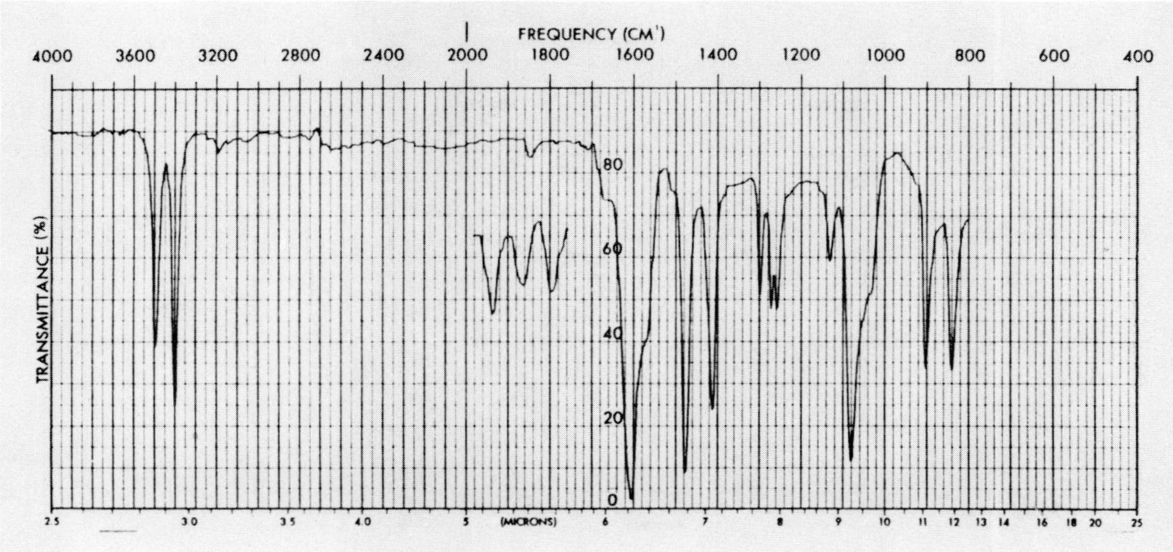

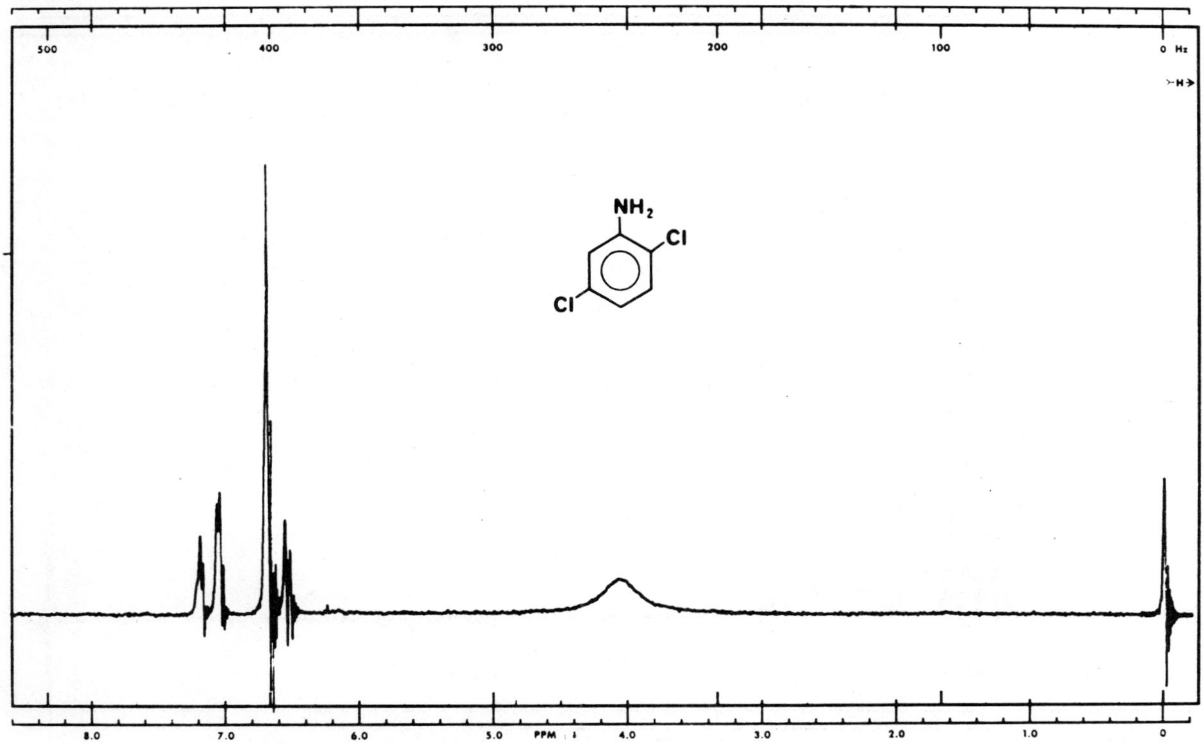

Figure 20.3 *Infrared spectrum (10% by weight in CCl_4, 0.1 mm cells) and ^1H NMR spectrum (in $CDCl_3$) of 2,5-dichloroaniline.*

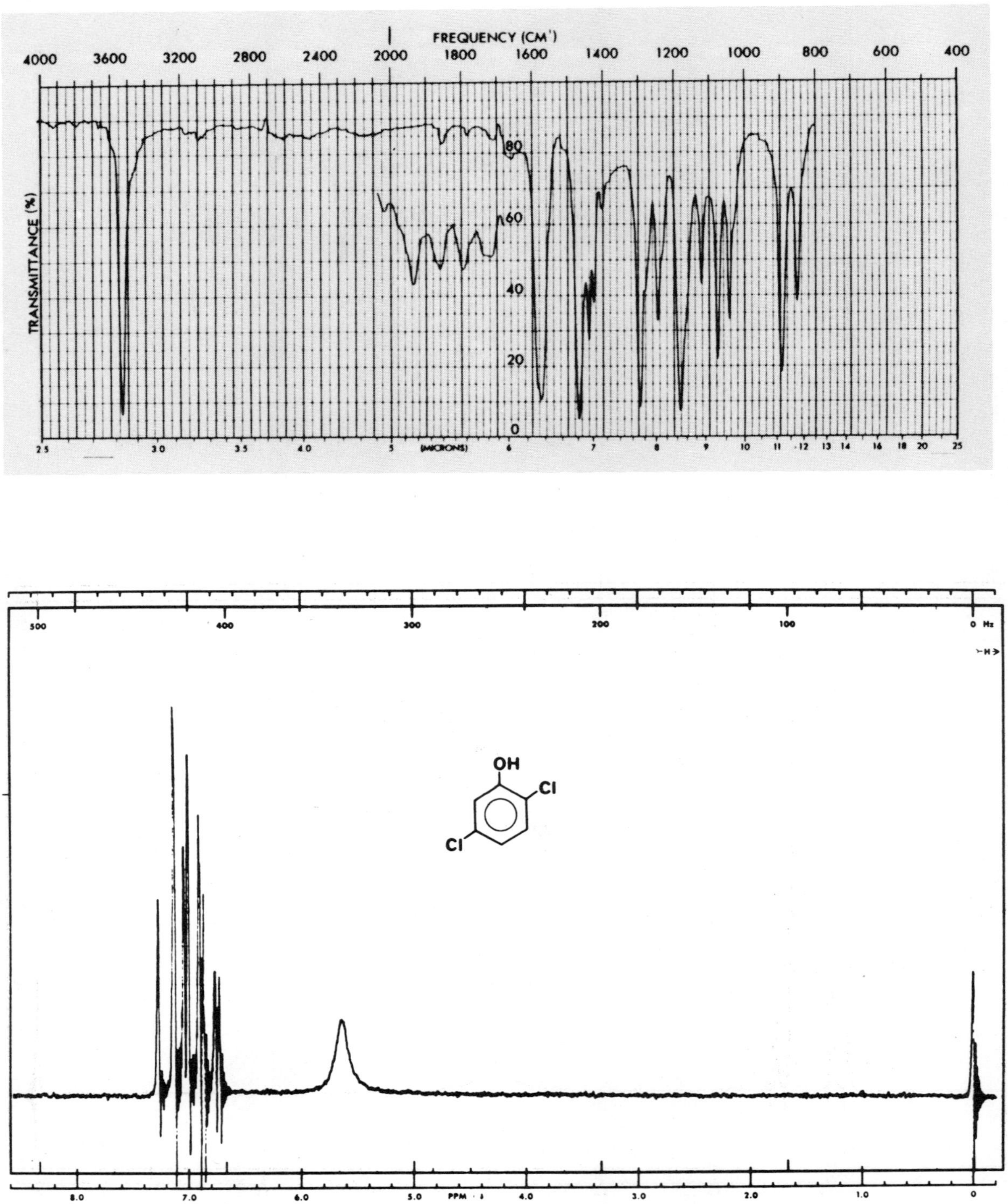

Figure 20.4 *Infrared spectrum (10% by weight in CCl_4, 0.1 mm cells) and ^1H NMR spectrum (in $CDCl_3$) of 2,5-dichlorophenol.*

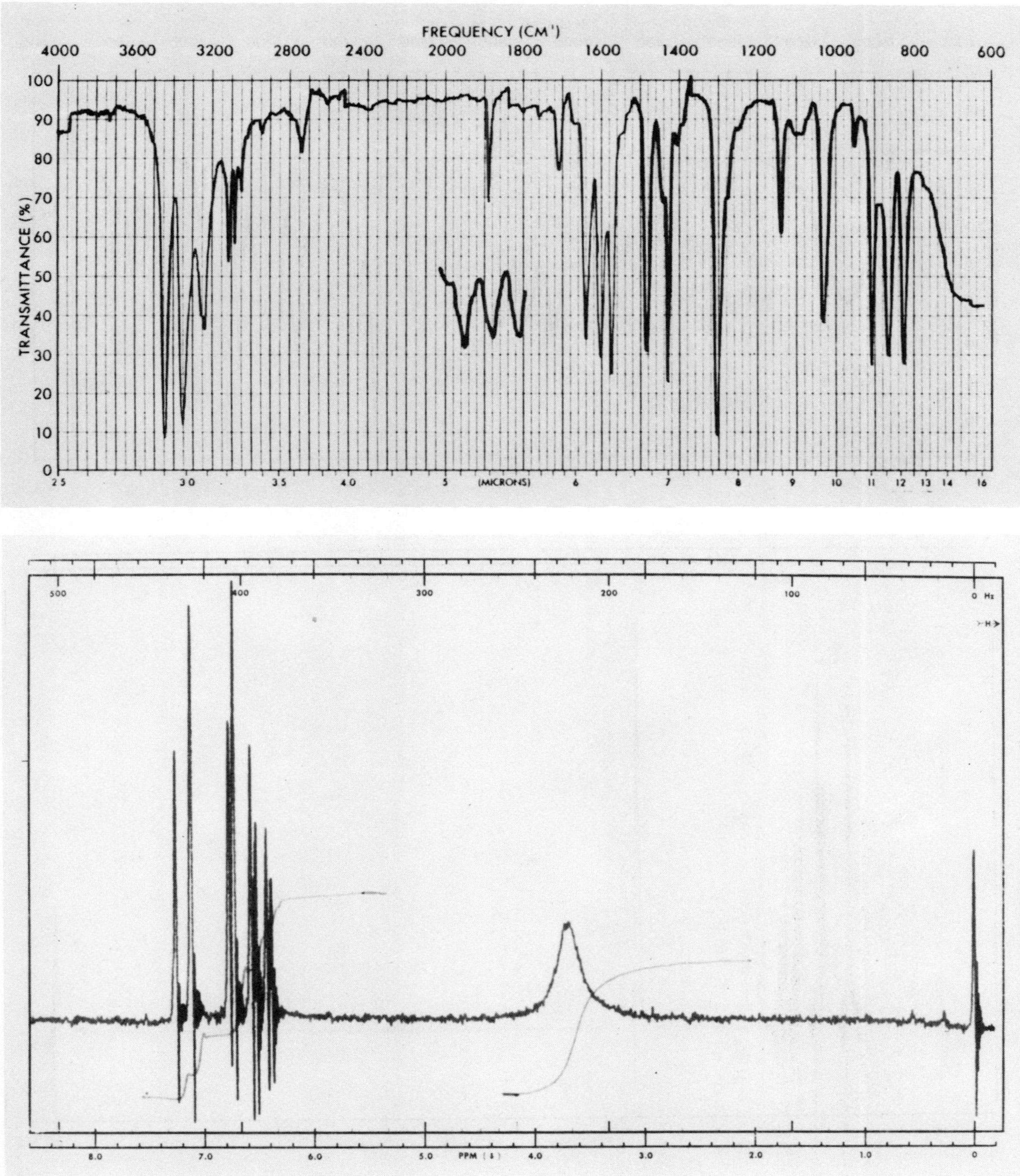

Figure 20.5 *Infrared spectrum (thin film) and ^1H NMR spectrum (in $CDCl_3$) of $C_6H_5Cl_2N$.*

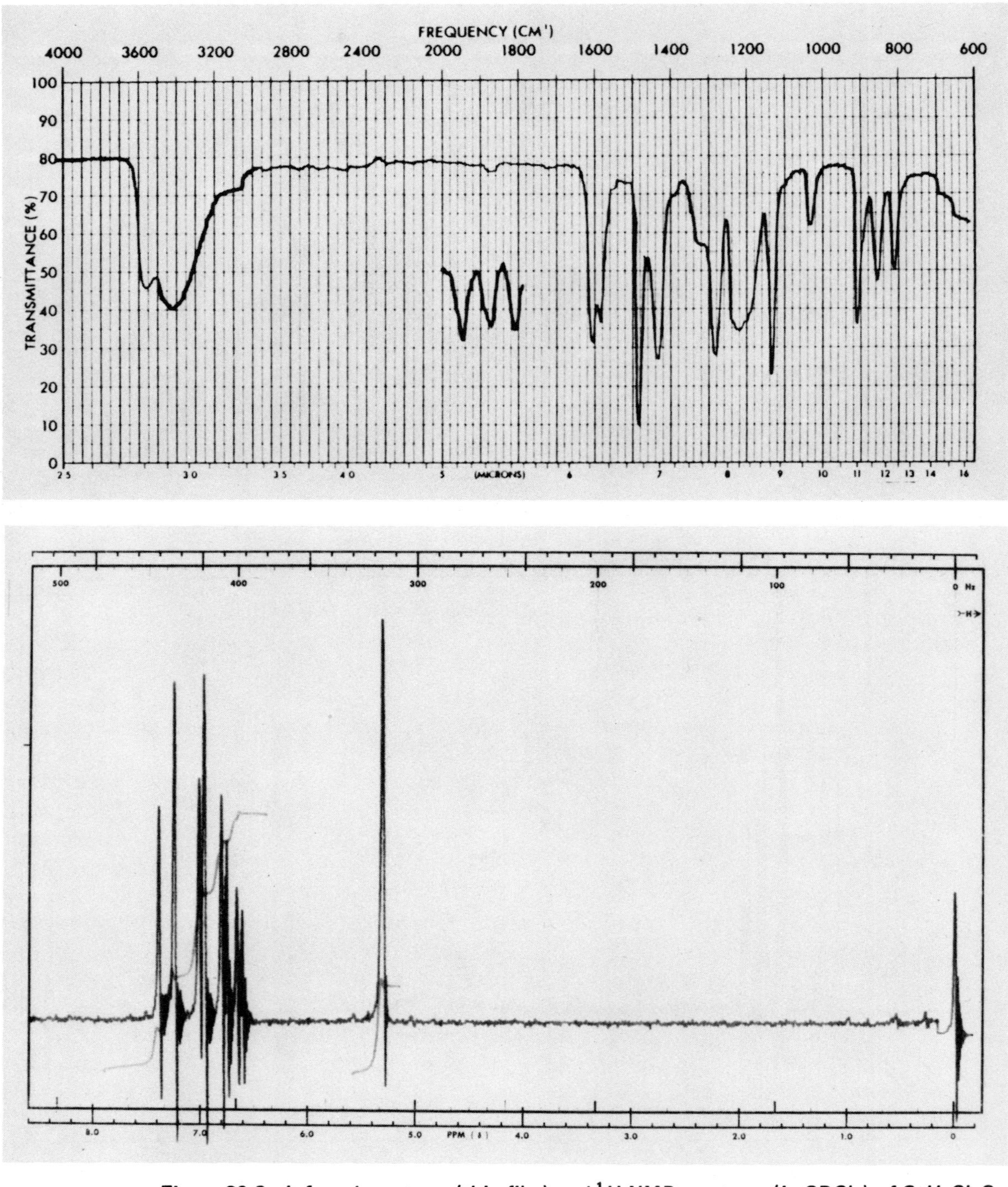

Figure 20.6 *Infrared spectrum (thin film) and ^1H NMR spectrum (in $CDCl_3$) of $C_6H_4Cl_2O$.*

Experiment Twenty-One

Identification of Organic Compounds Containing Carbon, Hydrogen, and Halogen

In Experiment 13 a systematic procedure for the identification of limited classes of organic compounds was introduced. This methodology will be expanded in the next four experiments to encompass the techniques necessary for the identification of all common classes of organic compounds containing the elements carbon, hydrogen, halogen, oxygen, and nitrogen.

In this experiment you will be asked to identify an unknown organic compound whose composition will be limited to combinations of the elements carbon, hydrogen, chlorine, bromine, and iodine. The specific chemical classes to which the unknown may belong are alkanes, alkenes, alkynes, aromatic hydrocarbons, alkyl halides, and aryl halides. Alkanes include all hydrocarbons that do not contain carbon-carbon double or triple bonds. Hydrocarbons that contain one or more carbon-carbon double bond are classified as alkenes. Alkynes are hydrocarbons that contain one or more carbon-carbon triple bond. Alkanes, alkenes, and alkynes are collectively referred to as *aliphatic hydrocarbons.* An alkyl halide is an aliphatic hydrocarbon in which one or more hydrogens have been replaced by halogen atoms. *Aromatic hydrocarbons* are those hydrocarbons that contain an aromatic ring (e.g., benzene); the aromatic ring may have aliphatic hydrocarbon substituents. Aryl halides are aromatic hydrocarbons in which one or more of the hydrogen atoms bonded directly to the aromatic ring have been replaced with halogen atoms.

The chemical and physical characterization of alkanes, alkenes, and alkyl halides has already been described in Experiment 13. These three classes of compounds are differentiated on the basis of their differing solubilities in concentrated sulfuric acid, elemental analyses, and reactions with bromine in carbon tetrachloride and with potassium permanganate.

Since both alkenes and alkynes are unsaturated hydrocarbons that react by addition to the π-bond, compounds in these two classes have the same solubility properties (i.e., soluble in ether and in concentrated sulfuric acid) and give the same results in the chemical characterization tests with bromine in carbon tetrachloride and with potassium permanganate (Experiment 7).

After a compound has been shown to be unsaturated by these two tests, the best method to distinguish between alkenes and alkynes is infrared spectroscopy. The infrared spectrum of a terminal alkyne will have a sharp absorption band of weak to medium intensity in the region of 2260 to 2100 cm^{-1} due to stretching of the carbon-carbon triple bond. Terminal alkynes will also have an absorption as a result of carbon-hydrogen stretching of the $-C{\equiv}C-H$ function in the region of 3330 to 3270 cm^{-1}. Unsymmetrically disubstituted carbon-carbon triple bonds show only a weak absorption band due to stretching vibrations in the 2260 to 2100 cm^{-1} spectral region; symmetrically substituted carbon-carbon triple bonds do not absorb in this spectral region. A weak to medium intensity absorption in the region of 1680 to 1620 cm^{-1} is observed in the spectrum of alkenes due to carbon-carbon double bond stretching. (See Experiment 12 and Appendix B for additional information on infrared spectroscopy.)

Since sulfonation of the aromatic ring of aromatic hydrocarbons and aryl halides by concentrated sulfuric acid at room temperature is relatively slow, these nonpolar compounds are insoluble in concentrated sulfuric acid. Like other nonpolar compounds, aromatic hydrocarbons and aryl halides are soluble in ether and insoluble in water. Compounds that contain an aromatic ring can be distinguished from aliphatic compounds by interpretation of their infrared spectra (Experiment 12) and by chemical characterization tests that involve selective reactions of the aromatic ring. Two chemical characterization tests for aromatic compounds, the aluminum chloride-chloroform test and the formaldehyde-sulfuric acid test, are described in this experiment.

The Aluminum Chloride–Chloroform Test

Aromatic hydrocarbons and aryl halides react with chloroform in the presence of aluminum chloride to produce colored products. The colors produced in this test are quite characteristic: benzene, alkyl substituted benzenes, and aryl halides produce an orange to red color, naphthalene and substituted naphthalenes produce a blue color, biphenyl and phenanthrene produce a purple color, and anthracene produces a green color. With time the colors tend to change to various shades of brown. Some reactive alkenes also produce a yellow or orange color upon reaction with $AlCl_3/CHCl_3$; therefore, it is important to demonstrate that the unknown is not an alkene before testing for aromatic character with the aluminum chloride in chloroform test. Many alkyl bromides and iodides also produce yellow-brown or purple colors in this test.

PROCEDURE

Dissolve 0.1 g (0.1 ml) of the compound to be tested in 1 ml of chloroform in a dry test tube. Tip the test tube to moisten the inner wall with solution, and then add 0.5 g of anhydrous aluminum chloride so that the powder strikes the wall of the tube. Note the color of the moist powder on the wall and the color of the solution. It is essential for the success of this test that the aluminum chloride is anhydrous. Aluminum chloride that has been exposed to the atmosphere or left in an unsealed bottle will rapidly react with atmospheric moisture and lose its activity.

The products formed in this characterization test are the result of aluminum chloride catalyzed Friedel-Crafts reactions between chloroform and the aromatic compound. For example, the Friedel-Crafts reaction of chloroform with an aromatic hydrocarbon produces a triarylmethane. The reaction pro-

$$\underset{\textit{Aromatic hydrocarbon}}{ArH} + CHCl_3 \xrightarrow{AlCl_3} \underset{\textit{Triarylmethane}}{(Ar)_3CH} + 3HCl$$

ceeds by way of intermediate carbonium ions, and disproportionation between the triarylmethane and the intermediate carbonium ions produces the relatively stable, highly colored triarylmethyl cation:

$$\underset{\textit{colorless}}{(Ar)_3CH} + R^+AlCl_4^- \rightleftharpoons \underset{\textit{yellow-orange}}{(Ar)_3C^+AlCl_4^-} + RH$$

The aluminum chloride-chloroform test is not reliable for detecting aromatic structure in compounds where the aromatic ring is substituted with a group that strongly deactivates the aromatic ring towards electrophilic substitution reactions.

The Formaldehyde–Sulfuric Acid Test

The formaldehyde-sulfuric acid test is useful in distinguishing aromatic compounds from nonaromatic compounds for those substances that are insoluble in concentrated sulfuric acid. Aromatic compounds react with this reagent to produce orange, red, pink, purple, blue, or green colors. With time these colors turn to brown or black. Color formation is due to the production of relatively stable carbonium ions similar to those formed in the aluminum chloride–chloroform test. Nonaromatic compounds that are insoluble in sulfuric acid produce either no color or a pale yellow color in the test reagent. Since many compounds that are soluble in concentrated sulfuric acid react to form colored products, this test should not be performed on those compounds that are soluble in concentrated sulfuric acid.

PROCEDURE

In a small test tube dissolve 30 to 50 mg (1 or 2 drops) of the compound to be tested in 1 ml of hexane. In a second small test tube add one drop of 37% formaldehyde to 1 ml of concentrated sulfuric acid. Add two drops of the hexane solution to the test tube containing the formaldehyde-sulfuric acid solution. Shake the tube gently and note the formation of color at the interface between the two immiscible layers and in the acid solution. The hexane used as the solvent for this test should always be tested with the reagent to confirm that the solvent does not contain aromatic impurities which cause misleading results.

Table 21.1 Summary of Chemical Tests[a]

Test	Alkane	Alkyl halide	Alkene	Alkyne	Aromatic hydrocarbon	Aryl halide	Reference for procedure
Solubility:							p. 82
ether	+	+	+	+	+	+	
water	–	–	–	–	–	–	
conc. H_2SO_4	–	–	+[b]	+[b]	–[c]	–[c]	
Elemental analysis:							
halogen	–	+	±[d]	±[d]	–	+	p. 159
Characterization tests:							
Br_2/CCl_4	–	–	+	+	–	–	p. 85
$KMnO_4$	–	–	+	+	–	–	p. 86
$AlCl_3/CHCl_3$	–	x	x	x	+	+	p. 252
H_2CO/H_2SO_4	–	–	x[e]	x[e]	+	+	p. 253
$AgNO_3/C_2H_5OH$	–	±[f]	±[f]	±[f]	–	–	p. 130

[a] A + indicates soluble or positive test. A – indicates insoluble or negative test. A ± indicates the test may be positive for some types of compounds of this class but not for others. x indicates that the test may give misleading results if applied to compounds of this class.
[b] Soluble with reaction and discoloration.
[c] Reacts and dissolves slowly at room temperature.
[d] Alkenes and alkynes may contain halogen.
[e] Not applicable to compounds that are soluble in conc. H_2SO_4.
[f] Only certain types of halohydrocarbons give a positive test (see Experiment 11).

Table 21.1 presents a summary of the chemical tests applicable to the unknown in this experiment and Figure 21.1 describes a general scheme for the systematic analysis of the unknown.

Refractive Index

Since hydrocarbons and halohydrocarbons are relatively unreactive compounds, it is not generally possible to prepare a solid derivative of one of these compounds in order to confirm an identification of an unknown. Therefore, in the identification of compounds in these classes it is important to obtain physical constants in addition to boiling and melting points that can be readily compared to those of known compounds. The refractive index of an organic liquid is a characteristic physical constant that can be determined accurately and quickly for use in the identification of an unknown.

The refractive index of a substance is the ratio of the velocity of light in air to the velocity of light in that substance. The refractive index is denoted by the letter n, with a superscript indicating the temperature at which the measurement was made and a subscript denoting the wavelength of light used. Normally, light of 589 nm, the sodium D line, and a temperature of 20°C are used for refractive index measurements. Thus, the refractive index of hexane is recorded as $n_D^{20} = 1.3749$.

The refractive index decreases as temperature increases. Although the varia-

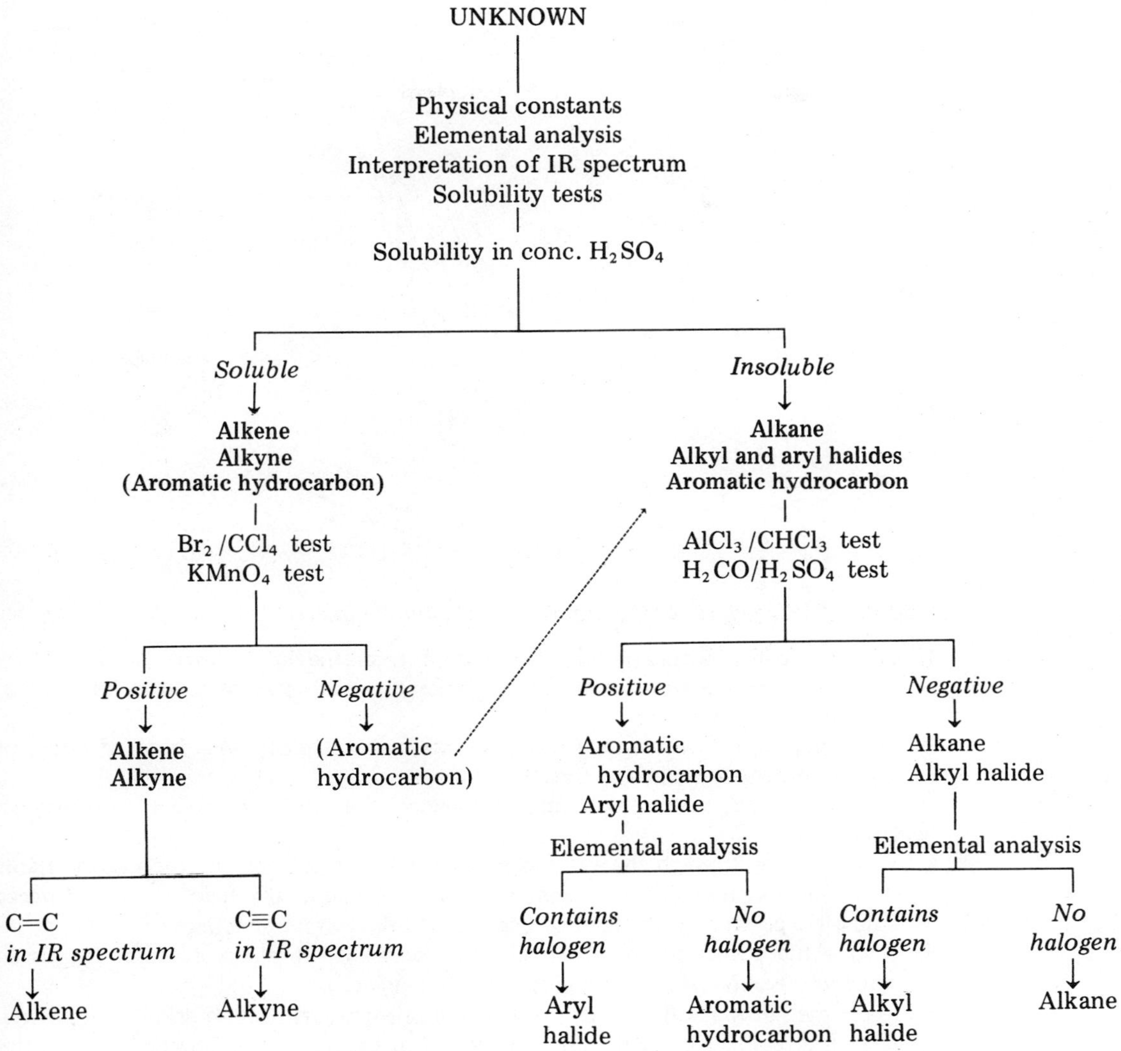

Figure 21.1 *General scheme for the classification of an organic compound that may contain only carbon, hydrogen, and halogen.*

tion in refractive index due to temperature change is slightly different for different organic compounds, measurements made near 20°C can be corrected to 20°C by assuming a decrease of 0.0004 in the refractive index per degree centigrade over 20°C. Most common organic compounds have refractive indices between 1.300 and 1.700. The refractive index of a compound is very sensitive to the presence of impurities. Unless the unknown has been carefully purified, it is unlikely that the refractive index of the unknown material will agree to within ±0.001 of the literature value of the refractive index for the known compound.

The refractive index of an organic liquid is measured by the use of a *refrac-*

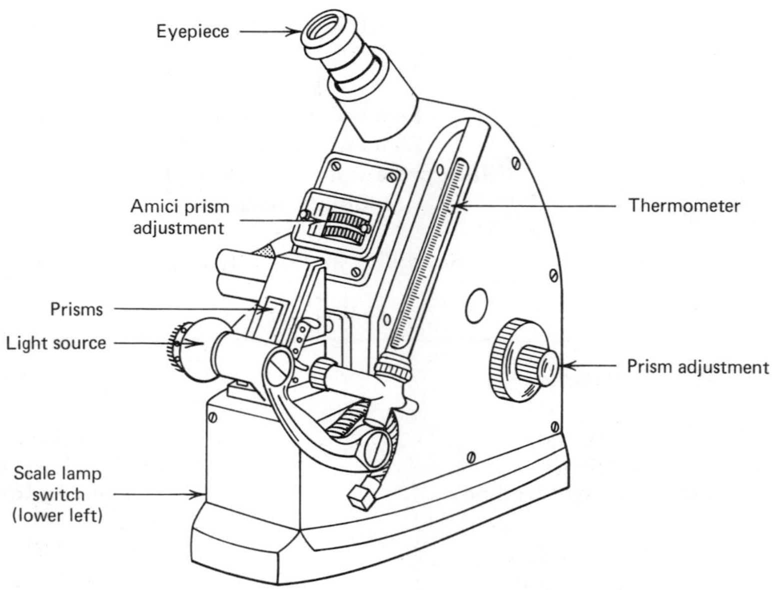

Figure 21.2 *Procedure for the use of the Abbe refractometer.*

1. With the switch located on the power cord, turn on the light source.
2. Record the temperature. A constant temperature bath may be used to keep the instrument at a set temperature.
3. Open the hinged prisms by lifting the top one and place two drops of liquid on the bottom prism. Close the top prism.
   CAUTION: The prisms must not be touched with any hard object since they scratch easily.
4. While looking through the eyepiece, rotate the prism adjustment knob (lower right side) until the boundary between light and dark appears in the field of view. If necessary, adjust the position of the light source for the brightest illumination.
5. Rotate the Amici prism adjustment (center front) to eliminate color fringes and to sharpen the boundary between light and dark areas.
6. Make the necessary fine adjustment of the prism adjustment to bring the light and dark areas into coincidence with the intersection of the cross hairs. The field of view through the eyepiece should appear like:

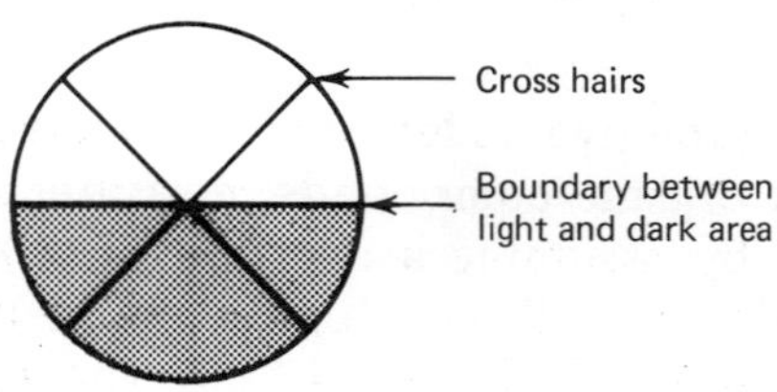

7. Push the switch (lower left side) to illuminate the scale and read the value of the refractive index.
8. Open the prism and gently clean the surfaces with a clean soft tissue dampened with ethanol. Close the prism when dry and turn off the light source.

tometer. The operation of the common laboratory refractometer is based on a simple optical principle: A beam of light that is passing at an angle across the interface between two transparent materials is bent toward the material in which it travels more slowly. The refractive index, n, is then given by

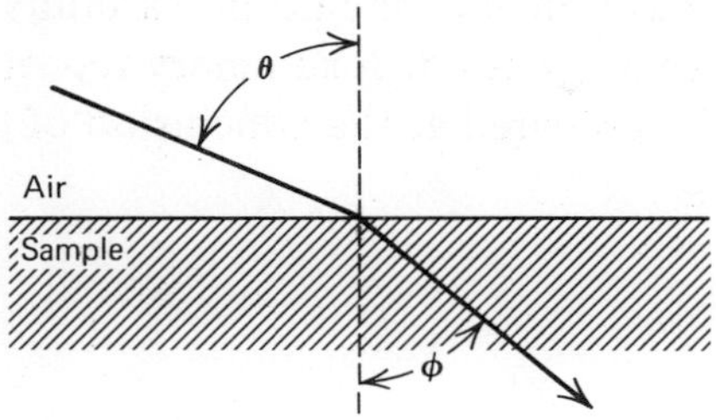

$$n = \frac{\sin\theta}{\sin\phi}$$

where θ is the angle of the incident beam and ϕ is the angle of the refracted beam. The operation of this type of refractometer, the Abbe refractometer, is described in Figure 21.2.

Prelab Questions

1. An unknown organic liquid (b.p. = 68–70°C, n_D^{20} 1.4652) is determined to be soluble in ether and in concentrated sulfuric acid but insoluble in water. Elemental analysis indicates that the compound contains bromine. The unknown reacts with bromine in carbon tetrachloride to give a colorless solution without evolution of hydrogen bromide. It also reacts with a dilute solution of potassium permanganate to give a dark brown precipitate. What is the probable identity of the unknown? Are there other compounds that fit these data?

2. An unknown organic liquid (b.p. = 158–160°C, n_D^{20} 1.520) is soluble in ether but insoluble in concentrated sulfuric acid and in water. Elemental analysis indicates that the compound contains chlorine. The unknown reacts with $AlCl_3/CHCl_3$ to produce a bright orange color. What is the probable identity of the unknown compound? If more than one compound is possible, what additional information would be necessary to differentiate between possibilities?

3. What chemical test could be used to distinguish between benzyl chloride (b.p. = 179°C) and *o*-dichlorobenzene (b.p. = 179°C)?

4. What additional data would be sufficient to distinguish between cyclopentane (b.p. = 49°C) and 2,2-dimethylbutane (b.p. = 50°C)?

5. Describe chemical tests that would distinguish between (a) 1-chloropropane and 3-chloropropene, (b) *p*-dichlorobenzene and hexachlorobenzene, (c) 3-chloro-1-butene and 2-chloro-1-butene.

6. How can infrared spectroscopy be used to distinguish between (a) 3-hexyne and 1-hexyne, (b) 1,1,2,2-tetrachloroethene and 1,1,1,2-tetrachloroethane, (c) *m*-xylene and *p*-xylene?

EXPERIMENTAL PROCEDURE

You will be expected to identify an unknown organic compound by making use of the procedures discussed in this experiment. The unknown compound will be either an alkane, alkene, alkyne, alkyl halide, aromatic hydrocarbon, or aryl halide. The compound that you are given may require purification before accurate physical data can be determined. Experiments should be carefully planned so that time and materials are not wasted. A summary report, as outlined in Experiment 13, should be prepared at the conclusion of the experiment.

Experiment Twenty-two

Identification of Organic Compounds Containing Carbon, Hydrogen, and Oxygen

Organic compounds that contain the elements carbon, hydrogen, and oxygen are commonly divided into three groups. The first group is composed of alcohols, phenols, and ethers. Aldehydes and ketones form the second group of compounds. The third group is made up of carboxylic acids and their derivatives. In this experiment the techniques employed for the characterization and identification of compounds in the first and second groups will be introduced. Methods for the analysis of carboxylic acids and carboxylic acid derivatives will be discussed in Experiment 24.

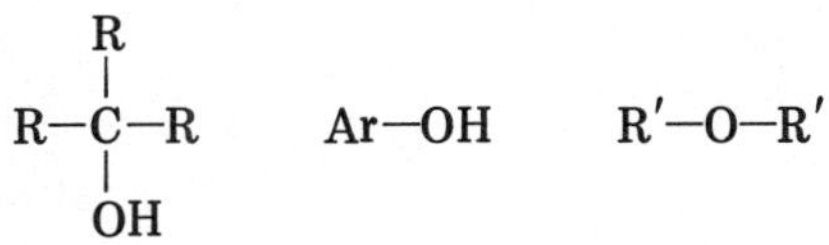

Alcohol *Phenol* *Ether*

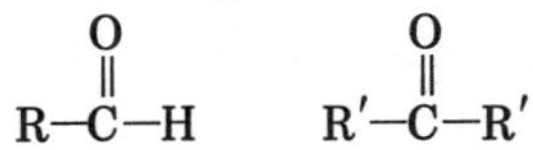

Aldehyde *Ketone*

R = hydrogen atom, or aliphatic or aromatic hydrocarbon group
R′ = aliphatic or aromatic hydrocarbon group
Ar = aromatic hydrocarbon group

The general procedure for the identification of an unknown, which was discussed in Experiment 13, should be followed in this experiment. Since compounds that are classified as alcohols, phenols, ethers, aldehydes, or ketones may also contain halogen substituents, elemental analysis for halogen should be performed on the unknown.

Solubility

Most alcohols, phenols, ethers, aldehydes, and ketones are soluble in ether. Since these compounds contain oxygen, they are Brønsted bases and react with concentrated sulfuric acid to form oxonium salts that are soluble in excess sulfuric acid. Monofunctional alcohols, aldehydes, and ketones of less than five carbons are generally soluble in water. Monofunctional compounds belonging to these classes that have more than five carbons are generally insoluble in water. The five-carbon monofunctional compounds that belong to these classes have varying solubility in water depending on their exact structure; the more highly branched isomers of a given class are more soluble than the straight-chain isomers.

Monofunctional ethers of less than four carbons are soluble in water. Four-carbon monofunctional ethers are marginally soluble in water (except tetrahydrofuran, which is miscible with water). Monofunctional ethers of more than four carbons are insoluble in water. A number of common polyfunctional ethers such as dioxane and 1,2-dimethoxyethane are soluble in water.

Phenol, C_6H_5OH, is only marginally soluble in water, and alkyl substituted and halogenated phenols are insoluble in water. Since phenols are weak acids ($K_a > 10^{-11}$), they react with 10% aqueous sodium hydroxide solution to form water-soluble phenoxide salts. However, some phenols with large hydrocarbon

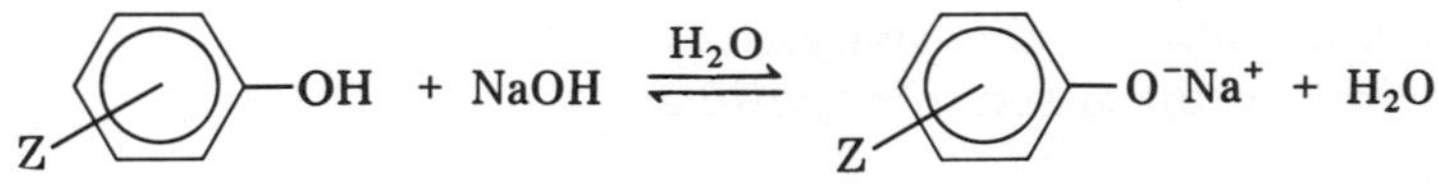

Substituted phenol [*water insoluble*] | *Phenoxide salt* [*water soluble*]

substituents are insoluble in 10% aqueous sodium hydroxide. Since aqueous sodium bicarbonate is not basic enough to react with most phenols, phenols are generally insoluble in 5% aqueous sodium bicarbonate—a characteristic that distinguishes them from carboxylic acids. Only phenols with acid-strengthening substituents such as *p*-nitrophenol are sufficiently acidic to dissolve in aqueous sodium bicarbonate solution.

Infrared Spectroscopy

The interpretation of the infrared spectrum of an unknown compound that contains carbon, hydrogen, and oxygen is useful in the determination of the chemical class of the unknown (Experiment 12 and Appendix B). Alcohols and phenols show a strong absorption in the range of 3650 to 3100 cm^{-1} due to the O—H stretching vibrations. Phenols also exhibit the absorptions characteristic for the aromatic ring in the region of 1670 to 1430 cm^{-1}. Ketones and aldehydes both absorb strongly in the 1750 to 1660 cm^{-1} spectral region, but these two classes can be distinguished from each other since aldehydes have an absorption near 2720 cm^{-1} due to C—H stretching vibrations of the aldehyde functional group. Ethers do not have characteristic functional group absorptions in the diagnostic portion of the infrared spectrum; only absorptions due to the hydrocarbon portion of the molecule are observed in this region. The characteristic absorptions due to carbon-oxygen single bond stretching occurs in the region of 1150 to 1090 cm^{-1} for aliphatic ethers and in the regions of 1280 to 1200 cm^{-1} and 1080 to 1020 cm^{-1} for aryl ethers.

Chemical Characterization Tests

Chemical characterization tests for alcohols, phenols, aldehydes, and ketones have been introduced in previous experiments (see Table 22.1). A systematic scheme for the application of these tests to the classification of the unknown in this experiment is described in Figure 22.1. Ethers are relatively unreactive compounds and give negative test results in the scheme. A compound is classified as an ether on the basis of its lack of reactivity, its solubility in concentrated sulfuric acid, and its infrared spectrum.

Preparation of Derivatives

The preparation of one or more derivatives is the last step in a rigorous systematic identification of an unknown organic substance. The successful con-

Table 22.1 Summary of Chemical Tests for Alcohols, Phenols, Ethers, Aldehydes, and Ketones[a]

Characterization test	Alcohol	Phenol	Ether	Aldehyde	Ketone	Reference for procedure
2,4-Dinitrophenylhydrazine	–	–	–	+	+	p. 180
Ceric ammonium nitrate	±[b]	±[b]	–	–	–	p. 163 and 245
Ferric chloride	–	+	–	–	–	p. 244
Chromic acid	±[c]	x	–	+	–	p. 164 and 181
Tollens' reagent	–	x	–	+	–	p. 182
Iodoform	±[d]	x	–	±[d]	±[d]	p. 182
Bromine-water	–	+	±[e]	x	x	p. 244
Lucas' reagent	±[f]	–	x	x	x	p. 165
Bisulfite reagent	–	–	–	±	–	p. 179

[a] A + indicates a positive test result. A – indicates a negative test result. A ± indicates some compounds of this class will give a positive result and other compounds will give a negative result. An x indicates that the test is not applicable to compounds in this class and misleading results may be obtained if the test is applied to compounds of this class.

[b] Alcohols give a red or orange complex and phenols give a brown complex.

[c] Tertiary alcohols do not react.

[d] Positive test results for $R-\underset{\underset{O}{\|}}{C}-CH_3$ and $R-\underset{\underset{OH}{|}}{\overset{\overset{H}{|}}{C}}-CH_3$, where R = alkyl, aryl, or hydrogen.

[e] Positive test results for some aryl ethers.

[f] Subclassification test for alcohols.

Figure 22.1 *Scheme for the classification of alcohols, phenols, ethers, aldehydes, and ketones by chemical characterization tests.*[a]

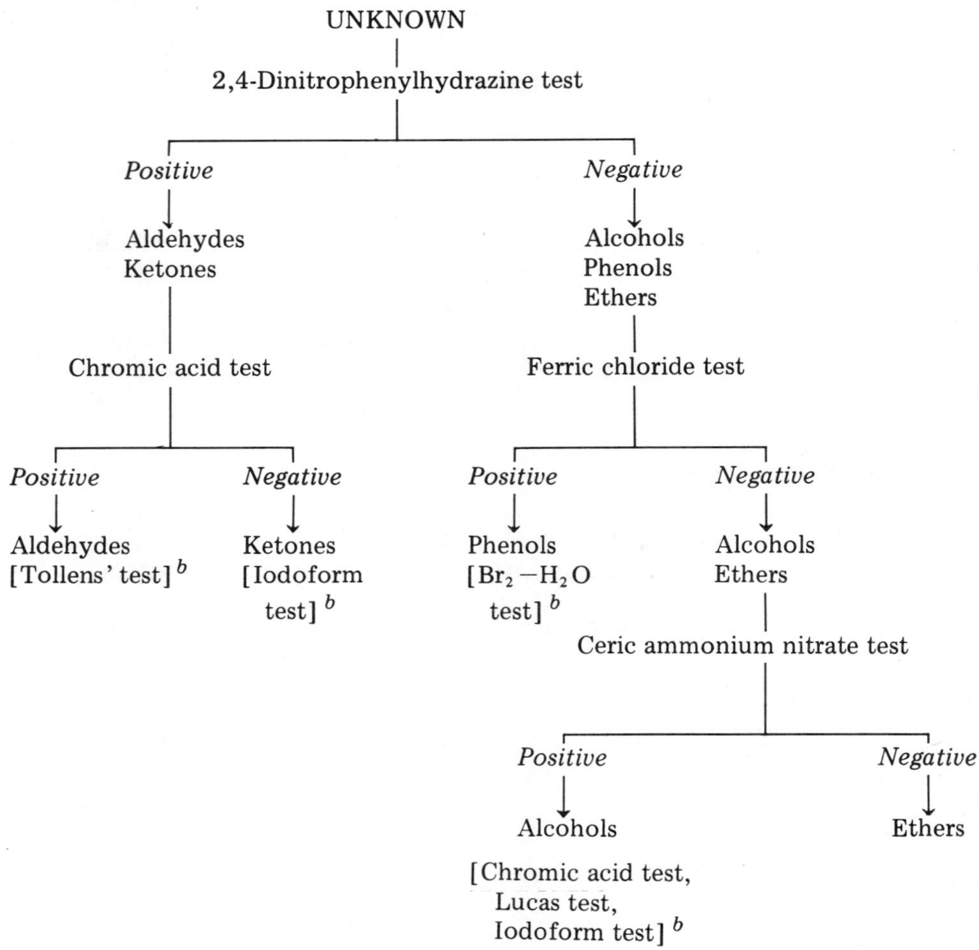

[a]Chemical characterization tests should be performed after the determination of physical constants, solubility tests, elemental analysis, and infrared spectrum.
[b]Confirming classification test or a subclassification test.

version of an unknown compound to a solid derivative provides evidence of chemical structure and an additional physical constant for comparison with literature data.

Before the preparation of a derivative is considered, the unknown should be fully characterized by physical and chemical techniques and a list of "possible compounds" that have similar properties should be prepared. A suitable derivative should fulfill the following criteria:

1. *The derivative should be a solid that melts, if possible, above 50°C and below 250°C.* Liquid derivatives are difficult to isolate and to purify. Compounds that melt below 50°C tend to separate as oils and are difficult to crystallize. If a compound melts much above 200°C, its melting point cannot be conveniently determined with accuracy.
2. *The derivatives of closely related compounds that are possibilities for the identification of the unknown should have melting points that differ by more than 5°C.*
3. *The derivative and the compound from which it is prepared should have different melting points.* A common error is to recover starting material and fail to recognize it as such.
4. *The reaction used to prepare the derivative should be complete in a short time and give a good yield of product with minimum interference from side reactions.*
5. *The derivative should be easily purified in good yield by recrystallization from a common solvent.*

If a melting point is not listed in the literature for a derivative of a possible compound, that derivative is not suitable for that compound. In some cases more than one derivative must be prepared in order to eliminate all but one compound from the list of possible compounds and to conclusively identify the unknown.

For most classes of organic compounds, many different types of derivatives are listed in the literature. Only preparations of the most generally suitable derivatives will be described in this text. The procedures described are designed to be applicable to most compounds and to yield 100 to 200 mg of the purified derivative. Since the preparation of derivatives is performed on a semimicro scale, it is important to maximize the amount of pure derivative that is isolated. Careful planning and good technique are necessary. The procedure should involve as few transfer operations as possible, and transfers should be quantitative. Reaction vessels should be as small as possible since the walls of large flasks will adsorb greater amounts of product. The reagents and solvents should be of high purity. It is often a good idea to make a "trial run" using a sample of known compound for the preparation and purification of a derivative before attempting to prepare a derivative from a limited amount of unknown.

Derivatives of Alcohols and Phenols

Carboxylic acid esters and urethanes are the most common types of derivatives prepared from compounds containing the hydroxyl group. Esters are most efficiently prepared by the reaction of the hydroxyl compound with an acid chloride. Nitroaromatic acid chlorides are most often employed because the

Nitroaromatic ester derivatives

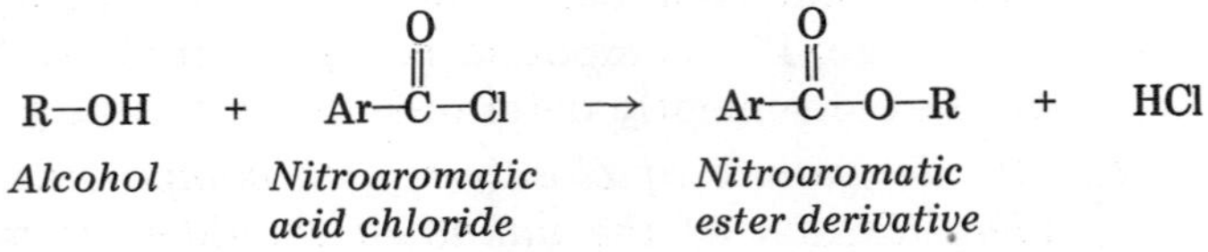

nitroaromatic esters are easily recrystallized solids. Urethanes are prepared by the reaction of a hydroxyl compound with an aromatic isocyanate, usually α-naphthyl isocyanate. The product is also called a carbamate.

Urethane derivatives

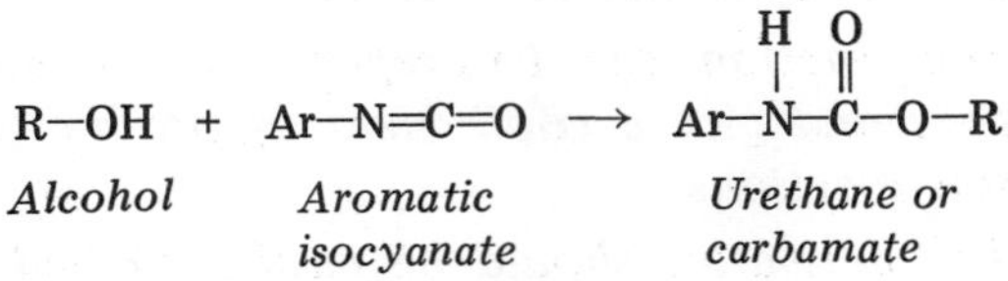

Derivatives of phenols can also be prepared by bromination of the aromatic ring (Experiment 20). The yields for this reaction are generally excellent.

3,5-Dinitrobenzoates and *p*-Nitrobenzoates. A small excess of the alcohol or phenol is heated with 3,5-dinitrobenzoyl chloride (or *p*-nitrobenzoyl chloride) to form the nitroaromatic ester. The product is pulverized and then extracted with a sodium carbonate solution to remove the acid and the acid anhydride formed in side reactions. The residue is then recrystallized from aqueous methanol or ethanol. Tertiary alcohols react slowly with the acid chloride and often undergo side reactions (substitution or elimination). For alcohols that do not readily form esters, it is advantageous to heat the acid chloride and alcohol dissolved in pyridine. Phenols also give better yields of ester derivatives if pyridine is used as the solvent. The pyridine serves as a catalyst and as a base to neutralize the hydrochloric acid produced in the esterification reaction.

The 3,5-dinitrobenzoyl chloride (m.p. = 72 to 74°C) or the *p*-nitrobenzoyl chloride (m.p. = 73 to 75°C) must be of high purity or the derivative will be formed in low yield. These reagents decompose slowly on exposure to the atmosphere. If the melting point of the reagent is below the range listed above, the reagent should be recrystallized from hexane before use in the following procedure.

PROCEDURE

(a) Place 1.0 g of 3,5-dinitrobenzoyl chloride and 0.1 g of alcohol in a dry test tube. Heat the tube gently for 5 minutes to form a melt at the bottom of the tube. Carefully add 10 ml of 5% sodium carbonate to the mixture and cool the test tube in an ice bath. By means of a glass rod or metal spatula, break up and pulverize the solid mass that forms so that no lumps remain. Filter the solid and wash the crystals with water.

Dissolve the crystalline mass in a minimum amount of hot ethanol (10 to 20 ml) and add water dropwise until the solution just turns cloudy. Allow the solution to cool slowly.

p-Nitrobenzoate esters are prepared by the same procedure through the use of *p*-nitrobenzoyl chloride in place of 3,5-dinitrobenzoyl chloride.

(b) Tertiary alcohols, phenols, and other hydroxyl compounds which fail to give a benzoate ester by procedure (a) generally react in a pyridine solvent. Add 5 ml of pyridine, 1 g of 3,5-dinitrobenzoyl

---

CAUTION: Pyridine has a nauseating odor and its fumes are moderately toxic. Work in a well-ventilated area or hood.

---

chloride (or *p*-nitrobenzoyl chloride), and 1 g of the hvdroxyl compound to a large size test tube. Heat the mixture gently on the steam bath for 15 minutes and then slowly pour the mixture with vigorous stirring into 20 ml of cold 10% aqueous hydrochloric acid. Allow the solid residue to settle to the bottom and decant most of the aqueous solution. Add 10 ml of 5% sodium carbonate to the precipitate, mix thoroughly, and isolate the solid derivative by filtration. Recrystallize the derivative as described in part (a).

α-Naphthylurethanes. α-Naphthylurethanes are formed by heating α-naphthyl isocyanate with a small excess of the hydroxyl compound. If the hydroxyl compound is a phenol, a small amount of pyridine is usually added to catalyze the reaction. The derivative is recrystallized from a hydrocarbon solvent such as hexane.

PROCEDURE

One gram of alcohol or phenol is placed in a large test tube, and 0.7 g of α-naphthyl isocyanate is added. If the reactant is a phenol, two drops

---

CAUTION: α-Naphthyl isocyanate is a powerful lachrymator. Work in a well-ventilated area or hood.

---

of pyridine are also added. After allowing any spontaneous reaction to occur, the mixture is heated on a steam bath for 5 minutes. The solution is then cooled in an ice bath, and the sides of the tube are scratched with a glass rod to induce crystallization. The solid residue is isolated by filtration and then dissolved in 10 ml of hot hexane. The hexane solution is filtered while hot to remove undissolved solid. The derivative is obtained by filtration after cooling the hexane solution in an ice bath.

Bromo Derivatives of Phenols. The aromatic ring of a phenol is very reactive in electrophilic aromatic substitution reactions. The bromine-water test for phenols (Experiment 20) can be modified to yield a solid product that in many cases is a useful derivative.

PROCEDURE

In a large test tube dissolve 0.5 g of phenol in 2 ml of methanol and add 2 ml of water. Add 1 ml of the bromine reagent* to the phenol solution

---

CAUTION: Bromine vapors are toxic and bromine causes serious burns upon contact with skin. Work in a hood and wear rubber safety gloves when handling molecular bromine.

---

and shake the mixture. Continue adding the bromine solution dropwise and shaking until the mixture retains a yellow color after shaking. Add 4 to 5 ml of water and shake the mixture vigorously. Cool the solution in ice to cause crystallization of the bromophenol. Isolate the derivative by filtration and wash the crystals thoroughly with water. If the derivative is impure, recrystallize the solid from hot methanol by the dropwise addition of water.

Derivatives of Aldehydes and Ketones

Derivatives of carbonyl compounds are conveniently prepared by substitution reactions with nitrogen nucleophiles. The most common derivatives of this type are the 2,4-dinitrophenylhydrazones [Z = 2,4-$(NO_2)_2C_6H_3NH-$], phenyl-

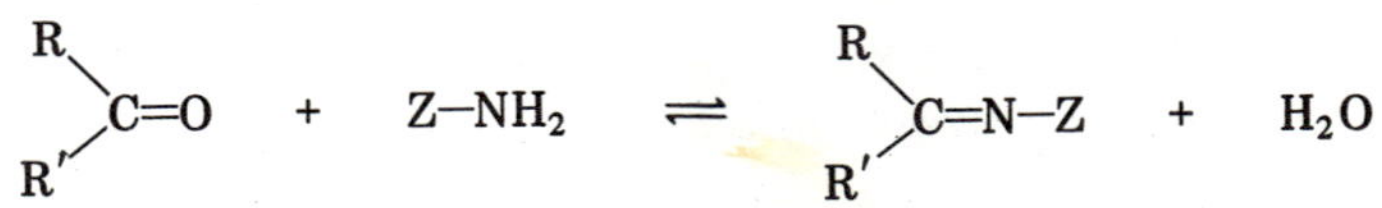

Carbonyl compound *Nitrogen nucleophile* *Derivative*

*The bromine reagent is prepared by dissolving 1 g of potassium bromide in 8 ml of water and CAREFULLY adding 0.20 ml of bromine.

hydrazones [$Z = C_6H_5NH-$], semicarbazones [$Z = H_2N\overset{O}{\overset{\|}{C}}NH-$], and oximes [$Z = HO-$].

2,4-Dinitrophenylhydrazones and Phenylhydrazones 2,4-Dinitrophenylhydrazones and phenylhydrazones are prepared by the same general method. However, the 2,4-dinitrophenylhydrazone deriviatives are generally easier to prepare and purify than the phenylhydrazone derivatives. Therefore, they are usually the derivative of choice, especially for lower molecular weight carbonyl compounds. A 2,4-dinitrophenylhydrazone derivative can often be isolated by filtration from the chemical characterization test employing this reaction (Experiment 14). If the crystalline derivative is carefully washed with cold 50% aqueous ethanol to remove excess 2,4-dinitrophenylhydrazine, it is usually not necessary to recrystallize the product.

PROCEDURE

To 10 ml of the 2,4-dinitrophenylhydrazine reagent (Experiment 14) in a 50-ml Erlenmeyer flask, add dropwise 0.2 g of the carbonyl compound and mix thoroughly. Warm the mixture on a steam bath and then allow the mixture to stand at room temperature for 5 minutes. If no precipitate is apparent, add water dropwise until the solution is cloudy. Cool the mixture and isolate the crystals by filtration. Wash the crystals carefully with two 5-ml portions of cold 50% aqueous ethanol. The derivative may be purified by recrystallization from ethanol-water.

A phenylhydrazone derivative is prepared by substituting a solution of 0.2 ml of phenylhydrazine and two drops of acetic acid dissolved in 10 ml of methanol for the 2,4-dinitrophenylhydrazine reagent in the above procedure.

CAUTION: Phenylhydrazine is toxic. Use care in handling this reagent.

Semicarbazones. Semicarbazones are prepared by reacting a slight excess of semicarbazide hydrochloride with the aldehyde or ketone in an aqueous solution buffered with sodium acetate. The derivatives of low molecular weight

Semicarbazone derivatives

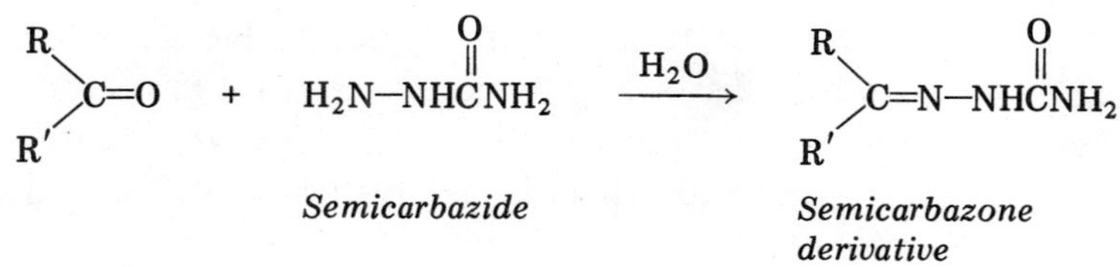

carbonyl compounds can be recrystallized from water. Higher molecular weight semicarbazones and semicarbazones of aromatic carbonyl compounds are recrystallized from aqueous ethanol or from ethanol.

PROCEDURE

Place 0.50 g of semicarbazide hydrochloride, 0.70 g of sodium acetate, 5 ml of water, and 5 ml of reagent grade ethanol in a 50-ml Erlenmeyer flask. Add 0.50 g of the carbonyl compound and warm the solution to 80°C on the steam bath. Continue heating for 10 minutes. If no precipitate has formed, add water dropwise until the hot solution turns cloudy or until 10 ml of water has been added. Cool the solution and filter the crystals. Wash the crystals with two 5-ml portions of water. The derivative may be recrystallized from a minimum amount of hot ethanol by the addition of water until the ethanol solution turns cloudy.

Oximes. Oximes are useful derivatives of many ketones, but they are generally not good derivatives of aldehydes. Oximes are prepared by the reaction of hydroxylamine hydrochloride with the ketone. The product is purified by recrystallization from water or aqueous ethanol.

Oxime derivatives

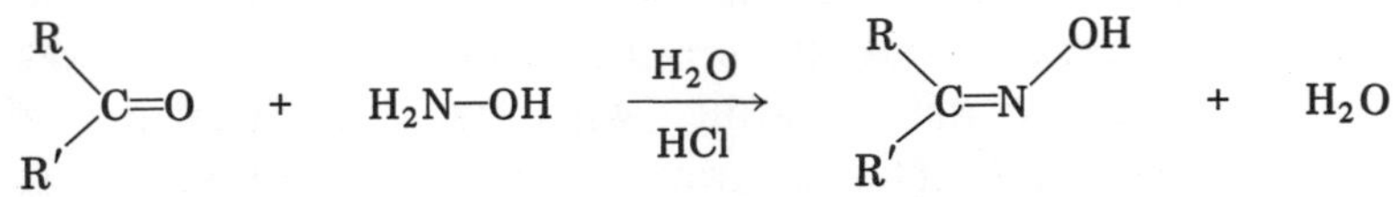

Hydroxylamine *Oxime derivative*

PROCEDURE

Place 0.5 g of hydroxylamine hydrochloride, 3 ml of water, 2 ml of 10% sodium hydroxide solution, and 0.3 g of ketone in a large test tube. If the ketone is not soluble in water, add just enough reagent grade ethanol to dissolve the ketone with warming. Heat the mixture on a steam bath for 15 minutes, and then cool the solution in an ice bath. Scratch the sides of the test tube with a glass stirring rod to promote crystallization. In some cases a few milliliters of water must be added to reduce the solubility of the oxime. The crystals are isolated by filtration and washed with a 5-ml portion of cold water. The derivative is purified by recrystallization. Lower molecular weight oximes recrystallize best from water. Higher molecular weight oximes are recrystallized by dissolving the derivative in a minimum volume of hot ethanol and adding water until the solution turns cloudy or until equal volumes of water and ethanol have been mixed. The solution is then cooled in an ice bath.

Derivatives of Ethers

Aliphatic ethers are relatively unreactive compounds and derivatives of these compounds are not commonly prepared. The identification of an unknown aliphatic ether is generally made on the basis of evidence from physical constants, characterization tests, and the interpretation of the infrared spectrum.

Derivatives of aromatic ethers may be prepared by taking advantage of the reactivity of the oxygen-substituted aromatic ring. For example, bromo derivatives of aromatic ethers are prepared by essentially the same procedure used to prepare bromo derivatives of phenols except that a solution of 2% bromine in glacial acetic acid is reacted with the aromatic ether that is also dissolved in glacial acetic acid. Water may be added to induce crystallization of the product.

PROCEDURE

A solution of 0.1 g of the aromatic ether in 10 ml of glacial acetic acid is prepared. To this solution is added dropwise a 2% bromine in glacial

---

CAUTION: Bromine vapors are toxic and bromine causes serious burns upon contact with the skin. Work in a hood and wear rubber safety gloves when working with bromine.

---

acetic acid solution until the solution retains a definite yellow color after shaking for 5 minutes. An equal volume of water is added to the acetic acid solution and the mixture is cooled. The bromo compound which separates is removed by filtration and recrystallized from aqueous ethanol.

Prelab Questions

1. An unknown organic solid (m.p. = 33 to 35°C) is found to contain no halogen on elemental analysis, to be insoluble in water and in 5% sodium bicarbonate solution, to be soluble in 10% sodium hydroxide solution, and to show a strong absorption at 3450 cm^{-1} in the infrared spectrum. To what chemical class does this compound belong? What chemical tests could be used to confirm this classification? What known compounds listed in the tables in Appendix A fit these properties?

2. What chemical tests could be used to distinguish between (a) hexanal and 2-hexanone, (b) *p*-cresol [p-$CH_3C_6H_4OH$] and anisole [$C_6H_5OCH_3$], (c) 3-phenyl-l-propanol and 1-decanol, (d) 2-heptanone and cyclohexanone, (e) *m*-cresol and benzyl alcohol?

3. What derivative would be most useful to distinguish between 4-methyl-2-pentanol and 3-methyl-l-butanol?

4. An unknown alcohol is found to boil over the range of 96 to 97°C. List the steps that

are necessary to identify this compound in the most efficient manner. (Assume that this boiling point is within 5°C of the literature value of the known compound.)

5. Only one primary alcohol and one aldehyde will yield iodoform upon treatment with a basic solution of iodine. What are these compounds?

6. An unknown compound having a boiling point of 153 to 155°C is found to be insoluble in 10% aqueous sodium hydroxide but is soluble in concentrated sulfuric acid. This compound gives a positive test with 2,4-dinitrophenylhydrazine (derivative m.p. = 161 to 162°C), is slowly oxidized by chromic acid (3 min), does not produce iodoform in the iodoform-formation test, and exhibits a strong absorption at 1715 cm^{-1}. The phenylhydrazone derivative formed for this unknown has a melting point of 75 to 76°C. What is the structure of this unknown compound?

7. What problems would you expect to encounter if the unknown that you are given is (a) salicaldehyde (*o*-HOC_6H_4CHO), (b) *p*-hydroxybenzyl alcohol, (c) 2-ethoxyethanol, (d) $CH_3COCH_2CH_2CHO$. For each compound list separately the expected results of solubility and chemical characterization tests.

EXPERIMENTAL PROCEDURE

You will be expected to identify an unknown alcohol, phenol, ether, aldehyde, or ketone by making use of the procedures discussed in this experiment. The compound that you are given may require purification before accurate physical data can be determined. Experiments should be carefully planned so that time and materials are not wasted. A derivative of the unknown should be prepared if a suitable derivative is listed in Appendix A. A summary report, as outlined in Experiment 13, should be prepared at the conclusion of the experiment.

Experiment Twenty-three

Identification of Organic Compounds Containing Carbon, Hydrogen, and Nitrogen

The techniques for the identification of an unknown amine, nitrile, or nitro compound are introduced in this experiment. Specific methods for the analysis of other classes of organic compounds that contain nitrogen, such as imines, nitrites, and diazonium salts, are not included here but can be found in specialized books on the characterization of organic compounds (see Useful References). Methods for the analysis of carboxylic acid amides, which are derivatives of carboxylic acids, will be discussed in Experiment 24. All organic compounds that contain nitrogen will give a positive test for nitrogen in the elemental analysis procedure described in Experiment 13.

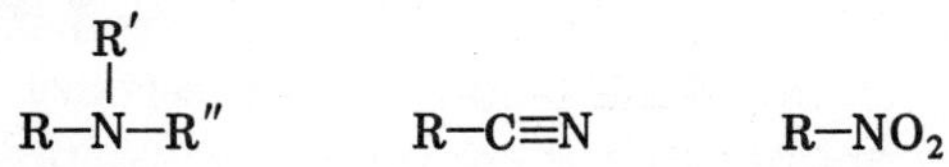

Amine *Nitrile* *Nitro compound*

Amines are divided into three subclasses on the basis of their chemical structure: Primary (1°) amines have one hydrocarbon substituent and two hydrogens attached to the nitrogen atom. Secondary (2°) amines have two hydrocarbon substituents and one hydrogen attached to nitrogen. In tertiary (3°) amines the nitrogen atom is bonded to three hydrocarbon substituents.

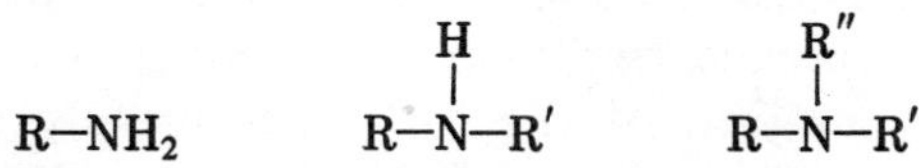

Primary amine *Secondary amine* *Tertiary amine*

Quaternary (4°) ammonium salts, compounds in which the nitrogen atom is bonded to four hydrocarbon substituents, will not be discussed in this experiment. Amines can also be classified as aliphatic amines, compounds in which the nitrogen is bonded directly to only saturated carbon atoms or hydrogen atoms, and aromatic amines, compounds in which nitrogen is bonded to one or more aromatic rings. Amines react as bases due to the nonbonded pair of electrons on nitrogen. Aliphatic amines are relatively strong bases ($K_b \cong 10^{-4}$).

Because aryl groups are electron withdrawing, aromatic amines are much weaker bases ($K_b \cong 10^{-10}$) than are aliphatic amines or ammonia.

Nitriles, also called *cyano compounds*, are structurally distinguished by the carbon-nitrogen triple bond. The $-C{\equiv}N$ functional group does not act as a base except with very strong acids such as concentrated sulfuric acid. Unlike the carbon-carbon triple bond, the carbon-nitrogen triple bond does not react with bromine in carbon tetrachloride or with dilute potassium permanganate.

Nitro compounds contain the $-NO_2$ functional group. This functional group is also not appreciably basic and only reacts as a base with very strong acids. Although some low molecular weight aliphatic nitro compounds such as nitromethane are readily available, the most commonly available nitro compounds are those formed by the nitration of aromatic compounds (Experiment 19).

Solubility

Amines, nitriles, and nitro compounds are soluble in ether and in concentrated sulfuric acid. Compounds belonging to these three classes that contain five carbons or less are generally soluble in water under the test conditions described in Experiment 7. Monofunctional nitriles and nitro compounds of six or more carbons are generally insoluble in water. The water solubility of monofunctional amines containing six or seven carbons is dependent on the structure of the amine. Branched-chain aliphatic amines are more soluble than straight-chain aliphatic amines or aromatic amines, and primary amines are more soluble than secondary or tertiary amines. Monofunctional amines of eight or more carbons are generally insoluble in water.

Primary, secondary, and tertiary aliphatic amines all react with dilute hydrochloric acid to form water soluble hydrochloride salts (Experiment 20). Therefore, most aliphatic amines are readily soluble in a 10% aqueous hydrochloric acid solution. Aromatic amines that have only one aryl group bonded to the amine nitrogen are generally soluble in 10% hydrochloric acid solution, but diarylamines and triarylamines are insoluble in 10% aqueous hydrochloric acid solution. In some cases, solid aromatic amines will react with dilute hydrochloric acid to form hydrochloride salts that are insoluble in water. By careful observation of the test, it is usually possible to determine that the solid amine is dissolving and that a new solid, the amine hydrochloride, is being formed.

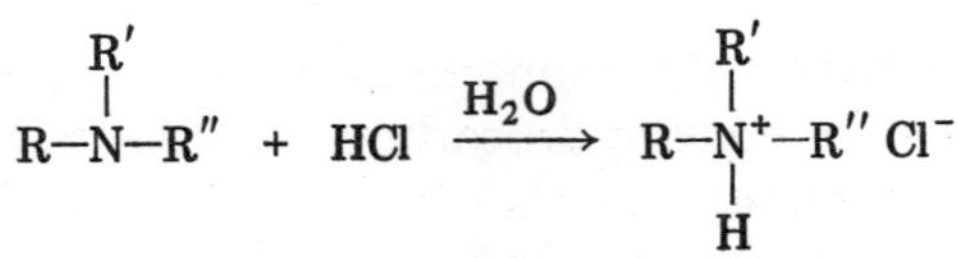

Amine *Amine hydrochloride*

Since nitriles and nitro compounds are not sufficiently basic to form hydrochloride salts in dilute hydrochloric acid, compounds that belong to these two classes are insoluble in 10% hydrochloric acid solution (if they are insoluble in water).

Infrared Spectroscopy

The interpretation of the infrared spectra of amines is discussed in Experiment 12 and in Appendix B. The most characteristic absorption of primary and secondary amines is due to the N—H stretching vibrations in the region of 3600 to 3200 cm^{-1}. Primary amines generally show two broad absorption bands in this region. Absorptions due to N—H bending vibrations of primary amines are observed in the 1650 to 1590 cm^{-1} region of the spectrum. The N—H bending absorption band of secondary amines is seldom observed due to its weak intensity. Aromatic amines display a strong absorption in the region of 1360 to 1260 cm^{-1} due to C—N stretching, but the corresponding absorptions for aliphatic amines are too weak to be useful for diagnostic purposes. Tertiary amines have no absorptions due to N—H stretching or bending vibrations.

The infrared spectrum of a nitrile exhibits a characteristic absorption band of medium intensity in the region of 2260 to 2220 cm^{-1} resulting from stretching vibrations of the carbon-nitrogen triple bond. Aliphatic nitriles absorb energy at the higher frequencies in this range (2260 to 2240 cm^{-1}), and aromatic and α,β-unsaturated nitriles absorb energy at the lower frequencies of this range (2240 to 2220 cm^{-1}).

Nitro compounds have two characteristic strong absorption bands in the infrared spectrum due to nitrogen-oxygen bond vibrations. One absorption band occurs in the range of 1570 to 1500 cm^{-1} and the second occurs in the range of 1380 to 1300 cm^{-1}. Although these two absorption bands will vary in relative intensity, both bands must be observed in the spectrum of a compound to indicate a nitro group in that compound.

Chemical Characterization Tests

Chemical tests for the detection of the amino, cyano, and nitro functional groups are not generally applicable to all types of compounds. Usually the presence of one of these functional groups in a molecule is first indicated on the basis of elemental analysis, solubility tests, and interpretation of the infrared spectrum. Characterization tests are then used to confirm the presence of the functional group.

Copper Sulfate Test

Low molecular weight amines that are water soluble will form a characteristic blue or blue-green complex with copper sulfate in water.

PROCEDURE

Add one drop of the amine to 1 ml of a 10% aqueous solution of copper sulfate. A blue or blue-green coloration or precipitate indicates an amine.

Hinsberg Test

The Hinsberg test is a subclassification test for distinguishing primary, secondary, and tertiary amines. The test is based on the fact that primary and secondary amines react with benzenesulfonyl chloride to form *N*-substituted benzenesulfonamides, but tertiary amines do not react with this reagent. Furthermore, since the benzenesulfonamides of primary amines are weak acids, they are generally soluble in dilute aqueous sodium hydroxide. The benzenesulfonamides of secondary amines are neutral compounds that are insoluble in sodium hydroxide solution. Tertiary amines do not react with benzenesulfonyl chloride to form a benzenesulfonamide, and the reagent is hydrolyzed by the basic aqueous solution to form the water-soluble sodium benzensulfonate. The unreacted tertiary amine is then observed as an oily layer on the top of the basic aqueous solution. The Hinsberg test is not reliable for distinguishing high molecular weight primary amines since the benzenesulfonamide salts of these compounds are not water soluble. The procedure described below for the characterization of primary, secondary, and tertiary amines also includes directions for the isolation of the benzenesulfonamide derivative of primary and secondary amines.

Hinsberg test

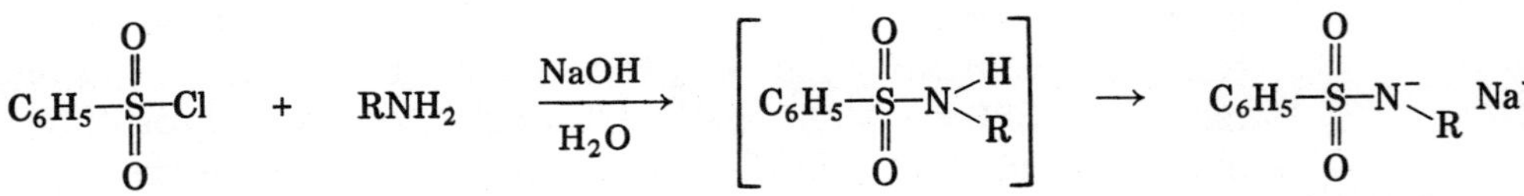

Benzenesulfonyl chloride — *1° Amine* — N-*Monosubstituted benzenesulfonamide* — *Sulfonamide salt [water soluble]*

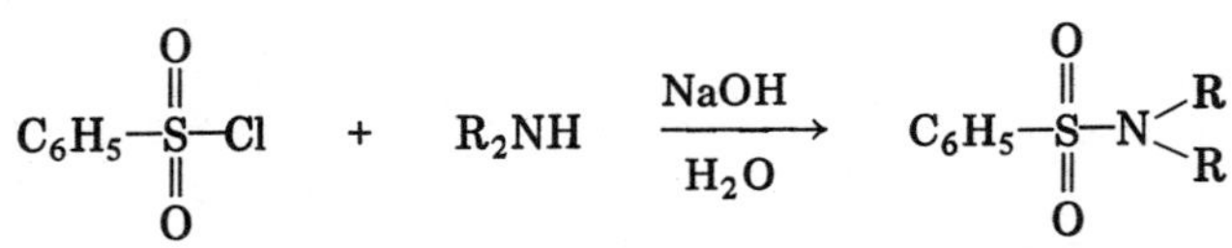

Benzenesulfonyl chloride — *2° Amine* — N,N-*Disubstituted benzenesulfonamide [water insoluble]*

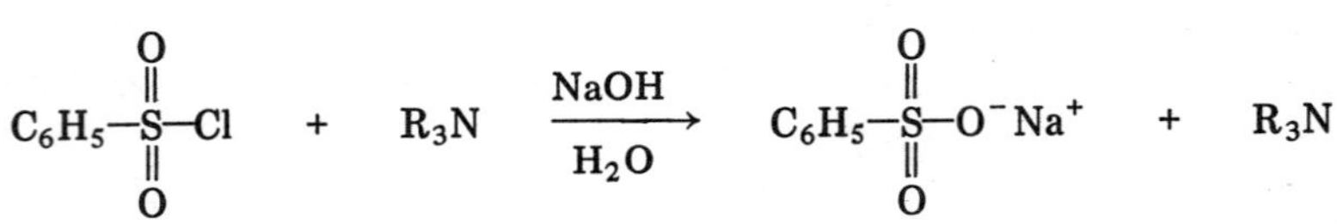

Benzenesulfonyl chloride — *3° Amine* — *Sodium Benzenesulfonate [water soluble]* — *3° Amine [oil]*

PROCEDURE

In a large test tube, add 2 ml of methanol, 0.1 g (five drops) of amine, and 0.4 g (eight drops) of benzenesulfonyl chloride. Warm the mixture over a steam bath for 5 minutes and then cool the solution. Add 10 ml of 10% aqueous sodium hydroxide and shake the solution for 10 minutes. If excess benzenesulfonyl chloride remains as a liquid globule at the bottom of the test tube, warm the solution to hydrolyze the excess sulfonyl chloride. Cool the solution to room temperature and carefully observe the mixture in the test tube.

If the unknown amine is a low molecular weight primary amine, the reaction mixture should be a homogeneous solution. If the unknown amine is a secondary amine, a heavy solid precipitate of the sulfonamide derivative is usually observed in the test tube. If the amine is a water-insoluble tertiary amine, a layer of the unreacted amine is observed on the top of the aqueous solution.

Carefully, with stirring, add 6*M* hydrochloric acid dropwise to the reaction mixture until the solution is acidic. The sulfonamide of a primary amine is insoluble in the acidic solution and will usually precipitate upon acidification of the reaction mixture. A tertiary amine will generally dissolve in the acidic aqueous solution. The solid sulfonamide of a secondary amine will remain unchanged during the addition of acid to the reaction mixture.

If a solid sulfonamide precipitate is present in the acidic reaction mixture, filter and then wash the solid with two 5-ml portions of 10% hydrochloric acid and one 5-ml portion of water. The isolated sulfonamide is used as a derivative of the primary or secondary amine. If necessary, the sulfonamide derivative is recrystallized from aqueous methanol.

If there is doubt about the subclassification of an amine in this test due to marginal solubility of the sulfonamide derivative in the basic reaction mixture, the following test is recommended. Transfer half of the isolated sulfonamide derivative to a small test tube and add 2 ml of 10% sodium hydroxide solution and 2 ml of water. Warm the mixture to 50°C and shake the solution vigorously. If the sulfonamide derivative dissolves, the original amine is primary. If the derivative is insoluble, a secondary amine is indicated. The sulfonamide of a primary amine can be recovered upon acidification of the alkaline solution.

Hydroxamate Test

Several different classes of organic compounds including nitriles and carboxylic acid derivatives can be converted into hydroxamic acids by reaction with hydroxylamine hydrochloride. The hydroxamic acid derivatives are easily detected since they form a deep red or violet complex with aqueous solutions of

Hydroxamate test

$$\underset{\textit{Nitrile}}{R{-}C{\equiv}N} + \underset{\textit{Hydroxylamine}}{H_2NOH} \xrightarrow[\text{HCl, heat}]{CH_3CH(OH)CH_2OH} \underset{\textit{Hydroxamic acid}}{R{-}\overset{\overset{\displaystyle O}{\|}}{C}{-}NHOH}$$

ferric chloride. Since compounds such as phenols will form a red or violet complex with ferric ion in aqueous solution, a compound that is to be tested by the hydroxamate test must first be tested to show that the compound itself does not give a colored complex with ferric ion.

PROCEDURE

Add 30 mg (one drop) of the unknown and 100 mg of hydroxylamine hydrochloride to 3 ml of a 0.3*M* solution of potassium hydroxide in propylene glycol. Add a boiling stone and gently heat the solution to boiling. Continue to boil the solution for 5 minutes and then cool the solution. Add 0.5 ml of a 5% aqueous solution of ferric chloride. The immediate formation of a red to violet color in solution is an indication that a hydroxamide acid is present and that the original compound is a nitrile (or carboxylic acid derivative).

Ferrous Hydroxide Test

Nitro compounds will oxidize ferrous hydroxide to ferric hydroxide with a corresponding color change from green to red-brown. Nitro compounds are the most common class of organic oxidizing agents. Less common classes of oxidizing agents are nitroso compounds, quinones, nitrates, and nitrites.

Ferrous hydroxide test

$$\underset{\textit{Nitro compound}}{R{-}NO_2} + \underset{\substack{\textit{Ferrous hydroxide}\\ \textit{[green precipitate]}}}{Fe(OH)_2} \xrightarrow[H_2O]{KOH} \underset{\textit{Amine}}{R{-}NH_2} + \underset{\substack{\textit{Ferric hydroxide}\\ \textit{[brown precipitate]}}}{Fe(OH)_3}$$

PROCEDURE

In a small test tube add 1 ml of a freshly prepared 5% aqueous solution of ferrous sulfate* and 10 mg of the compound to be tested. Then add 1 ml of 15% potassium hydroxide in aqueous ethanol.† Carefully

*The ferrous sulfate solution is prepared by dissolving 0.5 g of ferrous sulfate in 10 ml of distilled water and adding one drop of concentrated sulfuric acid. Excessive exposure of this reagent to air causes oxidation of the iron and misleading results.

†The 15% potassium hydroxide solution is prepared by dissolving 15 g of potassium hydroxide pellets in 15 ml of distilled water and adding this solution to 85 ml of ethanol.

flush the test tube with natural gas to exclude air. Stopper the test tube and shake vigorously. Note the color of the precipitate after 1 minute. A blank test without sample should always be run for this test to confirm the quality of the ferrous sulfate solution.

Derivatives of Amines

Primary and secondary amines are readily converted to amide or thiourea derivatives. Amide derivatives are prepared by treating the amine with an acid chloride in a basic solution. Benzamide and benzenesulfonamide derivatives are the most common types of amide derivatives. Thiourea derivatives are prepared by treating the amine with an aryl isothiocyanate, Ar—N=C=S. Phenyl isothiocyanate is employed for the preparation of the substituted phenylthiourea derivatives.

Derivatives of tertiary amines are prepared by treating the tertiary amine with an acid to form substituted ammonium salts. Salts of picric acid, 2,4,6-trinitrophenol, are the most common derivatives of tertiary amines.

Benzamides

Benzamide derivatives are prepared by reacting the primary or secondary amine with an excess of benzoyl chloride suspended in cold 10% sodium hydroxide solution. After vigorous shaking to ensure complete reaction, dilute hydro-

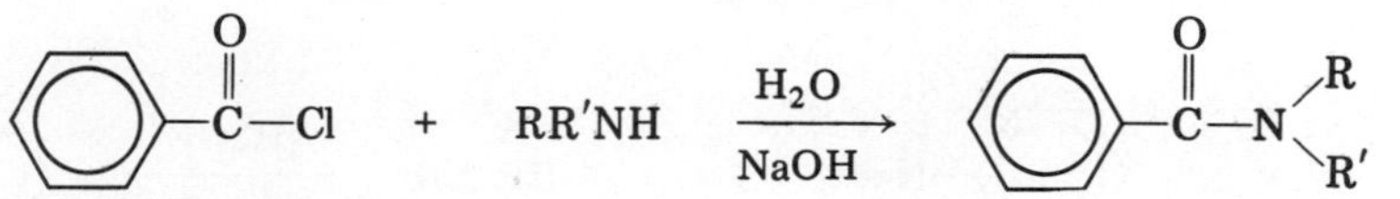

Benzoyl chloride *1° or 2° Amine* *Benzamide derivative*

chloric acid is added to the solution until the *p*H of the solution is in the range of 7 to 8. Some benzamide derivatives are appreciably soluble in strong alkaline solution, and if the solution is too acidic, benzoic acid, which is formed by hydrolysis of the excess benzoyl chloride, will crystallize from solution with the derivative. The derivative usually separates from solution as a solid mass and is isolated by filtration. Recrystallization from an alcohol-water mixture is usually necessary to obtain a pure derivative.

PROCEDURE

Add 0.20 g (0.20 ml) of the amine and 0.30 ml of benzoyl chloride to

---

CAUTION: Benzoyl chloride is a lacrymator. Use this reagent in a well-ventilated area or in a hood.

2 ml of cold 10% sodium hydroxide solution in a test tube. Stopper the tube and shake vigorously at intervals over a period of 5 to 10 minutes. After each period of shaking, carefully release the stopper on the test tube. After the reaction is complete, cool the solution in an ice bath. While mixing thoroughly, add 10% aqueous hydrochloric acid dropwise to the solution until the *p*H of the solution is in the range of 7 to 8. (Use *p*H test paper to check the *p*H after addition of every few drops of acid.) Filter the solid derivative from the solution and wash it with two 10-ml portions of cold water. Dissolve the solid in a minimum amount of boiling ethanol and add water until the solution turns cloudy. Cool the solution and isolate the crystalline derivative by filtration.

Sulfonamides

Sulfonamide derivatives are prepared by the procedure described for the Hinsberg test. The quantities of reagents used for the test can be doubled if larger amounts of derivative are desired. *p*-Toluenesulfonamide derivatives may be prepared by substituting *p*-toluenesulfonyl chloride for benzenesulfonyl chloride in the test procedure.

Phenylthioureas

Phenylthioureas are prepared by reacting a slight excess of the primary or secondary amine with phenyl isothiocyanate in ethanol. After heating the solu-

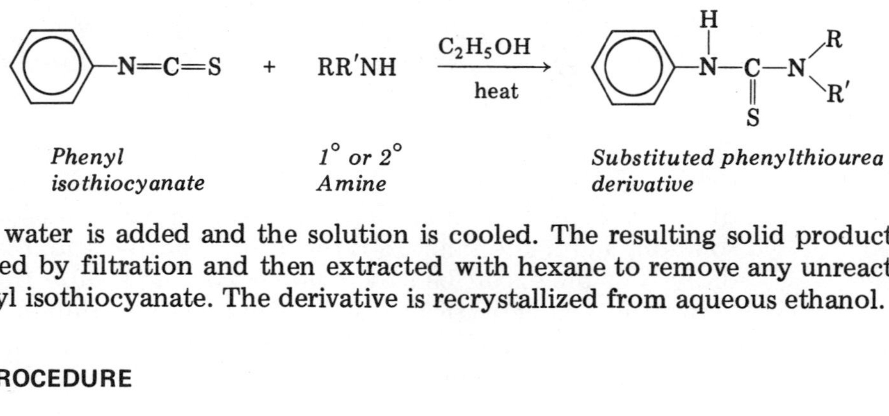

tion, water is added and the solution is cooled. The resulting solid product is isolated by filtration and then extracted with hexane to remove any unreacted phenyl isothiocyanate. The derivative is recrystallized from aqueous ethanol.

PROCEDURE

CAUTION: Phenyl isothiocyanate is a lacrymator. Use this reagent in a well-ventilated area or in a hood.

Add 0.20 ml (six drops) of phenyl isothiocyanate and 0.20 g (0.20 ml) of the amine to 2 ml of ethanol in a large test tube. Heat the mixture to boiling over a steam bath and continue to boil gently for 10 minutes. Add water dropwise to the warm mixture until it becomes cloudy and then cool the mixture in an ice bath. The derivative generally separates from solution as a thick oil that will solidify when carefully rubbed against the cold walls of the test tube with a glass rod. The solid derivative is isolated by filtration and dried by pressing between sheets of filter paper. (If the derivative is a gummy solid that cannot be filtered from solution, the solvent is decanted from the derivative, and the derivative is washed once with a 3-ml portion of cold ethanol before proceeding with the hexane extraction.) The solid derivative is placed in a large test tube and extracted with one 5-ml portion of hexane. The hexane is decanted from the solid. Save a small amount of the solid to use as seed crystals, and dissolve the rest of the derivative in a minimum amount of hot ethanol. Add water dropwise until the solution is cloudy, seed the solution with the solid material that was saved, and allow the solution to cool slowly to room temperature. Scratching the inner walls of the test tube with a glass rod may be necessary to induce crystallization. Cool the solution in an ice bath and isolate the product by filtration. Wash the product with one 5-ml portion of cold 50% aqueous ethanol.

Derivatives of Nitriles

Nitriles are hydrolyzed to carboxylic acids under acidic or basic conditions. High concentrations of acid or base and reflux temperatures are necessary to effect this conversion. If the carboxylic acid that is produced is a solid, it may be used as a derivative. Liquid carboxylic acids are not useful derivatives, but they can be converted into solid amide derivatives (Experiment 24) for characterization.

Hydrolysis of nitriles

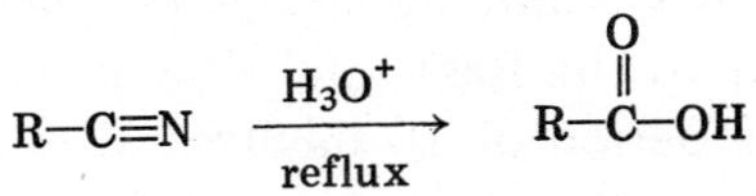

Nitrile *Carboxylic acid*

PROCEDURE FOR ACID HYDROLYSIS

To 4 ml of water in a 50-ml round-bottom flask, carefully add 8 ml of concentrated sulfuric acid and 0.1 g of sodium chloride. Add 1 g of the nitrile and several boiling stones to the flask and attach a reflux condenser. Reflux the solution for 30 to 60 minutes and then cool the solution in an ice bath. Add 10 g of cracked ice to the solution. If a solid precipitates from solution, it is isolated by filtration and recrystallized from ethanol or ether. If the carboxylic acid is a liquid (C_4 or larger), it is isolated by extraction with two 20-ml portions of ether.

The ether solution is then dried over anhydrous magnesium sulfate and concentrated to a small volume by distillation of the solvent on a steam bath. The residue, which is the carboxylic acid, is converted to a solid amide derivative by the procedures described in Experiment 24.

Derivatives of Nitro Compounds

Derivatives of nitro compounds are prepared by reducing the nitro compound to an amine by reaction with metals such as zinc or tin under acidic conditions. The amine is isolated by distillation or extraction after the reaction solution is made basic. The amine is usually converted to a solid amide derivative for characterization.

Reduction of nitro compounds

$$R{-}NO_2 \xrightarrow[HCl]{Sn} \xrightarrow{NaOH} R{-}NH_2$$

Aromatic nitro compounds can be reduced to amines, but often they can be more easily converted to solid derivatives by further nitration. Thus, mononitro aromatic compounds can be converted to dinitro compounds by treatment with nitric acid in sulfuric acid.

Reduction to Amines

Reduction of a nitro compound to an amine by the use of an excess of tin in hydrochloric acid is performed by the following procedure.

PROCEDURE

In a 50-ml round-bottom flask, place 4 ml of water and carefully add 4 ml of concentrated hydrochloric acid. Add 0.5 g of nitro compound and several boiling stones to the acid solution. Attach a reflux condenser to the flask, and slowly add 1 g of granular tin in small portions over a period of 10 minutes. Boil the mixture under reflux for 30 minutes (or longer if all of the nitro compound has not reacted and dissolved). Cool the mixture and slowly add 15 ml of 5*M* sodium hydroxide solution. If the amine product is a low molecular weight compound (C_5 or less), add fresh boiling stones, set up the flask containing the reaction mixture for distillation, and isolate the amine by distillation. If the amine contains more than six carbons, extract the amine with two 25-ml portions of ether, dry the ether over anhydrous magnesium sulfate, and distill the ether from the amine.

Liquid amines are converted to amide derivatives by one of the procedures described previously in this experiment. Solid amines may be used as derivatives after recrystallization from ethanol, or they may be converted to amides.

Nitration

Aromatic nitro compounds and many other substituted aromatic compounds can be converted into solid derivatives by further nitration of the aromatic ring.

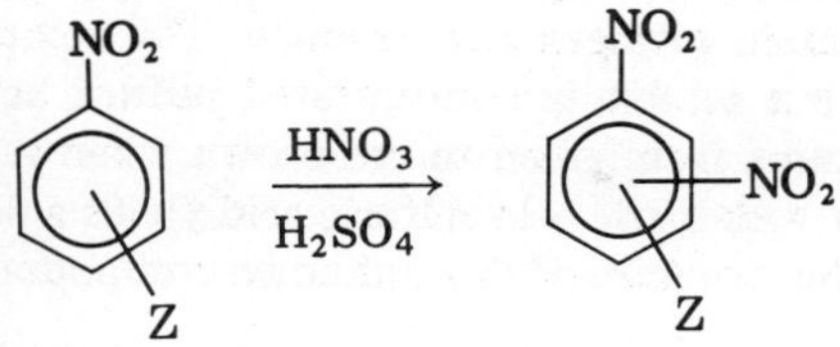

Several different procedures for nitration of aromatic compounds must be employed since different aromatic compounds require different reaction conditions to effect nitration without excessive side reactions. Compounds such as phenol, which are very reactive in nitration reactions, are best nitrated at low temperatures with dilute nitric acid. Less reactive compounds require more vigorous reaction conditions—higher temperatures and concentrated acids. The procedure for nitration described in this experiment is most useful for relatively unreactive substituted aromatic compounds such as nitro aromatic compounds.

PROCEDURE

Add 4 ml of concentrated sulfuric acid and 0.5 g of the aromatic compound to a large test tube. Cool the test tube in an ice bath and add dropwise with thorough mixing 4 ml of concentrated nitric acid. After all of the acid has been added, place the test tube in a beaker of water kept at a temperature of 50 to 60°C and heat the solution for 20 minutes at this temperature. After heating, the solution is cooled and poured into 25 ml of ice water in a beaker. The test tube is rinsed with 5 ml of cold water and the rinse is also added to the beaker. The aqueous mixture is cooled in an ice bath and the solid nitro derivative that separates from solution is isolated by filtration and washed with three 5-ml portions of cold water. The derivative may be purified by recrystallization from ethanol-water.

Prelab Questions

1. Write out a general scheme for the classification of an organic compound that may be an amine, a nitrile, or a nitro compound. (See Figure 21.1 for an example of a classification scheme.)

2. Describe what chemical tests would be useful to distinguish between: (a) *N*-methylaniline and *o*-toluidine [o-$CH_3C_6H_4NH_2$]; (b) pyridine and piperidine; (c) *o*-nitrotoluene and *o*-cyanotoluene; (d) 4-chlorobutanenitrile and pentanenitrile.

3. An unknown compound with a boiling point of 79 to 82°C is found by elemental analysis to contain nitrogen. The compound is soluble in water and in 10% hydrochloric acid solution but insoluble in 10% sodium hydroxide solution. The compound reacts with ben-

zenesulfonyl chloride to yield a solid derivative (m.p. = 92 to 93°C) that is insoluble in 10% sodium hydroxide solution. What is the structure of this unknown compound?

4. An unknown compound with a melting point of 39 to 42°C is found by elemental analysis to contain nitrogen and bromine. The compound is insoluble in 10% hydrochloric acid solution but soluble in concentrated sulfuric acid. Freshly prepared ferrous hydroxide precipitate turns from green to red-brown when reacted with the compound. Treatment of the unknown with nitric acid-sulfuric acid yields a derivative with a melting point of 70 to 72°C. What is the structure of this unknown compound?

5. An unknown organic compound, A (b.p. = 130 to 132°C), is found by elemental analysis to contain nitrogen. The unknown is soluble in ether and in water. The infrared spectrum of the unknown shows absorption bands at 2880, 1530, 1440, 1360, and 1340 cm^{-1} in the region above 1300 cm^{-1}. Unknown A was reacted with tin in hydrochloric acid. The product of this reaction, compound B, was isolated by distillation and treated with benzoyl chloride. The product of the reaction of compound B with benzoyl chloride was labeled compound C and had a melting point of 81 to 83°C. What are the structures of compounds A, B, and C?

EXPERIMENTAL PROCEDURE

You will be expected to identify an unknown amine, nitrile, or nitro compound by making use of the procedures discussed in this experiment. The compound that you are given may require purification before accurate physical data can be determined. Experiments should be carefully planned so that time and materials are not wasted. A derivative of the unknown should be prepared if a suitable derivative is listed in Appendix A. A summary report, as outlined in Experiment 13, should be prepared at the conclusion of the experiment.

Experiment Twenty-four

Identification of Carboxylic Acids and Carboxylic Acid Derivatives

The techniques for the identification of carboxylic acids, acid chlorides, acid anhydrides, esters, and amides are introduced in this experiment. Acid chlor-

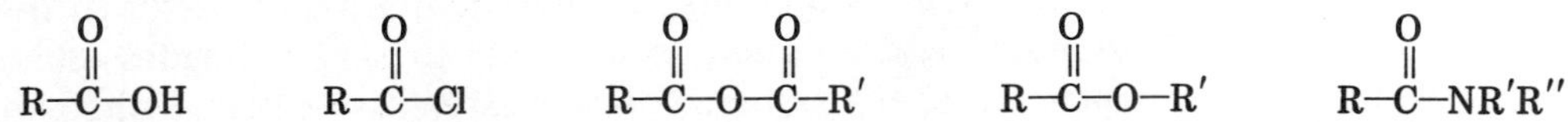

Carboxylic acid — *Acid chloride* — *Acid anhydride* — *Ester* — *Amide*

ides, acid anhydrides, esters, and amides are derived from carboxylic acids by the replacement of the hydroxyl group with chloride (—Cl), carboxylate (—OOCR), alkoxide (—OR′), and amide (—NR′R″) groups, respectively. Because of their derivative relationship to carboxylic acids, carboxylic acid derivatives are identified from their component functional groups.

Carboxylic acids are the most common class of organic acids. The acidity of carboxylic acids is dependent on the structure of the acid, but most carboxylic acids have K_a's in the range of 10^{-3} to 10^{-6}. Because they are more acidic than carbonic acid (H_2CO_3, $K_a = 4 \times 10^{-7}$), carboxylic acids react readily with sodium carbonate or sodium bicarbonate to evolve carbon dioxide.

Acid chlorides are highly reactive organic compounds that are hydrolyzed in aqueous solution to the corresponding carboxylic acid and hydrogen chloride. This reaction occurs rapidly at room temperature if the acid chloride is

Hydrolysis of acid chlorides

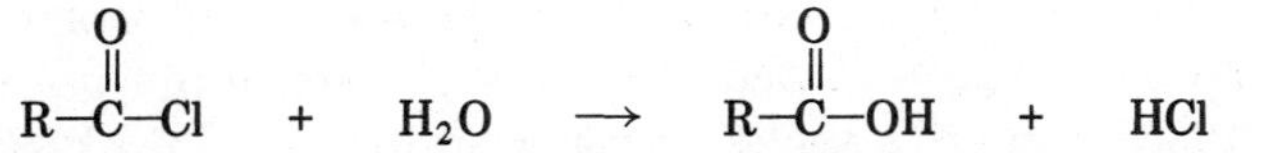

Acid chloride + *Water* → *Carboxylic acid* + *Hydrogen chloride*

soluble in the aqueous solution. Because they are generally toxic, these carboxylic acid derivatives should be handled with caution and in a well-ventilated hood. As you have already seen in Experiments 22 and 23, acid chlorides are quite useful in forming derivatives through reactions with alcohols to form carboxylate esters and with amines to form amides. Elemental analysis of all acid chlorides will indicate the presence of chlorine in the compound.

Carboxylic acid anhydrides are easily hydrolyzed in aqueous solution to the corresponding carboxylic acids. However, their reactivity with water is significantly less than that of their corresponding acid chlorides. Symmetric

Hydrolysis of acid anhydrides

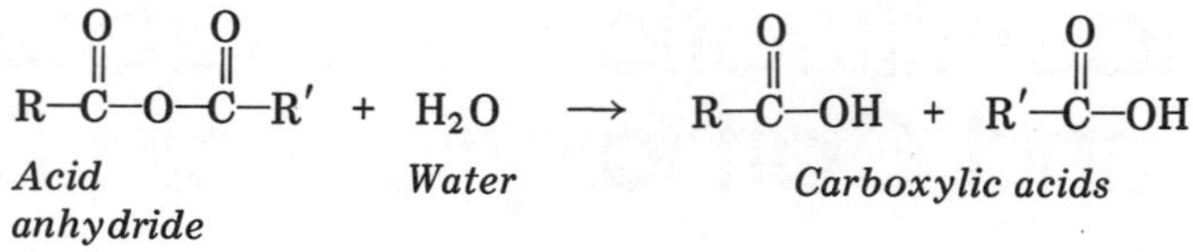

acid anhydrides, where R is identical to R′, are much more common than the unsymmetric acid anhydrides, where R is not identical to R′. Commercially available acid anhydrides are usually limited to symmetrical acid anhydrides or those produced from dicarboxylic acids that, like succinic acid and phthalic acid, can dehydrate to form a cyclic five-membered ring anhydride.

Carboxylate esters are not readily hydrolyzed in neutral aqueous solutions, but an acid or base will catalyze the hydrolysis of an ester to the corresponding carboxylic acid and alcohol. In contrast to acid chlorides and anhydrides that have distinctly sharp odors, volatile esters are distinguished by their pleasant odors. Many aliphatic esters have fruitlike odors.

Hydrolysis of esters

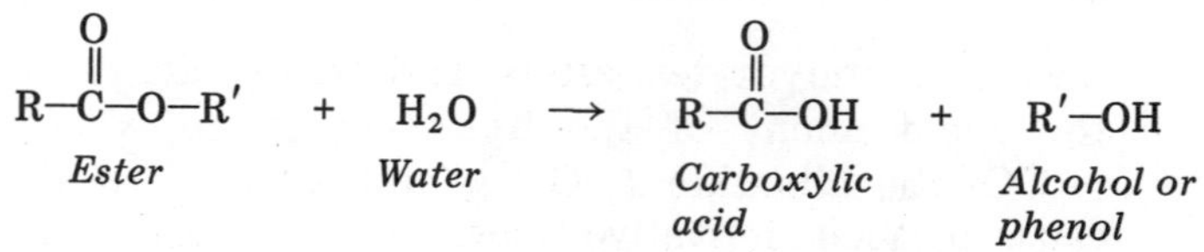

Carboxylic acid amides are classified as unsubstituted, monosubstituted, or disubstituted depending on the number of substituents attached to the nitrogen of the amide linkage. In contrast to esters, amides are very resistant to

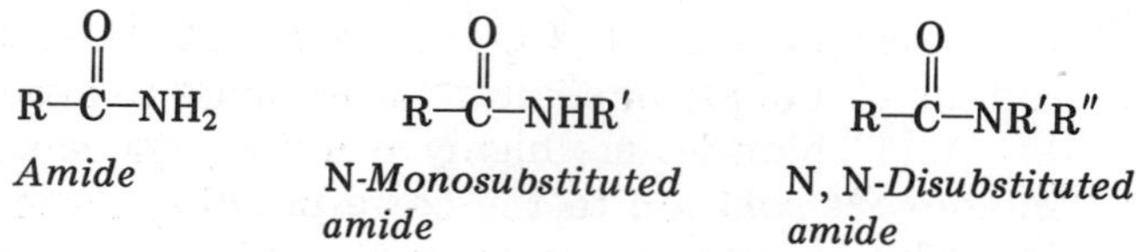

hydrolysis. High temperatures and high concentrations of acid or base catalysts are necessary to hydrolyze an amide to the corresponding carboxylic acid and amine. Unsubstituted amides yield ammonia upon basic hydrolysis. Monosubstituted amides yield primary amines and disubstituted amides yield secondary

Hydrolysis of amides

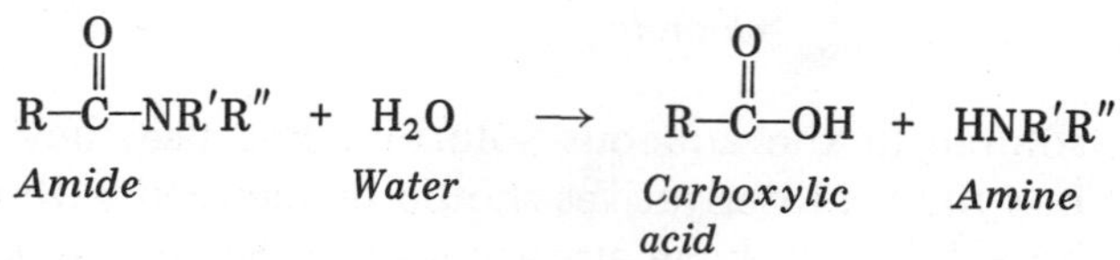

amines. Elemental analysis of all amides will indicate the presence of nitrogen in the compound. Most amides are solid organic compounds and because of their low volatility do not have characteristic odors.

Solubility

Carboxylic acids and carboxylic acid derivatives are soluble in concentrated sulfuric acid. Carboxylic acids of four carbons or less are soluble in water. The solubility of monofunctional carboxylic acids containing five carbons is marginal in water, and larger monofunctional carboxylic acids are insoluble in water. Carboxylic acids react with aqueous solutions of either sodium hydroxide or sodium bicarbonate to form water-soluble sodium carboxylate salts.

$$R{-}\overset{\overset{\displaystyle O}{\|}}{C}{-}OH + NaOH \rightleftharpoons R{-}\overset{\overset{\displaystyle O}{\|}}{C}{-}O^-Na^+ + H_2O$$

$$R{-}\overset{\overset{\displaystyle O}{\|}}{C}{-}OH + NaHCO_3 \rightleftharpoons R{-}\overset{\overset{\displaystyle O}{\|}}{C}{-}O^-Na^+ + H_2O + CO_2$$

Carboxylic acid — *Sodium carboxylate (water soluble)*

Acid chlorides and acid anhydrides hydrolyze to carboxylic acids when mixed with aqueous solutions. However, acid chlorides and acid anhydrides that contain more than five carbons are only slightly soluble (or insoluble) in aqueous solutions and, therefore, usually hydrolyze slowly. It may not be possible to visually observe any change in solubility for these compounds due to their slow rate of hydrolysis.

Carboxylate esters of less than four carbons are soluble in water. The solubility of monofunctional four-carbon esters is dependent on the structure of the ester. Monofunctional esters of five carbons or greater are insoluble in water. Since esters hydrolyze slowly at room temperature in 10% hydrochloric acid solution or in 10% sodium hydroxide solution, only water-soluble esters will be soluble in these test solutions.

Carboxylic acid amides of five carbons or less are soluble in water, and monofunctional amides of greater than five carbons are insoluble in water. Water insoluble amides are generally not soluble in 10% hydrochloric acid solution or in 10% sodium hydroxide solution.

Infrared Spectroscopy

The infrared spectra of all carboxylic acids and carboxylic acid derivatives exhibit strong carbonyl stretching absorptions. The frequencies of these absorptions, which are dependent on the nature of the groups bonded to the carbonyl groups, are indicative of the particular class of compound (Table 24.1).

The stretching frequency of the carbonyl group of carboxylic acids is in the range of 1720 to 1680 cm^{-1}. Saturated aliphatic acids absorb infrared radiation in the region of 1720 to 1705 cm^{-1}, and α,β-unsaturated acids and aromatic carboxylic acids absorb in the region of 1710 to 1680 cm^{-1}. The hydroxyl groups of carboxylic acid molecules are strongly hydrogen bonded even in

Table 24.1 Carbonyl Absorption Bands for Carboxylic Acids and Carboxylic Acid Derivatives

Structural Type	Frequency range (cm^{-1})	Usual frequency (cm^{-1}) for RCOZ R = alkyl	R = aryl
Acid chloride	1820–1770		
Acid anhydride (two bands)	1815–1775 and 1750–1720	1815 and 1750	1790 and 1730
Carboxylate ester	1775–1715	1735	1730
Carboxylic acid	1720–1680	1710	1700
Amide	1700–1630		

dilute solutions of nonpolar solvents. Consequently, the absorption band due to O—H stretching vibrations in a carboxylic acid is very broad and extends over the range of 3200 to 2500 cm^{-1} for a solid or liquid phase sample. Absorption bands due to C—O stretching vibrations in carboxylic acids result in a broad medium intensity band in the range of 1320 to 1280 cm^{-1}. Bands due to C—O—H bending vibrations of hydrogen-bonded carboxylic acids are observed in the range of 1440 to 1390 cm^{-1} and near 920 cm^{-1}.

The stretching frequency of the carbonyl group of acid chlorides is in the range of 1820 to 1770 cm^{-1}. Anhydrides display two bands in the carbonyl stretching region. Saturated noncyclic anhydrides absorb near 1815 and 1750 cm^{-1}. Conjugated noncyclic anhydrides exhibit absorption bands near 1775 and 1720 cm^{-1}. The higher frequency band is the stronger of the two absorptions. Due to ring strain, cyclic anhydrides with the anhydride linkage in a five-membered ring show the two carbonyl absorption bands at frequencies of 1865 and 1780 cm^{-1}. Anhydrides exhibit strong absorption bands in the region of 1300 to 1000 cm^{-1} due to carbon-oxygen stretching vibrations.

The carbonyl absorption band of saturated esters (except formates) occurs in the region of 1750 to 1735 cm^{-1}. The carbon-oxygen double-bond absorption bands of formates, α,β-unsaturated esters, and aryl esters occur at lower frequencies in the region of 1730 to 1715 cm^{-1}. The carbonyl stretching frequency is shifted to higher frequencies (1775 to 1750 cm^{-1}) in esters that have a carbon-carbon double bond in the alcohol portion of the molecule adjacent to the carbon-oxygen single bond. Esters also show several strong absorptions in the region of 1300 to 1000 cm^{-1} due to vibrations of the carbon-oxygen single bonds and combination vibrations of the $\mathrm{C{-}\underset{\underset{\displaystyle O}{\|}}{C}{-}O{-}C}$ group.

The carbonyl absorption of amides occurs in the region of 1700 to 1630 cm^{-1}. Most solid phase spectra of unsubstituted and monosubstituted amides exhibit absorption bands due to the carbonyl stretching vibrations near 1650 and 1640 cm^{-1}, respectively. Spectra of unsubstituted amides show two bands of moderate intensity near 3350 and 3200 cm^{-1} due to N—H stretching and a third band in the region of 1660 to 1620 cm^{-1} which often overlaps with the carbonyl absorption band and is due to bending vibrations of the $-NH_2$ group. Monosubstituted amides exhibit N—H stretching vibrations in the region of 3400 to 3000 cm^{-1} and bending vibrations in the region of 1570 to 1510 cm^{-1}. The absorption bands resulting from carbon-nitrogen stretching vibration of amides are generally of low intensity and are not useful for diagnostic purposes.

Chemical Characterization Tests

Acid chlorides, anhydrides, esters, and amides may be detected by conversion to hydroxamic acids. The hydroxamic acids are then treated with ferric chloride to produce a red-blue ferric hydroxamate complex. This test was described

Hydroxamate test

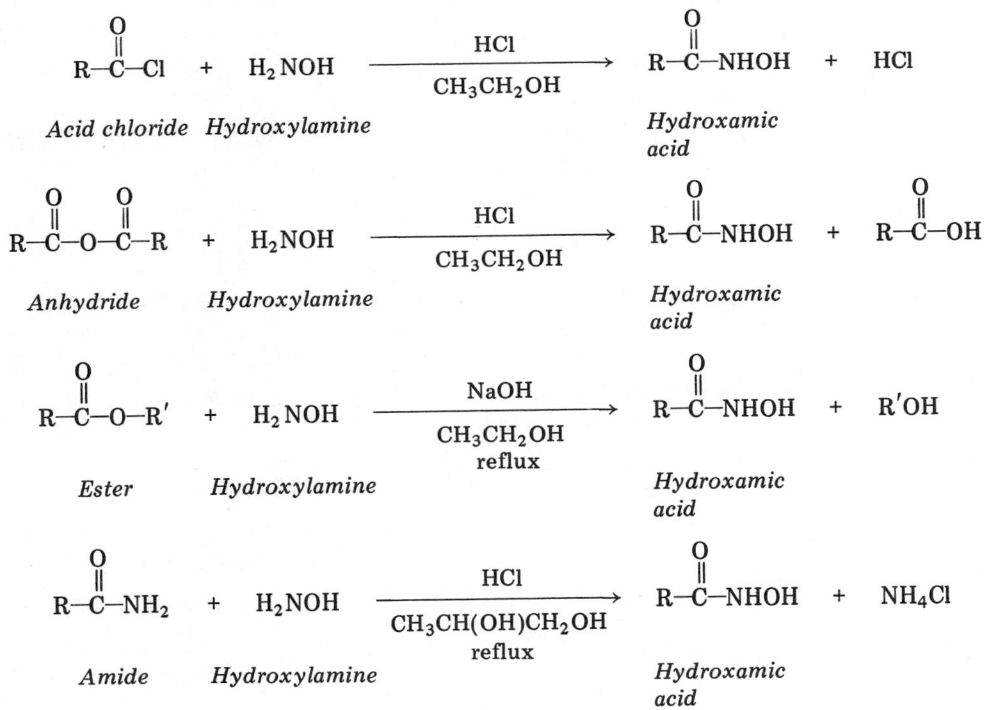

in Experiment 23 to detect nitriles. Acid chlorides and anhydrides are distinguished from esters since the acid chlorides and anhydrides react rapidly with hydroxylamine in an acidic solution. Esters do not react with hydroxylamine in acidic solution, but in a basic solution esters react rapidly with hydroxylamine to form hydroxamic acids. Amides are much less reactive with hydroxylamine than are acid chlorides, anhydrides, or esters. Amides require heating with hydroxylamine in a high boiling solvent such as propylene glycol (b.p. = 189°C) for conversion to hydroxamic acids.

Since compounds such as phenols will react with ferric chloride in aqueous solution, it is important to test any unknown with a solution of ferric chloride

(Ferric chloride test, Experiment 20) before performing the hydroxamate test. The ferric chloride test must show that the unknown compound does not give a colored complex with ferric ion or the observations from the hydroxamate test could be misleading.

Hydroxamate Test for Acid Chlorides and Anhydrides
Acid chlorides and anhydrides react rapidly with a warm acidic solution of hydroxylamine in ethanol to yield hydroxamic acids. Carboxylic acids, esters, and amides do not form these derivatives under these conditions.

PROCEDURE
Add 50 mg (two drops) of the compound to be tested to 1 ml of a 1*M* solution of hydroxylamine hydrochloride in ethanol. Add three drops of 6*M* hydrochloric acid to the solution and heat the solution to just boiling for a few seconds. Then allow the solution to cool to room temperature. Add two drops of 5% aqueous ferric chloride reagent and mix the solution. The immediate formation of a red-blue or violet color is an indication that a hydroxamic acid is present and that the original compound is an acid chloride or an anhydride.

Hydroxamate Test for Carboxylate Esters
Esters react with hydroxylamine under basic conditions in aqueous ethanol to form hydroxamic acids. Acid chlorides and acid anhydrides will also react under these conditions, but carboxylic acids and amides will not react with hydroxylamine under these test conditions to form hydroxamic acids.

PROCEDURE
Add 50 mg (two drops) of the unknown to 1 ml of 1*M* hydroxylamine hydrochloride in ethanol. Then add dropwise a 15% ethanolic solution of potassium hydroxide until the test solution is just basic to *p*H test paper. (Wet the test paper with distilled water before use.) Add five additional drops of the 15% ethanolic potassium hydroxide solution. Heat the mixture to boiling for 30 seconds and then allow it to cool to room temperature. With thorough mixing, add dropwise an aqueous solution of 10% aqueous hydrochloric acid until the *p*H of the mixture is approximately 3. Add two drops of the 5% aqueous ferric chloride and note the color. The immediate formation of a red-blue or violet color is an indication that the original compound is an ester.

Hydroxamate Test for Amides
Unsubstituted amides and most substituted amides will react with hyroxylamine in boiling propylene glycol to form hydroxamic acids. Since acid chlorides, anhydrides, and esters will also yield hydroxamic acids under these

conditions, the compound to be tested should have been shown to give a negative hydroxamate test for these classes of compounds.

PROCEDURE

In a large test tube, add 50 mg (two drops) of the compound to be tested to 3 ml of 1*M* hydroxylamine hydrochloride in propylene glycol. Add a boiling stone and heat the solution to boiling for 3 minutes. Cool the solution to room temperature and then add 1 ml (20 drops) of the 5% aqueous ferric chloride reagent. The immediate formation of a red-blue or violet color is an indication that the original compound is an amide.

Silver Nitrate Test

At room temperature acid chlorides will rapidly react with silver nitrate in ethanol to give a white precipitate of silver chloride (alcoholic silver nitrate test, Experiment 11). Tertiary alkyl halides, benzylic halides, and allylic halides will also give a precipitate under these conditions.

Neutralization Equivalent of Carboxylic Acids

A sample of an unknown carboxylic acid can be accurately titrated to the equivalence point with standardized sodium hydroxide to determine the number of equivalents of acid in the sample. From this value and the weight of the sample of unknown acid, the equivalent weight or neutralization equivalent of the acid is calculated. If the unknown is a monofunctional carboxylic

$$\text{Equivalents of acid in the sample} = \frac{\text{ml of NaOH} \times M \text{ of NaOH}}{1000}$$

$$\text{Neutralization equivalent} = \frac{\text{weight of sample in grams} \times 1000}{\text{ml of NaOH} \times M \text{ of NaOH}}$$

acid, the neutralization equivalent is equal to the molecular weight of the unknown. If the unknown is a dicarboxylic acid, the value of the neutralization equivalent is one-half the molecular weight of the acid. For best results at least two samples of acid should be titrated, and the results from the two titrations should agree to within 3%.

PROCEDURE

In a 125-ml Erlenmeyer flask, dissolve an accurately weighed sample (0.1 to 0.2 g) of the unknown acid in 30 ml of water or aqueous ethanol. If necessary, warm the solution to dissolve all of the sample. Add two drops of phenolphthalein indicator and titrate the solution to the phenolphthalein end point with standardized sodium hydroxide solution that is approximately $0.1M$ concentration. Measure the volume of sodium hydroxide added and calculate the neutralization equivalent of the acid.

Derivatives of Carboxylic Acids

Because most amides are solids with sharp melting points, carboxylic acids are usually converted to amide or substituted amide derivatives. The amide derivatives are prepared by first heating the carboxylic acid with thionyl chloride to obtain the carboxylic acid chloride, and then reacting the acid chloride with ammonia or an amine to obtain the corresponding amide. The most useful substituted amide derivatives are the anilides and the *p*-toluidides

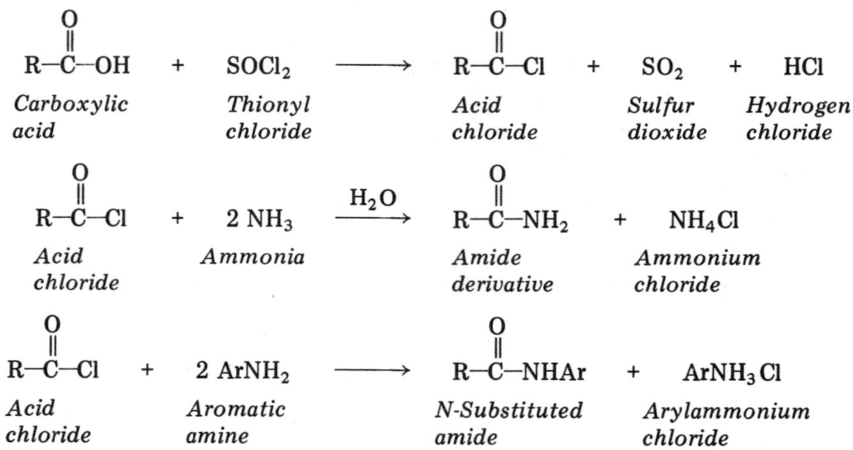

that are prepared from aniline ($C_6H_5NH_2$) and *p*-toluidine (*p*-$CH_3C_6H_4NH_2$), respectively.

PROCEDURE

To prepare the acid chloride, add 0.5 g of the acid and 2 ml of thionyl chloride to a 50-ml round-bottom flask and attach a reflux condenser.

CAUTION: Thionyl chloride is corrosive and gives off irritating, toxic fumes. This procedure should be performed in a hood.

Heat the mixture over a steam bath for 30 minutes.

(a) For the preparation of an unsubstituted amide, cool the carboxylic acid-thionyl chloride reaction mixture, and then carefully, with stirring, add the mixture into 15 ml of cold concentrated aqueous

CAUTION: The carboxylic acid–thionyl chloride reaction mixture will react vigorously with the aqueous ammonia. Since noxious fumes may be produced, this procedure should be performed in a hood.

ammonia solution ("ammonium hydroxide"). Cool the mixture. Filter the precipitated amide from solution and wash the precipitate with 5 ml of ice-cold water. The amide derivative is purified by recrystallization from water or aqueous ethanol.

(b) For the preparation of the anilide (or *p*-toluidide) derivative, dissolve 1.5 g of aniline (or *p*-toluidine) in 20 ml of hexane. Add the hexane solution of the amine to the cooled carboxylic acid-thionyl chloride reaction mixture. Reflux the resulting mixture on the steam bath for 10 minutes. Then cool the mixture to room temperature. Add 5 ml of cold water and 20 ml of ether. Pour the reaction mixture into a separatory funnel and separate the aqueous layer. Wash the organic layer successively with 5 ml of 10% hydrochloric acid, 5 ml of 10% sodium hydroxide, and then with 5 ml of water. Evaporate the organic solvent and recrystallize the resulting solid amide from aqueous ethanol.

Derivatives of Acid Chlorides and Anhydrides

Acid chlorides and anhydrides are converted directly to amide derivatives by reaction with ammonia, aniline, or *p*-toluidine. The procedures used for these reactions are the same as those described for the synthesis of amide derivatives from carboxylic acids in the previous section, except that the step involving reaction with thionyl chloride is omitted. A 0.5-g sample of the unknown acid chloride or anhydride is added directly to the ammonia or to the aniline (or *p*-toluidine) solution.

Derivatives of Carboxylate Esters

Carboxylate esters are generally characterized by preparation of derivatives of their carboxylic acid and alcohol (or phenol) constituents. The ester is first hydrolyzed, and then the carboxylic acid and alcohol are separated from the hydrolysis mixture and individually purified and characterized. Low molecular weight carboxylic acids or alcohols are difficult to successfully isolate from the hydrolysis mixture. Even with acids and alcohols that are insoluble in water, at least one gram of ester must be hydrolyzed to successfully obtain sufficient acid and hydroxy compound for characterization and derivative preparation.

The information obtained from the determination of physical constants and spectral properties of the unknown ester will give an indication of the probable structure of the ester. If it is clear from these data that both of the hydrolysis products will be water-soluble compounds, then the hydrolysis procedure will probably not be fruitful. However, if one or both of the components from the hydrolysis of the ester is insoluble in water, then the isolation of the water-insoluble components should be straightforward and this method of characterization should be attempted.

If an ester is not suitable for hydrolysis and isolation of the alcohol component, the alcohol portion of the unknown ester can be converted directly to a 3,5-dinitrobenzoic acid. This procedure is most successful with low molecular weight esters.

Hydrolysis of Carboxylate Esters

Esters are hydrolyzed by treatment with potassium hydroxide in hot aqueous ethylene glycol. After the hydrolysis is complete, water is added and the alcohol or phenol is extracted with ether. The aqueous solution is then acidified and the carboxylic acid component is isolated by filtration if it is a solid, or by extraction with ether if it is a liquid (Figure 24.1). After evaporating the ether, the hydroxy compound is converted to a derivative as described in Experiment 22. A derivative of the isolated carboxylic acid can be prepared by the procedures already described in this experiment.

PROCEDURE

In a 50-ml round-bottom flask, add 4 ml of ethylene glycol, 0.6 g (two pellets) of potassium hydroxide, and 0.5 ml of water. Warm the solution to dissolve the potassium hydroxide pellets. After cooling the hydroxide solution, add 1 g of the ester and several boiling stones. Attach a reflux condenser to the flask and heat the solution to a gentle reflux. After all of the ester has disappeared, continue to reflux the solution for 20 additional minutes. Cool the solution to room temperature, add 8 ml of water, and then add 6*M* hydrochloric acid dropwise to the aqueous solution until the *p*H of the solution is in the range of 7–8. Extract the resulting neutral solution with two 15-ml portions of ether to separate the water-insoluble alcohol or phenol. (Save the aqueous solution for the isolation of the carboxylic acid.) Combine the ether extracts and wash this combined solution with one 10-ml portion of water. Dry the ether solution over anhydrous magnesium sulfate and evaporate the ether to obtain the alcohol or phenol. If this compound is not a solid, a derivative of the compound is prepared according to the directions provided in Experiment 22.

Adjust the *p*H of the aqueous solution from the ether extraction to a *p*H of 2–3 by the dropwise addition of concentrated hydrochloric acid. If a precipitate of the solid carboxylic acid forms, it is isolated by filtra-

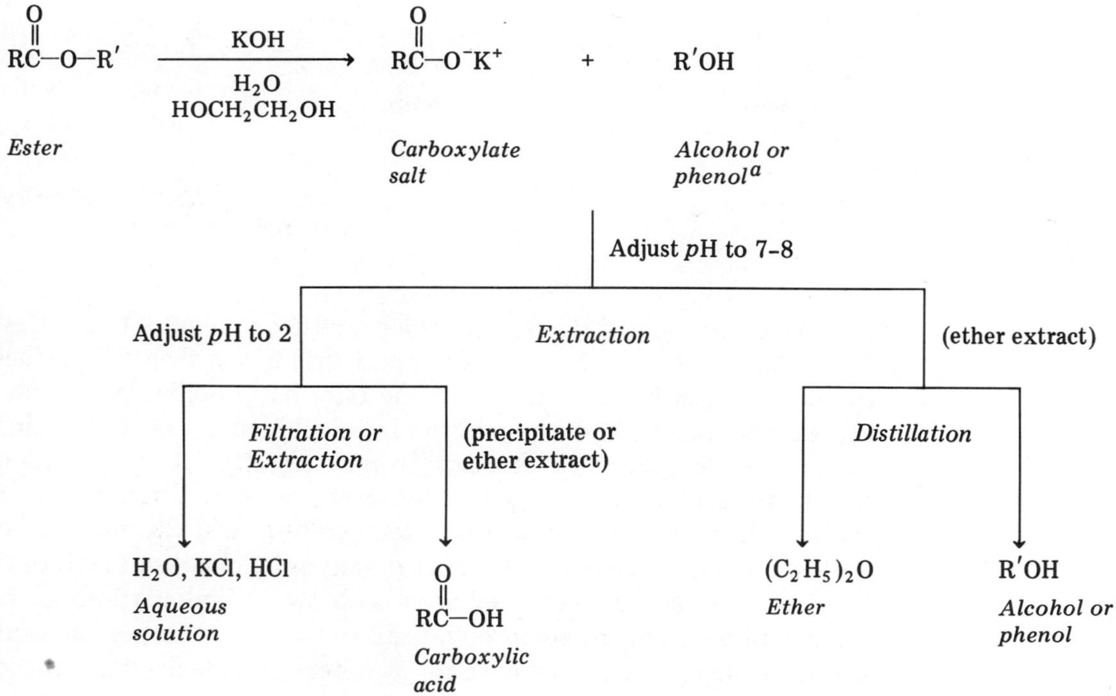

[a]A phenol will be ionized in the strong alkaline solution (ArO^-K^+). The *p*H is adjusted to 7–8 to obtain the neutral phenol (ArOH) before extraction. In this *p*H range the carboxylic acid remains a water-soluble salt.

Figure 24.1 *Hydrolysis of a carboxylate ester.*

tion, purified, and its melting point is determined. If the carboxylic acid is a water-insoluble liquid, it is isolated by extraction with two 15-ml portions of ether. Dry the ether extract over magnesium sulfate and then evaporate the ether to obtain the carboxylic acid. An amide derivative of the carboxylic acid residue is prepared according to the procedures described in this experiment.

Transesterification of Esters

The alcohol portion of an ester can be directly converted to a 3,5-dinitrobenzoate derivative by transesterification with 3,5-dinitrobenzoic acid. A strong acid such as sulfuric acid is necessary to catalyze the transesterification reaction. This method is applicable to simple esters, but it is not useful if the R′ group of the alcohol reacts with sulfuric acid. High molecular weight esters (M.w. > 250) usually fail to react.

PROCEDURE

In a small round-bottom flask, mix 2 g (2 ml) of the ester with 1.5 g of 3,5-dinitrobenzoic acid and add two drops of concentrated sulfuric

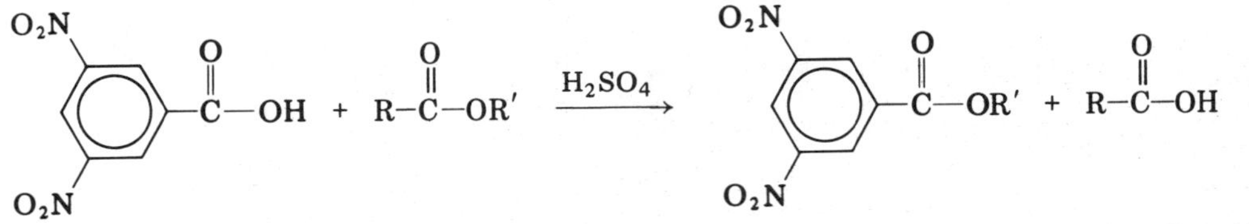

3,5-Dinitrobenzoic acid — *Ester* — *3,5-Dinitrobenzoate derivative* — *Carboxylic acid*

acid. Add two boiling stones and attach a reflux condenser to the flask. If the unknown ester has a boiling point that is less than 150°C, heat the mixture gently under reflux. If the ester has a boiling point that is higher than 150°C, heat the mixture in an oil bath at 150°C. Continue to heat the solution for at least 15 minutes after all of the 3,5-dinitrobenzoic acid has dissolved or for a maximum of 1 hour. After the solution has been cooled to room temperature, add 25 ml of ether. With the aid of a stirring rod, break up any solid chunks of material in the flask. Extract the ether solution with two 15-ml portions of 5% sodium bicarbonate to remove the sulfuric acid and carboxylic acids. Wash the ether solution with 15 ml of water and then dry this solution over anhydrous magnesium sulfate. Evaporate the ether and dissolve the residue in 8 ml of boiling ethanol. Add water dropwise to the boiling solution until the solution turns slightly cloudy and then cool the solution. (Scratching the inside walls of the container with a glass stirring rod will help to induce crystallization of the derivative.)

Derivatives of Amides

Like carboxylate esters, amides are generally identified by characterization of their carboxylate acid and amine constituents following hydrolysis. The best method for the preparation of a derivative of an amide is to hydrolyze the amide to the carboxylic acid and amine components. The acid and amine components are then separated and identified according to the procedures for these classes of compounds. The amide may be hydrolyzed under either acidic or basic conditions. However, alkaline hydrolysis is generally faster than acid hydrolysis.

To effect hydrolysis under alkaline conditions, the amide is refluxed with aqueous sodium hydroxide. If the amine component of the amide is low boiling, it may escape through the top of the condenser during the reflux period. This can be detected if a filter paper that has been soaked in a 10% cupric sulfate solution is placed near the opening at the top of the condenser. If ammonia or a volatile amine is produced, the filter paper will turn blue due to

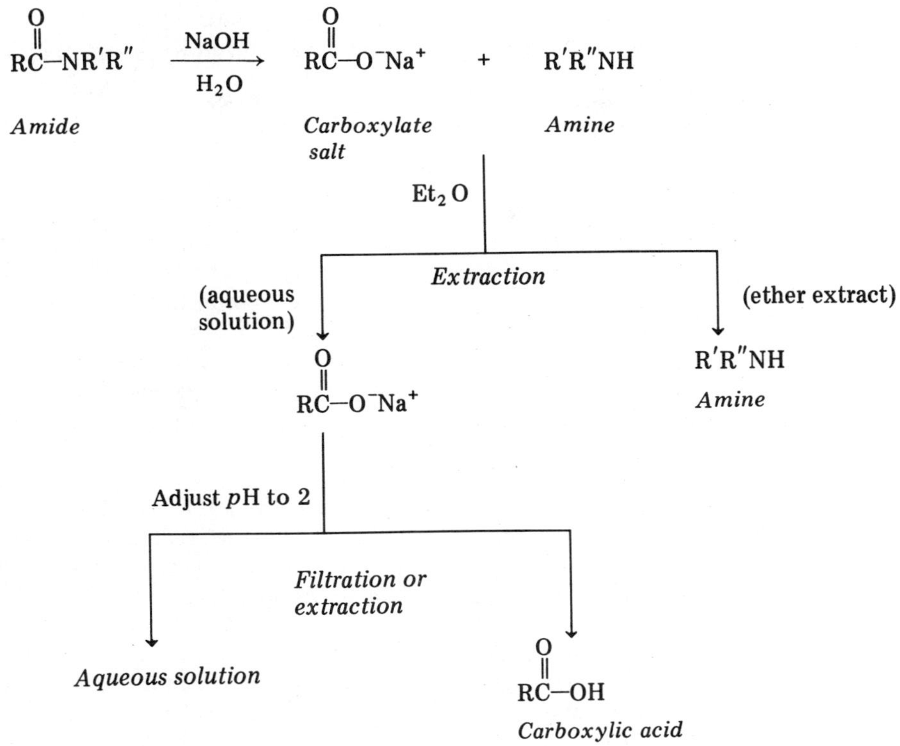

Figure 24.2 *Hydrolysis of an amide.*

the formation of a copper(II) amine complex. After hydrolysis, a nonvolatile amine is isolated from the basic aqueous solution by extraction with ether. The solution is then acidified and the carboxylic acid is isolated (Figure 24.2).

PROCEDURE

In a 50-ml round-bottom flask, add 20 ml of 10% aqueous sodium hydroxide solution, 1 g of the unknown amide, and a few boiling stones. Reflux the solution for 30 minutes. Test for the evolution of ammonia or a volatile amine by holding a piece of filter paper that has been soaked with a 10% cupric sulfate solution at the top opening of the reflux condenser during the reflux period. Cool the solution to room temperature. To isolate amines of limited water solubility, extract the reaction mixture with two 20-ml portions of ether. Dry the ether extract over anhydrous magnesium sulfate and isolate the amine by evaporation of the ether. A derivative of a liquid amine is prepared according to one of the procedures described in Experiment 23.

After extracting the amine, adjust the *p*H of the basic aqueous solution to a value of 2 to 3 by the dropwise addition of 6*M* hydrochloric acid. If solid acid precipitates, isolate the solid by filtration. If the acid is a water-insoluble liquid, extract the liquid with two 20-ml portions of ether. Dry the ether solution over anhydrous magnesium sulfate and evaporate the ether. The resulting acid is converted to a derivative by one of the procedures described in this experiment.

Prelab Questions

1. What chemical tests could be used to distinguish between (a) *p*-nitrobenzoic acid and *p*-toluic acid (*p*-$CH_3C_6H_4COOH$), (b) ethyl acetate and acetic anhydride, (c) benzamide and benzoic acid, (d) benzoyl chloride and *o*-chlorobenzoic acid, (e) *m*-nitrobenzoic acid and acetanilide?

2. What carboxylic acid and alcohol would be produced by the hydrolysis of (a) phenyl acetate, (b) ethyl phenylacetate, (c) phenyl carbonate, (d) butyl phthalate?

3. An unknown solid (m.p. = 155 to 157°C) was shown by elemental analysis to contain chlorine. Solubility tests indicated that the unknown was insoluble in water but soluble in 5% sodium bicarbonate solution. The unknown was refluxed with an excess of thionyl chloride and then reacted with aniline. The derivative synthesized by this procedure melted at 118 to 120°C. What is the structure of this unknown compound?

4. An unknown amide (m.p. = 140 to 142°C) was hydrolyzed in 10% sodium hydroxide solution. An unidentified volatile amine was produced in the hydrolysis. The acid component of the amide was isolated, purified, and determined to have a melting point of 138 to 140°C. What is the structure of the unknown amide?

5. An unknown compound (b.p. = 97 to 99°C) was shown by elemental analysis not to contain nitrogen or halogen. The compound was insoluble in water and in 10% sodium hydroxide solution. The unknown did not react with hydroxylamine hydrochloride under acidic conditions but yielded a hydroxamic acid when treated with hydroxylamine under alkaline conditions. Heating the unknown with 3,5-dinitrobenzoic acid and a few drops of sulfuric acid yielded a derivative (m.p. = 72 to 74°C). What is the structure of the unknown compound?

6. The hydrolysis of an unknown ester (b.p. = 214 to 216°C) produced a water-soluble alcohol that was not isolated and a solid carboxylic acid (m.p. = 185 to 187°C). What is the structure of the unknown ester?

7. An unknown solid (m.p. = 206 to 208°C) was shown to be soluble in 5% aqueous sodium bicarbonate. Elemental analysis confirmed the absence of halogen and nitrogen. The neutralization equivalent of this compound was found to be 82 ± 2 g/equiv. When the unknown solid was heated to 150°, water was evolved and a new compound (m.p. = 130 to 131°C) was formed. What is the structure of the unknown solid and its derivative?

EXPERIMENTAL PROCEDURE

In this experiment you will be expected to identify an unknown carboxylic acid, acid chloride, acid anhydride, carboxylate ester, or amide. The compound that you are given may require purification before accurate physical data can be determined. Since the unknown may contain halogen and nitrogen (i.e., amide, nitro, and nitrile) substituents, it is important to perform a sodium fusion and elemental analysis to determine the presence of halogen or nitrogen in the unknown compound. A summary report, as outlined in Experiment 13, should be prepared at the conclusion of this experiment.

Experiment Twenty-five

The Identification of a General Unknown

In this experiment you will be asked to identify an unknown organic compound that belongs to one of the chemical classes discussed in Experiments 21–24 (Table 25.1). Although the identification of this general unknown is more challenging than previous unknowns due to the number of potential chemical classes to which the unknown may belong, the same procedure used in previous experiments for the identification of unknown organic compounds is applicable to this problem.

A general unknown is identified by initially determining the chemical class to which the unknown belongs and then determining the actual structural identity of the unknown by the procedures described in the previous four experiments. Two techniques are commonly used to tentatively identify the chemical class of the unknown. The first technqiue, and probably the most powerful, is infrared spectroscopy. The second technqiue is classification on the basis of solubility tests and elemental analysis. Once the chemical class of the general unknown is tentatively identified by these methods, the chemical class is confirmed by specific characterization tests. The structure of the unknown is then identified on the basis of chemical and physical properties, and derivatives are prepared to confirm the identification.

Classification of Organic Compounds by Infrared Spectroscopy

Careful analysis of the infrared spectrum of a compound will allow you to determine what functional groups may be present *and what groups are not present* in an unknown compound. Specific details of the infrared absorption bands characteristic of different functional groups have been discussed (see Table 25.1 for references and Appendix B). The scheme outlined in Figure 25.1 provides a logical method for analyzing the infrared spectrum of an unknown. The first question that is asked in analyzing an infrared spectrum by this scheme is, "Does the spectrum contain a strong absorption in the region of 1820 to 1630 cm^{-1} due to the presence of a C=O group in the molecule?" If the answer to this question is "Yes," then proceed to the next question down the right-hand

Table 25.1 Chemical Classes and Functional Groups for the General Unknown

Chemical class	Functional group	Characterization tests in experiment	Infrared spectral analysis in experiment
Alkane	only C—C and C—H	21	12
Alkene	>C=C<	21	12
Alkyne	—C≡C—	21	12
Aromatic hydrocarbon[a]	Ar—H	21	12, 17
Alkyl halide[a]	R—X	21	12
Aryl halide[a]	Ar—X	21	12
Alcohol	—C(—)(—)—OH	22	12
Phenol[a]	Ar—OH	22	12, 20
Ether	—C(—)(—)—O—C(—)(—)—	22	12
Aldehyde	—C(—)(—)—C(=O)—H	22	12, 14
Ketone	—C(—)(—)—C(=O)—C(—)(—)—	22	12, 14
Amine	>N—	23	12, 20, 23
Nitrile	—C≡N	23	23
Nitro compound	$—NO_2$	23	23
Carboxylic acid	—C(=O)—OH	24	12, 24
Acid chloride	—C(=O)—Cl	24	24
Acid anhydride	—C(=O)—O—C(=O)—	24	24
Ester	—C(=O)—O—C(—)(—)—	24	24
Amide	—C(=O)—N<	24	24

[a]Ar = aryl; R = alkyl; X = Cl, Br, or I.

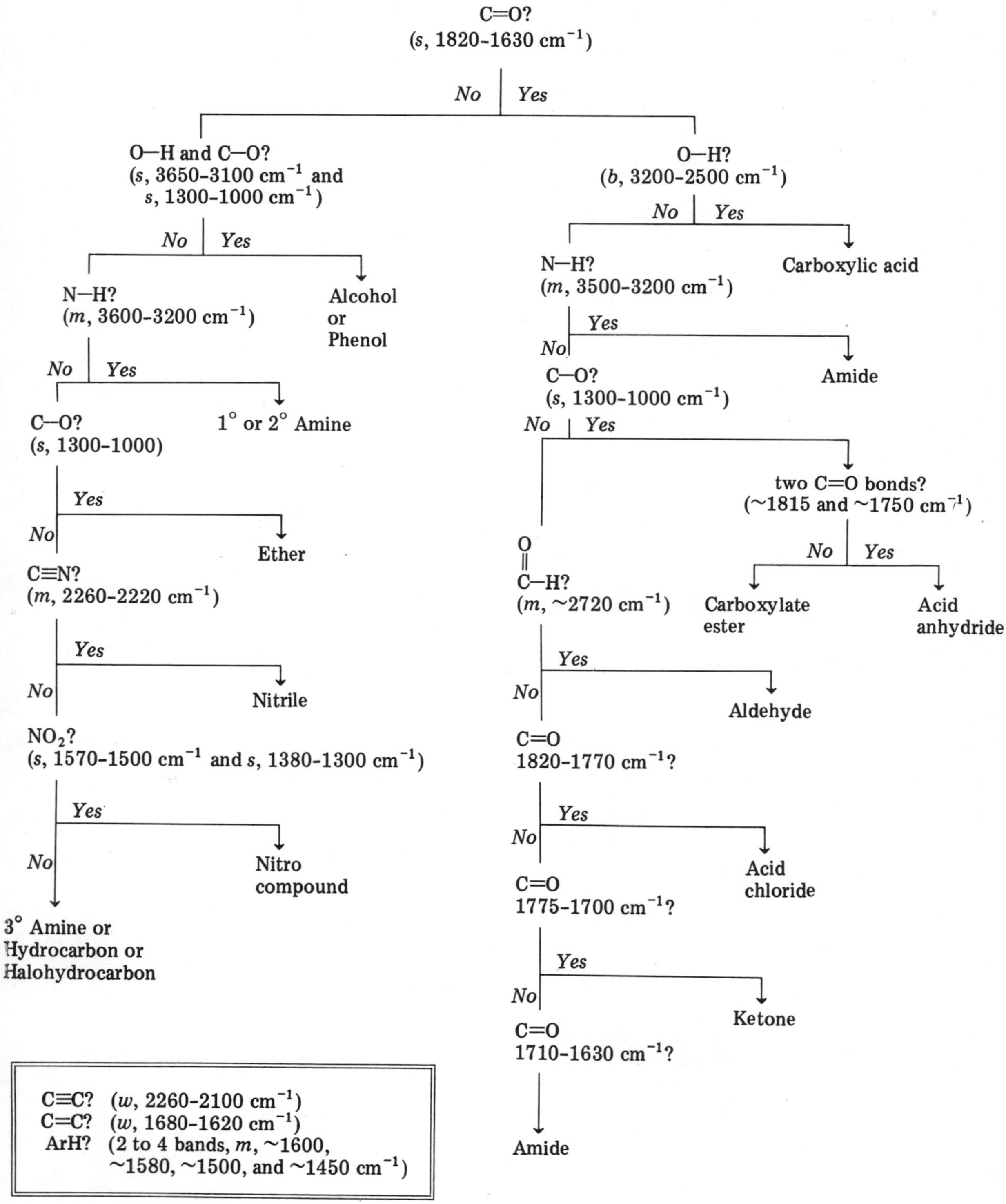

Figure 25.1 *Scheme for the analysis of an infrared spectrum.* (*s* = strong, *m* = medium, *w* = weak, *b* = broad).

branch of the scheme: "Does the spectrum contain a broad absorption in the region of 3200 to 2500 cm^{-1} due to the presence of a hydrogen-bonded O—H group?" If the answer to this question is "Yes," then the unknown may be a carboxylic acid and representative examples of infrared spectra of carboxylic acids should be examined for comparison. The scheme is followed in sequence until the presence of a particular functional group is indicated. All spectra should also be examined for absorption bands characteristic of carbon-carbon double bonds, carbon-carbon triple bonds, and aromatic rings.

The unknown may contain one or more additional functional groups listed in the scheme to the left or below carboxylic acids (e.g., a nitrobenzoic acid). The identification of the chemical class of the unknown must be considered to be tentative until the indicated functional group is confirmed by chemical tests. For example, an α-hydroxy ketone could be tentatively identified as a carboxylic acid by infrared spectral analysis since this difunctional compound exhibits a strong C═O absorption and a broad O—H absorption. However, chemical tests would clearly show that this compound is not an acid.

Classification of Organic Compounds by Solubility Tests and Elemental Analysis

The characteristic solubilities of the compounds in the various chemical classes have been discussed in previous experiments in this text (see Table 25.1). In this experiment you will use the solubility test scheme outlined in Figure 25.2 to classify an unknown into one of eight solubility classes. This scheme also makes use of the results of elemental analysis for classification. The eight different classes have the following characteristics:

Class S_1. This class includes compounds that are soluble in both water and ether. These compounds are polar.

Class S_2. This class includes compounds that are soluble in water and insoluble in ether. These compounds are very polar and may be ionic.

Class B. This class includes compounds that are insoluble in water, but soluble in 10% hydrochloric acid (1.2*M*). All compounds in this class contain nitrogen.

Class A_1. Compounds in this class are insoluble in water, but soluble in 10% aqueous sodium hydroxide solution and in 5% aqueous sodium bicarbonate solution. Compounds in this class are stronger acids than compounds in class A_2.

Class A_2. Compounds in this class are weak organic acids. They are soluble in 10% aqueous sodium hydroxide, but insoluble in 5% aqueous sodium bicarbonate.

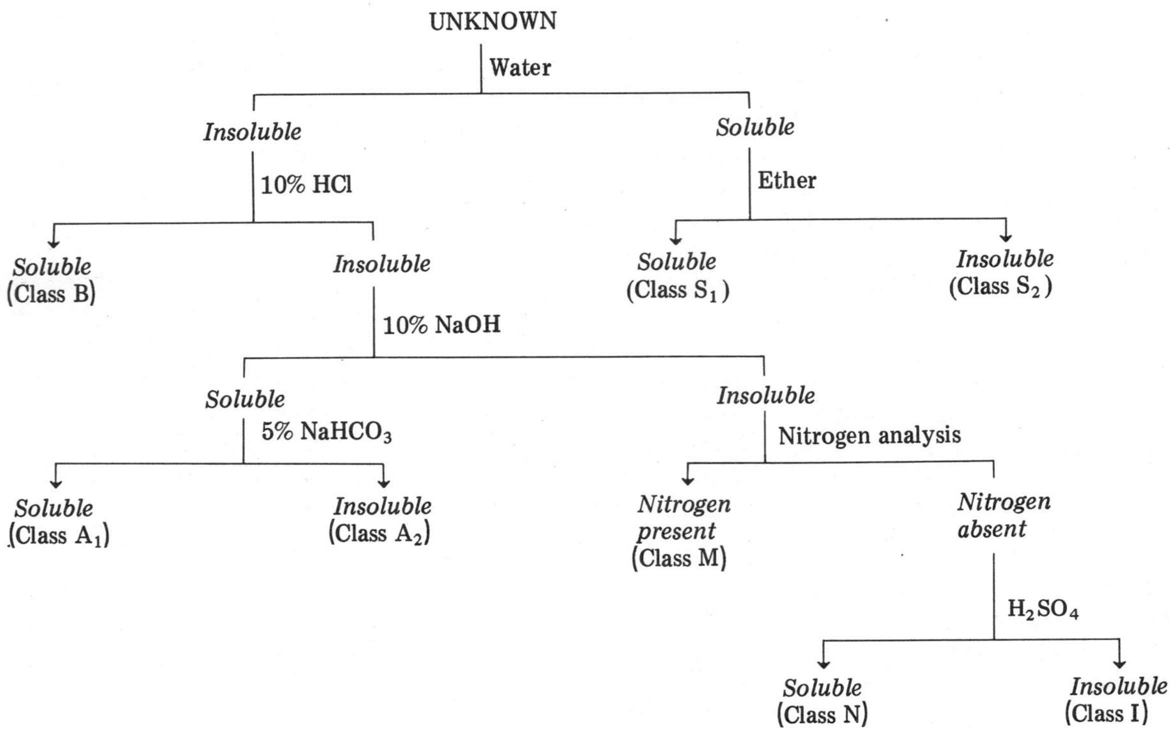

Figure 25.2 *Solubility classification scheme.*

Note: The procedure for the solubility tests is described in Experiment 7. The tests should be performed in sequence until the unknown is found to fit one of the classes.

Class M. This is a miscellaneous classification for compounds that contain nitrogen, but are not soluble in dilute acid or base. Very weak nitrogen bases such as diarylamines and nitriles belong to this solubility class.

Class N. Neutral compounds that do not contain nitrogen, but are soluble in concentrated sulfuric acid, are found in this class. Compounds in this class generally contain oxygen or they are unsaturated.

Class I. Compounds that are insoluble in all of the test solutions and that do not contain nitrogen are placed in this class. Compounds in this class are nonpolar and do not usually contain oxygen.

Table 25.2 gives a summary of the common chemical classes that may fall into a particular solubility classification. Only the most common polyfunctional compounds are listed in this table. Since the solubility of compounds that belong to a single chemical class may vary over a wide range, a given chemical class is often listed under several solubility classifications.

Table 25.2 Solubility Classification and Chemical Class

Class S_1

(a) Only C, H, and O present:
- Alcohols
- Aldehydes and ketones
- Anhydrides
- Esters
- Ethers
- Phenols

(b) Nitrogen present:
- Amides
- Amines
- Nitriles
- Nitroalkanes

(c) Halogen present:
- Halogen-substituted compounds of (a) or (b) above

Class B

- Amines
- Amino acids
- *N*,*N*-Dialkylamides

Class M

- Amides
- Amines
- Nitriles
- Nitro compounds
- Halogen substituted compounds of above classes

Class N

- Alcohols
- Aldehydes
- Ketones
- Esters
- Ethers
- Unsaturated hydrocarbons
- Anhydrides

Class I

- Hydrocarbons
- Alkyl halides
- Aryl ethers
- Aryl halides

Class S_2

(a) Only C, H, and O present:
- Polybasic carboxylic acids
- Polyhydroxy alcohols
- Polyhydroxy phenols
- Hydroxy acids

(b) Nitrogen present:
- Ammonium salts
- Amino acids and alcohols
- Amides
- Amines

(c) Halogen present:
- Halogen substituted compounds of (a) or (b) above

Class A_1

(a) Only C, H, and O present:
- Acids
- Anhydrides

(b) Nitrogen present:
- Amino acids
- Nitro and cyano acids
- Nitrophenols

(c) Halogen present:
- Acid halides
- Polyhalophenols
- Halo acids and anhydrides

Class A_2

(a) Only C, H, and O present:
- Phenols
- Some acids and anhydrides

(b) Nitrogen present:
- Amino acids
- Nitrophenols
- Cyanophenols
- Amides

(c) Halogen present:
- Halogen substituted compounds of (a) or (b) above

Prelab Questions

1. Predict the solubility classification for each of the following:

2-Propanol	Ethyl benzoate	*p*-Nitrophenol
1-Octene	Benzylamine	Toluene
Acetonitrile	Cyclohexanone	Limonene
p-Aminobenzoic acid	Benzamide	3-Chloropentane

2. Predict the characteristic infrared absorption bands that would be observed for each of the compounds listed in Question 1.

3. An unknown compound is found to belong in solubility class M. The infrared spectrum of this compound exhibits a strong absorption at 2250 cm^{-1}. To what chemical class does this unknown belong?

4. An unknown compound (m.p. = 108 to 110°C) is classified in solubility class A_1. The infrared spectrum of the compound shows strong absorptions in the region of 3200 to 2500 cm^{-1} and at 1690 cm^{-1}. A derivative (m.p. = 125 to 126°C) is prepared by reacting the unknown first with thionyl choride and then with aniline. What is the identity of this unknown?

5. A sweet-smelling natural product (b.p. = 228 to 230°C) was isolated by ether extraction of caraway seed. The compound has the characteristic solubilities of class N, and the infrared spectrum shows absorptions at 2800 to 3100 (*m*), 1680 (*s*), and 1450 (*m*) cm^{-1} in the diagnostic region. The compound reacts rapidly with aqueous $KMnO_4$ and with Br_2 in CCl_4. A derivative (m.p. = 187 to 188°C) was prepared by reacting the unknown with 2,4-dinitrophenylhydrazine reagent. What is the structure of this natural product?

EXPERIMENTAL PROCEDURE

You will be expected to identify an unknown that belongs to one of the nineteen chemical classes listed in Table 25.1. The compound you are given may require purification before accurate physical data can be determined. After determining the appropriate physical constants and the qualitative elemental composition of the unknown, you should employ the procedures discussed in this experiment to tentatively assign the unknown to a chemical class. This assignment must then be confirmed by the appropriate chemical tests. After searching the literature (Appendix A), you should prepare a list of compounds with similar chemical and physical properties. The identity of the unknown is then confirmed by further chemical and physical tests and by the preparation of one or more derivatives. A summary report, as outlined in Experiment 13, should be prepared at the conclusion of the experiment.

Experiment Twenty-six

Separation and Identification of the Components of a Mixture

The purification of organic compounds, which in reality involves the separation of mixtures, has been discussed extensively in this laboratory text. In this experiment these techniques will be employed to separate mixtures of organic compounds so that the purified components can be characterized and identified. All separations of mixtures are based on differences in the physical properties of the components at the time of separation. Separation techniques that are commonly used include: extraction (Experiments 1 and 3), which is based on differences in solubilities, fractional distillation and steam distillation (Experiments 3, 5, and 6), which are based on differences in vapor pressures, and column chromatography (Experiment 19), which is based on differences in the adsorption of the components on a solid surface. Separation techniques less commonly employed but possibly useful for certain mixtures include crystallization (Experiment 2), sublimation (Experiment 16), thin layer chromatography (Experiment 7), and gas-liquid partition chromatography (Experiment 10).

For many separations it is necessary to react one or more of the components in the mixture with a chemical agent in order to produce the required differences in physical properties. For example, in Experiment 1 benzoic acid in a toluene–benzoic acid mixture is converted into sodium benzoate by reaction with aqueous base to enable separation of the water-soluble benzoate salt from the water-insoluble toluene. After separation of the aqueous and organic layers, the aqueous solution is acidified to convert sodium benzoate to benzoic acid, which crystallizes from the solution. The technique of acid-base extraction is very useful for the rapid separation of many unknown mixtures on a sufficiently large scale to enable the isolation of the purified components for characterization and identification.

In this experiment you will separate a two-component mixture by one of the following techniques: extraction with nonreactive solvents, extraction with solvents that cause chemical change (acid-base extraction), or fractional distillation. You will then identify both of the purified components. Since a compound isolated from this mixture can belong to any chemical class discussed in Experiments 21–24, the identification of the compound should be performed as described for a general unknown (Experiment 25).

Preliminary Tests

If the mixture consists of more than one phase, the different phases should be separated and treated individually. In such situations, one compound may exist in more than one of the phases, and none of the phases may consist of a single compound. But often one phase will be a relatively pure component.

Small-scale solubility determinations should be made on representative samples of the mixture according to the procedures in Experiment 7. By careful observation of the amount of sample that dissolves, you may be able to determine which extraction solvent system (water, ether, water-ether, aqueous 10% hydrochloric acid-ether, or aqueous 10% sodium hydroxide-ether) can be used to separate the mixture. For example, if approximately one-half of the solubility test sample is soluble in aqueous 10% hydrochloric acid, partition of the mixture between aqueous 10% hydrochloric acid and ether should effect separation. *Concentrated sulfuric acid should not be used as a solvent for the separation of mixtures in this experiment.*

A small-scale distillation of a liquid mixture should be attempted using 2 to 3 ml of the mixture. The liquid in the distilling flask should not be heated over 150°C, and the mixture should be carefully observed for any evidence of decomposition (e.g., turning dark, or smoking). If decomposition is evident, the heat source should be immediately removed from the distilling flask. After distillation is complete, the residue remaining in the distilling flask and each fraction of distillate should be examined by redistillation or chromatography to determine if separation has been achieved. If the fractions have large differences in their boiling points, the components have been separated, and further analysis of the fractions is generally not necessary.

Once these preliminary solubility and distillation tests have been performed, you should decide what method of separation will be most effective and proceed directly to the large-scale separation of the mixture using that method. However, if it is not evident from the preliminary tests what single method can be used to separate the two-component mixture, the general scheme outlined below should be employed.

A General Procedure for the Separation of Mixtures Containing Two Components*

The general scheme outlined in Figure 26.1 can be employed for the separation of mixtures of two (or more) components. In separating mixtures that are known to contain only two components, stepwise progress through the scheme should be halted at the first point where it is evident that the two components have been successfully separated.

*This procedure was adapted from: N. D. Cheronis and J. B. Entrikin, *Identification of Organic Compounds* (New York, Interscience, 1963).

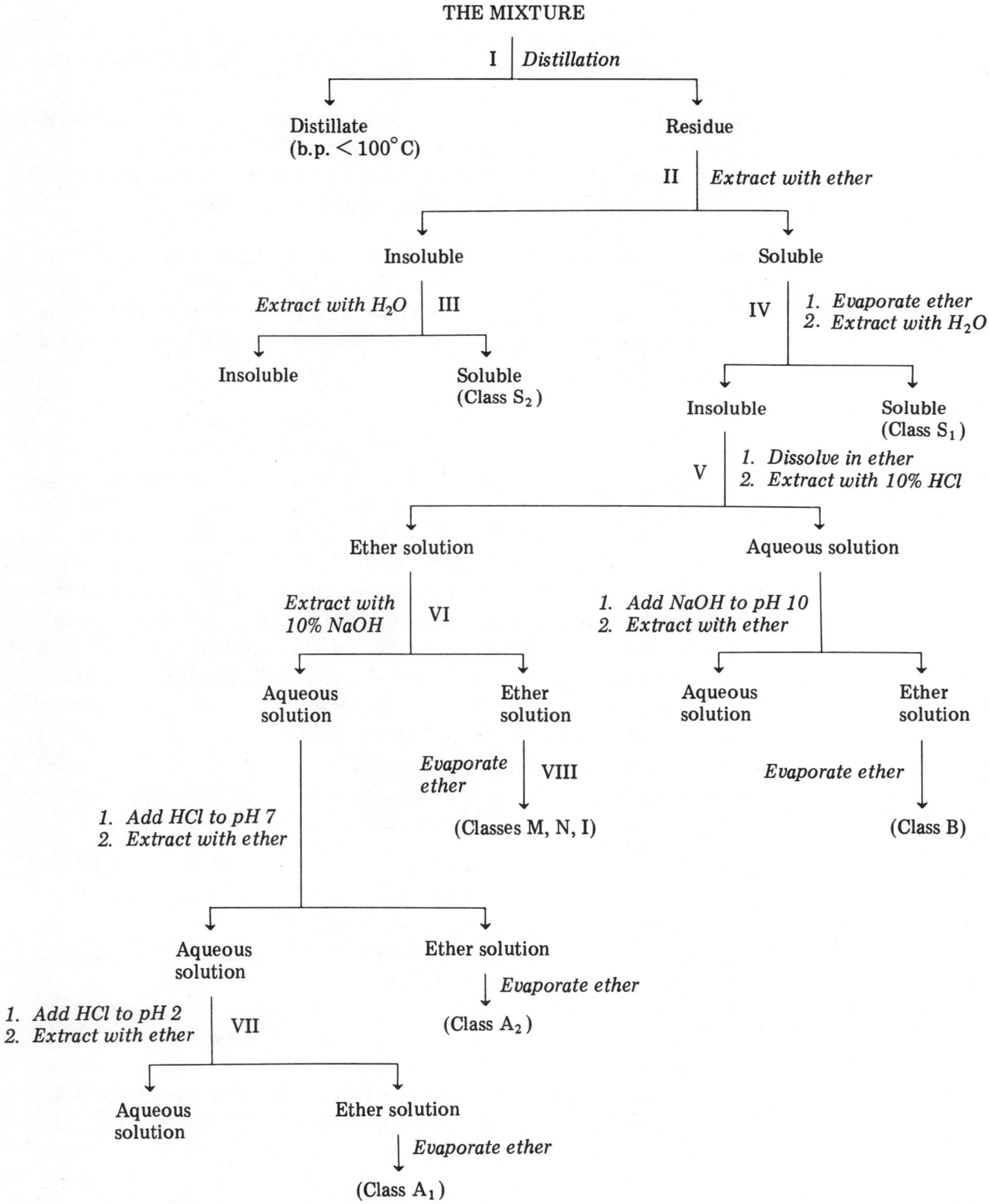

Figure 26.1 *A general procedure for the separation of mixtures.*

The Roman numerals refer to the number of the step in the procedure, and the compound classes refer to the solubility classes listed in Table 25.2.

PROCEDURE

If the preliminary distillation test showed decomposition or no distillate below 100°C or if the mixture is a solid, Step I should be omitted.

Step I. Place 5 to 10 ml of the liquid mixture in a 25-ml distilling flask and, using a well-cooled receiving flask, distill the mixture to remove any component that boils below 100°C. The distillate may contain low molecular weight compounds of almost any class.

Step II. The residue from Step I (or 5 to 10 g of the mixture if Step I was omitted) should be well mixed with 5 ml of ether per gram of mixture. After three minutes, separate any undissolved residue from the ether solution. Treat the residue by Step III, and save the ether solution for Step IV.

Step III. Allow all of the ether to evaporate from the residue obtained in Step II. Add 10 ml of water per gram of residue and shake vigorously. Separate the aqueous extract from any insoluble material. The aqueous solution will contain compounds that belong to solubility class S_2. Evaporate a small portion (~2 ml) of the aqueous extract. If little or no residue remains, no compounds of class S_2 are in your mixture. If appreciable residue remains after evaporating the water, test the *p*H of the aqueous extract. If the solution is basic, the salt of a weak acid may be present, and the extract should be acidified (to *p*H 2) with 6*M* hydrochloric acid to determine if the acid is insoluble in water. If the aqueous extract is acidic, the ammonium salt of an amine may be present, and 10% aqueous sodium hydroxide should be added (to *p*H 9) to the extract to determine if the amine is water insoluble. If the solution is neutral, the water should be evaporated under reduced pressure and the residue purified and characterized.

Step IV. Pour the ether solution from Step II into a distilling flask and distill the ether using a steam bath. Cool the residue and then extract it with 10 ml of water per gram of residue. Shake the solution well, and then separate any water-insoluble material from the aqueous solution. Save the water-insoluble material for Step V.

The aqueous solution will contain compounds that belong to solubility class S_1. Extract the aqueous solution with 5 ml of ether and discard the ether. Test the *p*H of the solution. If the solution is acidic, the aqueous layer contains a water soluble acid. Make the acidic solution strongly acidic (*p*H 2) by the addition of 6*M* sulfuric acid, and then distill the solution to remove any volatile acids. If the acid separates from solution when the sulfuric acid is added, isolate the acid by filtration or extraction. If the aqueous solution is not acidic, add solid potassium carbonate to the solution until it is saturated, and then extract the solution with an equal volume of ether. Distill the ether (steam bath), and the residue will be any nonacidic compounds belonging to class S_1.

Step V. Dissolve the water-insoluble residue from Step IV in 5 ml of ether per gram of residue. Extract the ether solution with two portions of aqueous 10% hydrochloric acid using 3 ml of aqueous acid per gram of residue for each portion. Save the ether solution for Step V. Make the combined, acidic, aqueous extracts alkaline (pH 9) by the addition of concentrated (~30%) sodium hydroxide solution, and then extract the basic, aqueous solution twice with equal volumes of ether. Dry the combined ether extracts over anhydrous potassium carbonate or magnesium sulfate. Then distill the ether (steam bath) to obtain any class M compound that was separated from the mixture.

Step VI. Extract the ether solution saved from Step V with two portions of aqueous 10% sodium hydroxide using a volume of aqueous sodium hydroxide equal to half the volume of ether for each extraction. Save the ether extract for Step VIII. Neutralize the combined alkaline extracts (pH 7) by the addition of 6*M* hydrochloric acid. Extract the neutral solution twice with equal volumes of ether to remove compounds that belong to solubility class A_2. Save the aqueous solution for Step VII. Dry the ether over anhydrous magnesium sulfate and distill off the ether (steam bath) to isolate class A_2 compounds. This fraction will contain phenols and other weak acids.

Step VII. Acidify the aqueous solution saved from Step VI (pH 2) by the addition of concentrated hydrochloric acid. Extract the acidic solution twice with equal volumes of ether. Wash the combined ether extracts with a portion of water equal to one-half the volume of ether, and then dry the ether over anhydrous magnesium sulfate. Distill the ether (steam bath) to obtain compounds belonging to solubility class A_1.

Step VIII. Wash the ether solution saved from Step VI twice with 10-ml portions of water and dry the ether over anhydrous magnesium sulfate. Distill the ether (steam bath) to obtain compounds that belong to solubility class M, N, or I.

This scheme should serve as a general guide for the separation of most mixtures. However, modifications may be necessary for a particular separation. Once the components have been separated, each component should be purified by recrystallization or distillation.

Prelab Questions

1. Outline a scheme for the separation of each of the following mixtures: (a) benzoic acid and ethyl benzoate, (b) aniline and nitrobenzene, (c) ethanol and ethyl acetate, (d) *p*-nitrobenzoic acid and *p*-aminobenzoic acid, (e) *o*-cresol and *o*-xylene, (f) anthracene and phthalic acid.

2. If the scheme outlined in Figure 26.1 were employed to separate the following mixtures, in which step would the two compounds be separated? (a) *p*-chlorophenol and *p*-dichlorobenzene, (b) cinnamic acid and cinnamaldehyde, (c) cyclohexane and bromocyclohexane, (d) cyclohexylamine and cyclohexanol, (e) *o*-chlorobenzonitrile and *o*-chlorobenzylamine, (f) acetone and dioxane.

3. Treatment of benzaldehyde with concentrated sodium hydroxide yields a mixture of equal amounts of benzyl alcohol and benzoic acid (as the sodium salt) in the aqueous alkali. Describe how this mixture could be separated and the components isolated.

4. Hydrolysis of *N*-phenyloctanamide in aqueous hydrochloric acid produces a mixture of octanoic acid and aniline (as the hydrochloride salt). Describe how this mixture could be separated and the components purified.

5. You are asked to determine the amount of phenacetin and aspirin in a combination headache remedy. How would you separate these two compounds?

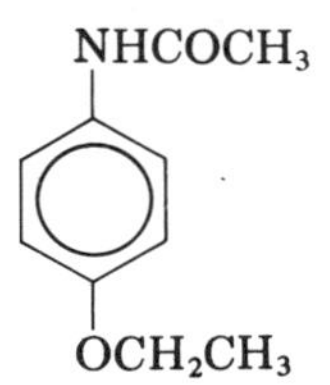

Phenacetin

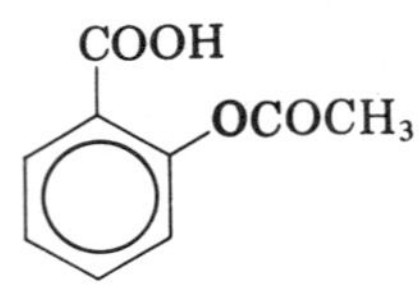

Aspirin

EXPERIMENTAL PROCEDURE

You will be expected to separate a two-component mixture by the procedures described in this experiment. Once the components of the mixture are separated, each compound should be purified and identified by making use of the procedures discussed in Experiments 21–25. A summary report that describes the methods used to separate the mixture and to identify each compound isolated should be prepared at the end of the experiment.

Experiment Twenty-seven

Phase Transfer Catalysis in the Synthesis of 7, 7-Dichlorobicyclo-(4.1.0)heptane

In this experiment you will prepare 7,7-dichlorobicyclo[4.1.0]heptane by reaction of chloroform and base with cyclohexene in the presence of a catalytic amount of benzyltriethylammonium chloride. This reaction involves the inter-

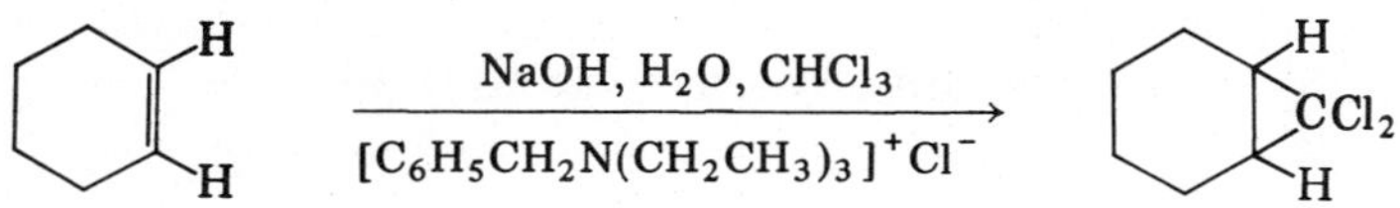

Cyclohexene
M.w. = 82.15 g/mol
b.p. = 83°C

7,7-Dichlorobicyclo-[4.1.0]heptane
M.w. = 165.1 g/mol
b.p. = 198°C

mediate formation of dichloromethylene, a *carbene*, and illustrates the use of a quaternary ammonium salt as a *phase transfer catalyst.*

Carbenes

Carbenes are reactive divalent carbon intermediates with two atoms or groups bonded to the carbon atom. Two nonbonded electrons reside on carbon. Therefore, a carbene has no net charge, but, because a carbene has an unfilled octet of electrons, it can react as an electrophile. The simplest carbene, methylene ($:CH_2$), can be generated by the decomposition of diazomethane:

$$H_2C{=}\overset{+}{N}{=}\overset{-}{N}: \xrightarrow[\text{or light}]{\text{heat}} H_2C: \quad + \quad :N{\equiv}N:$$

Diazomethane *Methylene* *Nitrogen*

Dichloromethylene, $:CCl_2$, is readily generated in the laboratory by two methods: the thermal decomposition of sodium trichloroacetate:

$$Cl_3C-\overset{\displaystyle O}{\overset{\|}{C}}-O^-Na^+ \xrightarrow{\text{heat}} :CCl_2 + CO_2 + NaCl$$

Sodium trichloroacetate — *Dichloromethylene*

and the reaction of chloroform with a strong base such as hydroxide or an alkoxide:

$$B:^- + Cl_3C-H \rightleftharpoons Cl_3C:^- + B-H$$

A base — *Chloroform* — *Trichloromethyl anion*

$$Cl_3C:^- \longrightarrow :CCl_2 + Cl^-$$

Dichloromethylene

The latter reaction involves proton transfer from chloroform to the base in an equilibrium step to form the trichloromethyl anion. This anion eliminates chloride ion to form dichloromethylene.

Because carbenes are very reactive, they are generated in the presence of the compound with which reaction is desired. With hydrocarbons, carbenes undergo either carbon-hydrogen bond insertion:

Carbon-hydrogen bond insertion

$$R-H + R'_2C: \longrightarrow R-\underset{R'}{\overset{R'}{C}}-H$$

A carbene

or addition to a carbon-carbon double bond:

Addition to carbon-carbon double bond

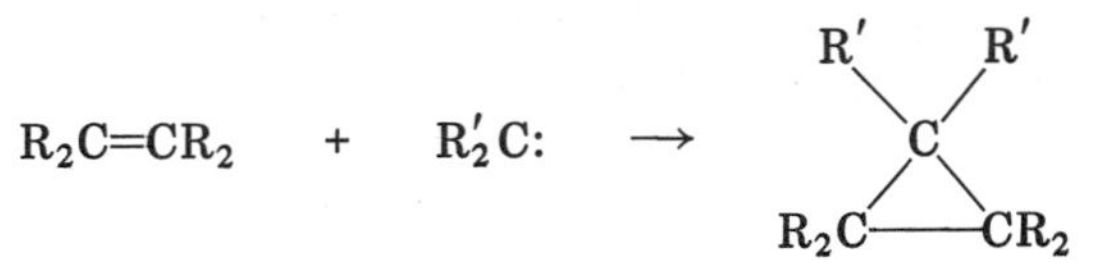

An alkene — *A carbene* — *A cyclopropane*

The carbene addition to an alkene yields a cyclopropane. Unlike the very reactive methylene, dichloromethylene will selectively undergo addition reactions with unsaturated hydrocarbons such as cyclohexene. Carbon-hydrogen insertion reactions are not characteristic of dichloromethylene.

Phase Transfer Catalysis

Normally, water must be carefully excluded from carbene generating reactions since carbenes react rapidly with water. However, in the late 1960s it was discovered that dichlorocarbene could be conveniently generated and reacted with an alkene in a two-phase, aqueous-organic system in which a quaternary ammonium salt, $R_4N^+X^-$, was present in catalytic amounts. In this procedure chloroform and the alkene compose the organic phase, and aqueous sodium hydroxide forms the second immiscible phase. The quaternary ammonium salt serves as a *phase transfer catalyst.* The phase transfer catalyst is employed to transfer the hydroxide ion into the organic phase where reaction between the hydroxide ion and chloroform generates dichloromethylene in the presence of the alkene. The sequence of reactions involved in this process is shown in Figure 27.1. The transfer of hydroxide ion into the organic phase occurs because the quaternary ammonium cation-hydroxide ion pair, $R_4N^+OH^-$, is soluble in the organic phase due to the large hydrophobic organic substituents on the quaternary ammonium cation. In the organic phase, reaction between chloroform and the quaternary ammonium hydroxide generates the organic-soluble quaternary ammonium salt of the trichloromethyl anion. This salt decomposes to dichloromethylene and the quaternary ammonium chloride salt. Thus, dichloromethylene is formed in the organic phase in the presence of cyclohexene, and reaction between dichloromethylene and cyclohexene results in the formation of the product, 7,7-dichlorobicyclo[4.1.0]heptane. The quaternary ammonium chloride salt is soluble in both the organic and aqueous phase and is distributed between these phases. Because the dichloromethylene is formed in the organic phase that is concentrated in cyclohexene, competing reactions of dichloromethylene such as reaction with water to form carbon monoxide or dimerization to form tetrachloroethane, $Cl_2C{=}CCl_2$, do not occur to a significant extent.

Many different quaternary ammonium salts have been tested as phase transfer catalysts. Of these salts, the best phase transfer catalysts have been

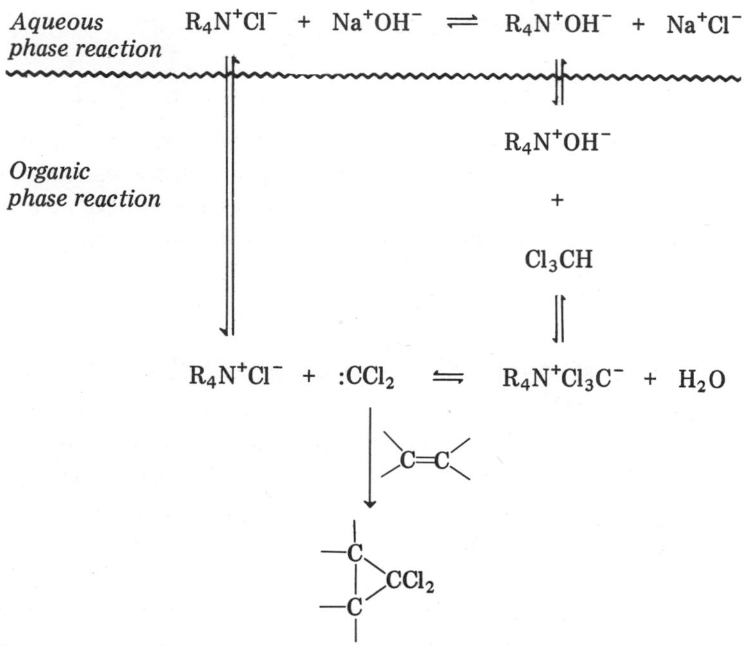

Figure 27.1 *Phase transfer catalysis in the generation of dichloromethylene.*

found to be large quaternary ammonium cations (M.w. > 150 g/mol) that are chloride or bisulfate salts. Small quaternary ammonium cations such as the tetramethylammonium ion are not good catalysts because they form salts of low solubility in organic phases. Benzyltriethylammonium chloride and tetrabutylammonium chloride are two efficient phase transfer catalysts which are commercially available.

Prelab Questions

1. Dichlorocarbene reacts with water to produce carbon monoxide and hydrogen chloride. Write a balanced equation for this reaction.

2. Two students performed this experiment and obtained quite different yields. The first student, who vigorously shook her reaction mixture throughout the reaction period, obtained a 68% yield of product. The second student shook his reaction mixture just a few times during the reaction period and was able to obtain only a 30% yield of product. Explain why this difference in yields occurred.

3. Why is only a catalytic amount of the quaternary ammonium salt necessary for phase transfer reactions?

4. Outline a flow diagram for the purification procedure in this experiment.

5. Only the *cis*-addition product is formed in this reaction. Could the *trans*-isomer be formed? Explain your answer.

EXPERIMENTAL PROCEDURE

CAUTION: Avoid skin contact with chloroform and breathing chloroform vapors. Skin contact with chloroform and breathing chloroform vapors is hazardous to your health. Perform this experiment in a hood or other well-ventilated area. Dispose of all chloroform containing fractions in the proper waste container.

Using your 250-ml round-bottom flask, Claisen head, addition funnel, and reflux condenser, set up the apparatus for reflux with addition (see Figure 17.2, omit the hydrogen chloride trap). To the 250-ml round-bottom flask add 8.2 g (0.10 mol) of cyclohexene, 30 ml (0.40 mol) of chloroform, and 0.5 g of benzyltriethylammonium chloride. Place 25 ml of 50% aqueous sodium hydroxide in the addition funnel. Add

---

CAUTION: Avoid skin contact with 50% aqueous sodium hydroxide. Wash any sodium hydroxide from your skin or clothes with copious amounts of water.

approximately 2 ml of the aqueous sodium hydroxide solution from the addition funnel into the reaction flask. Warm the solution on the steam bath and shake the apparatus to mix. It is important to keep the two phases well mixed during this reaction. Over the next 20 minutes add the sodium hydroxide solution in small portions, with shaking, to the refluxing reaction mixture. Continue to reflux the reaction mixture for one hour after addition is complete. Cool the solution and add slowly, with swirling, 25 ml of water and then 25 ml of hexane to the reaction mixture. Transfer the mixture to the addition funnel. Separate and discard the lower aqueous layer. Wash the organic layer twice with 25-ml portions of water, and then dry the cloudy organic layer with anhydrous magnesium sulfate. Distill the solvents from the dry organic layer using the steam bath. Dispose of the distillate in the waste container that is provided. Transfer the higher-boiling liquid residue to a small-scale distillation apparatus (Figure 17.3) and distill the product. The product is distilled under reduced pressure (b.p. = 96°C at 35 Torr) or at atmospheric pressure (b.p. = 198°C with only slight decomposition). Weigh the purified product and calculate the percentage yield. The collected product may be identified by IR spectroscopy (Figure 27.2) and its purity may be determined by GLC analysis on a 20% SE-30 column.

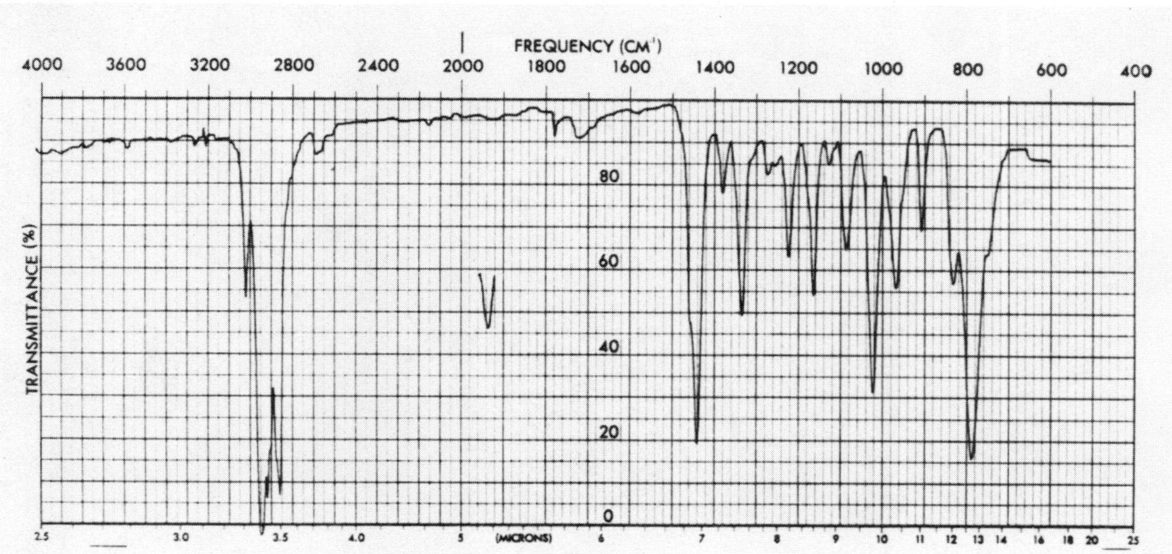

Figure 27.2 *Infrared spectrum of 7,7-dichlorobicyclo[4.1.0]heptane (thin film).*

Postlab Questions

1. Which reactant is the limiting reagent in this reaction?

2. What chemical tests could be used to confirm the identity of your product?

3. How could you determine if any unreacted cyclohexene remained in the reaction mixture after refluxing the two phase mixture?

4. The addition of dichloromethylene to alkenes is stereospecific. The addition of dichloromethylene to *trans*-alkenes yields *trans*-1,2-disubstituted-3,3-dichlorocyclopropanes, and the addition to *cis*-alkenes yields *cis*-1,2-disubstituted-3,3-dichlorocyclopropanes. Draw the structure of the dichloromethylene addition product formed by reaction with each of the following alkenes: (a) *cis*-3-hexene, (b) *trans*-3-methyl-2-pentene, (c) *cis*-1-chloro-2-bromoethene, (d) *trans*-3,5-dimethylcyclopentene.

Experiment Twenty-eight

Multistep Organic Syntheses: Preparation of Benzocaine

The synthesis of an organic compound from commercially available and inexpensive starting materials often requires more than one reaction. In a multistep synthesis the product from one reaction is employed as the reactant for subsequent chemical transformations. The preparation of *trans*-stilbene described in Experiment 16, for example, requires the reactant 1,2-diphenylethanol, which is produced from benzyl chloride and benzaldehyde by the Grignard reaction in Experiment 15. In turn, benzaldehyde may be synthesized from benzyl alcohol in an oxidation procedure (Experiment 14), and benzyl chloride can also be prepared from benzyl alcohol through a substitution reaction (as described in Experiment 8). Reaction processes commonly used for multistep organic syntheses will be described in this and subsequent experiments.

In this experiment you will prepare benzocaine (ethyl *p*-aminobenzoate) from the relatively inexpensive and commercially available *p*-nitrotoluene. Benzocaine (also known as *Anesthesin*, *Anesthone*, and *Parathesin*) is used medicinally as a local (usually external) anesthetic and is often an active component of ointments used for the treatment of sunburn. The synthesis of benzocaine from *p*-nitrotoluene is performed by three sequential reactions: oxidation of *p*-nitrotoluene,

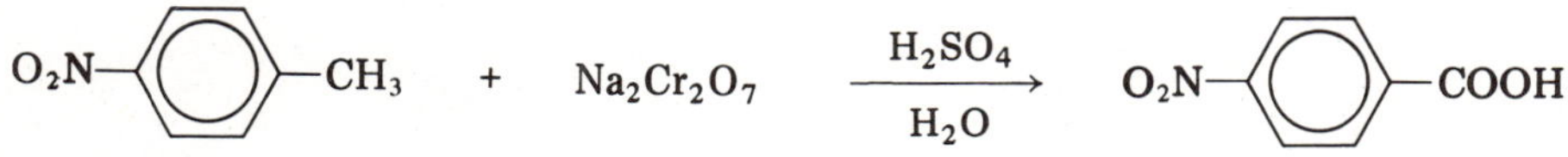

p-*Nitrotoluene*
M.w. = 137.1 g/mol
m.p. = 52°C
b.p. = 238°C

Sodium dichromate
M.w. = 298.0 g/mol
for $Na_2Cr_2O_7 \cdot 2H_2O$

p-*Nitrobenzoic acid*
M.w. = 167.1 g/mol
m.p. = 242°C

esterification of *p*-nitrobenzoic acid,

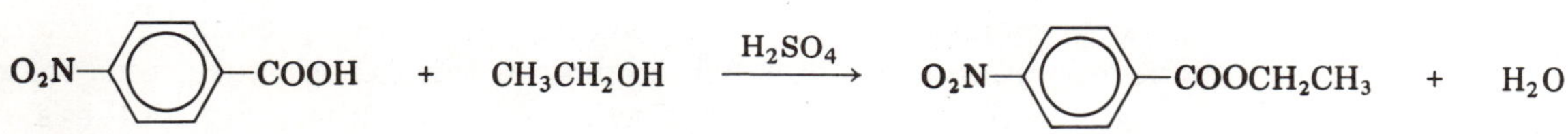

p-*Nitrobenzoic acid*

Ethanol
M.w. = 46.1 g/mol
b.p. = 78.5°C

Ethyl p-*nitrobenzoate*
M.w. = 195.2 g/mol
m.p. = 56°C

Water

and reduction of ethyl p-nitrobenzoate,

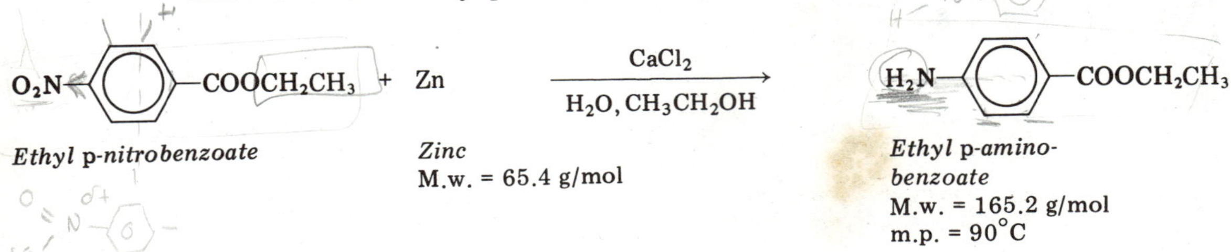

The order in which the three synthetic operations for the preparation of benzocaine is arranged is critical. Reduction of the nitro group may not precede oxidation since anilines undergo aromatic ring oxidation with strong oxidizing agents such as dichromate:

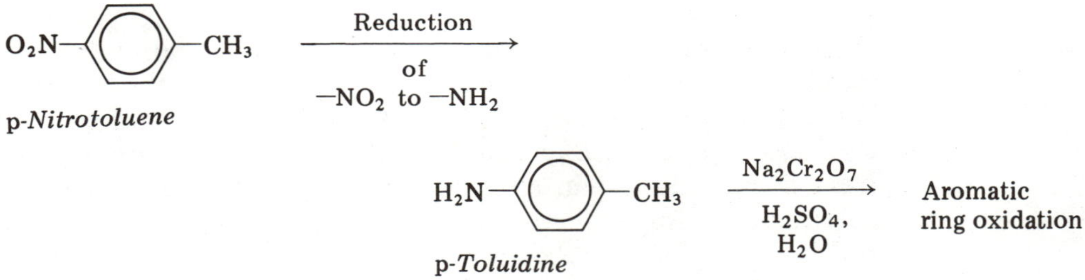

Similarly, an alternate procedure in which reduction of the nitro group precedes esterification presents isolation difficulties because p-aminobenzoic acid is soluble in both acid and base:

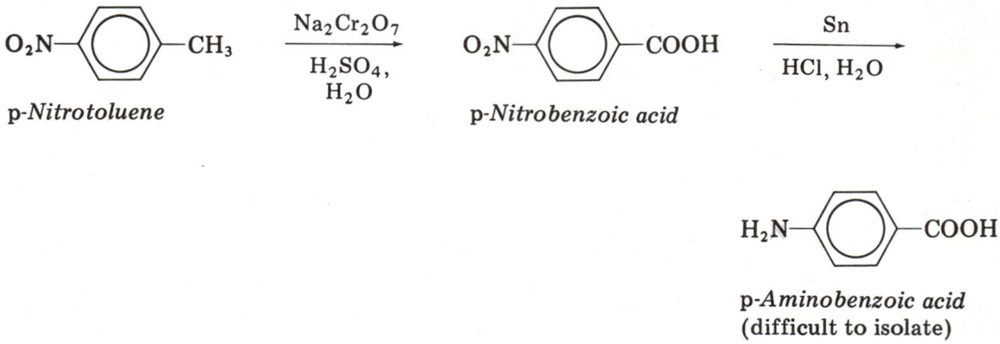

p-Aminobenzoic acid can be isolated only under precisely neutral conditions after removal of metal ions that form complexes with this product. The order of synthetic operations that you will employ in this experiment avoids these difficulties and represents a convenient scheme for the laboratory synthesis of benzocaine. For the synthesis of benzocaine on a larger scale, the zinc dust reduction of ethyl p-nitrobenzoate is replaced by a more economical catalytic hydrogenation.

Prelab Questions

1. Write balanced chemical reactions for each of the sequential steps in the conversion of *p*-nitrotoluene to benzocaine. Chromic sulfate, $Cr_2(SO_4)_3$, can be assumed to be the reduced product from the oxidation of *p*-nitrotoluene. Zinc hydroxide, $Zn(OH)_2$, may be assumed to be the oxidized product from the reduction of ethyl *p*-nitrobenzoate.

2. How many moles of sodium dichromate are required to oxidize one mole of *p*-nitrotoluene to *p*-nitrobenzoic acid?

3. Prepare a flow diagram that outlines the procedure for isolation of the desired organic product for each reaction in the preparation of benzocaine.

4. If you begin this sequence of reactions with 6.85 g of *p*-nitrotoluene and each step in the sequence affords a 60% yield of product, what amount of benzocaine would you expect to produce? What if the yield for each step was 80%?

EXPERIMENTAL PROCEDURE

Oxidation of p-Nitrotoluene. In a 250-ml 1-neck round-bottom flask dissolve 20 g (0.067 mol) of sodium dichromate dihydrate in 50 ml of water. With stirring, slowly add 25 ml of concentrated sulfuric acid to the chromic acid solution, cool the reaction solution to less than 50°C, and then add 6.8 g (0.050 mol) of *p*-nitrotoluene. After adding several boiling stones, attach the Claisen head to the round-bottom flask and place the thermometer adapter on the center connection of the Claisen head. Insert a thermometer through the adapter and into the reaction solution and attach a reflux condenser to the side connection of the Claisen head.

Heat the mixture to approximately 75°C with occasional swirling of the contents of the reaction flask. When the reaction temperature reaches approximately 75°C, an exothermic reaction will often be observed by a rapid increase in the reaction temperature. If this exothermic reaction begins, remove the heat source until the temperature begins to fall and then return the heat source. Reflux for 60 min, and plan to return solid *p*-nitrotoluene that enters the reflux condenser back into the reaction flask. Then allow the mixture to cool for 10 min, and subsequently pour the contents of the reaction flask onto approximately 100 g of ice in a 250-ml Erlenmeyer flask. Collect the resulting precipitate in a Büchner funnel by suction filtration and wash this solid with two 30-ml portions of water. Transfer the filtered solid to a 250-ml beaker and add 30 ml of water and 30 ml of 10% aqueous sodium hydroxide to dissolve *p*-nitrobenzoic acid. Warm the resulting mixture on a steam bath for 10 min to coagulate residual chromium salts as their insoluble hydroxides and then filter by suction. Add 1 g of decolorizing carbon to the resulting filtered solution; again heat on a steam bath for 10 min; and filter the mixture by gravity through a coarse filter paper. Prepare an aqueous acid solution by adding 20 ml of concentrated hydrochloric acid to 30 g of ice in a 250-ml beaker. Slowly and with stirring, pour the basic charcoal-decolorized solution into the aqueous acid solution. Test the resulting solution with *p*H paper after mixing is complete to ensure that this solution is strongly acidic. Suction filter the resulting precipitate and wash the collected solid with two 10-ml portions of cold water. Recrystallize the collected solid from ethanol, obtain the melting point and weight of the isolated product, and calculate the percentage yield of *p*-nitrobenzoic acid. Save a portion of your product for TLC and spectral analyses and perform these analyses if time is available to you.

Esterification of p-*Nitrobenzoic Acid.** Add 30 ml of absolute ethanol to 3.4 g (0.020 mol) of *p*-nitrobenzoic acid in a 100-ml round-bottom flask. Place several boiling chips in the flask, and attach a reflux condenser for heating under reflux. Slowly add 5 ml of concentrated sulfuric acid to the mixture through the condenser. Reflux the mixture for 1 hr or until all of the solid *p*-nitrobenzoic acid dissolves—whichever time is longer. Cool the reaction solution and then pour it into a mixture of 50 ml of 10% aqueous sodium hydroxide and approximately 50 g of ice. Isolate the precipitate by suction filtration and, if necessary, recrystallize the collected solid from ethanol-water. Obtain the melting point and weight of the isolated product and calculate the percentage yield of ethyl *p*-nitrobenzoate. Save a portion of your product for TLC and spectral analyses and perform these analyses if time is available to you.

Reduction of Ethyl p-*Nitrobenzoate.*† Dissolve 1.0 g of calcium chloride in 12 ml of water and then mix this solution with 55 ml of 95% ethanol. Pour the resulting solution into a 250-ml round-bottom flask that contains 2.5 g (0.013 mol) of ethyl *p*-nitrobenzoate, add 25 g of zinc dust, and attach a reflux condenser to the round-bottom flask. Reflux the reaction mixture for 2 hr and then cool to room temperature. Separate the unreacted zinc dust from the aqueous ethanol solution by suction filtration and wash the filtered solid with two 25-ml portions of ether. Extract the filtrate with 150 ml of water saturated with sodium chloride. (If necessary, divide the filtrate into two parts and extract each part with 75 ml of the saturated sodium chloride solution.) While retaining the organic layer for subsequent

*Adjust amounts of reagents to the molar amount of *p*-nitrobenzoic acid that you have prepared.

†Adjust amounts of reagents to the molar amount of ethyl *p*-nitrobenzoate that you have prepared.

treatment, wash the aqueous layer twice with 25-ml portions of ether. Combine the ether washings with the original organic layer and wash the combined ether solution with 50 ml of water. Dry the ether solution over anhydrous magnesium sulfate, filter, and then distill the ether on a steam bath to a residual volume of 10 to 15 ml. Transfer the ether residue to an Erlenmeyer flask and add 20 ml of pentane to precipitate the product. Recrystallize the isolated product from ether-pentane. Obtain the melting point and weight of the isolated product and calculate the percentage yield of ethyl *p*-aminobenzoate. If time is available, perform TLC and spectral analyses of this product.

Postlab Questions

1. Calculate the percentage yield of benzocaine from the initial reactant *p*-nitrotoluene.

2. Given that the cost of *p*-nitrotoluene is $3.50 per 100 g, of sodium dichromate is $7.00 per kg, of ethanol is $2.50 per liter, and of zinc dust is $11.00 per kg, calculate the materials cost for the production of benzocaine (per 100 g) from *p*-nitrotoluene. Assume that solvents and other reagents in this synthesis can be recovered.

3. *p*-Nitrobenzoic acid is priced at $11.75 per 100 g. Assuming a minimum wage for yourself, materials cost as described in Question 2, and a cost for utilities operation of $0.02 per gram of *p*-nitrotoluene, could you have produced *p*-nitrobenzoic acid at a lower cost? Assume that you are able to prepare 500 g of *p*-nitrobenzoic acid in the same amount of time that you spent for its preparation in this experiment.

4. Explain the function of each operation in the isolation and purification procedure for the synthesis of *p*-nitrobenzoic acid.

5. In the procedure for esterification of *p*-nitrobenzoic acid, what compound would you expect to precipitate after reacidification of the basic aqueous solution from which ethyl *p*-nitrobenzoate is filtered?

6. The usual method for the reduction of aromatic nitro compounds employs zinc dust in aqueous hydrochloric acid. Why is this procedure not applicable to the reduction of ethyl *p*-nitrobenzoate in this experiment?

7. Predict the characteristic differences in the infrared spectra of *p*-nitrotoluene, *p*-nitrobenzoic acid, ethyl *p*-nitrobenzoate, and ethyl *p*-aminobenzoate that would allow you to distinguish between these compounds.

8. Figure 28.1 provides the ^1H NMR spectra of the products formed in the multistep synthesis of benzocaine. Identify the characteristic absorptions and match each spectrum with the appropriate product structure. Why do the aromatic hydrogen absorptions for ethyl *p*-nitrobenzoate appear as a singlet while those for ethyl *p*-aminobenzoate are separated into two distinct sets of absorptions?

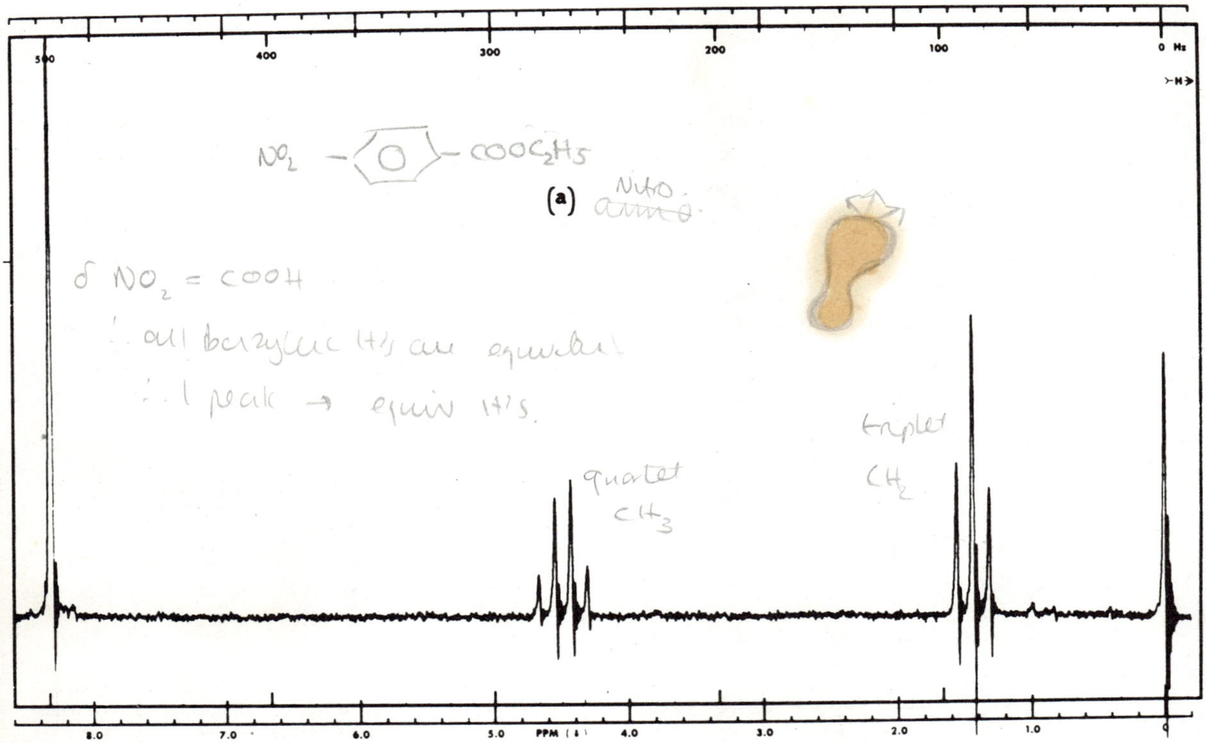

500
400
300
200
100
0 Hz
(a)
8.0
7.0
6.0
5.0
PPM (δ)
4.0
3.0
2.0
1.0
0

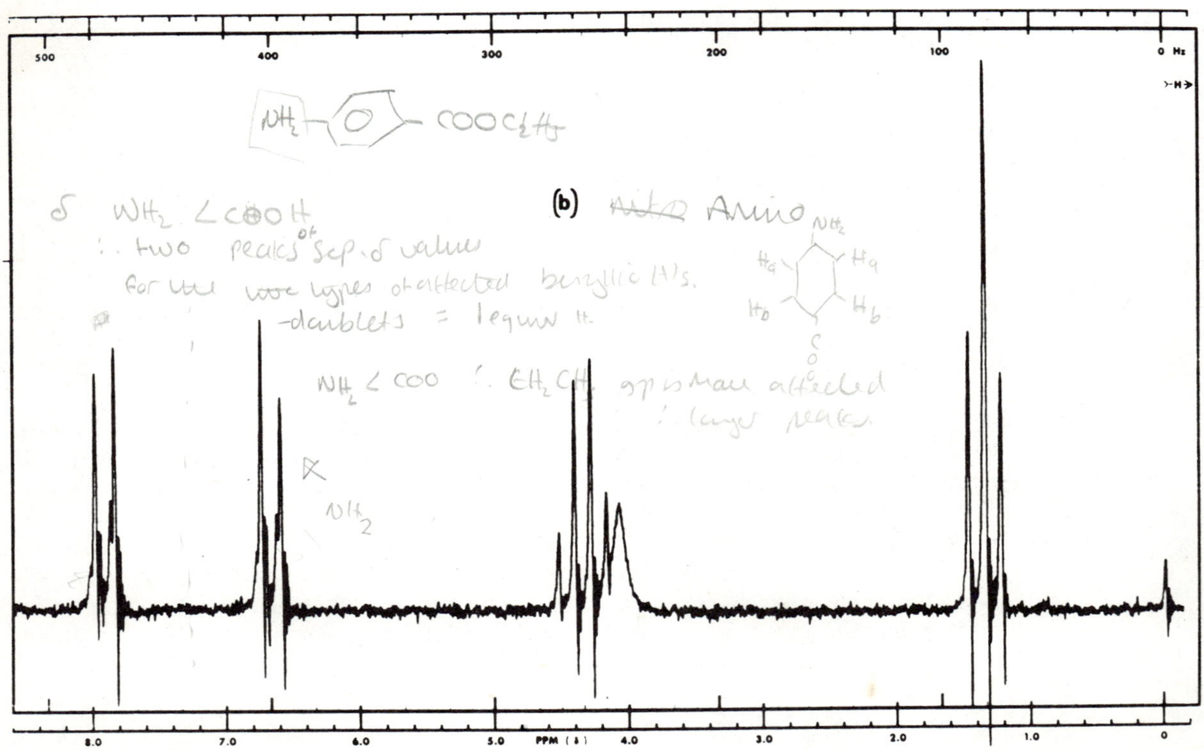

500
400
300
200
100
0 Hz
(b)
8.0
7.0
6.0
5.0
PPM (δ)
4.0
3.0
2.0
1.0
0

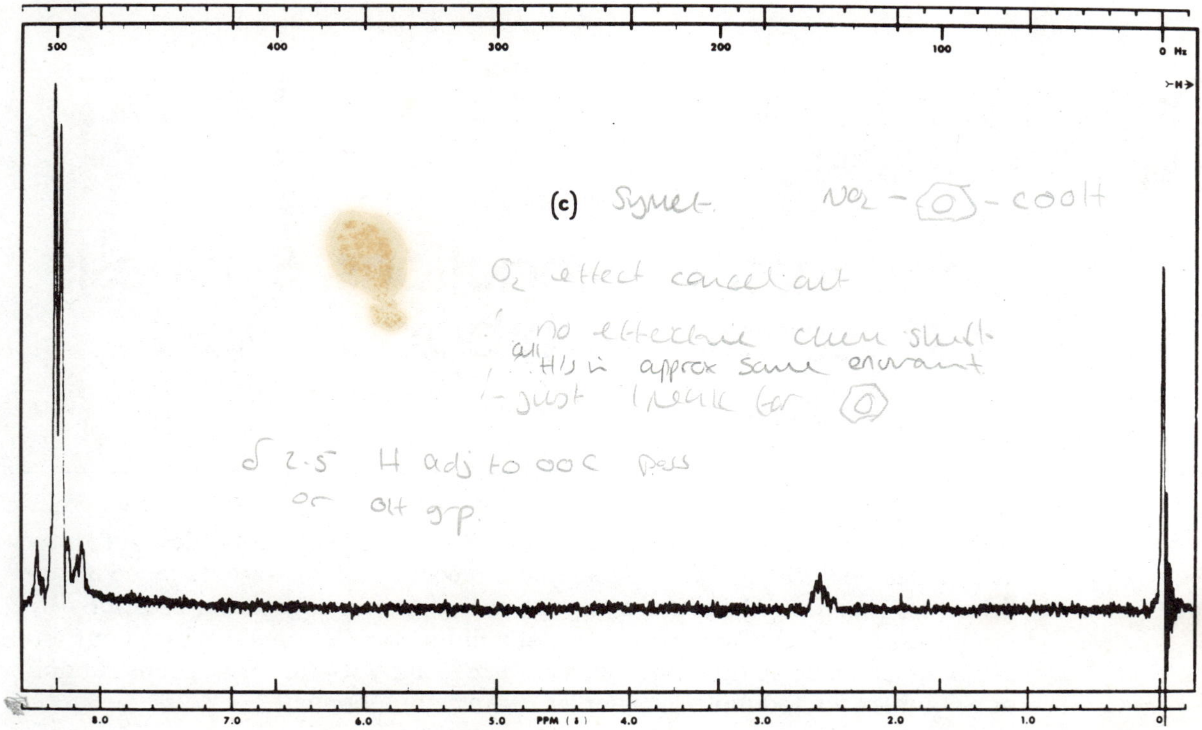

Figure 28.1 ^{1}H ***NMR spectra of products from multistep synthesis of benzocaine.***

(*a*) and (*b*) are recorded in $CDCl_3$ solution; (*c*) is recorded in DMSO-d_6 (the proton absorption at 2.6 δ is due to CD_3SOCD_2H).

ethyl ⌬ – have triplet & quartet ... i.e. 2 equiv + 3 equiv H's.

∴ must be a) & b)

Experiment Twenty-nine

Condensation Reactions: Synthesis of Coumarin-3-carboxylic Acid

A reaction resulting in the combination of two or more molecules to form a new molecule with the loss of a simple molecule such as water is called a *condensation reaction.* Carbonyl compounds undergo condensation reactions by the processes of addition and subsequent elimination. Normally carried out under basic conditions, these reactions are particularly useful for the construction of a complex compound from simpler organic compounds.

Carbonyl condensation reaction

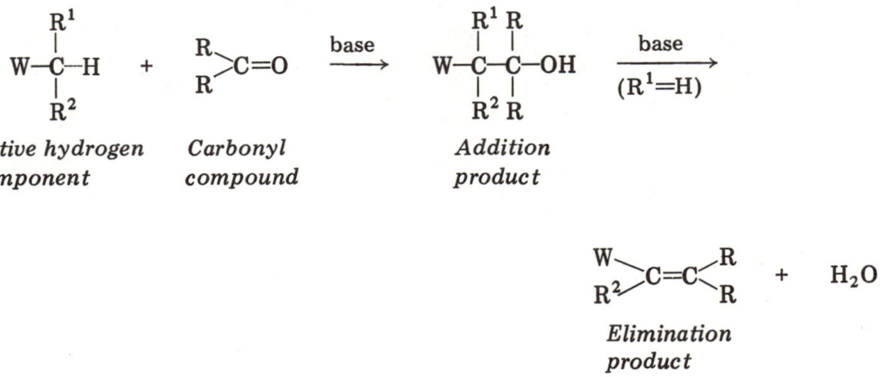

Initial combination occurs by the addition of the carbon and hydrogen of one molecule, the *active hydrogen component*, to the carbonyl group of another molecule. The W-group in the active hydrogen component is a strong electron withdrawing group such as a carbonyl group, a carboxy ester, a cyano group, or a nitro group. The electron withdrawing group of the active hydrogen component causes the hydrogen on the α-carbon to be more acidic and thus much more reactive in condensation reactions than other unactivated C—H groups. If the active hydrogen component has two electron withdrawing groups attached to the same carbon (R^1 or R^2 is also a W-group), then the acidity of the active hydrogen component is further increased. For example, diethyl malonate is 10^{12} times more acidic than is ethyl acetate.

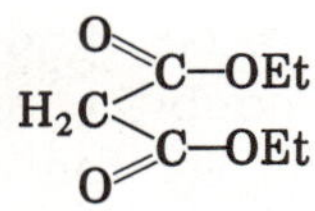

Diethyl malonate
$pK_a = 13$

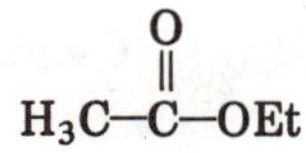

Ethyl acetate
$pK_a = 25$

Condensation reactions with carbonyl compounds involve nucleophilic addition. However, since the active hydrogen component is not itself sufficiently nucleophilic to add to the carbonyl group, base removal of a proton from the α-position of the active hydrogen component (the most acidic position) is required. The nucleophilic addition process can be considered to occur in three steps: (1) acid-base equilibrium reaction between the base and the active hydrogen component,

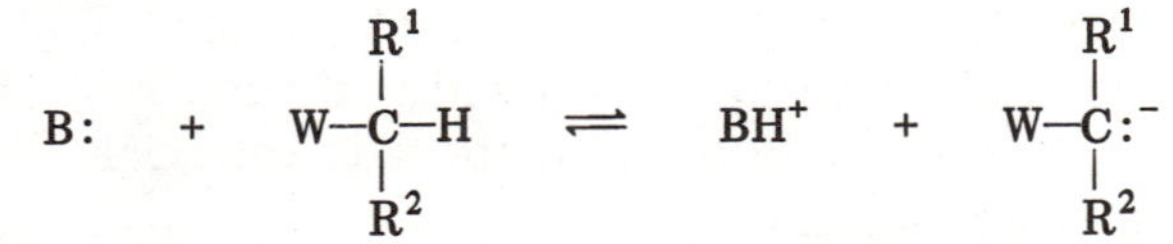

Base | *Active hydrogen component* | *Conjugate acid* | *Nucleophilic conjugate base*

(2) nucleophilic addition to the carbonyl compound,

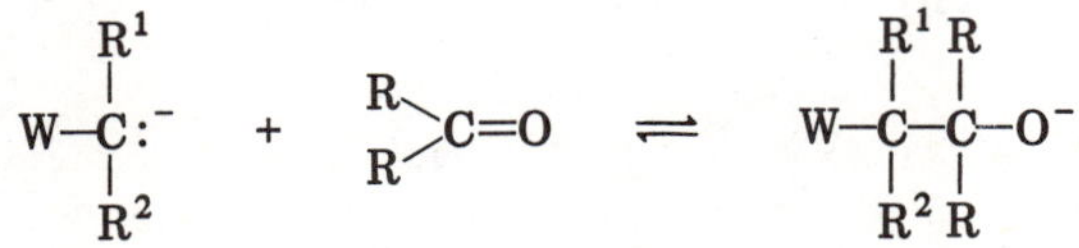

Carbonyl compound | *Alkoxide addition product*

and (3) protonation of the alkoxide by the conjugate acid,

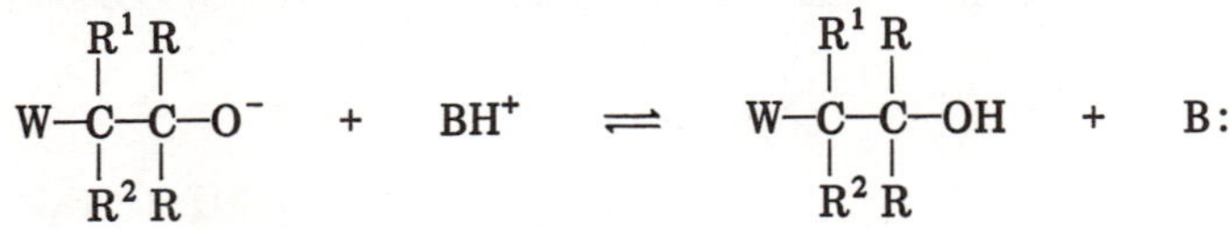

Addition product

In the last step the base is regenerated and is, therefore, a true catalyst in the reaction process. If the addition product contains an α-hydrogen (R^1 or $R^2 = H$), the elements of water may be elminated to form an unsaturated compound:

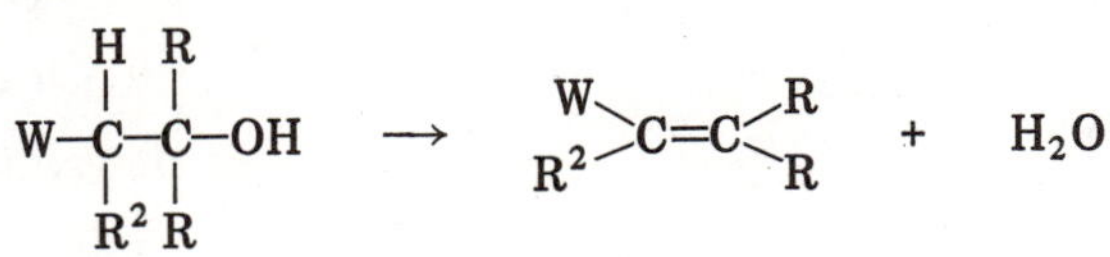

Addition product | *Elimination compound*

Elimination of water normally occurs under the reaction conditions of the condensation reaction when the double bond of the unsaturated compound will extend conjugation between the electron withdrawing group and an unsaturated R-group such as phenyl.

The choice of the base catalyst employed in carbonyl condensation reactions is important. The base chosen should be sufficiently basic to remove the α-hydrogen from the active hydrogen component but not basic enough to remove the hydrogen from the α-position of the carbonyl compound. Usually a base is chosen for condensation reactions so that the equilibrium concentration of the nucleophilic conjugate base of the active hydrogen component is quite low; that is, the K_a value for the conjugate acid of the base (BH^+) is 10^2 to 10^4 times greater than the K_a of the active hydrogen component. For example, piperidine (pK_a of $BH^+ = 11$) is often used in condensation reactions if the active hydrogen component is diethyl malonate ($pK_a = 13$).

In this experiment you will first prepare 3-carbethoxycoumarin by the condensation of diethyl malonate with salicylaldehyde using a catalytic amount of the base piperidine:

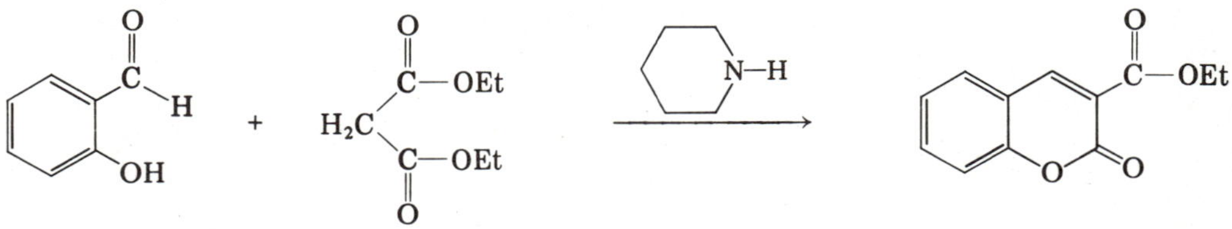

Salicylaldehyde	*Diethyl malonate*	*Piperidine*	*3-Carbethoxycoumarin*
M.w. = 122 g/mol	M.w. = 160 g/mol	M.w. = 85 g/mol	M.w. = 218 g/mol
b.p. = 197°C	b.p. = 199°C	b.p. = 106°C	m.p. = 93°C
d = 1.17 g/ml	*d* = 1.05 g/ml	*d* = 0.86 g/ml	

The first step in this reaction is the formation of the resonance stabilized diethyl malonate anion by reaction with the basic piperidine:

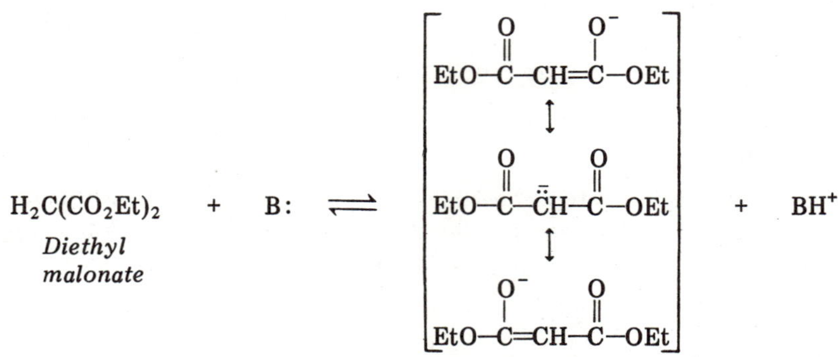

Diethyl malonate anion

The diethyl malonate anion adds to the carbonyl group of salicylaldehyde, and then the intermediate alkoxide is protonated to form the addition product:

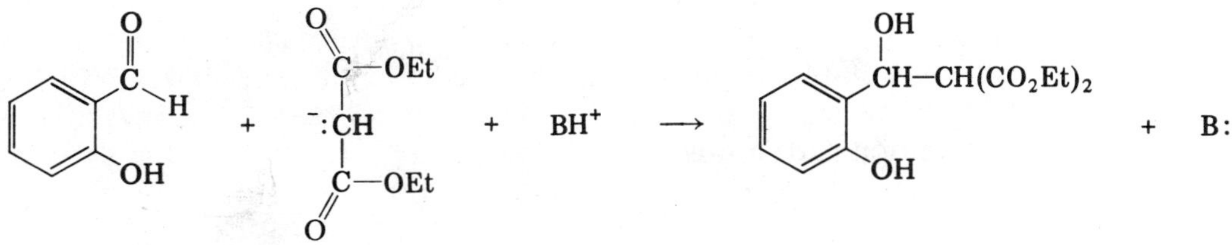

Salicyl-aldehyde *Diethyl malonate anion* *Addition product*

This intermediate addition product rapidly undergoes the loss of water and cyclization (via transesterification) to form the final product, 3-carbethoxy-coumarin:

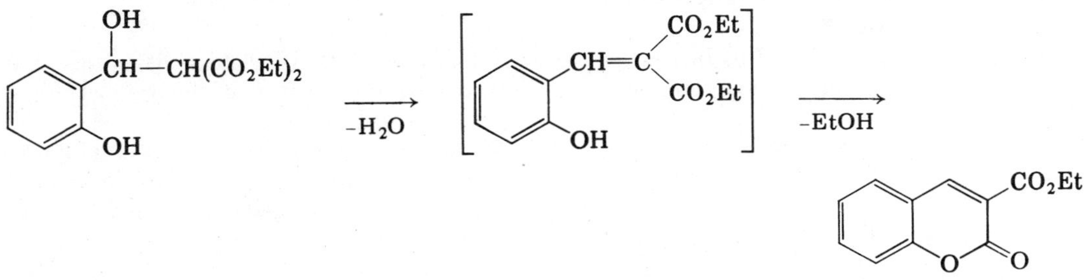

3-Carbethoxy-coumarin

The cyclization reaction may either follow the elimination of water, as illustrated, or it may precede the elimination of water.

To prepare coumarin-3-carboxylic acid, 3-carbethoxycoumarin is hydrolyzed by refluxing with a solution of potassium hydroxide in aqueous ethanol. During hydrolysis both the ethyl ester and the cyclic ester (lactone ring) are cleaved by the base to give the tri-potassium salt of *o*-hydroxybenzylidenemalonic acid.

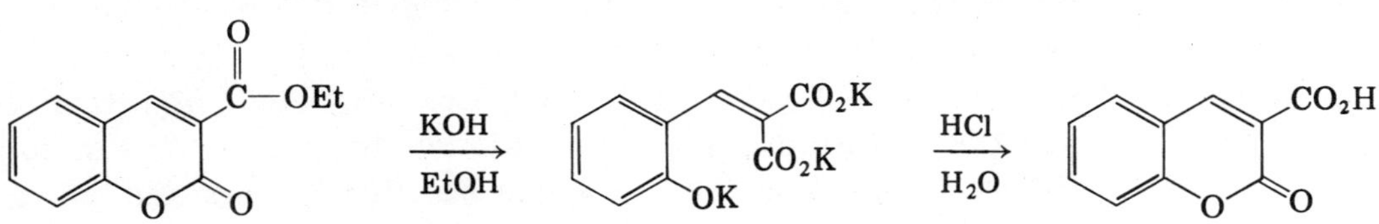

3-Carbethoxy-coumarin

Tri-potassium salt of o-hydroxybenzylidene-malonic acid

Coumarin-3-carboxylic acid
M.w. = 190 g/mol
m.p. = 188°C

This salt is not isolated, and upon acidification of the hydrolysis reaction mixture, rapid reformation of the lactone ring occurs yielding coumarin-3-carboxylic acid.

Coumarin compounds are widely distributed in nature. Coumarin is a constituent of many essential oils and extracts, including lavender oil and vanilla. Coumarin is also the volatile component of clover blossoms that is responsible for the characteristic odor of new-mown hay. Many 3-substituted derivatives of

4-hydroxycoumarin are powerful blood anticoagulants and are used as drugs to control blood clotting and as rodenticides (*warfarin*), which cause death by hemorrhage. Other substituted coumarin derivatives have unique ultraviolet absorption and fluorescence properties and are employed as dyes in lasers.

Prelab Questions

1. Why is it important to use absolute ethanol and not 95% ethanol as the solvent for this condensation reaction?

2. Calculate the number of moles of each reactant and reagent used in this experiment. What are the limiting reagents and the theoretical yields for the condensation reaction and for the hydrolysis reaction?

3. If 3-hydroxybenzaldehyde were used instead of salicylaldehyde as the starting material for this condensation reaction, what would be the structure of the product?

4. A drop of acetic acid added to the condensation reaction mixture improves the yield of 3-carbethoxycoumarin. This acid reacts with the excess base piperidine to form the salt piperidinium acetate. Explain how the presence of a small amount of this weakly acidic salt can improve the yield of 3-carbethoxycoumarin.

5. Predict the results that would be obtained for the solubility tests (Experiment 7), the ferric chloride test (Experiment 20), and the hydroxamate test for esters (Experiment 24) with each of the following compounds: salicylaldehyde, diethyl malonate, 3-carbethoxycoumarin, and coumarin-3-carboxylic acid.

EXPERIMENTAL PROCEDURE

Preparation of 3-Carbethoxycoumarin. In a dry 50-ml round-bottom flask place 5.0 g of salicylaldehyde (0.041 mol), 7.2 g of diethyl malonate (0.045 mol), 25 ml of absolute ethanol, 0.5 ml of piperidine, 0.02 ml (one drop) of glacial acetic acid, and three boiling stones. Equip the flask with a water-cooled condenser protected from atmospheric moisture with either a calcium chloride drying tube or a cotton plug. Reflux the solution over the steam bath for 2 hours, and then transfer the solution to a 250-ml Erlenmeyer flask. Add 35 ml of cold water and cool the solution in an ice bath. After crystallization is complete, filter the crystals and wash twice with 3-ml portions of ice-cold 50% aqueous ethanol. Determine the weight of the crude product, calculate the percentage yield, and obtain the melting point. This crude product is of suitable purity for the next step in the experiment, the preparation of coumarin-3-carboxylic acid. However, a purified sample of product should be obtained for characterization.

To obtain a sample of the product for characterization, dissolve 1.0 g of the crude product in 3 ml of hot ethanol and add 5 ml of hot water. Allow the solution to cool slowly to room temperature, and then cool

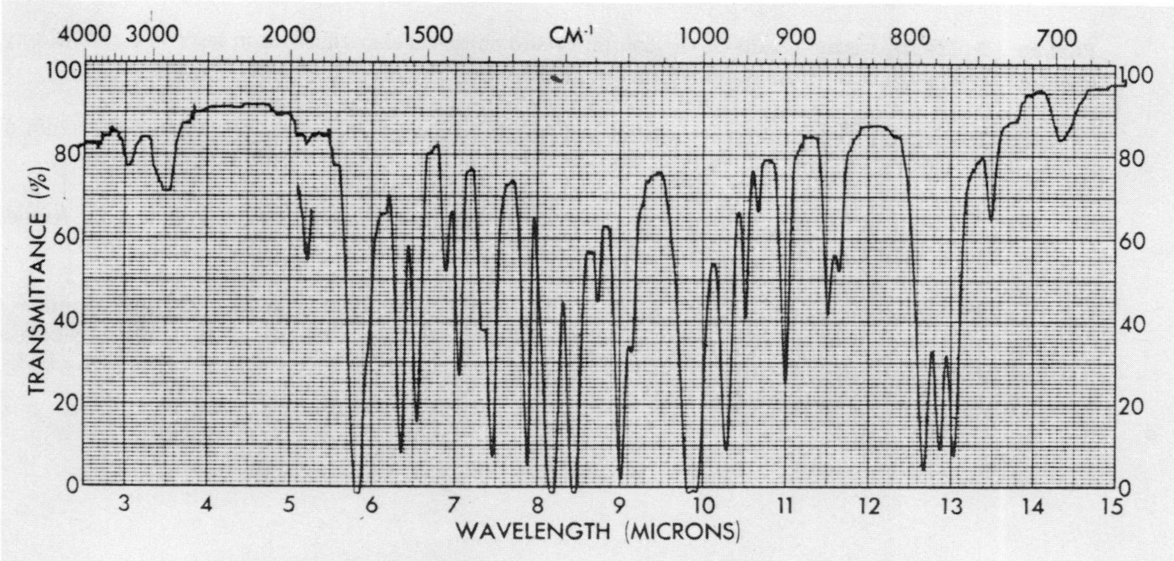

Figure 29.1 *Infrared spectrum of 3-carbethoxycoumarin (KBr pellet).*

the solution in an ice bath. Filter the solution and wash the crystals with a small volume of ice-cold 50% aqueous ethanol. Determine the weight of purified product and calculate the percentage recovery in the purification step. Obtain the melting point of the purified product. Characterize this product by solubility tests (Experiment 7), the ferric chloride test (Experiment 20), and the hydroxamate test for esters (Experiment 24). The collected product may be identified by IR spectroscopy (Figure 29.1).

Preparation of Coumarin-3-carboxylic Acid. Add 4.0 g of 3-carbethoxycoumarin, 20 ml of ethanol, 10 ml of water, and 4 g of potassium hydroxide pellets to a 125-ml Erlenmeyer flask. Add several boiling stones and heat the resulting mixture over the steam bath until the ester has dissolved. Continue heating the solution for an additional 15 minutes at a slow boil.

Prepare a dilute solution of hydrochloric acid by adding 10 ml of concentrated hydrochloric acid to 50 ml of water. Slowly, with stirring, pour the warm hydrolysis reaction mixture into the acid solution. Cool the resulting mixture in an ice bath and then isolate the product by filtration in a Büchner funnel. Wash the crystals with two 15-ml portions of ice-cold water. Allow the product to dry thoroughly and then determine the weight of the product and its melting point. Calculate the percentage yield. Characterize this product by the same chemical and solubility tests used to characterize 3-carbethoxycoumarin.

Postlab Questions

1. Compare the results of the solubility and chemical characterization tests on 3-carbethoxycoumarin with those obtained on coumarin-3-carboxylic acid. Explain these results.

2. What major differences would you predict to occur between the infrared spectrum of 3-carbethoxycoumarin and the infrared spectrum of coumarin-3-carboxylic acid?

3. Explain why the cyclic ester (lactone) of coumarin-3-carboxylic acid is rapidly formed upon acidification of the tri-potassium salt of the *o*-hydroxybenzylidenemalonic acid.

4. Give the structure of the product that would be isolated after base catalyzed condensation of each of the following pairs of compounds: (a) benzaldehyde with ethyl acetoacetate, (b) acetophenone with diethyl malonate, (c) benzaldehyde with acetone, (d) 2,4-dihydroxybenzaldehyde with diethyl malonate.

Experiment Thirty

Electronic Absorption Spectroscopy

Electronic absorption spectroscopy is the spectroscopic method for detection of electronic energy changes in molecules. Absorption of energy in the ultraviolet or visible region of the electromagnetic spectrum (Figure 12.1) causes electrons in the lower energy levels of molecules to be elevated to higher energy levels. The excitation of an electron in the lower energy orbital (*ground state*) to a higher energy orbital (*excited state*) occurs with the absorption of a discrete amount of energy (a *quantum* of energy). This electronic excitation energy

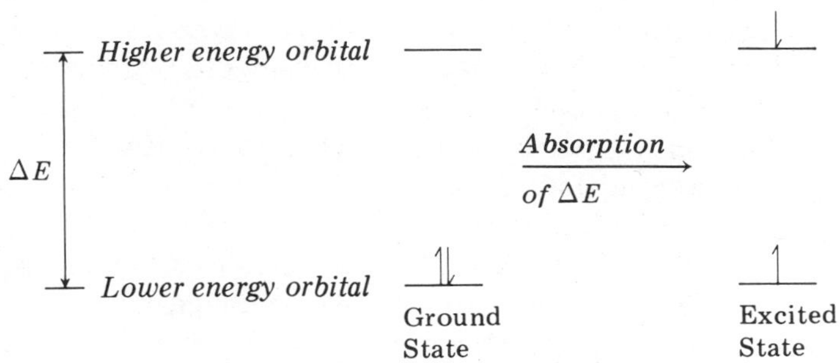

is the energy difference between the lower energy orbital and the higher energy orbital. The magnitude of the energy difference is dependent on the difference in energy between the two energy levels.

The electronic distribution in saturated hydrocarbons such as hexane only permits transitions from relatively low-lying σ-orbitals to high-energy σ^* antibonding orbitals. These transitions normally occur at high energies characterized by wavelengths less than 150 nm‡ and are not accessible with commercial spectrometers because oxygen in air also absorbs at wavelengths below 180 nm. Alkenes, on the other hand, absorb radiation at longer wavelengths (lower energy) due to the electronic transition from the π bonding orbital to a π^* antibonding orbital (a $\pi \rightarrow \pi^*$ transition). Thus a decrease in the energy difference between available orbitals in molecules provides a consequent increase in the

‡nm = nanometer. One nanometer equals 10^{-9} meter or 10^{-7} centimeter. In older chemistry literature you will find the use of millimicrons (mμ) and angstrom (Å) units to denote wavelength. One nm = one mμ. One angstrom = 10^{-10} meter = 0.1 nm.

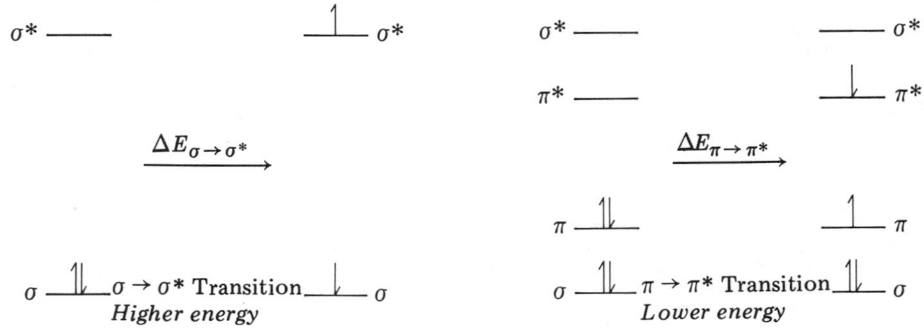

wavelength for absorption by these molecules. Since the primary electronic structural cause for the decrease in the difference in energy between molecular energy levels is conjugation, electronic absorption spectroscopy is employed primarily to detect and measure electronic transitions in conjugated molecules.

Ultraviolet Spectrometry

Ultraviolet (UV) or ultraviolet-visible (UV-Vis) spectrometry is the measurement of the quantity of ultraviolet or ultraviolet-visible radiation absorbed by a compound as a function of the wavelength of radiation. Thus, the data obtained from an ultraviolet absorption experiment is a graph of ultraviolet radiation wavelength (λ) verses the relative amount of ultraviolet radiation of that wavelength transmitted through a sample of the compound (Figure 30.1). This graph is called the *ultraviolet spectrum* of the compound.

The horizontal axis of the ultraviolet spectrum is linear in wavelength (λ). The ultraviolet region of the electromagnetic spectrum that is accessible on most UV spectrophotometers is radiation with wavelengths between 200 and 380 nm. The visible region of the spectrum extends up to 780 nm and is generally accessible on the same spectrophotometers. The vertical scale of the ultraviolet and visible spectrum is normally presented in units of *absorbance* (A) or *percent transmittance* ($\%T$), which are related by

$$\log_{10} \frac{100}{\%T} = A$$

When the percent transmittance is 100%, $A = 0.00$, and the sample is not absorbing radiation at that wavelength. Absorption of radiation by the sample is indicated by the upward deflection of the graph and a corresponding decrease in percent transmittance (increase in absorbance).

Electronic excitation of organic compounds in the ultraviolet region of the spectrum occurs with energies ranging from 72 kcal/mol (380 nm) to 125 kcal/mol (220 nm). These energies greatly exceed those required for vibrational excitation (<11 kcal/mol) and rotation. Consequently, the electronic absorption spectrum is a combination of electronic, vibrational, and rotational energy

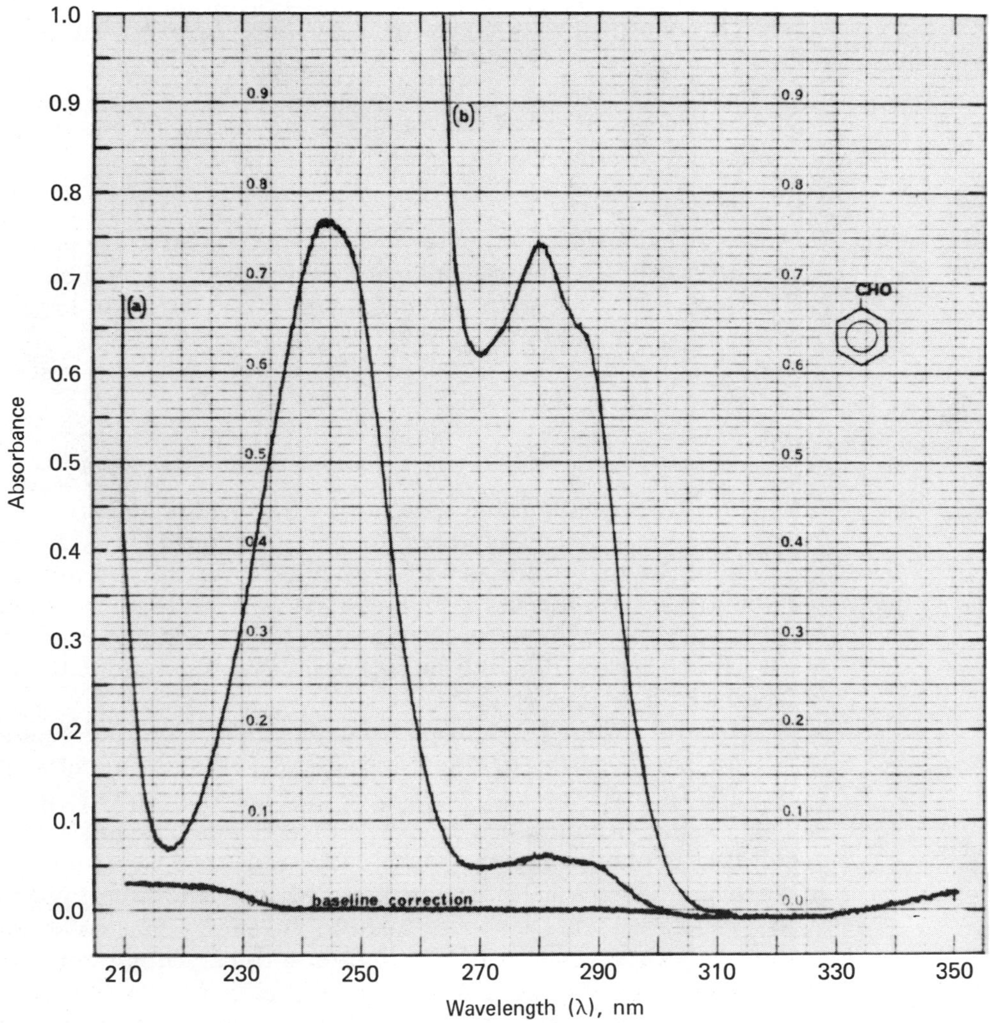

Figure 30.1 *Ultraviolet spectrum of benzaldehyde: (a) c = 6.00 × 10^{-5} M in 95% ethanol; (b) c = 6.00 × 10^{-4} M in 95% ethanol. Spectrum is transparent above 350 nm.*

absorptions that for all but the simplest organic compounds consists of broad absorption bands. The broad absorption bands do not usually contain the fine structure that is characteristic of an infrared spectrum and, therefore, the *wavelength of maximum absorption* (λ_{max}) is normally recorded as the characteristic spectral property in the UV–Vis spectrum.

The absorbance of a sample is dependent on its concentration in the path of the light (c in moles per liter), on the path length through the sample (l in centimeters), and on a physical constant characteristic of the absorbing sample (*molar absorptivity**, ϵ).

* In earlier literature molar absorptivity was referred to as the *molar extinction coefficient.*

This dependence is expressed in the *Beer–Lambert law* by

$$A = \epsilon cl$$

If the concentration of the sample is expressed in grams per liter, the Beer–Lambert law is defined by

$$A = acl$$

where a is the *absorptivity* ($\epsilon = a \times$ M.w.). The intensity of an absorption band is usually expressed as the molar absorptivity at the maximum absorption (ϵ_{max}). Thus, at the wavelengths of maximum absorption for benzaldehyde (Figure 30.1), ϵ_{max} = 1250 for λ_{max} = 280 nm and ϵ_{max} = 12,800 for λ_{max} = 244 nm.

The UV–Vis Spectrophotometer

The ultraviolet and visible spectra of a sample are measured on a UV–Vis spectrophotometer. The component design of a typical UV–V is spectrophotometer is similar to that described in Figure 12.3 for an infrared spectrometer. The radiation source for the visible region is a tungsten incandescent lamp that is replaced by a hydrogen discharge tube for determination of absorptions in the ultraviolet region. Light from the source is dispersed into its separate wavelengths by a monochromator.

As with infrared spectrophotometers, light from the source is divided into two radiation beams of equal intensity by the use of reflecting mirrors. One beam is focused through a cell that contains the sample dissolved in a solvent. The second beam, which is called the reference beam, is focused through a cell that contains only the solvent. The radiation beams emitted from the sample and reference cells are then passed into photomultiplier detection tubes that generate a voltage proportional to the energy that strikes the detector. (Because of these light-sensitive photomultiplier tubes, the detector area and entrance slits from the sample area are enclosed to protect them from exposure to external light sources such as from room light.) When the sample absorbs energy, there is a voltage decrease from the sample beam that is transmitted into a composite electrical signal and either recorded as an absorption peak on the spectrum or observed as a meter deflection on spectrophotometers without accessory recorders.

The cells employed for UV–Vis spectral analyses are normally 1.0 cm square and require approximately 3 ml of solution. However, cells of different sizes, 0.1 to 10 cm, may also be used. Glass cells may be used for determining the visible spectrum. However, because glass absorbs strongly at wavelengths less than 360 nm, quartz cells must be employed for the determination of spectra in the ultraviolet region.

Electronic Absorptions

Since the $\sigma \rightarrow \sigma^*$ transitions of organic compounds occur at wavelengths that are outside of the range of most spectrophotometers (λ < 200 nm are in the

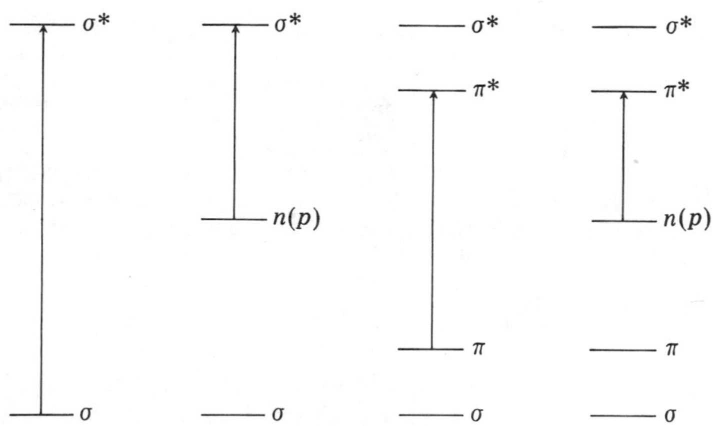

Transition	$\sigma \rightarrow \sigma^*$	$n \rightarrow \sigma^*$	$\pi \rightarrow \pi^*$	$n \rightarrow \pi^*$
Absorption intensity:	strong	weak	strong	weak
Absorption wavelength:	<150 nm	<250 nm	>160 nm	>250 nm
Examples:	C—C C—H	C—N: C—O: C-Halogen C—S:	C=C C=N C=O C=S	C=N: C=O: C=S:

Figure 30.2 *Classification of electronic transitions that are characteristic of organic compounds.*

far UV or vacuum UV region), the primary focus of UV-Vis spectroscopy is with the lower energy $\pi \rightarrow \pi^*$ and $n \rightarrow \pi^*$ transitions that occur at wavelengths above 200 nm. The $\pi \rightarrow \pi^*$ transitions result from excitation of electrons in π bonding orbitals to π^* antibonding orbitals, and their associated absorption bands are normally intense ($\epsilon_{max} > 10{,}000$). The $n \rightarrow \pi^*$ transitions result from excitation of electrons in nonbonding p-orbitals to π^* antibonding orbitals; these transitions are "forbidden" so that their corresponding absorption bands are weak ($\epsilon_{max} < 100$). The $n \rightarrow \sigma^*$ transitions, like $\sigma \rightarrow \sigma^*$ transitions, are higher in energy than $n \rightarrow \pi^*$ transitions, but, like these latter transitions, have absorption intensities that are characteristically weak. Figure 30.2 presents a classification of these transitions.

The functional groups responsible for electronic absorptions in the ultraviolet or visible region of the spectrum are called *chromophores.* Since electronic absorptions in the UV-Vis spectral region are generally the result of $\pi \rightarrow \pi^*$ and $n \rightarrow \pi^*$ transitions, chromophores are covalently unsaturated groups. The wavelengths of absorptions and the intensities of these absorptions are markedly affected by the electronic nature of atoms or groups bonded to the chromophore as well as by the nature of the atoms included in the chromophore. Factors that affect the wavelengths and intensities of absorptions are:

1. *The nature of atoms in the chromophore.* The C=C chromophore of alkenes absorbs strongly in the far UV (near 180 nm). Aldehydes and ketones

Table 30.1 Effect of Conjugation on λ_{max} and ϵ_{max} for C=C and C=O Chromophores

	$\pi \rightarrow \pi^*$			$\pi \rightarrow \pi^*$		$n \rightarrow \pi^*$	
Compound	λ_{max} (nm)	ϵ_{max}	Compound	λ_{max} (nm)	ϵ_{max}	λ_{max} (nm)	ϵ_{max}
$H_2C{=}CH_2$	175	15,000	$CH_3CH_2COCH_3$	185		279	16
$H_2C{=}CHCH{=}CH_2$	217	21,000	$H_2C{=}CHCOCH_3$	219	7080	324	21
$H_2C{=}CHCH{=}CHCH{=}CH_2$	258	52,500					
$H_2C{=}CHCH{=}CH$ (joined by a bond to) $H_2C{=}CHCH{=}CH$	304						

also exhibit a $\pi \rightarrow \pi^*$ transition near 180 nm. In contrast, the $n \rightarrow \pi^*$ transitions are affected by the nature of atoms in a chromophore; for example, the $n \rightarrow \pi^*$ transitions for aldehydes and ketones are in the region 270 to 300 nm, whereas those for thiocarbonyl compounds (C=S) are near 320 nm. In general, increasing the electronegativity difference between atoms of a chromophore results in a lower wavelength absorption for $n \rightarrow \pi^*$ electronic transitions.

2. *Conjugation.* The most important factor affecting the wavelengths and intensities of absorptions for organic compounds is conjugation of a chromophore with an unsaturated group. Conjugation lowers the energy difference between bonding (or nonbonding) orbitals and antibonding orbitals. Table 30.1 lists two sets of examples that illustrate the effect of conjugation.
3. *The nature of atoms or saturated groups bonded to the chromophore.* An atom or saturated group that, when bonded to a chromophore, alters both the wavelength and intensity of absorption is called an *auxochrome.* Auxochromes are usually electron withdrawing groups that decrease λ_{max} and increase ϵ_{max} relative to hydrogen. Examples of spectral changes due to the substitution of auxochromes are given in Table 30.2.

The Interpretation of Ultraviolet Spectra

The spectral properties employed for the interpretation of the ultraviolet spectra of organic compounds are the wavelengths at maximum absorption (λ_{max}) and

Table 30.2 Effect on $n \rightarrow \pi^*$ Transitions by the Substitution of Auxochromes on a Carbonyl Chromophore

Compound	Auxochrome	λ_{max} (nm)	ϵ_{max}
Acetaldehyde	—H	293	12
Acetyl chloride	—Cl	235	53
Acetamide	$—NH_2$	220(s)[a]	—
Ethyl acetate	$—OCH_2CH_3$	207	69
Acetic acid	—OH	204	41

[a] s = shoulder.

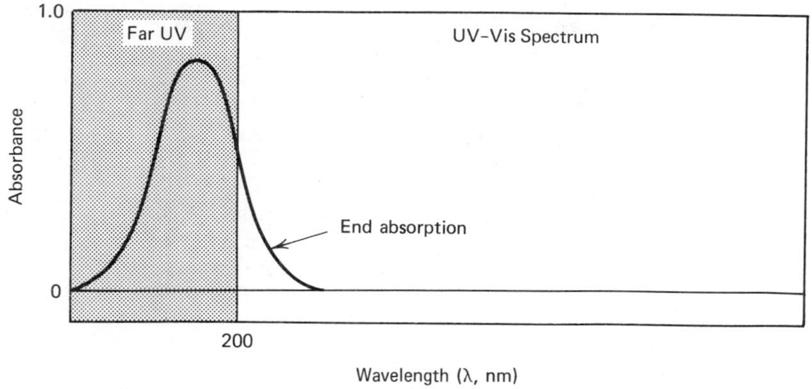

Figure 30.3 *End absorption for an organic compound.*

their corresponding molar absorptivities (ϵ_{max}). The λ_{max} of the absorption band characterizes the extent of conjugation in the organic compound and ϵ_{max} defines the electronic transition. Several classes of organic compounds, including alcohols, ethers, and alkyl chlorides, do not have a λ_{max} that is greater than 200 nm but do absorb out into the ultraviolet region; these compounds are said to have only "end absorption" or to be featureless above 200 nm (Figure 30.3).

The absorption band from a single electronic transition is symmetrical. The absorption band of benzaldehyde (Figure 30.1) centered at λ_{max} = 244 nm exemplifies this symmetry. However, the ultraviolet spectra of organic compounds may also exhibit unsymmetrical absorption bands like that at λ_{max} = 280 nm in Figure 30.1. In addition to the easily identified absorption maximum at 280 nm, there is a second absorption without a definable maximum at 287 nm. Absorption bands that occur on a primary absorption band and that are not resolved from the primary absorption band by an absorption maximum are referred to as a *shoulder*. In a compilation of absorption bands for a compound, a shoulder is listed with (s) following the wavelength at the position of maximum extension of the shoulder from the primary absorption. Thus a compilation of the ultraviolet data for benzaldehyde would include

λ_{max} (nm)	ϵ_{max}
244	12,800
280	1250
287 (s)	1100

In describing ultraviolet spectra two terms are commonly used to denote shifts in absorptions relative to a reference compound or reference solution. A *bathochromic shift* is an absorption shift to a longer wavelength (a red shift). A *hypsochromic shift* is an absorption shift to a shorter wavelength (a blue shift). The effects of conjugation, auxochrome substitution, and solvent on the wavelength of absorption are discussed in these terms.

Table 30.3 Absorption Characteristics of Saturated Compounds Containing Heteroatoms

Class of compound	General formula	λ_{max} (nm)
Alcohol	ROH	<200
Ether	ROR′	<200
Amine	R_3N, R_2NH, RNH_2	200–220
Alkyl chloride	RCl	<200
Alkyl bromide	RBr	200–220
Alkyl iodide	RI	250–270
Alkyl thiol	RSH	210–230
Alkyl sulfide	RSR′	200–220

Compounds Containing Only σ- and *n*-Electrons

As mentioned earlier, compounds that contain only σ-electrons—the alkanes—do not absorb in the ultraviolet region of the electromagnetic spectrum. Compounds that contain nonbonding electrons—those containing heteroatoms such as nitrogen, oxygen, halogen, and sulfur—absorb at higher wavelengths than do saturated hydrocarbons, and their absorptions are often detected by ultraviolet absorption spectroscopy (Table 30.3). However, most of these compounds only exhibit end absorptions in the ultraviolet region. Their absorptions are generally weak ($\epsilon_{max} < 1000$) and are of little diagnostic value relative to infrared spectral information.

Compounds Containing π-Electrons—The Ethylenic and Acetylenic Chromophores

Simple alkenes and alkynes do not absorb above 200 nm and are therefore transparent in the accessible ultraviolet region. However, as indicated by the data in Table 30.1, conjugation of the ethylenic or acetylenic chromophores increases the λ_{max} by 30 to 50 nm per additional double or triple bond (a bathochromic shift). For example, *trans*-β-carotene, a plant pigment found in numerous vegetables, has 11 double bonds in conjugation and therefore absorbs in the visible region. Two long wavelength absorptions are observed in the visible

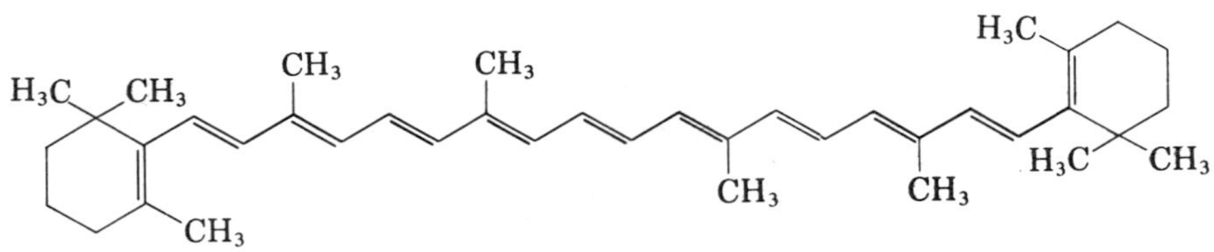

region, one at 483 nm ($\epsilon_{max} = 111{,}000$) and another at 453 nm ($\epsilon_{max} = 130{,}000$). Since these absorptions correspond to blue-green light, β-carotene appears orange. The absorption of blue-green light allows transmission of only red-yellow (orange) light through the sample.

The number of double bonds in conjugation determines the wavelength of maximum absorption and the characteristic color of that compound. A sample is colored when selected wavelength bands of visible light are absorbed by the compound. Absorption in the visible regions of the spectrum filters the ab-

Table 30.4 Characteristic Absorptions for Compounds Containing n- and π-Electrons

		λ_{max} (nm)	
Class of compound	General formula	$\pi \rightarrow \pi^*$ ($\epsilon_{max} > 1000$)	$n \rightarrow \pi^*$ ($\epsilon_{max} < 100$)
Aldehyde	RCHO	<200	290–310
Ketone	RCOR′	<200	275–295
Carboxylic acid	RCOOH	<200	200–220
Ester	RCOOR′	<200	200–220
Amide	$RCONR'_2$	<200	<200
Nitrile	RC≡N	<200	<200
Azo compound	RN═NR	<200	340–360
Nitrate	$RONO_2$	<200	260–280
Nitro compound	RNO_2	<200	260–280
Nitrite	RONO	200–230	340–360

sorbed bands from the white light and transmits the light that is not absorbed. However, only those compounds that absorb visible light are colored. Most organic compounds absorb light in the ultraviolet or far ultraviolet regions and are therefore colorless (liquids) or white (solids).

Compounds Containing n- and π-Electrons

Table 30.4 lists representative unsaturated functional groups that exhibit both $\pi \rightarrow \pi^*$ and $n \rightarrow \pi^*$ transitions. Generally, however, compounds with nonbonding electrons on nitrogen do not exhibit identifiable $n \rightarrow \pi^*$ transitions unless nitrogen is π-bonded to oxygen or nitrogen. As with the ethylenic chromophore, λ_{max} increases with conjugation (Table 30.1). Substitution of auxochromes on the carbonyl chromophore generally results in a hypsochromic shift (Table 30.2).

The Benzene Chromophore—Effect of Conjugation and of Auxochromic Substitution

Compounds containing the benzene ring exhibit characteristic absorption bands. As in the ultraviolet spectrum of benzaldehyde (Figure 30.1) two absorption bands are generally observed: a high intensity band ($\epsilon_{max} > 5000$) at the lower wavelength and a low intensity band ($\epsilon_{max} < 2000$) that occurs at the higher wavelength. The high intensity, lower wavelength absorption band is referred to as the *K-band* (or E_2-band) and the low intensity, higher wavelength absorption band is called the *B-band*. Table 30.5 provides examples of the ultraviolet absorption data for benzene and selected monosubstituted benzenes.

The effect of auxochromic substitution on the benzene ring is most pronounced in the B-band. Electronegative auxochromes such as —Cl, —OCH_3, and —OH increase the wavelength of absorption of the B-band by at least 10 nm relative to benzene (bathochromic shift). Relative to these electronegative substituents, the effect of alkyl substitution is small.

Conjugation affects both the K- and B-band. However, the bathochromic shift of the K-band is usually greater than that for the B-band when chromo-

Table 30.5 Ultraviolet Spectral Data for Benzene and Monosubstituted Benzenes

Compound	Benzene substituent	K-band		B-band	
		λ_{max}	ϵ_{max}	λ_{max}	ϵ_{max}
Benzene		204	7900	255	230
Styrene	$-CH{=}CH_2$	244	12,000	282	450
Phenylacetylene	$-C{\equiv}CH$	236	12,500	278	650
Benzoic acid	$-COOH$	230	10,000	270	800
Acetophenone[a]	$-COCH_3$	240	13,000	278	1100
Nitrobenzene[b]	$-NO_2$	252	10,000	280	1000
Toluene	$-CH_3$	206	7000	261	225
Chlorobenzene	$-Cl$	210	7600	265	240
Anisole	$-OCH_3$	217	6400	269	1480
Phenol	$-OH$	210	6200	270	1450
Aniline	$-NH_2$	230	8600	280	1430

[a] $n \rightarrow \pi^*$ transition at 319 nm (ϵ_{max} = 50).
[b] $n \rightarrow \pi^*$ transition at 330 nm (ϵ_{max} = 125).

phores such as $-CH{=}CH_2$, $-COR$, and $-NO_2$ are substituted onto the benzene ring. Thus the λ_{max} for K- and B-bands of a benzene derivative can usually be employed to differentiate between substituents that are auxochromes or conjugated chromophores.

Solvents for UV-Vis Spectroscopy

Commercially available spectral-grade solvents are available for UV-Vis spectroscopy. Such solvents are free of impurities that absorb in the ultraviolet and can be employed to obtain the spectrum of a sample in those spectral regions in which the solvent itself does not absorb. Table 30.6 lists solvents commonly employed for electronic absorption spectroscopy and the range of wavelengths through which these solvents do not absorb. Normally, end absorption by these solvents limits their range.

Table 30.6 Solvents for UV-Vis Spectroscopy

Solvent	Transparent above λ (nm)	Relative solvent effect (nm)[a]
Hexane	>200	+11
Cyclohexane	>210	+11
Chloroform	>250	+1
Ethyl ether	>210	+7
Acetonitrile	>190	+6
1,4-Dioxane	>220	+5
95% Ethanol	>210	0
Methanol	>210	0
Water	>190	-8

[a] Relative to spectrum of a polar sample in 95% ethanol.

The solvent employed for UV-Vis spectroscopy of a polar compound has a measureable effect on the wavelength for maximum absorption. Polar compounds such as α,β-unsaturated carbonyl compounds and nitro compounds are strongly affected by the solvent. Table 30.6 lists the solvent effect on λ_{max} for absorptions of a polar compound relative to 95% ethanol. The λ_{max} of nonpolar compounds such as conjugated dienes and alkylbenzenes generally do not vary with the changing polarity of the solvent.

Spectra of Acids and Bases

Changing the pH of a solution of an organic acid or base has a marked effect on the UV–Vis spectrum of that compound. For example, in water, aniline exhibits two absorption maxima, the K-band at 230 nm and the B-band at 280 nm; however, in aqueous acid where aniline is completely protonated these same bands are observed with a hypsochromic shift to 203 nm and 254 nm, respectively. The opposite effect is observed with phenol. A bathochromic shift characterizes the conversion of phenol to the phenoxide anion.

$$C_6H_5NH_2 + H^+ \rightleftharpoons C_6H_5NH_3^+$$

Aniline

λ_{max}	ϵ_{max}
230	8600
280	1430

Anilinium cation

λ_{max}	ϵ_{max}
203	7500
254	160

$$C_6H_5OH + OH^- \rightleftharpoons C_6H_5O^- + H_2O$$

Phenol

λ_{max}	ϵ_{max}
210	6200
270	1450

Phenoxide anion

λ_{max}	ϵ_{max}
235	9400
287	2600

When the spectra of pure acid and pure conjugate base or of pure base and pure conjugate acid cross at a point in the UV-Vis spectrum, the spectra of all solutions containing ratios of these two species must also pass through this point, called the *isobestic point.* The isobestic point provides verification that one is dealing with a simple equilibrium between two absorbing species. To determine the isobestic point or isobestic points, the sum of the concentrations of both species must be constant and the absorption coefficients of both species must be insensitive to changing pH and to other nonabsorbing species, such as the buffer. Figure 30.7 illustrates these properties.

Sample Preparation

Quantitative analytical methods are employed to determine the electronic absorption spectrum of a compound. The sample must be pure, and the solvent in which the sample is to be dissolved must be free of any absorbing impurities. In preparing spectral solutions, the sample is accurately weighed and then dissolved in the solvent using a volumetric flask to obtain an accurate concentration measurement.

Spectrophotometers usually record absorbance from 0.0 to 1.0; as illustrated in Figure 30.1, the most accurate determinations are made when the absorbance is greater than 0.5 for a maximum absorption. If 1.0-cm cells are employed, an absorbance of 1.0 will be recorded for the maximum absorption dependent on the molar absorptivity and concentration of absorbing species:

For ϵ_{max} =	C =
10	$0.10M$
100	$0.010M$
1000	$1.0 \times 10^{-3}M$
10,000	$1.0 \times 10^{-4}M$

An appropriate initial concentration of the solution can be determined by estimating the ϵ_{max} for the highest intensity absorption band expected in the spectrum. Aliquots of solution are then removed from the volumetric flask and additional dilutions are made until the desired concentration has been obtained. The cells used for spectral measurement must be thoroughly cleaned and monitored for absorption readings with the pure solvent prior to use with the sample solution.

Prelab Questions

1. Only the K- and B-bands are recorded in the ultraviolet spectrum of benzaldehyde (Figure 30.1). The $n \rightarrow \pi^*$ transition for benzaldehyde occurs at 328 nm with ϵ_{max} = 20. Calculate the absorbance at 328 nm for a solution that exhibits an absorbance of 0.75 at 280 nm (ϵ_{max} = 1,250).

2. Why do solvents such as 95% ethanol and hexane affect the absorption characteristics of a polar compound differently?

3. Describe a series of spectral measurements that would confirm that the Beer–Lambert law is obeyed for a compound.

4. Predict how the ultraviolet spectra of (a) salicylaldehyde and (b) coumarin-3-carboxylic acid should differ from the spectrum of benzaldehyde.

5. If an absorbance of 0.01 is detectable, what is the minimum concentration of benzaldehyde that can be identified by UV spectroscopy if (a) a 1.0-cm cell is used and (b) if a 10.0-cm cell is employed?

EXPERIMENTAL PROCEDURE

The ultraviolet spectra of salicylaldehyde, 3-carbethoxycoumarin, and coumarin-3-carboxylic acid are given in Figures 30.4, 30.5, and 30.6, respectively. List the λ_{max} and corresponding ϵ_{max} for all absorption bands and discuss the factors that affect the spectral bands relative to benzaldehyde.

Figure 30.7 presents the ultraviolet spectrum of coumarin-3-carboxylic acid and its conjugate base. List the λ_{max} and corresponding ϵ_{max} for all absorption bands and identify the isobestic points in the spectrum.

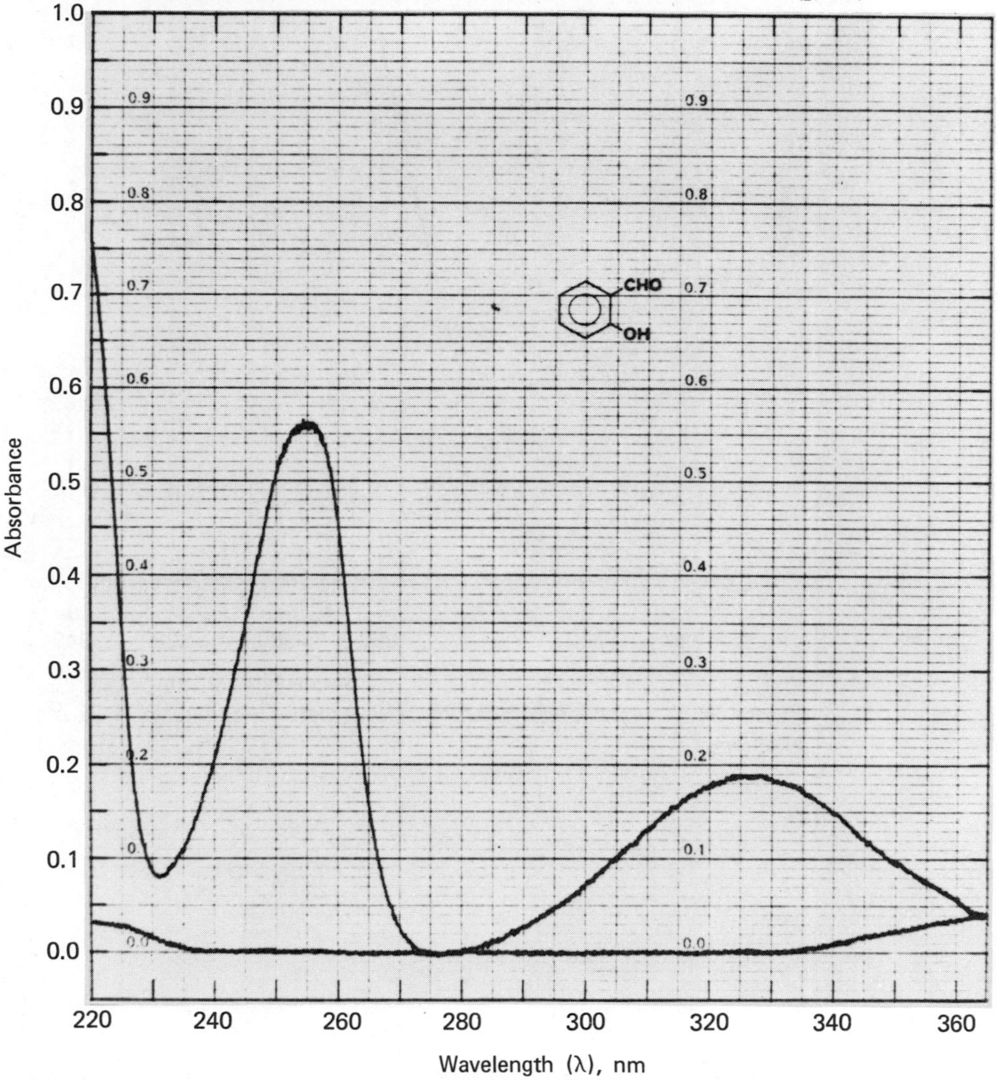

Figure 30.4 *Ultraviolet spectrum of salicyladehyde (0.613 mg in 100.0 ml of 95% ethanol). Spectrum is transparent in visible region.*

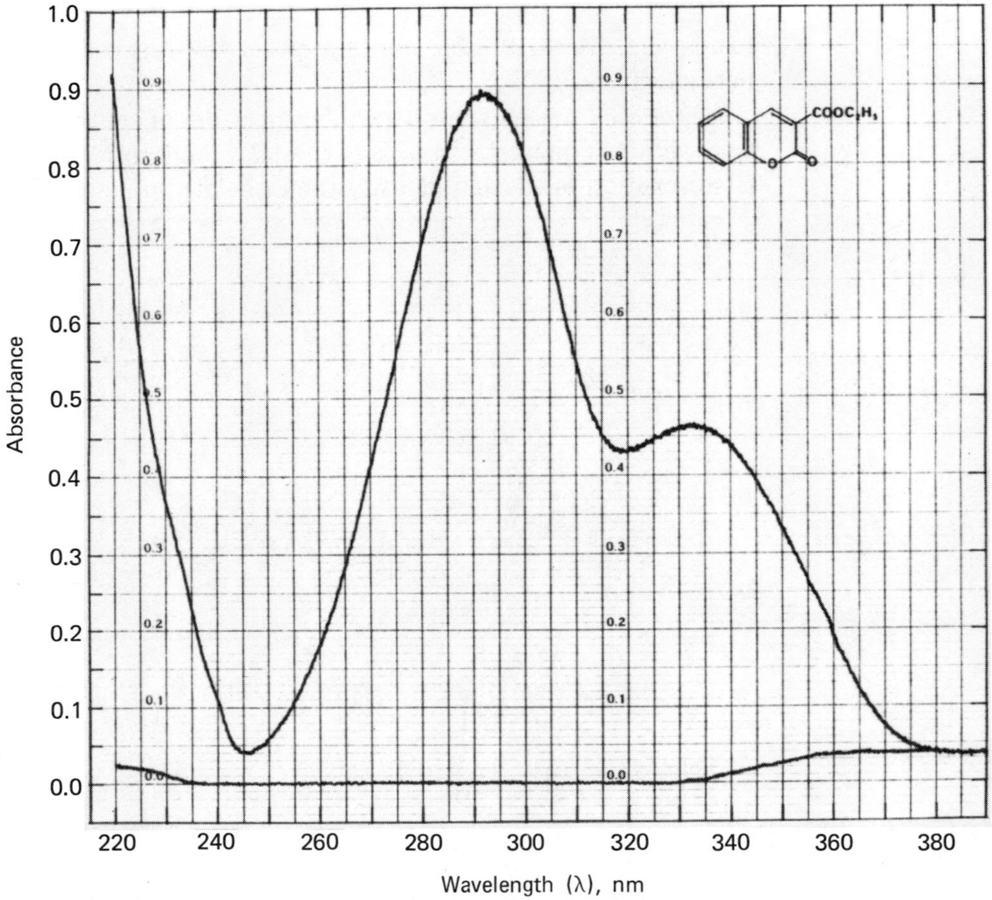

Figure 30.5 *Ultraviolet spectrum of 3-carbethoxycoumarin (1.32 mg in 100.0 ml of 95% ethanol).*

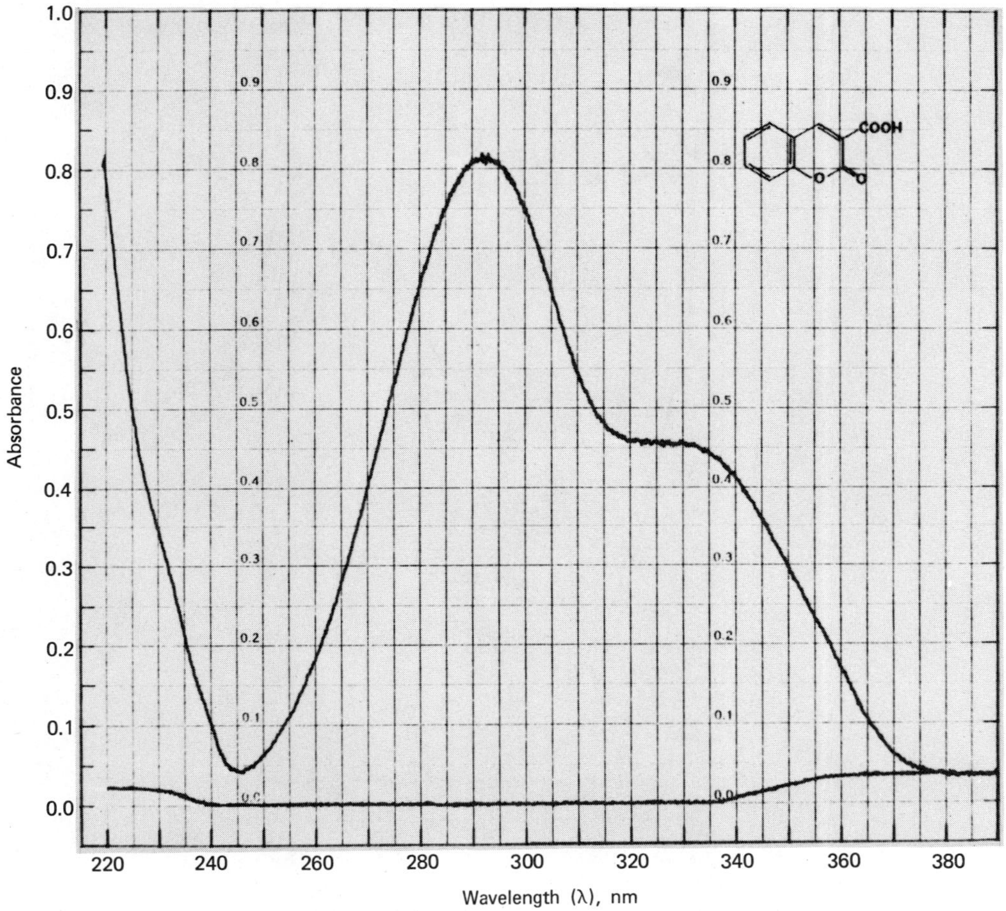

Figure 30.6 *Ultraviolet spectrum of coumarin-3-carboxylic acid (1.21 mg in 100.0 ml of 95% ethanol).*

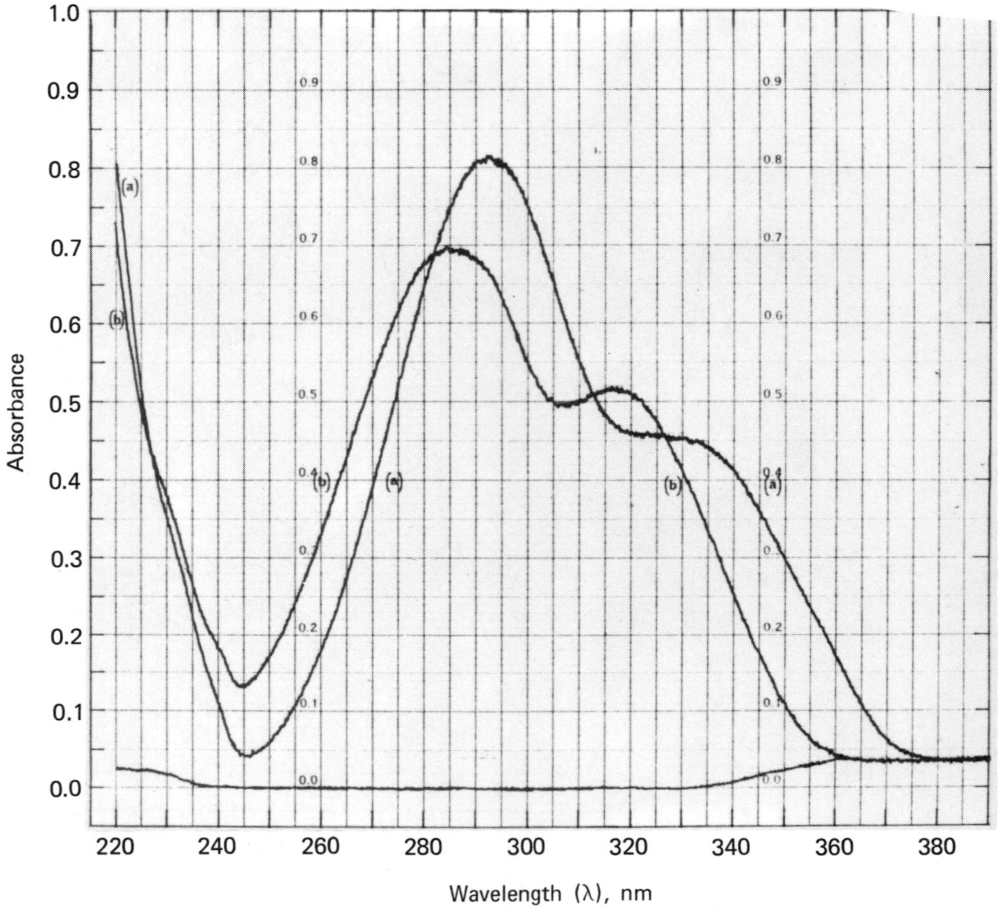

Figure 30.7 (*a*) *Ultraviolet spectrum of coumarin-3-carboxylic acid* (*1.21 mg in 100.0 ml of 95% ethanol*). (*b*) *Ultraviolet spectrum of sodium cormarin-3-carboxylate* (*1.35 mg in 100.0 ml of 95% ethanol*).

Experiment Thirty-one

Use of Protective Groups in Organic Synthesis: Preparation of 4,4-Diphenyl-3-buten-2-one

Organic compounds that possess two or more similar functional groups can often be chemically modified so that only one of the functional groups will react with a particular reagent. The chemical modification occurs by a reaction that is specific to one of the functional groups present in the molecule. This reaction masks the identity of the functional group so that a subsequent chemical reaction can be selectively performed on the unmodified functional group. The *masked functional group* is thus protected from involvement in the subsequent reaction.

The use of *protective groups*—functional groups that result from chemical modification to protect them from subsequent chemical reactions—is an important process in the construction of complex organic compounds. To be useful protective groups must be easily transformed back to their unmodified state.

The synthesis of 4,4-diphenyl-3-buten-2-one from ethyl acetoacetate by a selective Grignard reaction at the ester functional group with phenylmagnesium bromide is made possible with the use of a protective group for the ketone

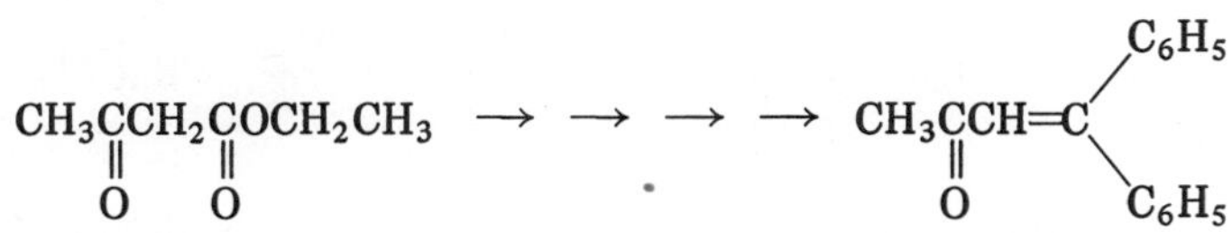

Ethyl acetoacetate *4,4-Diphenyl-3-buten-2-one*

functionality. Both the ketone and ester functional groups react with Grignard reagents. However, selective chemical modification of the ketone functional group through formation of a ketal masks the ketone functional group and permits selective reaction of the Grignard reagent with the ester functional group. Ketals are unreactive towards basic reagents such as phenylmagnesium bromide but are readily transformed to the ketone in aqueous acid.

$$R_2C{=}O \quad + \quad 2\,R'OH \quad \overset{H^+}{\rightleftharpoons} \quad R_2C(OR')_2 \quad + \quad H_2O$$

Ketone *Alcohol* *A ketal*

$$RCOOR'' \quad + \quad R'OH \quad \xrightarrow{H^+} \quad \text{No reaction under the same reaction conditions}$$

Ester *Alcohol*

Ethylene glycol will be used in this experiment to mask the ketone functional group from subsequent reaction with phenylmagnesium bromide. Ketones react with ethylene glycol to preferentially form the cyclic 2,2-disubstituted-1,3-dioxolane structure with negligible formation of the open chain structure $R_2C(OCH_2CH_2OH)_2$. Since ketal formation is an equilibrium process

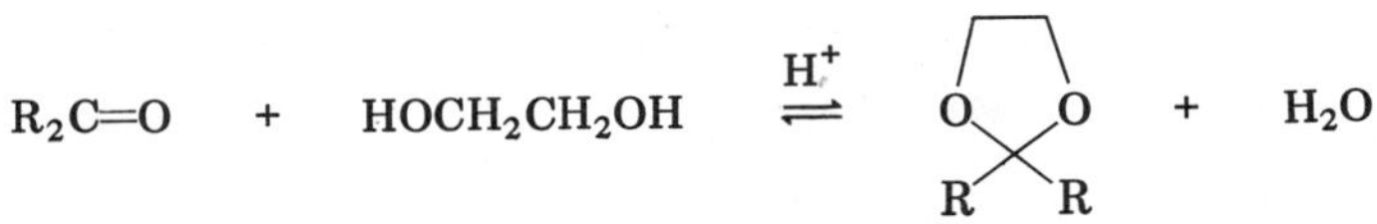

Ethylene glycol

A 2,2-disubstituted-1,3-dioxolane (an ethylene ketal)

that favors the reactants in acidic media, water is removed from the reaction medium as it is formed in order to drive the reactants to product.

In this experiment you will prepare 4,4-diphenyl-3-buten-2-one in three stages: (1) selective formation of the ethylene ketal of ethyl acetoacetate,

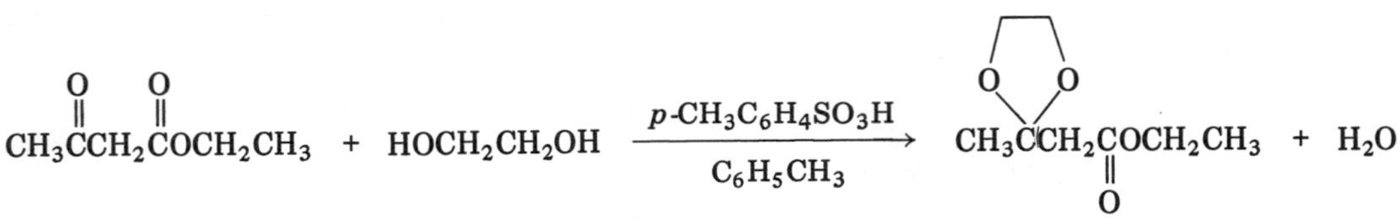

Ethyl acetoacetate M.w. = 130.1 g/mol b.p. = 180°C

Ethylene glycol M.w. = 62.1 g/mol b.p. = 198°C

Ethyl 2-methyl-1,3-dioxolane-2-acetate M.w. = 174.2 g/mol b.p. = 102°C at 23 Torr (137°C at 50 Torr)

Water (removed by azeotropic dehydration)

(2) Grignard reaction of the masked substrate with phenylmagnesium bromide,

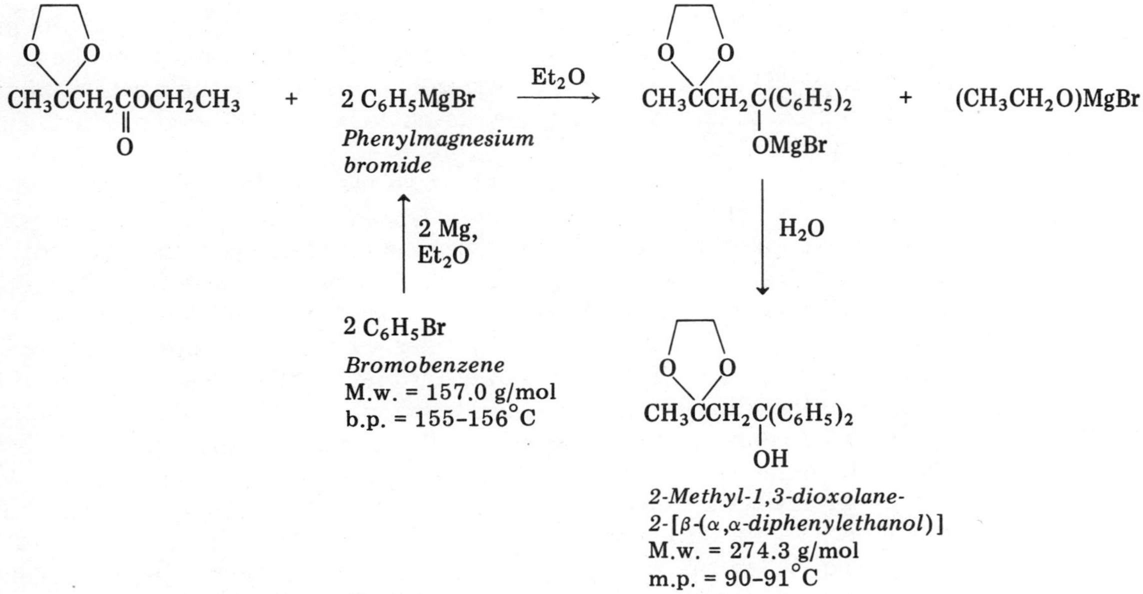

and (3) acid catalyzed hydrolysis and dehydration of the Grignard reaction product that both removes the protective group and dehydrates the alcohol,

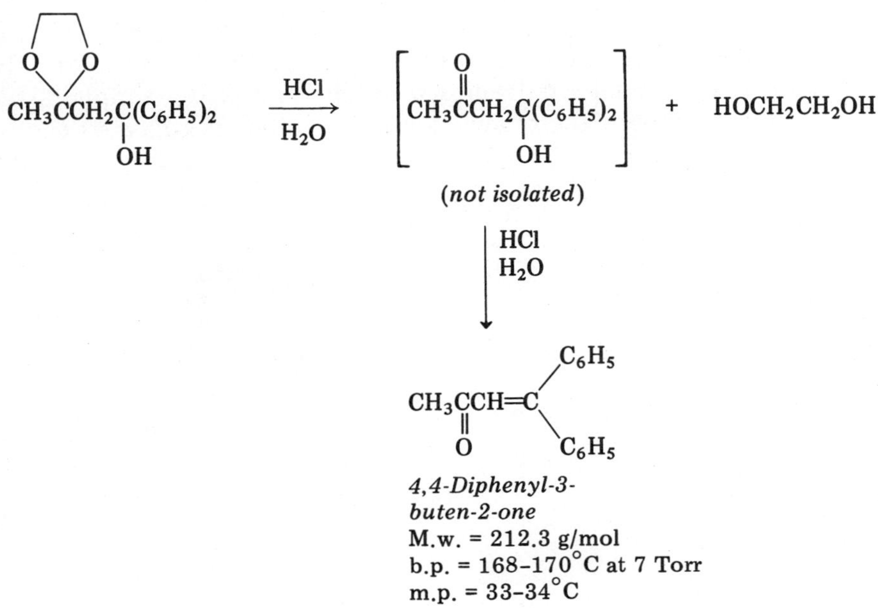

Azeotropic Dehydration

When water is an undesirable by-product of a chemical reaction, its removal from the reaction mixture as it is formed is an essential experimental operation.

In the ketal formation reaction that is employed in this experiment the presence of water in the reaction mixture causes the reverse reaction—hydrolysis of the ketal—to occur. However, if water is removed from the reaction mixture as it is formed, the reverse reaction is prevented, and the desired product can be formed in high yield.

The most commonly used and effective method for water removal from these reaction solutions is *azeotropic dehydration*—the codistillation of water as an azeotropic mixture with an appropriate immiscible organic solvent (see Experiment 6). As the azeotrope condenses, water separates from the cooled organic solvent as a separate liquid phase. If the density of water is greater than that of the immiscible organic solvent, water will form the lower liquid layer of the collected mixture. The volume of water that is collected is a measure of the progress of the water-forming chemical reaction.

A water collecter, known as a *Dean-Stark trap*, is commonly used for the collection and measurement of water. The apparatus construction that employs this water collector consists of a reaction flask, the Dean-Stark trap, and a condenser (Figure 31.1). The reaction solution contained in the reaction flask is refluxed. When the azeotrope of water and the organic solvent condenses, water separates from the less dense, immiscible organic solvent and falls to the bottom of the trap. The upper organic layer overflows back into the reaction flask. Reflux is continued until all of the water formed in the reaction solution has been collected.

The Dean-Stark trap is a convenient apparatus for water collection. This device automates the process for azeotropic dehydration. However, water may also be removed from a reaction solution by simple distillation of the water azeotrope if the organic solvent distillate is collected and returned to the reaction flask.

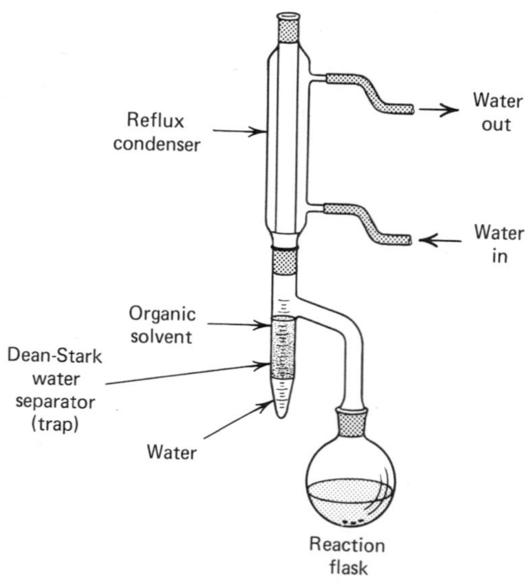

Figure 31.1 *Use of a Dean-Stark trap for azeotropic dehydration.*

Benzene (b.p. = 80°C) has been traditionally employed as the organic solvent for azeotropic dehydration. With a density less than that of water, it forms an azeotropic mixture with water that boils at 69°C and is composed of 91% benzene and 9% water. However, because of the health hazards attributed to benzene, toluene (b.p. = 111°C) is alternatively employed in this experiment. Toluene forms an azeotropic mixture with water that is composed of 80% toluene and 20% water. Upon condensation the toluene layer consists of 99.95% toluene and 0.05% water, and the water layer includes a similarly limited amount of toluene. However, unlike benzene, the higher boiling toluene also forms an azeotropic mixture with ethylene glycol, which necessitates the use of greater than one molar equivalent of ethylene glycol for ketal formation in this experiment.

Prelab Questions

1. Calculate the volume of water that will be collected if the ketal formation reaction of this experiment proceeds to completion.

2. The *p*-toluenesulfonic acid that is employed as the acid catalyst for the formation of the ethylene ketal of ethyl acetoacetate is the monohydrate *p*-$CH_3C_6H_4SO_3H \cdot H_2O$. Dehydration of this monohydrate occurs in refluxing toluene. How much water should be collected from dehydration of the acid catalyst?

3. Ethylene glycol and water codistill with toluene as an azeotropic mixture. Upon condensation of the azeotropic mixture only two liquid phases are observed. In which phase, toluene or water, would you expect to find the greater amount of ethylene glycol?

4. Describe three methods by which you could selectively detect the presence of ethyl acetoacetate in the distillate obtained by fractional vacuum distillation of ethyl 2-methyl-1,3-dioxolane-2-acetate in this experiment.

5. What product or products will be formed from reaction of ethyl acetoacetate with an excess of phenylmagnesium bromide? What will be the product or products after acid catalyzed dehydration?

EXPERIMENTAL PROCEDURE

Preparation of Ethyl 2-Methyl-1,3-dioxolane-2-acetate. Add 15 g of ethyl acetoacetate (0.12 mol) and 15 g of ethylene glycol (0.24 mol) to a 250-ml round-bottom flask followed by 75 ml of anhydrous toluene, 0.20 g of *p*-toluenesulfonic acid monohydrate, and several boiling stones. One of two procedures—use of a Dean-Stark trap or simple distillation—may be employed for the azeotropic dehydration process.

Azeotropic Dehydration With a Dean-Stark Trap. Attach the reaction flask to the Dean-Stark trap and the reflux condenser as described in Figure 31.1. Fill the water collector of the Dean-Stark trap with toluene and proceed to heat the reaction solution to reflux. The reaction solution is refluxed until the theoretical volume of water is collected (1 hr).

Azeotropic Dehydration By Simple Distillation. Attach the reaction flask to a simple distillation apparatus in which a Claisen adapter is inserted between the reaction flask and the distillation head (similar to the apparatus described in Figure 14.3). Place an additional funnel containing 10 ml of toluene onto the center connection of the Claisen adapter so that distilled toluene can be readded to the reaction flask. Slowly distill the toluene solution at approximately 2 ml per minute. Collect the distillate in a 10-ml graduated cylinder and, at regular intervals, transfer the toluene layer to the addition funnel. Add toluene from the addition funnel to the reaction flask at a rate that compensates for the volume of distilled toluene. Distillation is continued until the theoretical volume of water is collected (1 hr).

Upon completion of the reaction, cool the reaction solution to room temperature and extract with 30 ml of 10% aqueous sodium hydroxide and then sequentially with two 30-ml portions of water. Dry the resulting toluene extract over anhydrous magnesium sulfate. Filter the dried toluene solution into a 250-ml round-bottom flask and distill the toluene at atmospheric pressure using a simple distillation apparatus. Dispose of the collected toluene in the container provided for waste toluene. Transfer the residue remaining after removal of toluene to a side-arm distillation flask or to a 50-ml round-bottom flask and fractionally distill this residue under aspirator pressure (Figure 14.2). Collect the fraction containing the ethylene ketal of ethyl acetoacetate (b.p. = 100 to 105°C at 23 Torr, 135 to 140°C at 50 Torr). Record the boiling point range for your distilled product and the pressure at which distillation occurred. Obtain the weight of this product and calculate its percentage yield. The collected product may be identified by ^1H NMR and IR spectroscopy (Figure 31.3) and its purity may be determined by GLC analysis on a 20% SE-30 or 20% Carbowax 20*M* column.

Preparation of 2-Methyl-1,3-dioxolane-2-[β-(α,α-diphenylethanol)]. Prepare phenylmagnesium bromide from 2.7 g of magnesium turnings (0.11 mol) and 15.7 g of bromobenzene (0.10 mol) by the procedure described for the preparation of benzylmagnesium chloride in Experiment 15. Cool the flask that contains the Grignard reagent in a ice-water bath and add 25 ml of anhydrous ether to the reaction solution through the addition funnel. Dissolve 8.7 g of the ethylene ketal of ethyl acetoacetate (0.050 mol) in 25 ml of anhydrous ether and pour this solution into the addition funnel used in the preparation of phenylmagnesium bromide. Add the ether solution containing the ethylene ketal dropwise to the reaction flask with continual agitation of the mixture. Moderate refluxing of the ether should be maintained during the dropwise addition. After addition is complete, continue to agitate the reaction mixture for 30 min at room temperature then slowly add a mixture of 50 g of ice in 50 ml of water to the frequently agitated reaction mixture. After the ice has melted, stopper the reaction flask and shake the contents of the flask to ensure thorough mixing. At this point the reaction mixture should be composed of two liquid layers and a white precipitate of magnesium salts. Decant the liquid layers into a separatory funnel (or filter the reaction mixture using a coarse filter paper), separate the two layers, and wash the water layer with 50 ml of technical grade ether. Combine the original ether layer with the ether wash, wash the combined ether solution with 50 ml of water, and then dry the ether solution over anhydrous magnesium sulfate. Filter the ether solution into a 250-ml round-bottom flask and distill the ether on a steam bath. Dispose of the collected ether in the container provided for waste ether. Add 20 ml of hexane to the residue remaining after distillation of ether and continue heating the flask to remove the last traces of ether. While still hot, transfer the hexane solution to an Erlenmeyer flask and then allow the solution to cool. Collect the solid product and obtain its melting point. Reduce the volume of hexane after filtration to obtain a second crop of crystals. Recrystallize the product if the melting point is less than 85°C. Weigh the collected colorless product and calculate its percentage yield. The isolated product may be identified by ^1H NMR and IR spectral analyses.

Preparation of 4,4-Diphenyl-3-buten-2-one. With the aid of a long-stemmed funnel, pour 1.5 ml of concentrated hydrochloric acid and

---

CAUTION: Concentrated hydrochloric acid will burn the skin upon contact. Wash acid from your skin at once with generous amounts of water.

3.0 ml of water into a 100-ml round-bottom flask that has been set up for reflux as shown in Figure 4.2. Dissolve 5.5 g of the ethylene ketal of 4,4-diphenyl-4-hydroxy-2-butanone (0.020 mol) in 50 ml of acetone and then add this solution and several boiling stones to the acid solution. Reflux the resulting solution on a steam bath for one hour and then allow this solution to cool to room temperature. Dilute the cooled reaction solution with 100 ml of water and extract with two 50-ml portions of ether. Wash the combined ether extract with 50 ml of

5% aqueous sodium bicarbonate and 50 ml of water and then dry the resulting ether solution over anhydrous magnesium sulfate. Filter the ether solution into a 250-ml round-bottom flask and remove the ether by simple distillation. Weigh the crude product and calculate its percentage yield. Perform TLC analyses on the crude product using silica gel as the solid adsorbant and both 50:50 pentane:ether and absolute ether as the developing solvents. Based on your results from TLC analyses, purify 1.0 g of the crude product by column chromatography (Experiment 19) using 50 g of silica gel. The isolated light yellow product may be identified by ^{1}H NMR and IR spectral analyses.

Postlab Questions

1. Calculate the overall yield of 4,4-diphenyl-3-buten-2-one from ethyl acetoacetate.

2. Outline an alternate method for the preparation of 4,4-diphenyl-3-buten-2-one from commercially available reactants. Discuss the relative advantages or disadvantages of the alternate method.

3. In the procedure employed for the isolation of 1,2-diphenylethanol by the Grignard synthesis, dilute acid is added to the reaction mixture following addition of benzaldehyde (Experiment 15). Why is a similar procedure—use of aqueous sulfuric acid—not employed for the Grignard synthesis in this experiment?

4. Why is acetone and not ethanol employed as the solvent in the acid catalyzed hydrolysis-dehydration of the Grignard reaction product (contrast with the dehydration procedure used in Experiment 16).

5. If ethanol was used to form the diethyl ketal of ethyl acetoacetate in the first reaction of this experiment, what procedural changes in the ethylene ketal formation reaction are required? Could ethanol simply replace ethylene glycol?

6. Identify the characteristic infrared absorptions from Figures 31.2 and 31.3 that would allow you to distinguish between ethyl acetoacetate and its ethylene ketal.

7. Assign the individual NMR absorptions of Figures 31.2 and 31.3 to the protons that produce these signals.

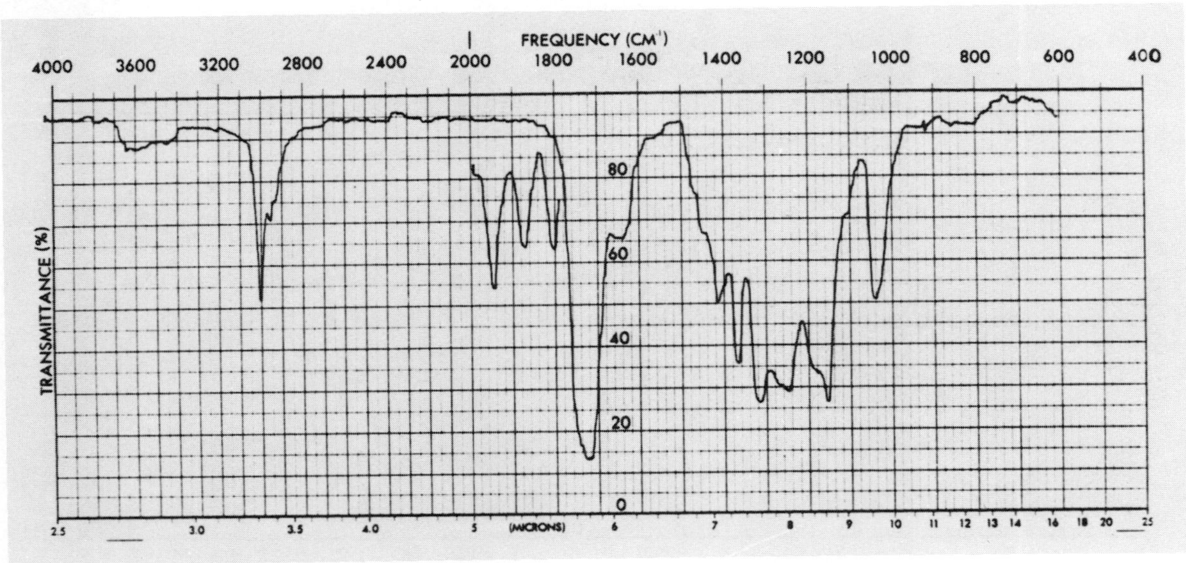

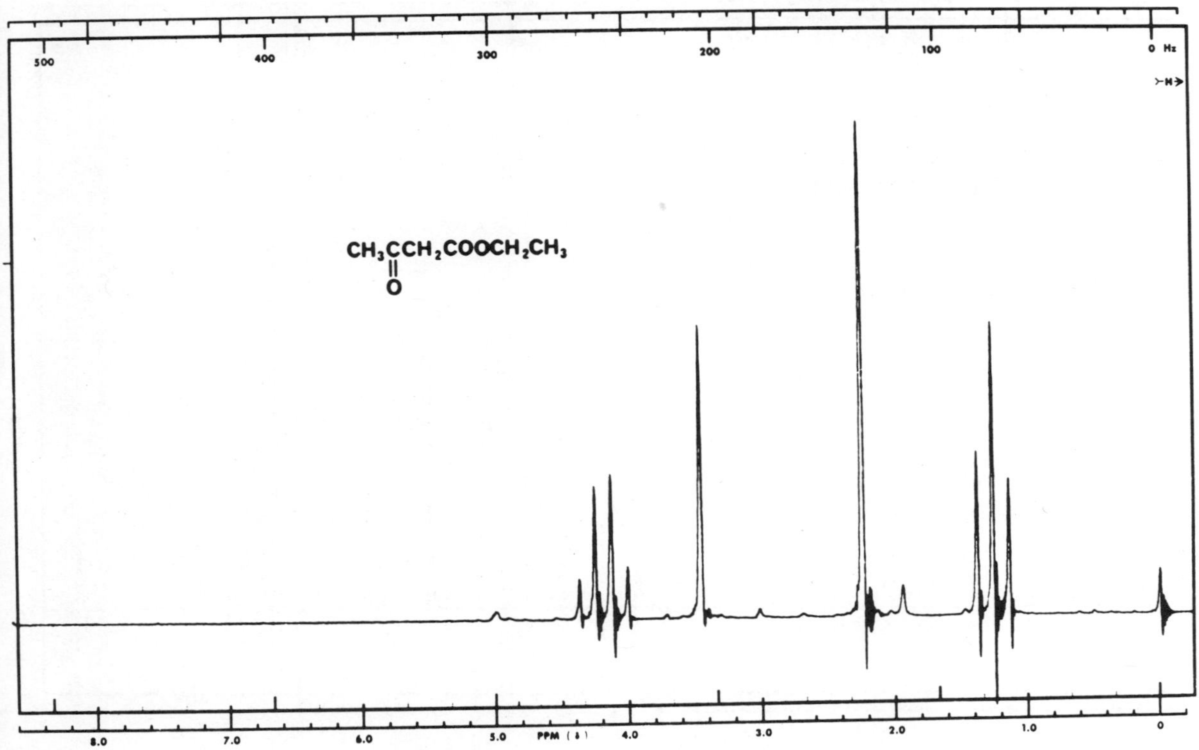

Figure 31.2 *IR (thin film) and NMR (in CCl_4) spectra of ethyl acetoacetate.*

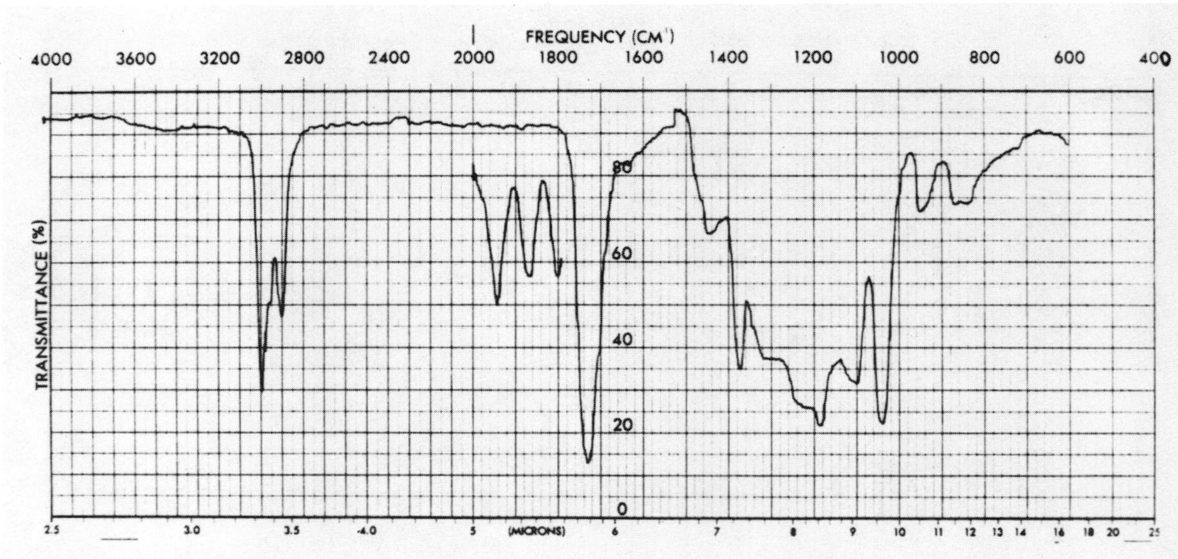

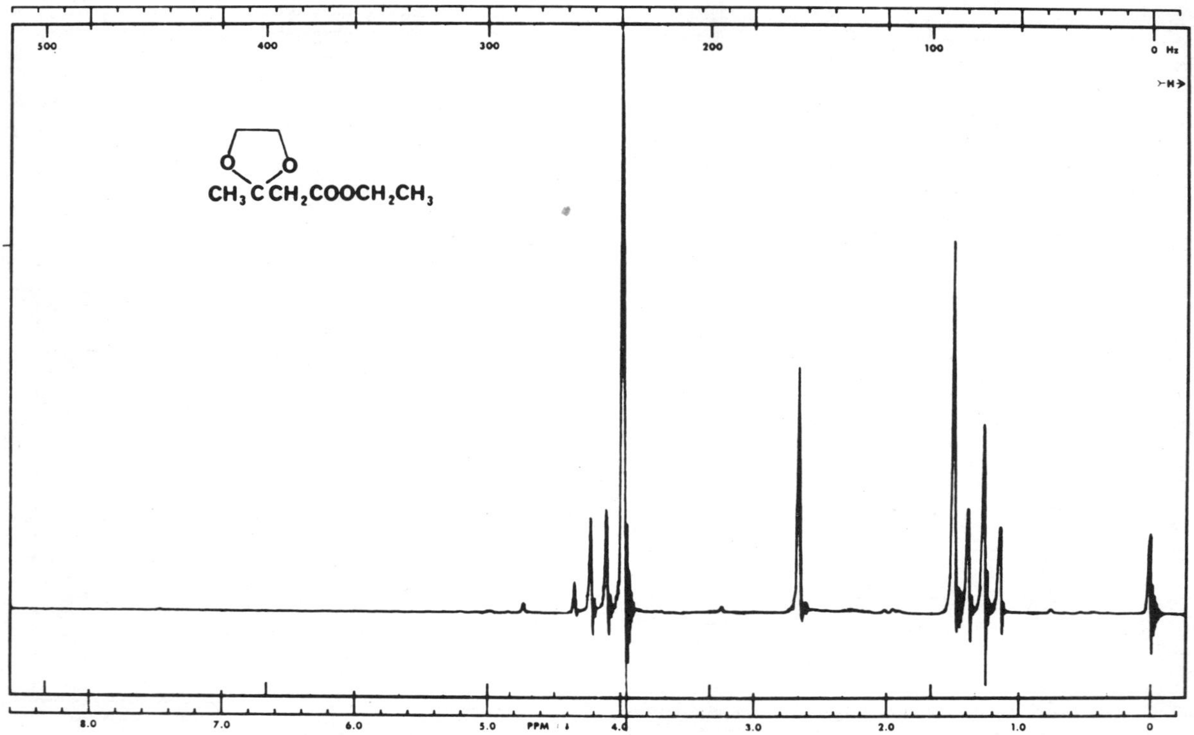

Figure 31.3 *IR (thin film) and NMR (in CCl_4) spectra of 2-methyl-1,3-dioxolane-2-acetate.*

Experiment Thirty - two

Stereoselectivity in Reductions of 3-Methylcyclohexanone

In previous experiments you have performed reactions that have resulted in the predominant formation of one of two possible stereoisomers. In Experiment 9, for example, dehydrohalogenation of 3-methyl-3-chloropentane produces a mixture of olefinic compounds in which *trans*-3-methyl-2-pentene predominates over its stereoisomeric *cis*-isomer. Similarly, *trans*-stilbene is selectively formed by dehydration of 1,2-diphenylethanol in Experiment 16; *cis*-stilbene, if produced at all, is only a minor constituent of the isolated product. Reactions that yield predominantly one of two or more possible stereoisomers are called *stereoselective reactions.* In this experiment you will examine stereoselectivity in more detail by quantitatively determining the relative yields of stereoisomeric alcohols formed in chemical reductions of 3-methylcyclohexanone by two different reagents.

The reduction of carbonyl compounds is an important synthetic route to alcohols. Aldehydes and ketones are reduced by a wide variety of reducing agents to their respective alcohol products. With 3-methylcyclohexanone two stereoisomeric alcohols, *cis*- and *trans*-3-methylcyclohexanol, are formed upon reduction. The *cis*-isomer, in which both the methyl and hydroxyl substituents

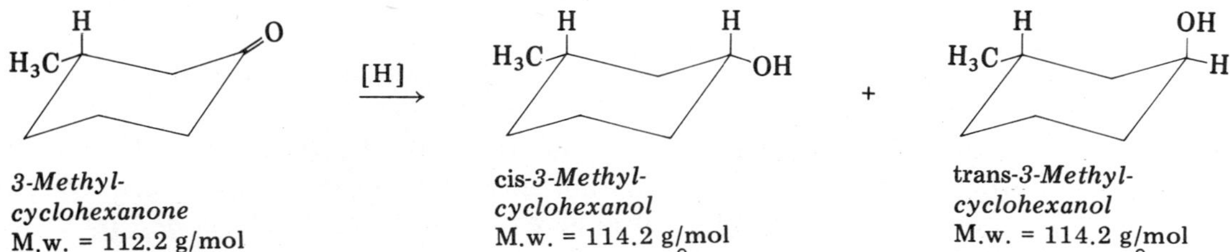

3-Methyl-cyclohexanone	cis-*3-Methyl-cyclohexanol*	trans-*3-Methyl-cyclohexanol*
M.w. = 112.2 g/mol	M.w. = 114.2 g/mol	M.w. = 114.2 g/mol
b.p. = 169–170°C	b.p. = 174–175°C	b.p. = 174–175°C
$n_D^{20°}$ = 1.4430	$n_D^{20°}$ = 1.4573	$n_D^{20°}$ = 1.4583

can exist in equatorial positions, is more stable than the *trans*-isomer by 1.1 kcal/mol at 25°C. If the 1.1 kcal/mol difference in stability between *cis*- and *trans*-3-methylcyclohexanol is reflected in the transition state leading to products, the *cis*-isomer should be the predominant product. Under reaction conditions in which the two alcohol isomers are in equilibrium at 25°C, the equilibrium amount of *cis*-3-methylcyclohexanol is 87% of the total 3-methylcyclohexanol.

In this experiment you will investigate the stereoselectivity of reductions by reducing agents that are *hydride donors*—sodium borohydride and triethylsilane. Hydride reducing agents formally transfer a hydride ion to the carbon atom of the carbonyl group in the rate-limiting step of the reduction process.

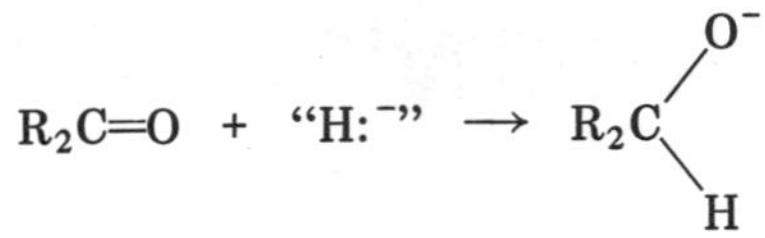

Hydride addition to the carbonyl group of 3-methylcyclohexanone may occur from either of two sides: attack from the axial side leading to the *cis*-alcohol isomer and attack from the equatorial side resulting in the *trans*-alcohol isomer.

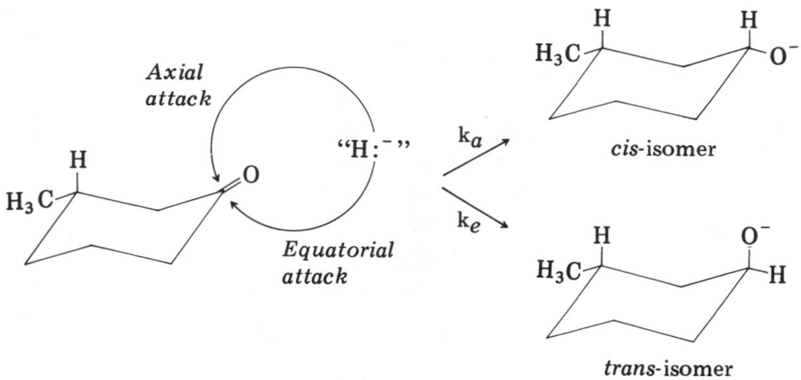

The rate for axial attack is given by

$$\frac{d[\textit{cis}\text{-isomer}]}{dt} = k_a[\text{3-Methylcyclohexanone}]\,[\text{“H:}^-\text{”}]$$

and that for equatorial attack is

$$\frac{d[\textit{trans}\text{-isomer}]}{dt} = k_e[\text{3-Methylcyclohexanone}]\,[\text{“H:}^-\text{”}]$$

Comparison of the relative yields of the isomeric 3-methylcyclohexanols formed from reductions of 3-methylcyclohexanone will allow you to determine the relative rates for axial and equatorial attack (k_a/k_e) by different reducing agents.

$$\frac{\%\ \textit{cis}\text{-3-methylcyclohexanol}}{\%\ \textit{trans}\text{-3-methylcyclohexanol}} = \frac{k_a}{k_e}$$

Your results may be compared with those in Table 32.1 to determine the relative abilities of sodium borohydride and triethylsilane to selectively reduce 3-methylcyclohexanone.

Table 32.1 Relative Yields of Alcohol Isomers from Reductions of 3-Methylcyclohexanone

	Relative % yield	
Reducing agent	*cis*-3-Methyl-cyclohexanol	*trans*-3-Methyl-cyclohexanol
$LiAlH_4$ in Et_2O, 25°C	84	16
H_2, Pt in CH_3COOH, 25°C	73	27
Li(*sec*-Butyl)$_3$BH in THF[a],		
0°C	15	85
-78°C	6	94

[a]THF = tetrahydrofuran.

Prelab Questions

1. Define "relative yield" for the products, *cis*- and *trans*-3-methylcyclohexanol, formed in this experiment. How does "relative yield" differ from the actual or theoretical yield?

2. Reduction of 4-methylcyclohexanone by hydride reducing agents also forms two isomeric alcohol products. Name the two products and state which of the two isomers is more stable.

3. 3-Methylcyclohexanone is capable of optical activity. If you reduce 3-methylcyclohexanone with either sodium borohydride or triethylsilane, would you expect the alcohol products to be optically active? What if you used optically active (+)-3-methylcyclohexanone in these reductions? Would you expect the stereoselectivity in reductions of (+)-3-methylcyclohexanone by sodium borohydride to differ from that obtained from reduction of (−)-3-methylcyclohexanone?

4. Which reactant, 3-methylcyclohexanone or sodium borohydride, is the limiting reagent in the experimental procedure for sodium borohydride reduction in this experiment?

5. What ether soluble products will be contained in the product mixture following triethylsilane reduction of 3-methylcyclohexanone?

6. Can the relative rates for axial and equatorial attack (k_a/k_e) be determined for reductions of cyclohexanone? Explain.

Reduction of 3-Methylcyclohexanone by Sodium Borohydride

Sodium borohydride, $NaBH_4$, is a white crystalline ionic compound that is widely employed for reductions of aldehydes and ketones to alcohols. All four hydrogens of $NaBH_4$ are reactive in carbonyl group reductions. A major

$$4\ R_2C{=}O \quad + \quad NaBH_4 \quad + \quad 4\ H_2O \quad \longrightarrow \quad 4\ R_2CHOH \quad + \quad NaB(OH)_4$$

Sodium borohydride

advantage of this reducing agent is its unreactivity with the vast majority of common organic functional groups including the C=C and C≡C groups, esters, and alkyl halides.

Although stable in water and alcohol solvents, sodium borohydride reacts rapidly with strong acids to form diborane (Experiment 36). Consequently, reductions are normally performed in neutral aqueous or alcoholic media.

EXPERIMENTAL PROCEDURE

CAUTION: Careful handling of sodium borohydride is recommended. Avoid contact of this reagent with acids and with your skin.

In a 50-ml round-bottom flask dissolve 0.10 g of sodium borohydride (2.6 mmol) in 2 ml of water and add to this solution 0.65 g of 3-methylcyclohexanone (5.3 mmol) dissolved in 10 ml of ethanol. Fit the round-bottom flask with a reflux condenser and add several boiling stones. Reflux the reaction solution for 30 minutes over a steam bath, then add 2 ml of 10% aqueous sodium hydroxide and continue to reflux the solution for an additional 15 minutes. Cool the reaction solution and then pour the contents of the reaction flask into a separatory funnel containing 50 ml of water. Wash the aqueous solution twice with 20-ml portions of ethyl ether and dry the combined ether solution over anhydrous magnesium sulfate. After filtering, remove the ether solvent by simple distillation, and save the residue for GLC analysis. Dispose of the collected ether in the container provided for waste ether.

Reduction of 3-Methylcyclohexanone by Triethylsilane

Unlike sodium borohydride, triethylsilane is a nonpolar liquid (b.p. = 106°C) that is only active as a reducing agent in acidic solutions. Like sodium borohydride, however, triethylsilane can be employed for the selective reduction of

$$\underset{}{R_2C{=}O} + \underset{\textit{Triethylsilane}}{(CH_3CH_2)_3SiH} + H_2O \xrightarrow{H^+} R_2CHOH + \underset{\textit{Triethylsilanol}}{(CH_3CH_2)_3SiOH}$$

carbonyl compounds. Reduction occurs through hydride transfer from silicon to carbon with resulting formation of triethylsilanol. Triethylsilanol may be converted under the reaction conditions to hexaethyldisiloxane,

$(CH_3CH_2)_3SiOSi(CH_2CH_3)_3$. Both triethylsilanol and hexaethyldisiloxane as well as triethylsilane are soluble in organic solvents and are, consequently, isolated with the alcohol products in this experiment.

EXPERIMENTAL PROCEDURE

In a 50-ml Erlenmeyer flask dissolve 0.65 g of 3-methylcyclohexanone (5.3 mmol) and 0.70 g of triethylsilane (6.0 mmol) in 1.0 ml of ethyl ether. Prepare a 70% aqueous sulfuric acid solution by carefully adding

---

CAUTION: Sulfuric acid will burn the skin upon contact. Wash any acid off the skin at once with large volumes of water.

0.7 ml of concentrated sulfuric acid to 0.3 ml of water and, after cooling to room temperature, transfer the acidic solution dropwise to the ether solution. Stopper the Erlenmeyer flask and swirl the flask to ensure thorough mixing of the reactants. After 30 minutes of regular mixing, cool the reaction solution in an ice bath and add 20 ml of 10% aqueous sodium hydroxide and 20 ml of ether to the reaction solution. Pour the contents of the reaction flask into a separatory funnel and, following thorough mixing of the two layers, remove the water layer. (Save the ether extract.) Wash the water layer with 20 ml of ether, combine this ether wash with the ether extract, and dry the combined ether solution over anhydrous magnesium sulfate. After filtering, remove the ether solvent by simple distillation, and save the residue for GLC analysis. Dispose of the collected ether in the container provided for waste ether.

Product Analysis

Although the physical properties of *cis*- and *trans*-3-methylcyclohexanol are nearly identical, these isomeric alcohols can be conveniently separated, identified, and quantitatively analyzed by gas-liquid partition chromatography. The *trans*-isomer elutes first from the column; the *cis*-isomer has the longer retention time. The detector sensitivities for these stereoisomeric compounds are identical so that peak areas can be used directly as a measure of the relative percentage yield:

$$\left(\frac{\text{Area}_{cis}}{\text{Area}_{cis} + \text{Area}_{trans}}\right) 100 = \text{Relative \% } cis\text{-3-methylcyclohexanol}$$

EXPERIMENTAL PROCEDURE

Refer to Experiment 10 for the operational details of GLC analyses. A 5-ft 25% glycerol column operated at 100°C (maximum operating temperature = 100°C) is employed for the separation and analysis of *cis*- and *trans*-3-methylcyclohexanol. Determine the retention times of 3-methylcyclohexanone and of the commercial mixture of 3-methylcyclohexanol isomers. If the resolution of the column for separation of the alcohol isomers is less than $R = 1.0$, optimize the resolution by adjustments in the column temperature and flow rate. Determine the relative percentage of *cis*-3-methylcyclohexanol in the commercial mixture. Analyze the product mixtures obtained by reductions of 3-methylcyclohexanone, assign the ketone (if present) and alcohol peaks through retention time comparisons with authentic samples under identical conditions, and calculate the relative yields of isomeric 3-methylcyclohexanols.

Postlab Questions

1. Describe how you could determine that the alcohol products formed by reduction of 3-methylcyclohexanone were not equilibrated under the reaction conditions employed in these experiments.

2. In reductions by sodium borohydride all four hydrogens are active. Describe an experiment that would allow you to determine if the reduction stereoselectivity from the first hydride transfer is the same as that from transfer of the fourth hydride.

3. Calculate the theoretical plate numbers for the GLC analysis of *cis*- and *trans*-3-methylcyclohexanol.

4. Explain why *cis*-3-methylcyclohexanol has a longer retention time on a 25% glycerol column than its *trans*-isomer. Does your explanation also explain the observation that *trans*-4-methylcyclohexanol has a longer retention time than its *cis*-isomer on the same column?

5. Calculate k_a/k_e for each of your product analyses.

Experiment Thirty - three

The Association of Ceric Ion with Benzyl Alcohol: The Equilibrium Expression and Equilibrium Constant

In Experiment 14 you observed an immediate color change to bright red upon mixing the orange ceric ammonium nitrate solution with colorless benzyl alcohol. The intensity of the colored solution decreased with time as benzyl alcohol was converted to benzaldehyde. The end result of this oxidation was a colorless solution. Since the cerium product from this oxidation, cerium (III), is colorless and benzaldehyde has no visible absorption of light, the change from orange to red cannot be due to a change in the oxidation state of the cerium(IV) oxidant. Instead, this color change must be due to some interaction between cerium(IV) and the alcohol, a color-changing interaction that is the basis for the use of cerium(IV) in a colorimetric test for alcohols.

The use of cerium(IV) as a colorimetric reagent for the detection of alcohols has been known for over 30 years. Cerium(IV) is orange-colored in $1M$ aqueous solutions (λ_{max} at approximately 315 nm). When an alcohol and this reagent are mixed, orange cerium(IV) instantly changes to dark red. Since the alcohol is colorless and since it can be determined that the color change represents a bathochromic shift in the λ_{max} for cerium(IV) (Experiment 30), the color change is reasonably assumed to be due to the formation of a complex between cerium(IV) and the alcohol:

$$n\,\text{ROH} + \underset{\text{(orange)}}{m\,\text{Ce(IV)}} \overset{K_{eq}}{\rightleftharpoons} \underset{\text{(red)}}{\text{Complex}}$$

$$K_{eq} = \frac{[\text{Complex}]}{[\text{ROH}]^n\,[\text{Ce(IV)}]^m}$$

Equilibrium between the alcohol, cerium(IV), and the complex is rapidly attained and measurement of the equilibrium constant, K_{eq}, is relatively simple.

Ceric ammonium nitrate (CAN), $Ce(NH_4)_2(NO_3)_6$, the cerium reagent that is most commonly employed as a colorimetric reagent, has six nitrate ions surrounding each cerium in a bidentate fashion:

Bidentate Cerium-Nitrate Coordination

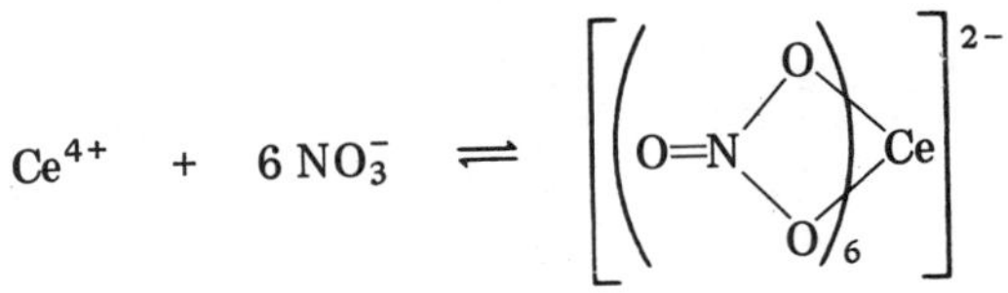

Association of ceric ammonium nitrate with an alcohol molecule will replace one of the Ce—O bonds of the cerium nitrate with a new Ce—O bond formed to the alcohol:

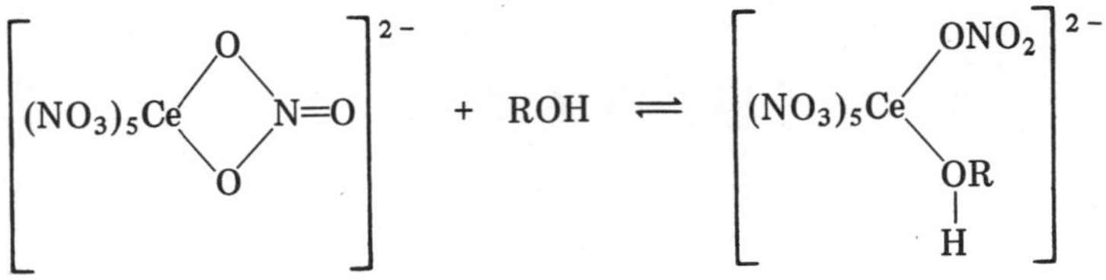

The number of alcohol molecules associated with one cerium ion and the equilibrium constant for complex formation will be determined in this experiment.

Spectrophotometric Analyses

Since association of an alcohol with cerium(IV) involves a color change, the equilibrium expression and the equilibrium constant can be determined by spectrophotometric methods. Spectrophotometric methods are among the most useful analytical procedures since measurements by these methods are accurate and can be made at low concentrations (Experiment 30). The experimental basis for spectrophotometric measurements is the Beer-Lambert law,

$$A = \epsilon c l$$

which equates the absorption of light by a solution (A) with the length (l) of a cell (in cm), the molar concentration of the colored substance (c), and the molar absorptivity of that substance (ϵ). The Beer-Lambert law is obeyed if the absorption of a solution at a particular wavelength is directly proportional to the concentration of the absorbing substance over the range of concentrations used in the study. In this experiment the Beer-Lambert law is obeyed throughout the range of concentrations that will be used.

Absorption measurements for this experiment can be made with a single-beam *Spec 20* Visible Spectrophotometer. Both 0 and 100% adjustments are made on the instrument prior to each measurement of absorption from the sample. The meter reading is adjusted to 0% transmittance when there is no sample in the cell compartment. The 100% adjustment is made with a reference sample in the cell compartment. The meter reading for the sample can then be recorded either as percent transmittance (%T) or absorption (A). The percent transmittance is related to the absorption by the equation

$$\log_{10}(100/\%T) = A$$

Calculation of K_{eq} for Cerium(IV)-Alcohol Complex Formation

For the present we do not know how many alcohol molecules are included in the cerium(IV)-alcohol complex, nor do we know the nature of the data that we must obtain to determine the equilibrium constant. To acquire this information we must analyze the general equilibrium expression,

$$M + nS \overset{K_{eq}}{\rightleftharpoons} MS_n$$

in terms of the absorption and concentration measurements that we can make. In the equilibrium expression M = cerium(IV), S = benzyl alcohol, and n is an integer.

Since both M and MS_n are colored, the measured absorption (A_T) is the sum of the individual absorbtions for M [$A_m = C_m(1 - X)\epsilon_m$] and MS_n [$A_c = C_m X \epsilon_c$, where $C = MS_n$ complex]

$$A_T = A_m + A_c = C_m(1 - X)\epsilon_m + C_m X \epsilon_c$$

where C_m is the total concentration of cerium(IV) in solution and X is the mole fraction of complexed cerium(IV). The mole fraction of cerium(IV) not complexed with alcohol is $(1 - X)$.

The equilibrium expression for complex formation is

$$K_{eq} = \frac{[MS_n]}{[M][S]^n}$$

which can be rearranged (how?) into an equation for the mole fraction of MS_n,

$$\text{Mole fraction of complexed cerium(IV)} = X = \frac{K_{eq}[S]^n}{1 + K_{eq}[S]^n}$$

Substituting this equation into that for A_T gives

$$A_T = \frac{C_m K_{eq}[S]^n \epsilon_c}{1 + K_{eq}[S]^n} + C_m \epsilon_m - \frac{C_m K_{eq}[S]^n \epsilon_m}{1 + K_{eq}[S]^n}$$

Since $C_m \epsilon_m = A_0$, the absorption due to cerium(IV) in the absence of alcohol,

$$A_T - A_0 = \Delta A = \frac{\Delta\epsilon C_m K_{eq}[S]^n}{1 + K_{eq}[S]^n}$$

where $\Delta\epsilon = \epsilon_c - \epsilon_m$. Inverting this equation for ΔA gives

$$\frac{1}{\Delta A} = \left(\frac{1}{[S]^n}\right)\left(\frac{1}{\Delta\epsilon C_m K_{eq}}\right) + \frac{1}{\Delta\epsilon C_m} \qquad (1)$$

which is conveniently in the form of the linear equation, $y = mx + b$.

If $1/\Delta A$ is plotted verses $1/[S]^n$, the slope of the line is

$$\text{slope} = \frac{1}{\Delta\epsilon C_m K_{eq}}$$

and the intercept is

$$\text{intercept} = \frac{1}{\Delta\epsilon C_m}$$

from which you can see that

$$\frac{\text{intercept}}{\text{slope}} = K_{eq} \qquad (2)$$

It is now a simple matter to see what you are required to measure in order to determine n and K_{eq} for the association of cerium(IV) with benzyl alcohol: (1) measure A_0, (2) measure A_T at constant [Ce(IV)] with various amounts of added benzyl alcohol, (3) plot $1/\Delta A$ versus $1/[C_6H_5CH_2OH]^n$ and accurately determine both the intercept and the slope of the best straight line, and (4) calculate K_{eq}.

Prelab Questions

1. What will be the equation for $1/\Delta A$ if $n = 1$?

2. If the concentration of CAN is found to be in error by 50%, will the experimentally determined equilibrium constant also be in error by 50%?

3. If $n = 1$, will the calculated value of K_{eq} be higher, lower, or the same value if you find the alcohol concentration to be 2.0*M* instead of 1.0*M* and you have used the 1.0*M* value in your calculation of K_{eq}?

4. Does this experiment require that you determine the molar absorptivity of the cerium(IV) and the cerium(IV)-alcohol complex? Why?

5. Why must the total concentration of cerium(IV) be known for this experiment? How would you determine the wavelength that should be used for absorbance measurements in this experiment?

EXPERIMENTAL PROCEDURE

Prepare a stock solution that is 0.067*M* in ceric ammonium nitrate (CAN) and 1.6*M* in nitric acid by accurately measuring 2.1 ml of the 1.6*M* CAN stock solution into a 50.0-ml volumetric flask, followed by 25.0 ml of the 3.2*M* nitric acid stock solution, and by filling to the 50.0-ml mark with distilled water. Label this solution as the *CAN Solution.* This *CAN Solution* cannot be stored for longer than four hours and should be used immediately after preparation for best results.

Prepare a 1*M* solution of benzyl alcohol in acetonitrile using a 50.0-ml volumetric flask by accurately weighing the appropriate amount of benzyl alcohol and quantitatively transferring this liquid to the volumetric flask using acetonitrile in this transfer; after thorough mixing of acetonitrile and benzyl alcohol fill to the mark with acetonitrile.

CAUTION: Be careful when using acetonitrile solutions. Do not breathe the vapor and avoid contact of acetonitrile with the skin.

The concentration of benzyl alcohol in this solution, labeled as the *Alcohol Solution*, must be accurately determined to three significant figures. Do not discard the *Alcohol Solution* after you complete this experiment; you will use this solution again in Experiment 34.

In your notebook prepare a table listing the source (chemical supplier) and approximate purity of each of the stock reagents that you employ. Record the weight and the number of moles of benzyl alcohol that you use to prepare the *Alcohol Solution*. The data in the following table should be completed before you begin your experimental measurements. You should be prepared to defend the precision of your data by detailing how each solution was prepared.

Solution	Conc. CAN (*M*)	Conc. HNO_3 (*M*)	Conc. ROH (*M*)	Vol. Soln. (ml)	Solvent
CAN stock solution		—	—		H_2O
HNO_3 stock solution	—		—		H_2O
CAN Solution			—	50.0	H_2O
Alcohol Solution	—	—		50.0	CH_3CN

Spectral measurements begin with the determination of A_0, the absorbance of cerium(IV) without added alcohol. Accurately pipet 3.00 ml of the *CAN Solution* into an Erlenmeyer flask and mix thoroughly with 7.00 ml of acetonitrile. Fill the spectrophotometer cell with this solution and record its absorbance at 520 nm (A_0) as rapidly as possible. Use a solution of 70% acetonitrile–30% water (70% aqueous acetonitrile) in the reference cell for the 100% adjustment in this experiment.

To measure the change in absorbance due to complexation of CAN with benzyl alcohol, the concentration of CAN is kept constant and increasing amounts of the *Alcohol Solution* are added. Pipet 3.00 ml of the *CAN Solution* into an Erlenmeyer flask, then add 6.00 ml of acetonitrile and 1.00 ml of the *Alcohol Solution*. Mix thoroughly and note the time of mixing, then transfer this solution to the spectrophotometer cell and record its absorbance at 520 nm (A_T). Since oxidation of the alcohol by cerium(IV) produces a slow decrease in absorbance with time, it is essential that you take absorbance readings rapidly and at intervals of 1.0, 2.0, and 3.0 min after the noted time of mixing. Record the temperature of the solutions and do not leave your solution cells in the spectrophotometer. When left in the sample chamber solutions increase in temperature. Variations in temperature will produce changes in absorbance measurements. Repeat these measurements by using 2.00, 3.00, 4.00, and 5.00 ml of the *Alcohol Solution* and decreasing the volume of acetonitrile so that the total volume of added reagents is 10.00 ml. To record your measurements prepare a table that lists the volume of the *CAN Solution*, the volume of the *Alcohol Solution*, the volume of acetonitrile, the time of measurement from mixing, A_T, and the temperature of the reaction solution.

If there was a decrease in the absorbance of the cerium(IV)-alcohol solution with time,* the absorbance value (the A_T value) to be used in experimental calculations should be determined by extrapolating the absorbance to zero time. Assume that absorbance decreases linearly with time.

For each of the solutions that you examined tabulate the concentrations of benzyl alcohol, CAN, and nitric acid. Calculate 1/[ROH], $[ROH]^2$, $1/[ROH]^2$, A_T-A_0, and $1/\Delta A$. Plot $1/\Delta A$ versus 1/[ROH] and $1/\Delta A$ versus $1/[ROH]^2$ on separate sheets of graph paper. Observe the distribution of data points. By evaluating which value of *n* gives the best straight line plot, determine whether your data support a 1:1, 1:2 or some other ratio for the cerium(IV)-alcohol complex. Calculate K_{eq} using Equation (2) and estimate the errors in this experimental analysis.

Postlab Questions

1. Estimate the error in your reading of absorbance. How does this error compare with those in the preparation of volumetric solutions? What is the source of the greatest error in this experiment?

2. Estimate the percent variance in K_{eq} from your error analysis in Question 1.

3. No change in absorbance is observed when benzyl alcohol is added to ceric ammonium sulfate, $Ce(NH_4)_2(SO_4)_3$. Suggest a reason for this observation.

*Absorbance readings that increase with time usually result from temperature changes of the reaction solution.

4. Suggest an explanation for the use of 70% aqueous acetonitrile as the solvent for this experiment rather than water.

5. Using the K_{eq} value that you calculated in this experiment, determine the percent of the total cerium(IV) that exists at equilibrium as the cerium(IV)-alcohol complex for each of your reaction solutions.

Experiment Thirty - four

Ceric Ion Oxidation of Benzyl Alcohol: The Rate Law and Rate Constant

---

Observation of the changes that occur during the ceric ammonium nitrate oxidation of benzyl alcohol to benzaldehyde suggests a mechanism for this reaction. Not only is there a color change when ceric ammonium nitrate is mixed with benzyl alcohol, there is also a slow decrease in the intensity of this color with time. The end result is a colorless solution that is indicative of the formation of cerium(III), an ion that has no visible absorption. These observations suggest that a rapid reaction between cerium(IV) and benzyl alcohol forms an intermediate complex and that this intermediate complex undergoes oxidative electron transfer to cerium in a subsequent slow step to produce cerium(III) and an organic product:

$$\text{Ce(IV)} + C_6H_5CH_2OH \overset{K_{eq}}{\rightleftharpoons} \text{Complex} \underset{\text{slow}}{\overset{k_2}{\longrightarrow}} \text{Ce(III)} + \text{Product}$$

Your results from Experiment 14 indicate that in order to completely oxidize benzyl alcohol to benzaldehyde two moles of cerium(IV) per mole of benzyl alcohol are required:

$$2\,\text{Ce(IV)} + C_6H_5CH_2OH \longrightarrow 2\,\text{Ce(III)} + C_6H_5CHO + 2\,H^+$$

Benzaldehyde is formed by a two-electron oxidation of benzyl alcohol. The conversion of cerium(IV) to cerium(III) is a one-electron reduction. The balanced reaction describes the stoichiometric relationship between reactants and products and is the essential starting point for investigations of the reaction mechanism.

In Experiment 33 you examined the nature of the complex formed between cerium(IV) and benzyl alcohol, and you obtained the equilibrium constant for association, K_{eq}. In this experiment you will determine the rate law and obtain the rate constant for oxidation of benzyl alcohol by cerium(IV). By combining the data that you obtain from Experiment 33 with that from this experiment, you can begin to describe a detailed mechanism for the oxidation of benzyl alcohol by cerium(IV).

Determination of the Rate Expression and Rate Constant

The rate of reaction for the oxidation of benzyl alcohol by cerium(IV) can be obtained by several techniques. We could observe the rate of formation of products or the rate of loss of reactants. The choice of a method for rate measurement depends on how accurately and easily the concentration of a particular species can be determined. Consideration of both of these factors for the oxidation of benzyl alcohol by cerium(IV) suggests that we should follow the rate of loss of cerium(IV), the only colored species in solution.

Let us assume that the rate of oxidation of an alcohol by cerium(IV) is dependent only on $[\mathrm{Ce(IV)}]^m$ and $[\mathrm{ROH}]^n$, where m and n are integers that represent the number of each chemical species in or prior to the transition state for electron transfer. The general equation for the rate of loss of cerium(IV) with time is

$$\frac{-d[\mathrm{Ce(IV)}]}{dt} = k[\mathrm{Ce(IV)}]^m[\mathrm{ROH}]^n \tag{1}$$

The constant k is the specific reaction *rate constant* and is independent of [Ce(IV)] and [ROH] but is dependent on such reaction variables as the solvent and the temperature.

If the concentration of alcohol is large compared to the concentration of cerium(IV), say by at least a factor of 10, the alcohol concentration does not measurably change during the reaction and is, therefore, a constant. At such a high alcohol concentration the equation for the rate of loss of cerium(IV) (Equation(1)) with time becomes

$$\frac{-d[\mathrm{Ce(IV)}]}{dt} = k'[\mathrm{Ce(IV)}]^m \tag{2}$$

where $k' = k[\mathrm{ROH}]^n$. Thus if the alcohol concentration is kept constant, we can determine the dependence of the rate on the cerium(IV) concentration and obtain a value for k'. If we now double the alcohol concentration and again measure the rate of reaction, the observed rate constant in this experiment should be $2k'$ if $n = 1$ and $4k'$ if $n = 2$. Equation (2) is known as a *pseudo* first-order rate expression.

If $m = 1$, Equation (2) can be integrated to give

$$\ln[\mathrm{Ce(IV)}]_t = -k't + \ln[\mathrm{Ce(IV)}]_0 \tag{3}$$

where $[\mathrm{Ce(IV)}]_t$ is the ceric ion concentration at time t and $[\mathrm{Ce(IV)}]_0$ is the initial ceric ion concentration. Since we will measure the concentration of cerium(IV) spectrophotometrically, and $[\mathrm{Ce(IV)}] = A_{\mathrm{Ce(IV)}}/l\,\epsilon_{\mathrm{Ce(IV)}}$ from the Beer-Lambert law, Equation (3) can be rewritten in terms of absorbance A,

$$\ln A_{\mathrm{Ce(IV)}_t} = -k't + \ln A_{\mathrm{Ce(IV)}_0} \tag{4}$$

or, converting to $\log_{10}$

$$\log_{10} A_t = \frac{-k'}{2.303} t + \log_{10} A_0 \tag{5}$$

A plot of $\log_{10} A_t$ versus time should yield a straight line with a slope of $-k'/2.303$ and an intercept of $\log_{10} A_0$ if the rate law is first ordered in cerium(IV).

If $m = 2$, Equation (2) can be integrated to give

$$\frac{1}{[\text{Ce(IV)}]_t} = k't + \frac{1}{[\text{Ce(IV)}]_0} \tag{6}$$

and rewritten in terms of absorbance,

$$\frac{1}{A_t} = \frac{k't}{l\epsilon_{\text{Ce(IV)}}} + \frac{1}{A_0} \tag{7}$$

A plot of $1/A_t$ versus time should yield a straight line with a slope of $k'/l\epsilon_{\text{Ce(IV)}}$ and an intercept of $1/A_0$ if the rate law is second order in cerium(IV).

To obtain the rate law and the rate constant, k, for the oxidation of benzyl alcohol by ceric ammonium nitrate two experiments are required: (1) measurement of $A_{\text{Ce(IV)}_t}$ versus time at a relatively high alcohol concentration, and (2) measurement of $A_{\text{Ce(IV)}_t}$ versus time at twice the alcohol concentration in (1). However, if the rate of solvent oxidation is comparable to the rate of alcohol oxidation, the rate constant for solvent oxidation must also be obtained and subtracted from the rate constant for alcohol oxidation in that solvent. For this experiment teams of two or three students can obtain the necessary information in the time allotted. In a team of three students one student is to measure the rate of oxidation at the low alcohol concentration, one student is to measure the rate of oxidation at the high alcohol concentration, and one student is to measure the rate of solvent oxidation. In a team of two students, each student is to measure the rate of oxidation at one alcohol concentration and the rate of solvent oxidation.

Prelab Questions

1. Derive Equation (4) from Equation (3) by assuming that the Beer-Lambert law, $A_{\text{Ce(IV)}} = \epsilon_{\text{Ce(IV)}} l [\text{Ce(IV)}]$, is obeyed.

2. To obtain the rate law and rate constant for the oxidation of benzyl alcohol by cerium(IV), how accurate must be your measurement of the initial cerium(IV) concentration?

3. To obtain the rate law and rate constant for the oxidation of benzyl alcohol, how accurate must be your measurement of the initial benzyl alcohol concentration?

4. Does experimental determination of the rate constant require that you measure A_0 accurately?

EXPERIMENTAL PROCEDURE

Prepare a stock solution that is $0.128M$ in ceric ammonium nitrate (CAN) and $2.94M$ in nitric acid by accurately measuring 4.00 ml of the

1.6*M* CAN stock solution into a 50.0-ml volumetric flask, and then add the stock 3.2*M* aqueous nitric acid solution to fill the volumetric flask to the 50.0-ml mark. Label this solution as the *CAN Solution.* This *CAN Solution* cannot be stored for longer than four hours at room temperature and should be used immediately after its preparation for best results. Only one student of the experimental team of two or three students should prepare this reagent since this will be a sufficient quantity for the team.

Heat a water bath to 55°C and hold that temperature. The water bath may be a large beaker of water heated at the required temperature by a Bünsen burner. Since temperature fluctuations will produce inaccuracies in your measurements, monitor the bath temperature at frequent intervals. A variation of ±1°C is acceptable for these rate measurements.

Rate of Oxidation at Low Alcohol Concentration. Transfer 10.0 ml of the 1.0*M* benzyl alcohol solution that you prepared in Experiment 33 to a 250-ml Erlenmeyer flask followed by 65 ml of acetonitrile and 20.0 ml of 3.2*M* nitric acid. Place a rubber stopper loosely over the flask. Place this loosely-stoppered flask in the constant temperature bath and allow five minutes for the temperature of the aqueous acetonitrile solution to come to equilibrium at 55°C. Carefully pipet 5.00 ml of the *CAN Solution* into the Erlenmeyer flask, mix the reaction solution thoroughly, and remove a 5.00-ml aliquot from the reaction solution. Transfer the aliquot to a clean, dry test tube or Erlenmeyer flask, and quickly cool the aliquot to or below room temperature. Note the time of addition, usually when one-half of the *CAN Solution* has been added; this time is t_0. At regular intervals, usually between five and ten minutes, remove additional 5.00-ml aliquots from the reaction solution, transfer the aliquots to clean, dry test tubes or Erlenmeyer flasks, and cool to or below room temperature. Lowering the temperature of the aliquot to room temperature effectively minimizes further oxidation. Record the time at which the aliquot was removed from the reaction solution. Time measurements are taken from t_0 in seconds and must be consistent for each aliquot; recording the time at which one-half of the aliquot was removed, for example, gives consistent average time for quenching.

After an aliquot in an Erlenmeyer flask has reached room temperature, transfer the sample to a cuvette and measure its absorbance at 460 nm. Be certain that the temperature of each solution whose absorbance is measured is the same. The absorbance measurement should be made within five minutes after the removal of each aliquot from the reaction solution. The reference cell for absorbance readings at 100% transmittance in this experiment contains 75% aqueous acetonitrile. Follow the rate of reaction until aliquot absorbance readings are less than 0.15. Take a final absorbance reading after oxidation is complete (greater than five half-lives).

In your notebook tabulate for each aliquot the time (in seconds) at which the aliquot was removed from the reaction solution, the temperature at which absorbance measurements were made, the absorbance reading (A_t), $\log_{10} A_t$, and $1/A_t$. In addition, calculate the initial concentrations of cerium(IV), benzyl alcohol, and nitric acid in the 75% aqueous acetonitrile reaction solution.

Rate of Oxidation at High Alcohol Concentration. Transfer 20.0 ml of the 1.0*M* benzyl alcohol solution that you prepared in Experiment 33 to a 250-ml Erlenmeyer flask followed by 55 ml of acetonitrile and 20.0 ml of 3.2*M* nitric acid. Follow the procedure for the oxidation at low alcohol concentration. Since the rate of oxidation is expected to be dependent on the concentration of benzyl alcohol, aliquots should be removed more frequently in this experiment—at intervals of five minutes or less.

Rate of Solvent Oxidation. Transfer 75 ml of acetonitrile to a 250-ml Erlenmeyer flask followed by 20.0 ml of 3.2*M* nitric acid. Follow the procedure for the oxidation at low alcohol concentration. Remove aliquots every ten minutes in order to follow the rate of solvent oxidation for a time period of approximately 90 min.

For each of your rate studies plot $\log_{10} A_t$ versus time (in seconds) and plot $1/A_t$ versus time (in seconds) on separate sheets of graph paper. Observe the distribution of data points and determine whether the oxidation is first- or second-order in cerium(IV). On the appropriate graph draw the best straight line through the data points and calculate k'. Subtract k' for the solvent oxidation from the k' value for each of the alcohol oxidations. Determine the order in benzyl alcohol of the cerium(IV) oxidation from the k' values that are corrected for solvent oxidation. Calculate k, the reaction rate constant for oxidation of benzyl alcohol by ceric ammonium nitrate at 55°C in 75% aqueous acetonitrile.

The treatment that we have used in Experiments 33 and 34 also applies to enzyme-catalyzed biochemical reactions (Experiment 35). In enzyme-catalyzed reactions an enzyme (E) forms a complex with a substrate molecule (S). The formation of the enzyme-substrate complex (ES) is usually rapid and is followed by the conversion of the ES complex to product (P) plus the regenerated enzyme:

$$\mathrm{E + S \rightleftharpoons ES \longrightarrow P + E}$$

When the enzyme-substrate complex is formed rapidly, the only difference in our treatment of the cerium(IV)-benzyl alcohol oxidation and that required for investigations of enzyme-catalyzed reactions is time. The benzyl alcohol–cerium(IV) complex, as we have observed, has a reasonably long lifetime; the ES complex has a very short lifetime by comparison.

Postlab Questions

1. Estimate the errors in your reading of absorbance and in your solution preparations. Describe the sources of other errors that contribute to inaccuracy in your determination of the rate law and rate constant for the cerium(IV) oxidation of benzyl alcohol.

2. Estimate the precision in the value for k from the experimentally determined values for k'.

3. You have previously determined in Experiment 33 that a rapid and reversible reaction between benzyl alcohol and cerium(IV) occurs prior to the oxidation step. The observed rate constant k is actually a combination of reaction constants. Write k in terms of k_2 and K_{eq}.

4. Describe two other experimental methods by which we could measure the rate of cerium(IV) oxidation of benzyl alcohol. Discuss the merits of these methods and compare them to the spectrophotometric procedure that you used in this experiment.

5. From the rate law and equilibrium expression that you determined in Experiments 33 and 34 propose a stepwise mechanism for the oxidation of benzyl alcohol by cerium(IV). The mechanism that you write must be consistent with the rate law and with complex formation, and must account for the stoichiometric requirement of two cerium(IV) ions per benzyl alcohol.

6. What is the purpose of taking a final absorbance reading after the oxidation reaction is complete?

Experiment Thirty - five

An Enzyme - Catalyzed Reaction: Hydrolysis of Urea

Enzymes are *protein* molecules that catalyze specific biochemical reactions. Most of the chemical reactions that occur in a living organism are mediated by enzymes. The enzyme *urease*, for example, catalyzes the hydrolysis of urea, but it does not catalyze the hydrolysis of simple amides or esters. In this experiment you will study the catalytic action of urease for the hydrolysis of urea.

Hydrolysis of urea

$$\underset{\textit{Urea}}{H_2N{-}\overset{\overset{\displaystyle O}{\|}}{C}{-}NH_2} + H_2O \xrightarrow{\text{Urease}} \underset{\substack{\textit{Carbon}\\ \textit{dioxide}}}{CO_2} + \underset{\textit{Ammonia}}{2\,NH_3}$$

Urease, which is conveniently isolated from the jack bean, was the first enzyme to be obtained in crystalline form. The purification and crystallization of urease was accomplished by J. B. Sumner in 1926, and in later studies Sumner showed that this enzyme had a molecular weight of almost 500,000. Sumner received the Nobel Prize in Chemistry for his pioneering work in the study of enzymes.

More recent studies of jack bean urease have led to the discovery that the urease molecule is composed of four or five equal structural subunits (M.w. = 105,000), each of which contains two atoms of bound nickel. The nickel atoms are necessary for the hydrolytic activity of this enzyme. Acetohydroxamic acid [$CH_3CONHOH$] forms very stable complexes with nickel(II) and will inhibit the catalytic activity of urease. Metal ions, such as the nickel ions in urease, or small organic molecules that are necessary for the catalytic function of enzymes, are called *cofactors*.

Active Site of the Enzyme

Although the mechanism of the urease-catalyzed hydrolysis of urea has not been determined in detail, there are a number of facts about the reaction process that are well known. The enzyme and the substrate, urea, first combine together in a reversible reaction to form an *enzyme-substrate complex* (Figure

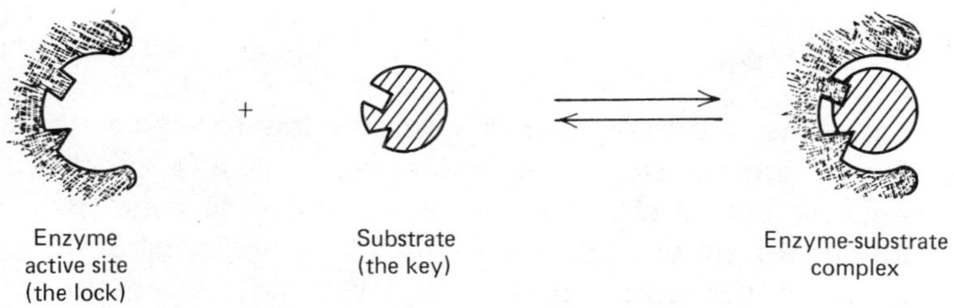

Figure 35.1 *A schematic picture of the lock-and-key theory for enzyme-substrate complex formation.*

35.1). The substrate binds to the enzyme at a specific site called the *active site*. The specificity that is characteristic of enzyme reactions is due to the selective binding of the substrate in the active site of the enzyme. Considerations regarding specificity have usually been based on the *lock-and-key theory*. According to this theory the substrate (the key) and the active site (the lock) are structurally complementary. This allows for the selective binding of the substrate into the active site of the enzyme.

The active site in an enzyme is formed by the folding of the protein chain (three-dimensional structure). In urease the active site is known to contain several sulfhydyl groups (–SH) as well as the nickel ions. Each of the sulfhydryl groups is present at the active site because the residue from the amino acid cysteine [$HSCH_2CH(NH_2)COOH$] is folded into the active site. Oxidation of these sulfhydryl groups to form disulfide linkages results in a reversible inactivation of the enzyme as illustrated below:

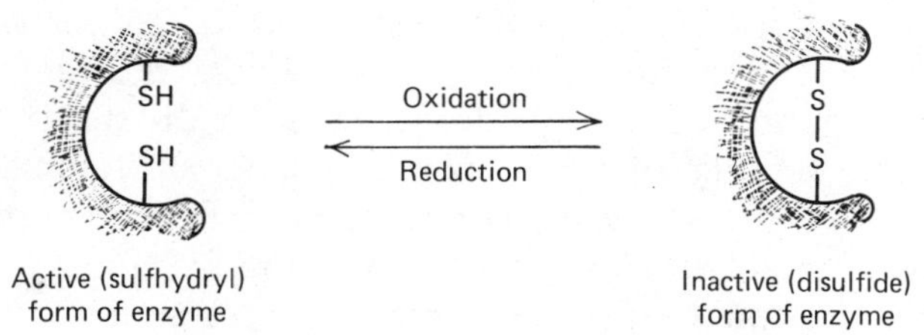

The inactive, disulfide form of the enzyme can be reactivated (reduced) by reaction with other sulfhydryl compounds such as the free amino acid cysteine. Reaction of certain heavy metals such as Hg^{2+}, Ag^{+}, or Cu^{2+} with the sulfhydryl groups of the enzyme will also cause inactivation of the enzyme through the formation of a metal sulfide.

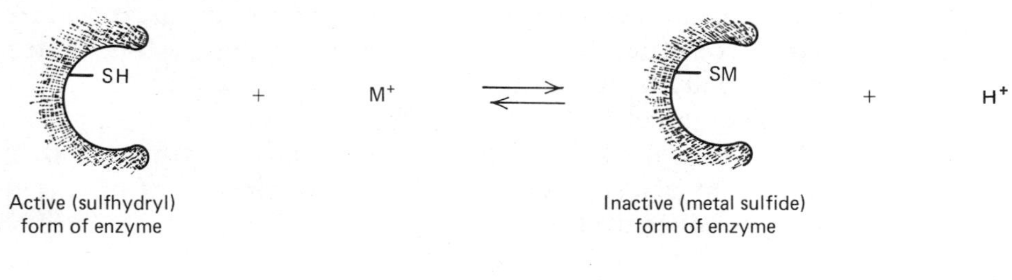

Turnover Number

Once the enzyme-substrate complex has formed with the active enzyme, functional groups from the amino acid residues of the protein chain and any enzyme-bound cofactors are in a position to react with the substrate. The close alignment of the substrate with the specific catalytic groups in the active site provides the catalysis that is observed. After the reaction products have formed, they are released from the active site of the enzyme and the catalytic cycle is complete. The enzyme molecule is then free to complex with another substrate molecule and to continue the catalytic process. The maximum number of substrate molecules converted to product per active site of enzyme per second is called the *turnover number* of the enzyme. Turnover numbers of enzymes vary from one to over a million, but many enzymes have turnover numbers of about 100 substrate molecules per active site per second. The turnover number of urease cannot be accurately determined since the number of active sites per molecule of enzyme is unknown. However, urease is one of the more reactive enzyme catalysts, and a turnover number of approximately 10^4 /sec is the best available estimate for the pure enzyme.

In this experiment the rate of hydrolysis of urea by urease will be compared to the rate of the hydrolysis reaction without enzyme. Quantitative evaluations of the rate accelerations brought about by enzymes compared to nonenzymatic reactions are difficult to make since in many cases the nonenzymatic reaction does not proceed at a measurable rate in the absence of enzyme under the experimental conditions in which enzyme catalysis is observed. Furthermore, an enzyme-catalyzed reaction is likely to proceed through a reaction mechanism that is entirely different from that of the predominant nonenzymatic reaction. However, it is still important to understand that both enzyme and nonenzyme catalysts cause rate acceleration of a reaction by lowering the free energy of activation for the reaction.

The activity of an enzyme is dependent on experimental conditions such as the *p*H of the solution and the temperature. A particular enzyme usually has a maximum activity over a narrow *p*H range, and a higher or lower hydrogen ion concentration results in a drastic decrease in enzyme activity. The optimum *p*H for urease is near neutrality. Enzyme-catalyzed reactions are run in a *p*H-buffered solution to maintain a constant *p*H.

The rate of enzyme activity generally increases with temperature. However, enzyme molecules are not stable in solution at high temperature; the complex three-dimensional structure of the enzyme is destroyed as the protein chain unfolds at elevated temperatures and the enzyme is said to be *denatured.* Most enzymes in aqueous solution are rapidly denatured at temperatures above 50°C. Reactions with urease are generally run at temperatures between 20 and 37°C (body temperature).

The activity of urease is determined by measuring the number of μmoles (10^{-6} moles) of ammonia that are produced from the hydrolysis of urea per minute at *p*H 7.0 and 30°C. The concentration of urea is very large compared to the concentration of enzyme. The number of μmoles of ammonia produced is determined by titrating the ammonia with standard aqueous hydrochloric acid.

Prelab Questions

1. It has been estimated that the first-order rate constant for the uncatalyzed hydrolysis of urea in water at 20°C is about 3×10^{-10} sec^{-1} and that the hydrolysis of urea bound to the enzyme urease takes place at 20°C in neutral aqueous solution with a first-order rate constant of 3×10^{4} sec^{-1}. What is the rate acceleration for the hydrolysis of urea by the enzyme urease?

2. Why is it essential to use distilled water to prepare the buffer solutions for the enzyme assays?

3. Ethylenediaminetetraacetic acid (EDTA) is a reagent that forms very stable complexes with many different metal ions in solution including copper(II) and nickel(II). Small amounts of EDTA in solution can prevent the deactivation of urease by metals, but larger amounts of EDTA will deactivate the enzyme. Explain why EDTA will deactivate urease.

4. Explain why urease will not catalyze the hydrolysis of *N,N,N′,N′*-tetramethylurea.

5. Urease is very widely used in clinical laboratories as a catalyst for the quantitative determination of urea. Propose how urease could be employed for the quantitative determination of urea in physiological fluids.

EXPERIMENTAL PROCEDURE

In this experiment you will use 18 × 150 mm test tubes for the enzyme reactions. Each tube should contain the reagents as listed in Table 35.1. Set up a large beaker to be used as a 30°C water bath. Fill the beaker with water, heat the water to 30°C, and adjust the rate of heating to maintain a constant 30°C (±1°C) temperature.

Tube number 1 is the "blank" tube. Add all of the reagents listed in Table 35.1 except for the urea solution to the test tube. Add four drops of a 1% $CuSO_4$ solution to the test tube to completely inhibit the enzyme, and then add the urea solution. Mix the solution, transfer a 10.0 ml aliquot of the solution to a 50-ml Erlenmeyer flask, and add two drops of methyl red indicator. Using a 50-ml burette, titrate the solution to the first pink color with standard 0.05*M* hydrochloric acid. Record the volume of acid used. This volume is the amount of acid necessary to titrate the buffer and reagents to the methyl red end point and is subtracted from all of the other titration volumes.

Now add the reagents, *except for the 0.3M urea solution*, to tubes 2, 3, 4, 5, and 6. Place these tubes in the water bath at 30° for 10 minutes. The reaction is initiated by the addition of the 0.3*M* urea in 0.05*M* Tris buffer to each test tube, and each tube is kept in the 30° water bath for 20 minutes. At the end of 20 minutes, four drops of 1% $CuSO_4$ are added with mixing to tubes 2–5 to stop the enzyme reaction. Titrate 10-ml aliquots from each tube using the methyl red indicator as previously done for the blank. The volumes of standard acid

Table 35.1 Urease Enzyme Assays

Reagents[a]	Tube Number 1	2	3	4	5	6
0.05*M* Tris buffer (*p*H 7.0) (ml)	10	10	10	10	10	11
Urease enzyme solution (ml)	1	1	1	1	1	
$1 \times 10^{-2} M$ cysteine (ml)			1			
$1 \times 10^{-2} M$ copper(II) sulfate (ml)[b]				1		
$1 \times 10^{-2} M$ acetohydroxamic acid					1	
Distilled water (ml)	1	1				1
0.3*M* Urea in 0.05*M* Tris buffer, (*p*H 7.0) (ml)	10	10	10	10	10	10
Titration Data and Calculations						
Volume of 0.05*M* HCl in titration						
Corrected volume of titrant						
Total μmoles of ammonia produced						

[a]The reagents are prepared by your instructor using glass-distilled water for all solutions according to the following procedures:

0.05*M* Tris buffer (*p*H 7.0) is prepared by dissolving 6.05 g of tris(hydroxymethyl)-aminomethane in 800 ml of distilled water and titrating to *p*H 7.0 at 30° C (*p*H meter) with 10% hydrochloric acid. Dilute this solution to a total volume of 1000 ml.

Urease enzyme solution is prepared fresh each day and kept cold (~4° C) until it is to be used. Suspend 10 g of crude jack bean meal in 500 ml of 0.05*M* Tris buffer (*p*H 7.0) at room temperature. After stirring the mixture for 10 to 15 min, cool the solution and allow the solids to settle to the bottom. Use the cloudy supernatant liquid for the enzyme assays. Commercially available urease may also be used as the enzyme for this reagent.

Prepare $1.0 \times 10^{-2} M$ cysteine by dissolving 1.75 g of L-cysteine hydrochloride hydrate in 1000 ml of glass-distilled water. Prepare this reagent fresh each day.

Prepare $1.0 \times 10^{-2} M$ copper(II) sulfate solution by dissolving 2.50 g of $CuSO_4 \cdot 5\,H_2O$ in 1000 ml of water.

Prepare $1.0 \times 10^{-2} M$ acetohydroxamic acid solution by dissolving 0.75 g of acetohydroxamic acid in 1000 ml of glass-distilled water.

Prepare 0.3*M* urea in 0.05*M* Tris buffer (*p*H 7.0) by dissolving 18 g of urea in 1000 ml of previously prepared 0.05*M* Tris buffer (*p*H 7.0).

[b]$1.0 \times 10^{-2} M$ Mercury(II) chloride may be substituted.

required to titrate these aliquots should be at least as large as the volume used to titrate the blank solution.

The total number of μmoles of ammonia produced in the assay per minute at 30°C is given by the following calculation:

$$\mu\text{moles of } NH_3 \text{ per min} = \frac{(X\text{-}A) \times \text{molarity of HCl} \times 1000}{20 \text{ min}} \times \frac{22 \text{ ml}}{10 \text{ ml}}$$

where *A* is the volume of standard acid used to titrate the blank reaction solution (tube 1), *X* is the total volume of standard acid used to titrate the enzyme reaction solution, and *X-A* is, therefore, the volume of standard acid necessary to titrate the ammonia produced by the hydrolysis of the urea. The 20 minutes represents the total time of the assay, and the 10 ml represents the size of the aliquot that was titrated from the total reaction volume of 22 ml. Calculate the number of μmoles of ammonia produced per minute in each of these assays.

Postlab Questions

1. Compare the amount of ammonia produced in the uncatalyzed reaction (tube 6) with the amount produced in the enzyme-catalyzed reaction (tube 2). In which reaction was the urea hydrolyzed most rapidly?

2. What was the effect on the activity of the enzyme of the added cysteine, the added $CuSO_4$, and the added acetohydroxamic acid? Explain the cause of each effect that was observed.

3. Calculate the number of μmoles of *urea* that were hydrolyzed per minute in each of the enzyme assays.

4. Explain why the factor of 1000 is necessary in the equation for the calculation of the number of μmoles of ammonia produced per minute.

5. What other methods besides adding excess copper(II) ions could be used to stop the enzyme activity at the end of the assay time period?

6. If the turnover number of urease is 10^4/sec under the assay conditions used in this experiment, how many μmoles of enzyme are necessary to hydrolyze 1 μmole of urea per minute (assume one active site per molecule)? How many grams of enzyme would this be if the weight of urease is estimated to be 500,000 g/mol?

Experiment Thirty-six

Polarimetry: Hydroboration–Oxidation of (+)-α-Pinene

Hydroboration-oxidation is a widely used synthetic procedure for the preparation of alcohols. *Hydroboration* is an electrophilic addition reaction in which borane (BH_3) or alkylboranes (R_2BH, R = alkyl, H) add to carbon-carbon multiple bonds. *Oxidation* by hydrogen peroxide in aqueous base converts alkylboranes formed by hydroboration to alcohols.

Hydroboration

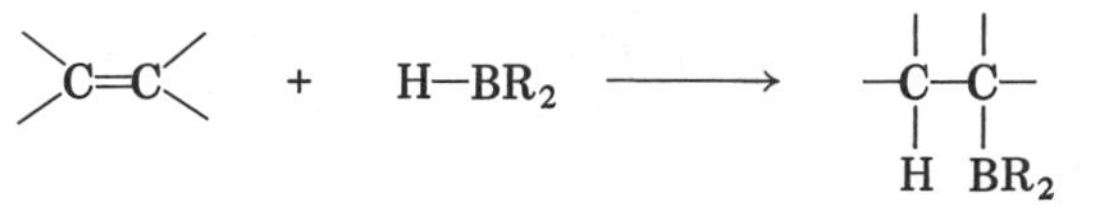

An alkene · *A borane* · *An alkylborane*

Oxidation

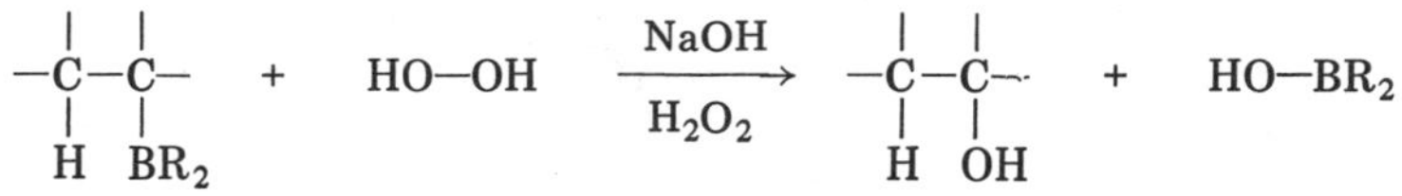

Hydroboration occurs by *cis*-addition: that is, boron and hydrogen add to the same side of the carbon-carbon multiple bond. Addition occurs in a Markovnikov fashion (Experiment 11) with the more positive end of the polar borane molecule, boron, forming a bond to the carbon possessing the greater number of hydrogens. Oxidation of the alkylborane occurs with retention of configuration at carbon; the C—OH bond is formed from the alkylborane without a change in the configuration at carbon. For example, hydroboration-

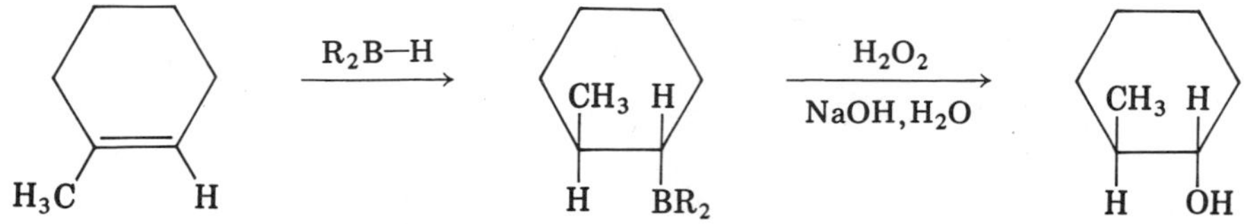

1-Methylcyclohexene

Hydroboration (*cis*-addition, Markovnikov orientation)

Oxidation (retention of configuration)

trans-*2-Methylcyclohexanol*

oxidation of 1-methylcyclohexene produces *trans*-2-methylcyclohexanol.* In this experiment you will use the stereospecific hydroboration-oxidation procedure to prepare optically active isopinocampheol from (+)-α-pinene.

α-Pinene occurs widely in nature and is the major constituent of turpentine obtained by steam distillation of the resinous components of various conifers. Hydroboration of α-pinene occurs stereoselectively from the less hindered side of the carbon-carbon double bond to form an alkylborane that

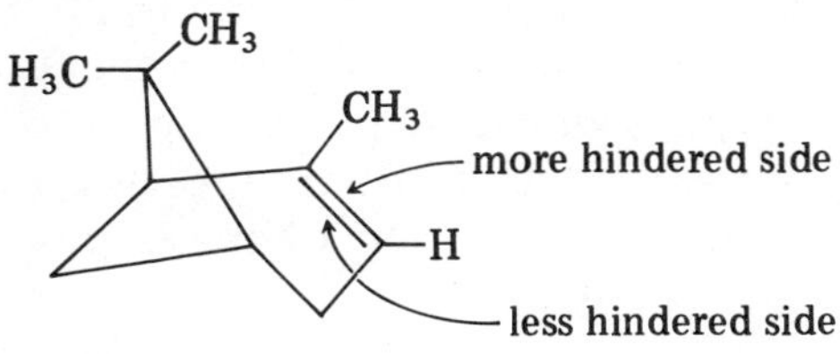

α-Pinene

is converted to isopinocampheol by oxidation with hydrogen peroxide. The borane reagent, either as diborane (B_2H_6) or as a solvent-borane complex (Sol-BH_3), is generated in the presence of the olefin by treating sodium borohydride with trifluoroacetic acid in tetrahydrofuran (THF).

$NaBH_4$	+	CF_3COOH	+	Sol:	⟶	Sol—BH_3	+	H_2	+	CF_3COONa
Sodium borohydride M.w. = 37.83 g/mol m.p. = 400°C (dec.)		*Trifluoro-acetic acid* M.w. = 114.0 g/mol b.p. = 72.4°C d_{20}°C = 1.535 g/ml		*Solvent*		*Solvent-borane complex*		*Hydrogen*		*Sodium trifluoro-acetate*

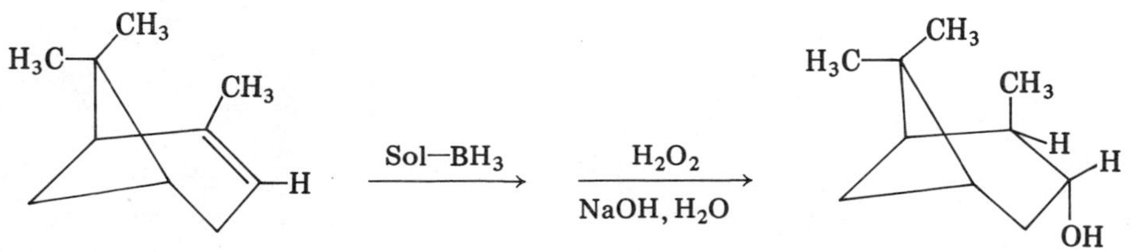

α-Pinene
M.w. = 136.2 g/mol
b.p. = 155°C
$n_D^{20°}$ = 1.4658
$d_{20°}$ = 0.858 g/ml

Isopinocampheol
M.w. = 154.2 g/mol
b.p. = 216°C
m.p. = 55–57°C
$n_D^{20°}$ = 1.4865

The method of generation of the borane reactant in this experiment takes advantage of the reactivity of borane compounds with olefins and utilizes the strong Lewis acidity of the borane monomer. In the absence of solvents such as

*Hydroboration-oxidation of 1-methylcyclohexene may be employed as an alternate experiment. 2-Methylcyclohexanol can be obtained by fractional distillation (b.p. = 167 to 168°C), and the relative yield of the *trans*-isomer can be determined by GLC analysis using the columns and procedures described in Experiment 32.

tetrahydrofuran that can serve as Lewis bases, borane is unstable and dimerizes to diborane. Diborane is a gas at room temperature that is explosive in air. When generated in the tetrahydrofuran reaction solution (*in situ*), however, borane forms a solvent complex that rapidly adds borane to the olefin. The formation of the solvent complex and the subsequent rapid addition of this reagent to the olefin prevents accumulation of the dangerous diborane reagent.

Since the orientation and stereochemistry of the hydroboration-oxidation reactions are known, use of optically active (+)-α-pinene having a known absolute configuration will allow you to define the absolute configuration of the product isopinocampheol. Since hydroboration-oxidation is a stereospecific reaction and addition does not occur at a chiral center of (+)-α-pinene, the absolute configuration of the isopinocampheol product is directly related to that of (+)-α-pinene.

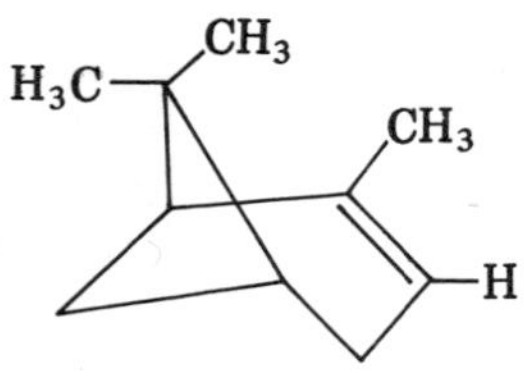

(+)-*α-Pinene*
$[\alpha]_D^{22°C} = +46.5°$

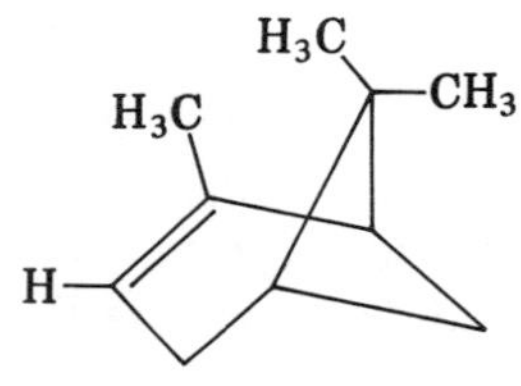

(−)-*α-Pinene*
$[\alpha]_D^{22°C} = -46.5°$

↓ Hydroboration-oxidation

↓ Hydroboration-oxidation

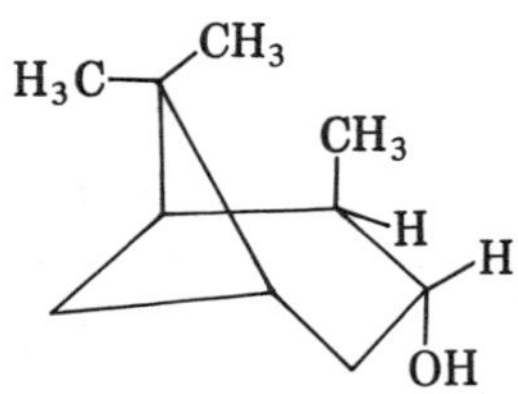

Isopinocampheol
$[\alpha]_D^{20°C} = +32.8°$ or $-32.8°$

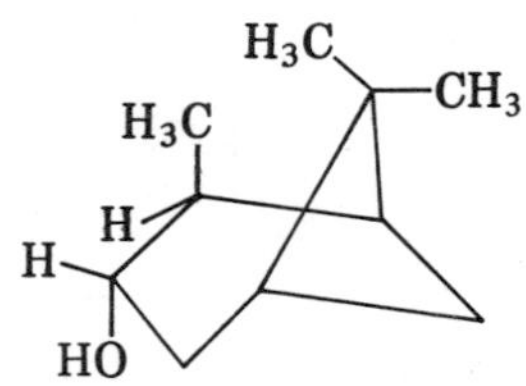

Isopinocampheol
$[\alpha]_D^{20°C} = +32.8°$ or $-32.8°$

Polarimetry

Chiral compounds such as (+)- and (−)-α-pinene have identical physical and chemical properties in an achiral (symmetric) environment. For example, the boiling points, densities, IR and NMR spectra, and rates of reaction with common achiral reagents for (+)- and (−)-α-pinene are exactly the same. However, chiral compounds that, like the α-pinene stereoisomers, are related as enantiomers differ in their effect on plane-polarized light. Interaction of plane-polarized light—light in which the wave vibrations are all in one plane—with a chiral

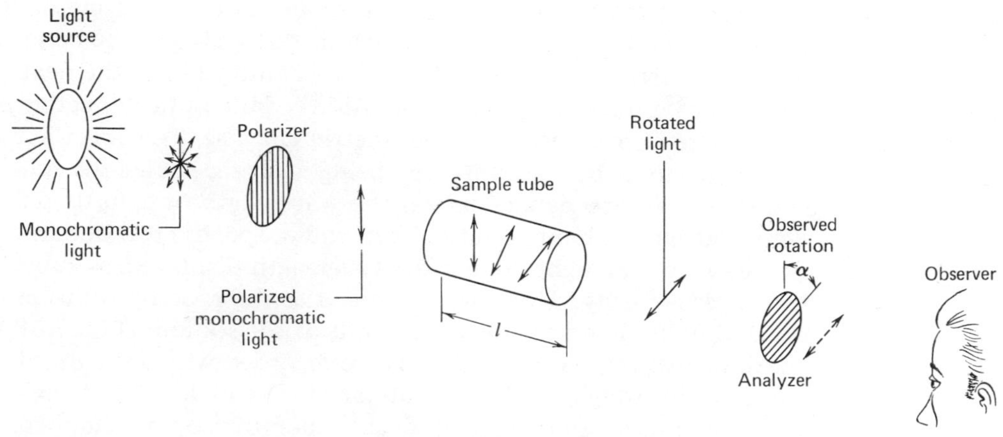

Figure 36.1 *A schematic description of a polarimeter.*

object causes rotation of the plane of polarization. *Enantiomers rotate the plane of polarized light in equal but opposite directions.*

The instrument used to detect and measure the rotation of polarized light is called a *polarimeter.* A polarimeter is a relatively simple instrument (Figure 36.1) in which polarized light from a monochromatic light source is passed through a sample tube containing the optically active substance with subsequent detection of the direction and degree of rotation (α). The light source normally used in a polarimeter is that from a sodium lamp (sodium D line, $\lambda = 589$ nm). The polarizer, either a Polaroid lens (film) or a Nicol prism, allows only light emanating from the source in one plane to pass. After the polarized light has passed through a glass sample tube containing the optically active substance, a second polarizer that serves as the analyzer is rotated to determine the change in rotation of polarized light. Rotation of polarized light to the right (clockwise) is a positive rotation and is recorded with a plus sign before the numerical value for the degree of rotation. Rotation of polarized light to the left (counterclockwise) is a negative rotation and is recorded with a negative sign before the numerical value for the degree of rotation.

The degree of rotation, the angle α, is dependent on the extent of interaction of the polarized light with the optically active compound and on the number of optically active molecules encountered by the polarized light in the sample tube. Thus the degree of rotation is proportional to a specific rotation constant, $[\alpha]$, that is a physical property of the substance being examined, to the concentration of the optically active substance (d or c), and to the pathlength of the sample tube (l) that contains the optically active substance. These relationships are defined by the equations:

For the neat liquid:

$$[\alpha] = \frac{\alpha(\text{in degrees})}{l(\text{in decimeters}) \times d(\text{in g/ml})}$$

For solutions:

$$[\alpha] = \frac{\alpha(\text{in degrees}) \times 100}{l(\text{in decimeters}) \times c(\text{in g/100 ml})}$$

The angle of rotation, α, is measured by use of the polarimeter. The pathlength of the sample tube (in dm, where 1 dm = 10 cm) and the concentration of the optically active compound, either as the density (d) of the neat liquid or as the concentration (c in g/100 ml of solution) of the optically active compound in a solvent, are determined for polarimetric analysis. The specific rotation, $[\alpha]$, is then calculated for the substance being analyzed. Since the specific rotation of a substance is also dependent on the wavelength of light used for the measurement and on the temperature of the liquid, specific rotations are recorded with temperature and light source or wavelength displayed as superscript and subscript, respectively. Thus $[\alpha]_D^{20°C}$ specifies the specific rotation of a substance at 20°C determined at the wavelength of the sodium D line (589 nm). Specific rotations determined for solutions are recorded with the solvent and concentration (usually weight in 100 ml of solution) listed in parentheses after the rotation value; for example, one enantiomer of isopinocampheol has a specific rotation

$$[\alpha]_D^{20°C} = -32.8° \text{ (1 g/100 ml ethyl ether)}$$

A variety of polarimetric instruments, ranging from manually operated instruments that require direct readings by the operator to automated spectropolarimeters, are available. Manually operated instruments are constructed so that the analyzer can be rotated relative to the polarizer zero position to observe the maximum (or minimum) light intensity. Whichever instrument is used, however, the observed rotation of an optically active sample must be corrected by subtracting the rotation of the sample tube (when neat liquids are used) or the sample tube that contains only the solvent (when solutions are used).

Sample Preparation for Polarimetric Analysis

Polarimetric analyses are performed on pure compounds. Prior to sample preparation you must establish that the material whose optical activity you will measure is homogeneous. The physical properties of the material (b.p. or m.p., for example) or its spectral properties should be evaluated to determine that the sample is not contaminated with solvent, starting materials, or by-products.

The method employed for sample preparation depends on whether polarimetric analysis is performed on a neat liquid or a solution. In either case, however, the sample tube and its end caps must be clean and dry. The transparent parts of the end caps should be wiped with a soft lens paper to remove opaque substances, particularly fingerprints. The sample to be analyzed must be clear and, when filled, the sample tube must be free of air bubbles in the liquid that would be exposed to the polarized light. The end caps to the sample tube should fit tightly so that the sample does not leak, but strain on the caps that causes apparent rotation of polarized light must be avoided.

Neat Liquids
The sample tube is filled directly with the neat liquid having a known density.

Solutions
The approximate volume of the sample tube is measured by filling the tube with the solvent that is to be employed and then pouring the solvent from the sample tube (polarimeter cell) into a graduated cylinder. A weighed quantity of the optically active sample is dissolved in a minimal volume of the solvent and transferred with rinsing into a volumetric flask of appropriate size. Additional solvent is added to fill to the mark with mixing to ensure homogeneity. The sample tube is then filled with the solution of known concentration.

Polarimetric Analysis

For analysis of neat liquids, determine the optical rotation for the empty sample tube and, if different from 0.00°, subtract this value from the observed rotation of the optically active liquid. For analyses of solutions, determine the optical rotation for the sample tube containing the pure solvent and, if different from 0.00°, subtract this value from the observed rotation of the optically active solution. The zero-point reading should be obtained as an average of several different determinations.

When manually operated instruments are employed, readings should be taken in a consistent fashion. The analyzer should be rotated to the observed rotation from the same direction, and the rotation readings should be taken several times from the same sample. The optical rotation of samples determined with manually operated instruments can be read at two different positions: $\alpha \pm 180°$. If the actual rotation cannot be determined with certainty from one measurement, a second polarimetric measurement with a solution of one-half the concentration will normally alleviate any confusion.

Optical Purity

The optical purity of a compound is a measure of the excess of one enantiomer over another in an optically active sample. An optically pure compound consists of only one enantiomer. Since enantiomers have equal but opposite optical rotations, an equal mixture of enantiomers (a racemic mixture) exhibits zero rotation and has 0% optical purity. An optically active sample consisting of 0.75 mol of one enantiomer and 0.25 mol of the other is 50% optically pure: 50% of the sample is a racemic mixture and 50% is one enantiomer in excess of the racemic modification.

The optical purity of a sample is determined by comparing the specific rotation of a sample to that of the pure stereoisomer. *Percent optical purity is the ratio of the specific rotations of the sample and pure stereoisomer multiplied by 100:*

$$\% \text{ Optical purity} = \frac{[\alpha]\text{Sample}}{[\alpha]\text{Pure stereoisomer}} \times 100$$

Thus, if the specific rotation of the isopinocampheol that you determine in this experiment is found to be -16.4°, the percent optical purity of that sample is 50%.

Prelab Questions

1. Why must opaque substances be removed from the transparent parts of the end caps of the sample tube used for polarimetric analysis? What is the effect of turbidity in the liquid sample on polarimetric analysis of that sample?

2. If the optical purity of the α-pinene that you use in this experiment is only 75% and you determine that the isopinocampheol obtained by hydroboration-oxidation has a specific rotation of -24.60°, what percentage of optically active α-pinene has undergone racemization during the reaction process?

3. Calculate the specific rotation and optical purity of an isopinocampheol sample if 0.50 g of that sample, dissolved in 50 ml of ethyl ether and placed in a 1.0 dm sample tube, has an observed rotation of -0.25° at 20°C using the sodium D line as the light source.

4. Can you predict the sign for rotation of polarized light for the isopinocampheol enantiomer expected from hydroboration-oxidation of (+)-α-pinene? Why?

5. Predict the effect on the calculated specific rotation for isopinocampheol (increase or decrease) (a) if the sample contains solvent impurities and (b) if the sample contains unreacted (+)-α-pinene. (Two answers, dependent on the sign of rotation of isopinocampheol are required for part (b).)

EXPERIMENTAL PROCEDURE

For the hydroboration reaction, all equipment must be thoroughly dried and all reactants and solvents must be anhydrous. The glassware that is employed is identical to that described for the Grignard synthesis in Experiment 15. The equipment is thoroughly dried as directed in that experiment. Sodium borohydride may be dried in the reaction flask if care is taken to heat the flask at temperatures that are well below the decomposition point of sodium borohydride (400°C). The tetrahydrofuran solvent and (+)-α-pinene reactant are dried over molecular sieves.

CAUTION: Careful handling of sodium borohydride is recommended. Avoid contact of this reagent with acids and with your skin. Do not weigh out sodium borohydride in the vicinity of trifluoroacetic acid.

Trifluoroacetic acid containing 5% (by volume) trifluoroacetic anhydride is employed in this experiment; trifluoroacetic acid is kept anhydrous by the anhydride.

CAUTION: Exposure of the relatively low boiling trifluoroacetic acid to the air results in fuming. The acid strength of trifluoroacetic acid is comparable to that of mineral acids. Use trifluoroacetic acid only in well-ventilated areas and keep this reagent well away from sodium borohydride until you begin the hydroboration experiment. Wash any acid off the skin with large volumes of water.

Obtain the specific rotation of (+)-α-pinene as a neat liquid. Add 2.0 g of sodium borohydride (0.052 mol) to a 500-ml three-neck round-bottom flask that is fitted with an addition funnel, reflux condenser, and a glass stopper. Pieces of cotton or drying tubes containing anhydrous calcium sulfate should be positioned at the openings of the equipment to the air. After drying the reaction flask and when the reaction flask has cooled to room temperature, add 5.5 g of (+)-α-pinene (0.040 mol) followed immediately by 80 ml of anhydrous tetrahydrofuran. Place the reaction flask in a water–ice bath to maintain the reaction temperature below 15°C throughout the experimental procedure.

Dissolve 3.0 g (2.0 ml) of the 95% trifluoroacetic acid–5% trifluoroacetic anhydride solution in 20 ml of anhydrous tetrahydrofuran and pour this solution into the addition funnel. Carefully add this solution dropwise into the reaction flask over a 30-min period and monitor the change in the undissolved sodium borohydride. Agitate the reaction solution at regular and frequent intervals during the addition. Continue agitation of the reaction solution for one hour after addition is complete, and then slowly add 20 ml of 10% aqueous sodium hydroxide from the addition funnel, followed by 8.0 ml of 30% aqueous hydrogen peroxide. Continue agitation of the reaction solution for an additional hour, then replace the separatory funnel and condenser with stoppers, and store the reaction solution until your next laboratory period. During this time the reaction mixture may separate into two distinct liquid phases. If phase separation has not occurred, add sodium chloride to form a salt-saturated solution. Then add 25 ml of ether.

Pour the contents of the reaction flask into a separatory funnel. If a solid is present in the reaction flask, filter the solution into the separatory funnel. Separate the organic layer from the aqueous layer and save the organic layer. Wash the aqueous phase three times with 25-ml portions of ether. Combine the ether washings with the original tetrahydrofuran organic layer, wash the combined solution with 50 ml of water, and dry the combined solution over anhydrous magnesium sulfate. Filter the organic solution and distill the ether and tetrahydrofuran solvents on a steam bath. Dispose of the collected solvents in the container provided for waste solvent. Remove the last traces of solvent by distillation under reduced pressure (aspirator). Cool the residue in an ice bath and collect the crystalline solid. Recrystallize the product from a minimal volume of pentane. Obtain the melting point range for the recrystallized product and the weight of the isolated compound. Calculate the percentage yield of the recovered isopinocampheol. When the melting point range of the isolated product is within ±2°C of the reported value for isopinocampheol, obtain the specific rotation of the product in an ethyl ether solution (10.0 g/100 ml solution). Calculate the optical purity of the isopinocampheol and determine the percentage of (+)-α-pinene that has undergone racemization during the reaction process.

Postlab Questions

1. Describe an alternate method for the purification of isopinocampheol. Will this method lead to racemization of optically active isopinocampheol?

2. Why does the tetrahydrofuran reaction mixture remain homogeneous during the addition of aqueous base and hydrogen peroxide? What factor(s) causes the eventual separation of the two phases?

3. Trifluoroacetic acid undergoes addition to alkenes. Trifluoroacetate esters are, however, readily hydrolyzed to alcohols and trifluoroacetate in aqueous base. Predict the product that would be formed if trifluoroacetic acid underwent addition to α-pinene at a faster rate than its reaction with sodium borohydride.

4. (a) How can you determine if α-pinene is a contaminant of your isolated product? (b) If a considerable amount of α-pinene is unreacted, the isopinocampheol will not crystallize. Would the method that you described in Question 1 separate α-pinene from isopinocampheol? If not, describe an effective separation method.

5. Predict the product(s) from hydroboration-oxidation of (a) 1-hexene, (b) limonene, (c) *trans*-stilbene, (d) *trans*-3-methyl-2-pentene.

Experiment Thirty - seven

Diels Alder Reactions: Synthesis of cis - 4 - Methyl - 4 - cyclohexene - 1, 2 - dicarboxylic Acid Anhydride

An important and versatile method for the synthesis of cyclic organic compounds having six-membered rings is the Diels-Alder reaction, a 1,4-addition reaction of an alkene to a diene. Since Diels-Alder reactions involve the addi-

Diels-Alder reaction

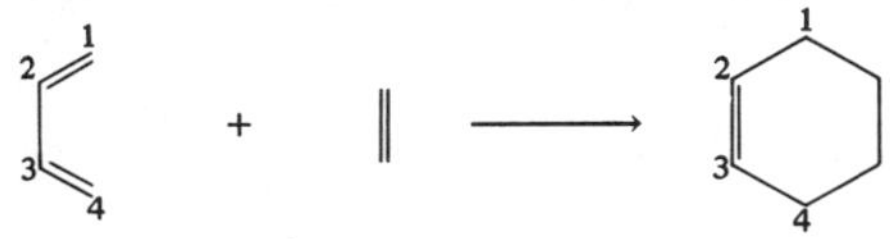

Diene *Alkene (Dienophile)* *A cyclohexane*

tion of a 2-π electron system (the alkene or *dienophile*) to a 4-π electron system (the *diene*), these reactions are also referred to as [4 + 2] *cycloaddition reactions.* The simplest example of this reaction is the addition of ethylene to 1,3-butadiene:

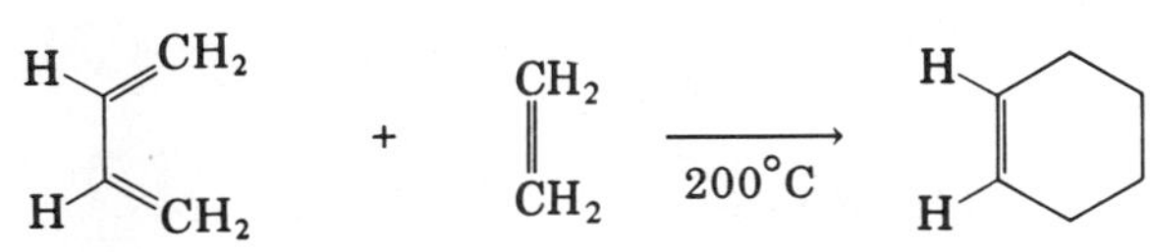

1, 3-Butadiene *Ethylene* *Cyclohexene*

However, this particular example requires high temperatures and pressures to form the Diels-Alder product, cyclohexene. The Diels-Alder reaction is facilitated by electronic withdrawing groups such as carbonyl groups, carboxylate ester groups, or cyano groups substituted on the dienophile and by electron donating groups such as alkyl groups substituted on the diene. Alkynes can also act as dienophiles. Generally no solvent is used in Diels-Alder reactions; the reactants are simply mixed together and heated.

The Diels-Alder reaction occurs in a single step and does not involve the formation of a reactive chemical intermediate. The transition state for this reaction involves overlap of the π-orbitals of the diene and dienophile to form the two new σ-bonds of the cyclic system, as illustrated in Figure 37.1. This

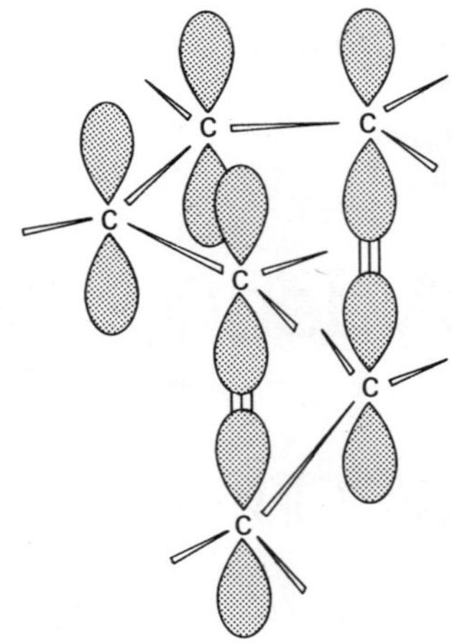

Figure 37.1 *The transition state of the Diels-Alder reaction.*

one-step, or concerted, reaction mechanism has stereochemical consequences; the reaction must occur by *cis*-addition, both with respect to the diene and the dienophile. The following example illustrates this stereochemistry:

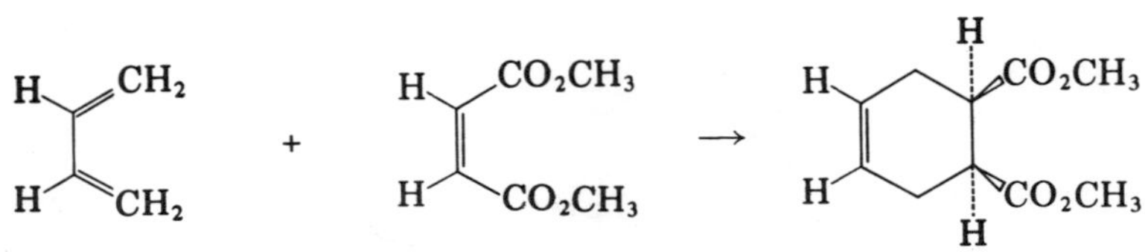

In this experiment you will prepare *cis*-4-methyl-4-cyclohexene-1,2-dicarboxylic acid anhydride by the cycloaddition reaction of maleic anhydride to 2-methyl-1,3-butadiene (isoprene):

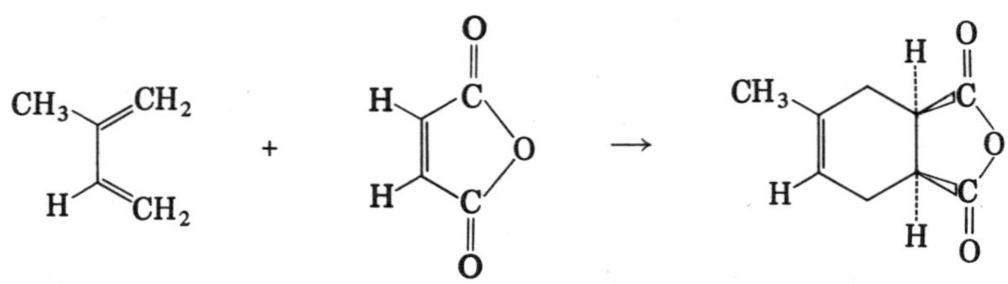

2-Methyl-1,3-butadiene	*Maleic anhydride*	cis-*4-Methyl-4-cyclohexene-1,2-dicarboxylic acid anhydride*
M.w. = 68.1 g/mol	M.w. = 98.1 g/mol	M.w. = 166.2 g/mol
b.p. = 34°C	m.p. = 56°C	m.p. = 63-64°C
d_{20} = 0.68 g/ml		

Since the dienophile, maleic anhydride, is substituted with two activating groups, this reaction takes place rapidly at the reflux temperature of the reaction mixture.

You will prepare the 2-methyl-1,3-butadiene for the Diels-Alder reaction by the dehydration of 2-methyl-3-buten-2-ol using phosphoric acid as the dehydration catalyst:

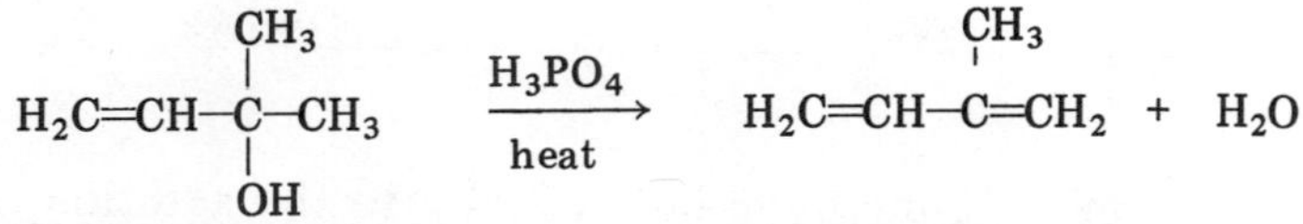

2-Methyl-3-buten-2-ol
M.w. = 86.1 g/mol
b.p. = 99°C
d_{20} = 0.89 g/ml
n_D^{20} = 1.4160

2-Methyl-1,3-butadiene
n_D^{20} = 1.4219

The experimental procedures and scope of dehydration reactions have been discussed previously in Experiments 9 and 16. In this experiment, better yields of the diene are obtained when the easily dehydrated allylic alcohol is reacted with the weaker phosphoric acid catalyst rather than with the stronger sulfuric acid catalyst that was used in Experiment 16. If sulfuric acid is used as a catalyst for the dehydration of 2-methyl-3-buten-2-ol, a tarry polymeric material is the major product. To reduce the amount of polymer formation, even when using the phosphoric acid catalyst, the 2-methyl-3-buten-2-ol is slowly added to the hot acid and the diene is removed from the acid solution as it is formed by distillation. Even with these precautions, the average percentage yield of 2-methyl-1,3-butadiene is only about 30%.

Prelab Questions

1. Why must the receiving flask for the distillation of 2-methyl-1,3-butadiene be cooled in an ice bath?

2. Draw a portion of the structure of the expected polymeric side products formed in the dehydration of 2-methyl-3-buten-2-ol.

3. If the 2-methyl-1,3-butadiene is not dried prior to reaction with maleic anhydride, *cis*-4-methyl-4-cyclohexenedicarboxylic acid is produced. Explain how this material is formed.

4. What chemical tests could you perform to confirm the structure and identity of each product prepared in this experiment?

5. How much maleic anhydride should be used to react with 6.3 g of 2-methyl-1,3-butadiene?

EXPERIMENTAL PROCEDURE

Preparation of 2-Methyl-1,3-butadiene. Using a 500-ml, three-neck, round-bottom flask as the distilling flask, set up the apparatus for fractional distillation as shown in Figure 6.4. Add 60 ml of concentrated phosphoric acid (86% by weight) and several boiling stones to the three-neck flask. Attach an addition funnel to one neck of the three-neck flask, and stopper the third neck of the flask. Place 30

CAUTION: Phosphoric acid will burn the skin. Wash acid from your skin with generous amounts of water.

---

ml of 2-methyl-3-buten-2-ol into the addition funnel. Place an ice bath in position to cool the 50-ml round-bottom flask used to collect the distillate. Heat the phosphoric acid in the three-neck flask over the steam bath until the acid is hot, and then start to add the 2-methyl-3-buten-2-ol dropwise (approximately 1 to 2 ml/min) to the hot acid solution. After all of the alcohol has been added, continue to heat the reaction mixture for 15 minutes. During the addition and heating processes, collect all of the low boiling distillate that condenses. Dry the distillate by the addition of less than one gram of anhydrous magnesium sulfate, and then filter the product directly into your 100-ml round-bottom flask that has been pre-weighed with a glass stopper. Stopper the flask with the same glass stopper, and re-weigh the flask, stopper, and product. Determine the weight and percentage yield of 2-methyl-1,3-butadiene. If this product is not used at once in the following reaction, it should be stored at 0°C.

Preparation of cis-*4-Methyl-4-cyclohexene-1,2-dicarboxylic Acid Anhydride.* Calculate the number of moles of 2-methyl-1,3-butadiene contained in the 100-ml round-bottom flask. Add several boiling stones and equip the flask with a reflux condenser. Cool the flask in an ice bath, and then add an equal number of moles of maleic anhydride to the flask. Remove the ice bath, but keep it nearby to moderate the reaction as required. If necessary, initiate the reaction by warming the solution on a steam bath. If vigorous reflux occurs, cool the flask to prevent the volatile 2-methyl-1,3-butadiene from evaporating out of the condenser. After the initial spontaneous reflux subsides (10 to 15 min), heat the solution on the steam bath for 20 minutes. Mild boiling should occur at first, but reflux should not be evident at the end of this heating period. Allow the mixture to cool, and then dissolve the semi-solid residue in a minimum volume of hot toluene heated on a steam bath. Filter the solution to remove any solid that does not readily dissolve in the hot toluene. Add an equal volume of hexane to the toluene solution, allow the solution to cool to room temperature, and finally cool the mixture in an ice bath. Isolate the product by filtration, and wash the crystals with a small portion of cold toluene-hexane (1:1). Weigh the dry product and determine its melting point. Calculate the percentage yield. Confirmation of the structure of your product may be obtained by infrared spectroscopy.

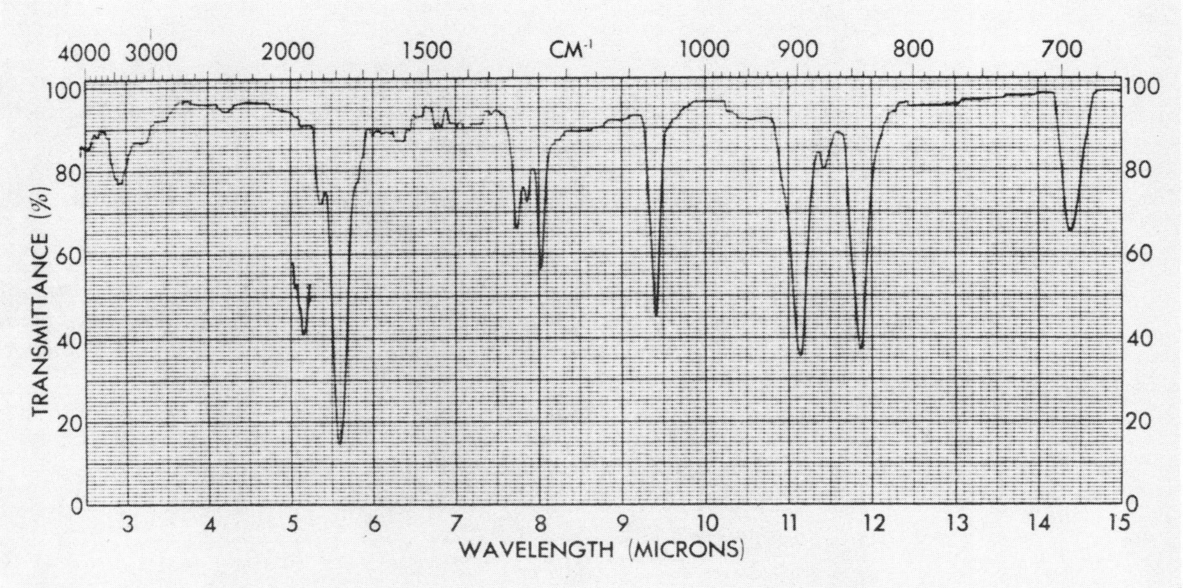

Figure 37.2a *Infrared spectrum of maleic anhydride (KBr pellet).*

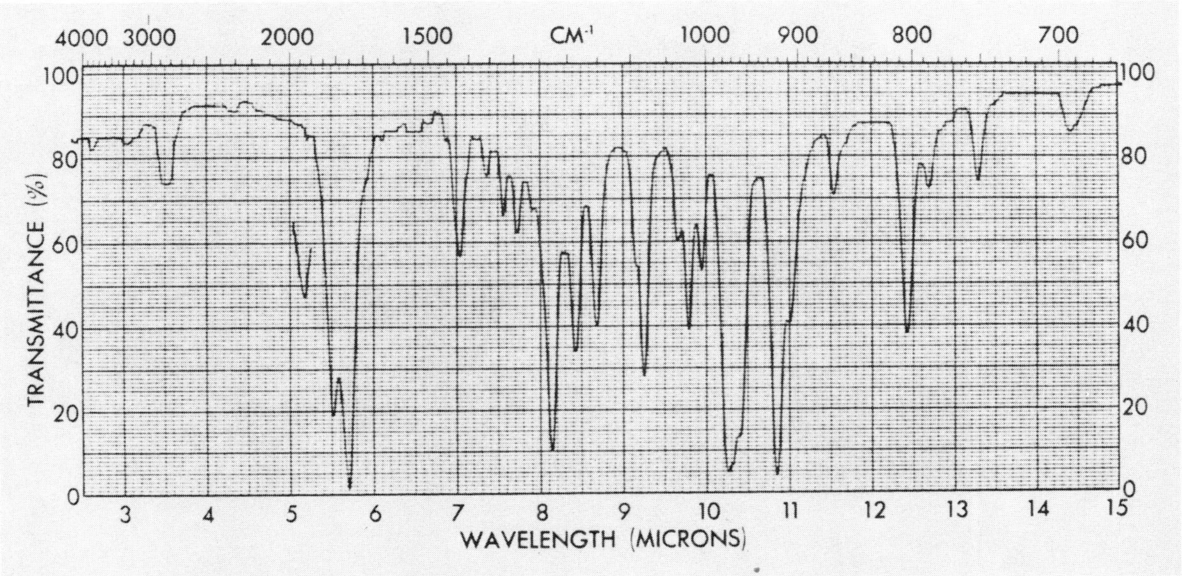

Figure 37.2b *Infrared spectrum of* cis-*4-methyl-4-cyclohexene-1,2-dicarboxylic acid anhydride (KBr pellet).*

Postlab Questions

1. Assign the major peaks observed in the infrared spectra of maleic anhydride and of *cis*-4-methyl-4-cyclohexene-1,2-dicarboxylic acid anhydride (Figure 37.2*a* and 37.2*b*) to the corresponding functional groups in the two molecules.

2. Why does reflux stop near the end of the heating period during the preparation of the *cis*-4-methyl-4-cyclohexene-1,2-dicarboxylic acid anhydride?

3. Give the structure of the Diels-Alder reaction products that would be formed from the following dienophiles reacting with 1,3-butadiene: (a) dimethyl fumarate (*trans*-butenedioic acid dimethyl ester), (b) dimethyl acetylenedicarboxylate (butynedioic acid dimethyl ester), (c) 3-buten-2-one.

4. Why does the slow addition of 2-methyl-3-buten-2-ol to the hot phosphoric acid result in a higher yield of 2-methyl-1,3-butadiene than mixing the alcohol and acid rapidly?

Experiment Thirty - eight

Polymers and Polymerization

Polymers are high molecular weight compounds that are constructed by the covalent combination of a large number of low molecular weight molecules (*monomers*). Many polymers have molecular weights greater than 10^6 and

$$n \text{ Monomer} \longrightarrow (\text{Monomer})_n$$

Low molecular weight molecule	*High molecular weight polymer*

are constructed from more than 10^4 monomer units per polymer molecule. Because of their large molecular size, polymers are often called *macromolecules*. The unique physical and chemical properties of macromolecules have led to numerous applications of these materials in our modern society. Table 38.1 provides a brief list of common polymers, their chemical structure, and their typical applications.

Polymerization is the process of forming polymer molecules by linking together monomers in a continuous chain of repeating units. A polymerization process involves a sequential series of chemical reactions: monomer (M) reactants form a dimer,

$$2\,M \longrightarrow M{-}M$$

the dimer combines with another monomer to form a trimer,

$$M{-}M + M \longrightarrow M{-}M{-}M$$

and this sequential process continues under favorable conditions until a high molecular weight polymer is formed.

Addition Polymers

Polymers are divided into two main classes, *addition polymers* and *condensation polymers*, depending on the method of polymerization. Addition polymers are formed by addition of monomer units in a sequence of reactions that involve initiation, chain-propagation, and chain termination. The structural unit of the addition polymer has the same molecular formula as the monomer. The monomer used to form an addition polymer generally contains one or more

Table 38.1 Synthetic and Natural Polymers

Name	Monomer(s)	Chain structure	Use
Polyethylene	$H_2C{=}CH_2$	$\sim CH_2{-}CH_2{-}CH_2{-}CH_2{-}CH_2{-}CH_2\sim$	Films, bags, electrical insulation, bottles
Polypropylene	$H_2C{=}CH(CH_3)$	$\sim CH_2{-}CH(CH_3){-}CH_2{-}CH(CH_3){-}CH_2{-}CH(CH_3)\sim$	Fibers, packaging films, molded products, bottles
Poly-*cis*-1,4-isoprene (Natural rubber)	$H_2C{=}CH{-}C(CH_3){=}CH_2$	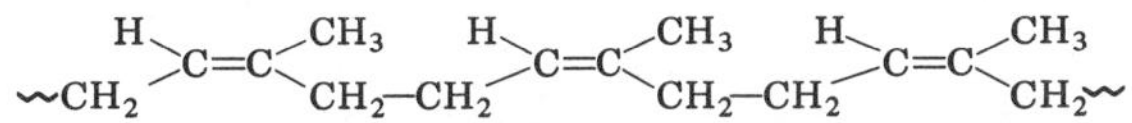	Polymer is crosslinked with sulfur (vulcanization) for tires, footwear
Polychloroprene (Neoprene)	$H_2C{=}CH{-}C(Cl){=}CH_2$	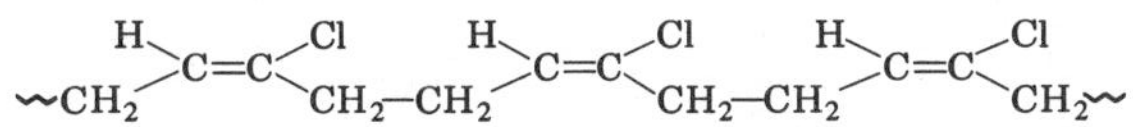	Weather-resistant rubber products
Polyvinylchloride (PVC)	$H_2C{=}CH(Cl)$	$\sim CH_2{-}CH(Cl){-}CH_2{-}CH(Cl){-}CH_2{-}CH(Cl)\sim$	Plastic pipe, floor tile, phonograph records, furniture covering
Polytetrafluoroethylene (Teflon)	$F_2C{=}CF_2$	$\sim CF_2{-}CF_2{-}CF_2{-}CF_2{-}CF_2{-}CF_2\sim$	Nonstick coatings, chemically-resistant labware
Polymethyl methacrylate (Plexiglas, Lucite)	$H_2C{=}C(CH_3)(CO_2CH_3)$	$\sim CH_2{-}C(CH_3)(CO_2CH_3){-}CH_2{-}C(CH_3)(CO_2CH_3){-}CH_2{-}C(CH_3)(CO_2CH_3)\sim$	Substitute for glass, molded products
Polystyrene (Styrofoam)	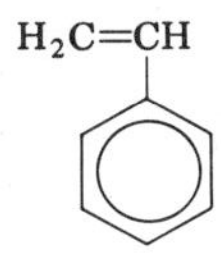	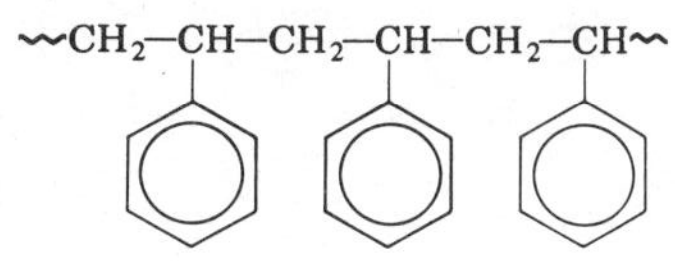	Molded products, foams for insulation
Polyacrylonitrile (Orlon)	$H_2C{=}CH(CN)$	$\sim CH_2{-}CH(CN){-}CH_2{-}CH(CN){-}CH_2{-}CH(CN)\sim$	Fiber, fabrics

Nylon 6.6	$H_2N(CH_2)_6NH_2$ $HO_2C(CH_2)_4CO_2H$	$\sim NH(CH_2)_6NHC(=O)(CH_2)_4C(=O)NH(CH_2)_6NHC(=O)(CH_2)_4C(=O)\sim$	Fiber, molded products
Poly(ethylene terephthalate) (Dacron)	$HOCH_2CH_2OH$ 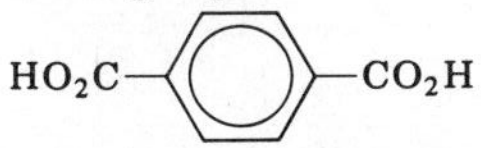	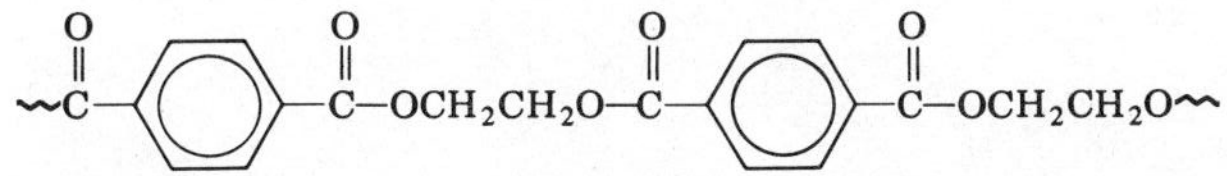	Fiber, fabric, films
Polycarbonate (Lexan)	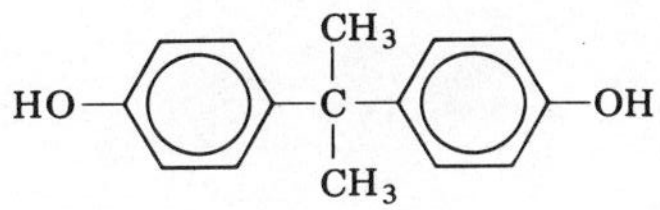 and $O{=}CCl_2$	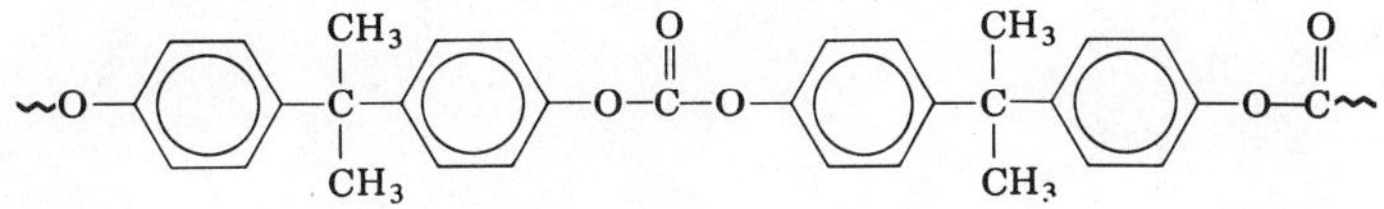	Clear bottles, molded products, substitute for glass
Cellulose (Natural fiber)	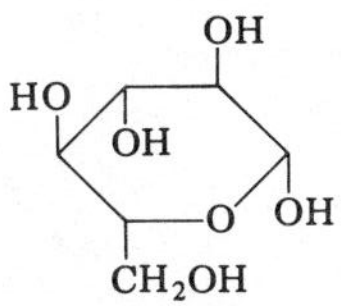 (Glucose)	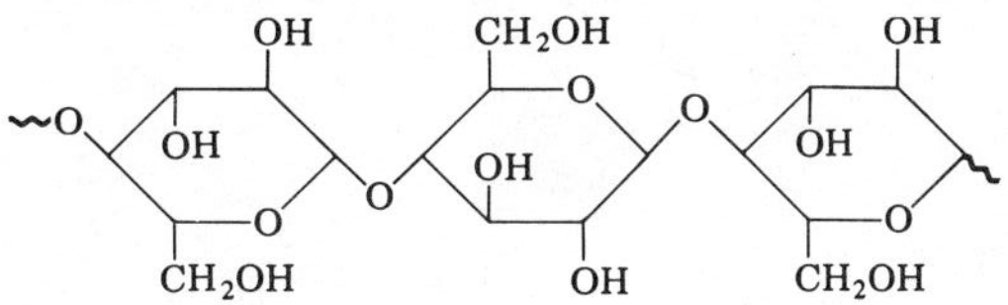	Cotton, wood
Phenol-Formaldehyde Resin (Bakelite)	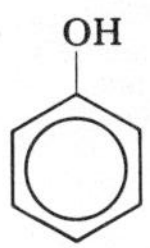 and $O{=}CH_2$	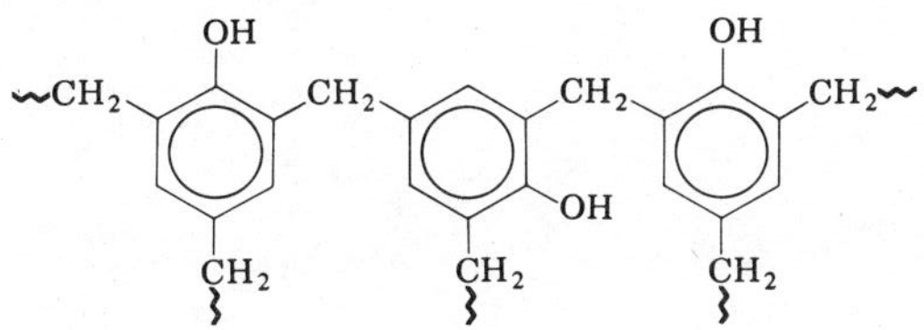	Thermosetting molded products, adhesives
Poly(dimethylsiloxane) (Silicone)	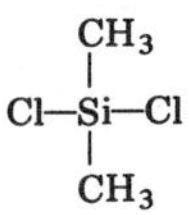 and H_2O	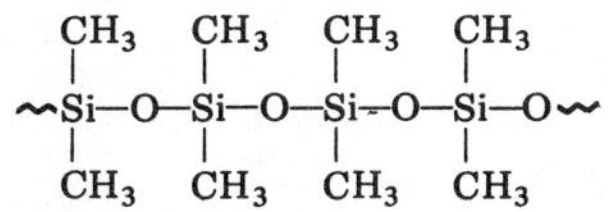	High-temperature fluids, rubber products

$$n\ \mathrm{A{=}B} \longrightarrow \mathrm{-\!\!\!+A{-}B+\!\!\!-}_n$$

Monomer — *Addition polymer*

$$n\ \mathrm{H_2C{=}CH_2} \longrightarrow \mathrm{-\!\!\!+CH_2{-}CH_2+\!\!\!-}_n$$

Ethylene monomer — *Polyethylene polymer*

multiple bonds. Familiar examples of addition polymers include polyethylene and polystyrene.

Addition polymerizations are chain reactions involving a chain-propagating species that is either an ion or a radical. Polymers formed from vinyl monomers are the most common examples of this type of polymer. A typical example of an addition polymerization involving a radical chain-propagating species is the benzoyl peroxide initiated polymerization of methyl methacrylate. The benzoyl

Initiation

Propagation

Termination

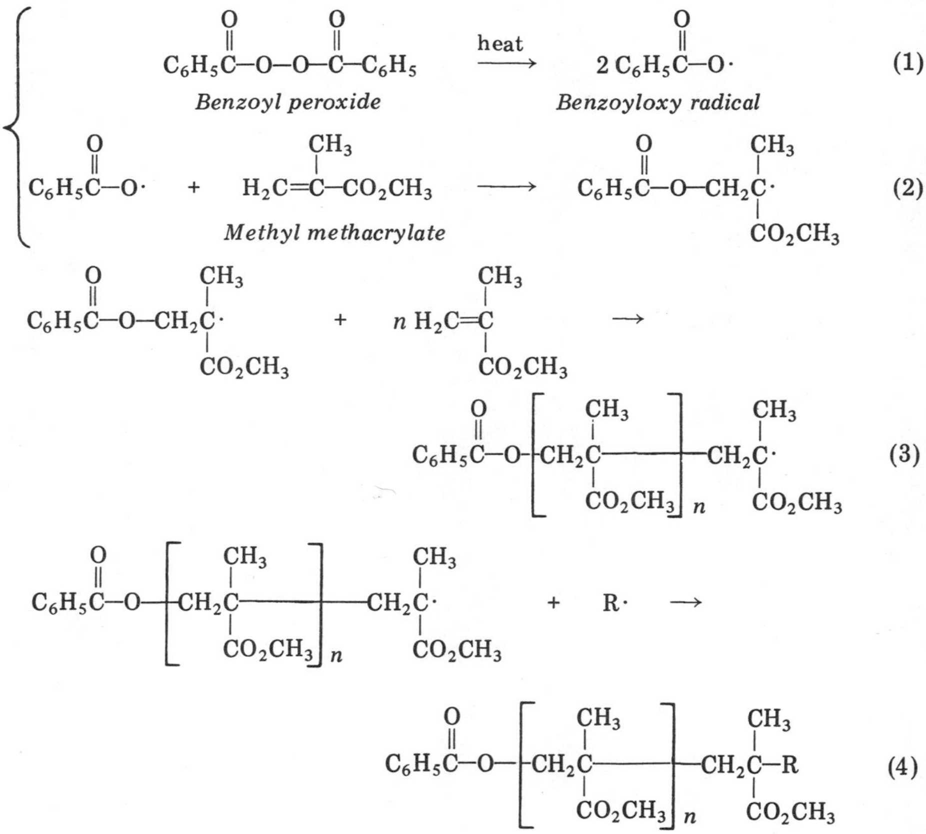

peroxide initiator is decomposed thermally (80 to 100°C) to form a reactive benzoyloxy radical (Reaction (1)). This radical adds to the carbon-carbon double bond of a methyl methacrylate monomer to form a new reactive free radical species (Reaction (2)). This carbon radical then adds to another monomer in the chain-propagating step and this process produces a growing polymer chain. Growth continues until the chain reaction is terminated, typically by two growing chains reacting with one another by radical combination to form a polymer molecule (Reaction (4)). In this experiment you will prepare polymethyl methacrylate and polystyrene by addition polymerization reactions.

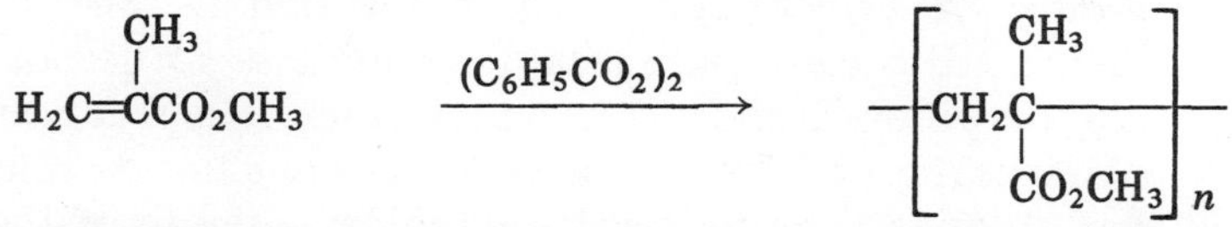

Methyl methacrylate
M.w. = 100.1 g/mol
b.p. = 100°C
d_{20} = 0.94 g/ml

Benzoyl peroxide
M.w. = 242.2 g/mol
m.p. = decomposes

Polymethyl methacrylate

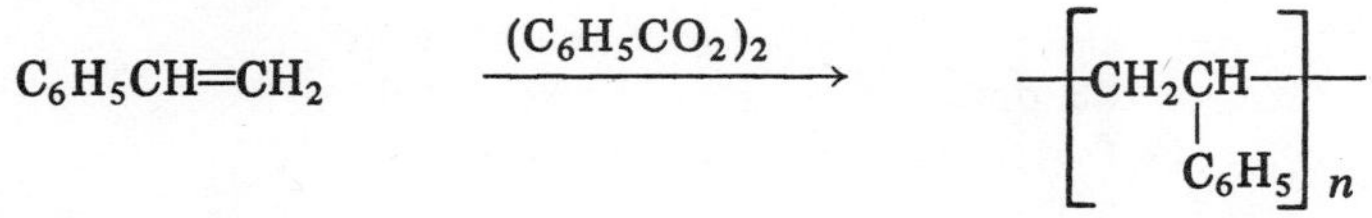

Styrene
M.w. = 104.2 g/mol
b.p. = 145°C
d_{20} = 0.90 g/ml

Polystyrene

Condensation Polymers

Condensation polymers are formed by reactions of bifunctional or polyfunctional molecules with the elimination of some small molecule (such as water, methanol, ammonia, or hydrogen chloride) in the polymerization process. The covalent bonds formed between monomer units are commonly the result of the formation of ester, amide, or acetal linkages between monomer units. Nylons and polyesters are familiar examples of polymers that are synthesized by condensation reactions. The synthesis of a polyester, poly(ethylene terephthalate), is illustrated in the following equation:

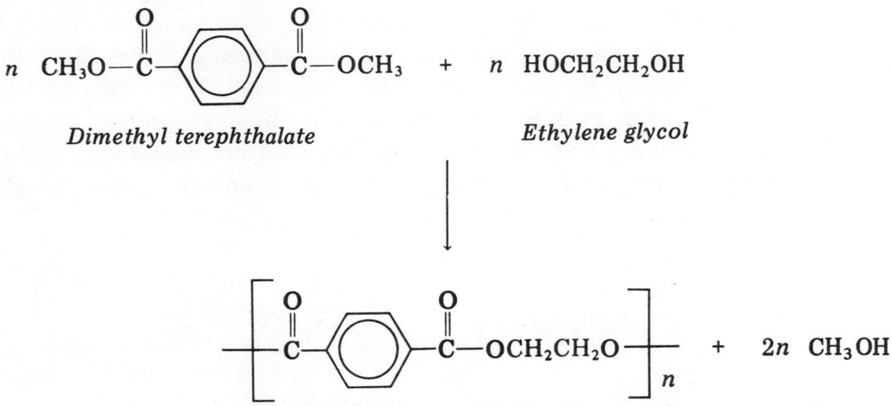

Dimethyl terephthalate — *Ethylene glycol* — *Poly(ethylene terephthalate)* — *Methanol*

Condensation polymerizations are stepwise reactions, and the polymer chain growth proceeds progressively by discrete condensation reactions—first forming low molecular weight materials and then forming higher molecular weight materials by further condensation reactions of the terminal functional groups. A typical condensation reaction is the formation of nylon 6.6 (so designated because each monomer contains six carbons) by the condensation of adipoyl chloride and 1,6-diaminohexane. The reaction proceeds by the stepwise condensation of the amine with the acid chloride resulting in the loss of hydrogen chloride. The terminal acid chloride group of the polymer chain can react

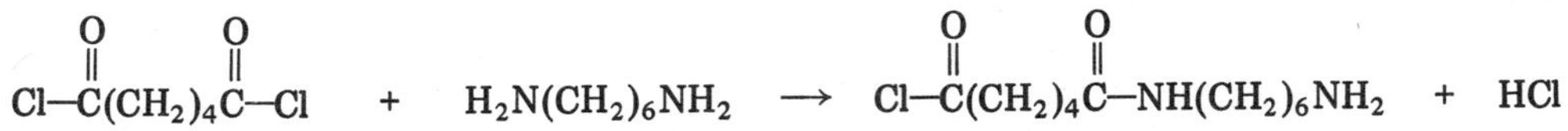

Adipoyl chloride
M.w. = 183 g/mol
b.p. (12 Torr) = 126°C

1,6-Diaminohexane
M.w. = 116 g/mol
b.p. = 204°C

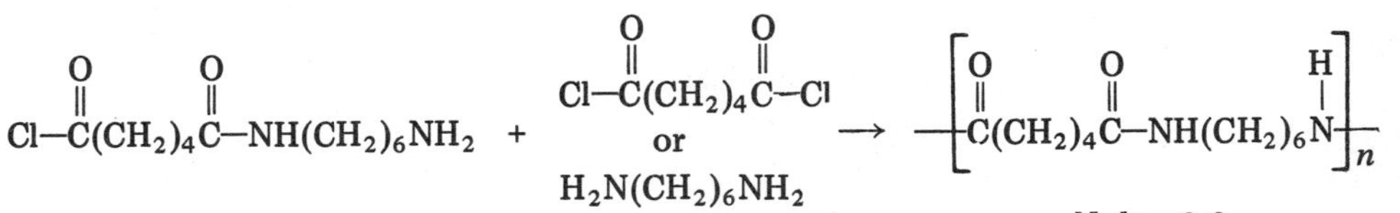

Nylon 6.6

with an amino group either from 1,6-diaminohexane or from another polymer chain. Polymer chain growth stops when the concentration of either reactive terminal group becomes too low or when side reactions, such as hydrolysis of the acid chloride terminal group, prevent further condensation. In this experiment you will prepare nylon 6.6 by the condensation polymerization of adipoyl chloride with 1,6-diaminohexane. In addition, you will prepare two different polyesters by the condensation of phthalic anhydride with either ethylene glycol or with glycerol. Since glycerol has three reactive hydroxyl groups a much more complex polymer structure will be formed.

Polymer molecules can be formed into either linear chains of monomer units, branched chains of monomer units, or the polymer chains may be cross-linked to form a complex molecular network (See Figure 38.1). The cross-linked polymers are generally higher melting, more rigid, and less soluble in solvents than linear or branched polymers. The condensation of phthalic anhydride with ethylene glycol forms a linear polymer, but the condensation of phthalic anhydride with glycerol forms a cross-linked polymer. You will compare the physical properties of the linear and the cross-linked polyesters in this experiment.

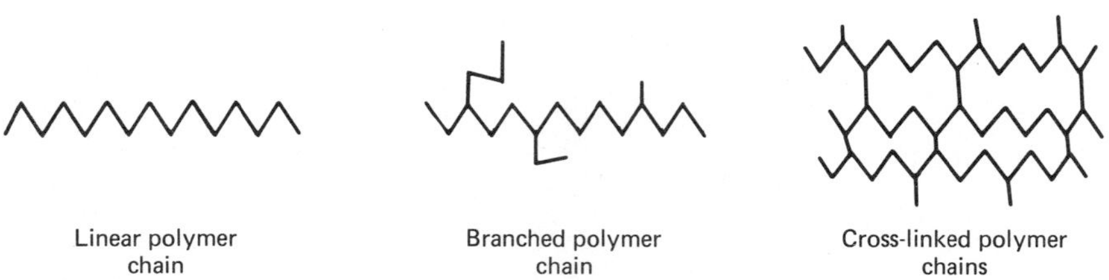

Figure 38.1 *Linear, branched, and cross-linked polymers.*

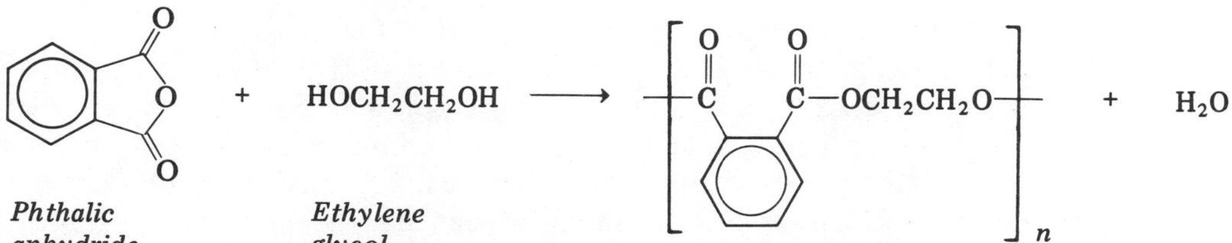

Phthalic anhydride M.w. = 148 g/mol m.p. = 132°C; *Ethylene glycol* M.w. = 62.1 g/mol b.p. = 198°C; *Poly(ethylene phthalate)*; *Water*

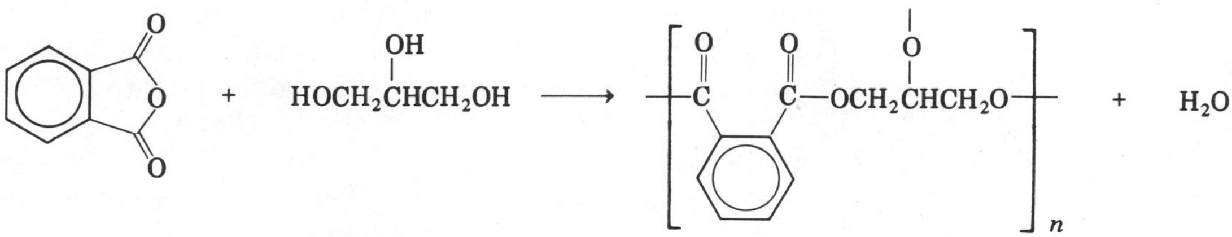

Glycerol M.w. = 92.1 g/mol b.p. = 290°C (decomp.); *Glyptal polymer*

Polymers that may be softened or melted by heat and reformed (molded) into another shape are called *thermoplastics.* Thermoplastics are linear or branched chain polymers, and polyethylene and nylon are familiar examples of thermoplastic materials. *Thermosetting polymers* are materials that permanently harden upon heating in the final stage of polymerization through the formation of cross-links between the polymer chains. These cross-linked polymers cannot be softened by heating and remolded without destroying the polymer.

Analysis of Polymers

The physical properties of polymers are dependent on the average molecular weight of the polymer molecules, the molecular weight distribution of the polymer molecules, the method of preparing the polymer, the degree of cross-linking or branching of the polymer chains, as well as the structure of the monomer units used to prepare the polymer. Thus, characterizing and analyzing polymers can be much more complex than the analysis of low molecular weight compounds. Many polymers do not normally crystallize in the solid state (a very highly ordered solid structure) but rather form an amorphous glass (a disordered solid structure). A true melting point cannot be determined for these glasses, but a transition from the hard, brittle solid to soft semi-fluid liquid can be observed over a temperature range of 5 to 20°C. This temperature range is called the *softening range* of the polymer. The softening range of amorphous polystyrene is 100 ± 10°C, for example, but the melting point of crystalline polystyrene is 230°C.

Infrared spectroscopy is very useful for characterizing polymers. Many polymeric materials can be formed into a thin (~ 0.01–0.03 mm) film, and in this form the infrared spectrum of the polymer may be obtained by placing the film directly in the spectrometer sample beam in a suitable holder. The infrared spectrum of polystyrene (Figure 12.20) was obtained by this technique. Polymer films for infrared analysis can be obtained by the use of a molding press, but a more general technique is to cast them from solution onto a sodium chloride plate. To do this the polymer is dissolved in any reasonably volatile solvent; the solution is then poured onto the salt plate; and the solvent is evaporated by gentle heating or under reduced pressure. A reasonably uniform polymer film is formed on the plate, and a spectrum of the polymer can be obtained by mounting the salt plate in the spectrometer with the holder shown in Figure 12.14. The solvent chosen for casting thin films by this technique should be able to dissolve the polymer (a 1 to 5% solution is sufficient) and be readily evaporated with gentle heating. Heating may be necessary to dissolve the polymer. Acetone, 2-butanone, or tetrahydrofuran are useful solvents for polymers such as polyesters that are high in oxygen content. Vinyl polymers like polystyrene can usually be dissolved in solvents such as 2-butanone, cyclohexane, or toluene. Polyamides (nylons) are insoluble in most common organic solvents, but will dissolve in boiling *N,N*-dimethylformamide. However, this solvent is not very volatile (b.p. = 178°C), and evaporation of the solvent must be performed under reduced pressure. Care must be taken when warming the salt plates to evaporate the solvent since rapid heating will cause the salt plate to crack.

Infrared spectra of solid polymers can also be obtained by using the potassium bromide pellet technique that is described in Experiment 12. This method of sample preparation is especially useful for polymeric materials that are insoluble in common solvents. However, when employing this technique, care must be taken to grind the polymer and potassium bromide to a fine powder before pressing the pellet; otherwise, the particles of polymer will scatter the infrared radiation.

Prelab Questions

1. Write equations showing the chain of reactions involved in the benzoyl peroxide polymerization of styrene.

2. Draw the structure of the polyester formed in the reaction of phthalic anhydride with glycerol showing at least three or four polymer units to illustrate the network structure of the polymer.

3. Prepare a table listing the number of moles of each reagent used in each of the suggested procedures in this experiment. (a) What is the limiting reagent in each experiment? (b) Why is the mole ratio of phthalic anhydride to ethylene glycol different than the mole ratio of phthalic anhydride to glycerol?

4. Why is sodium hydroxide added to the aqueous solution of 1,6-diaminohexane in the preparation of nylon 6.6?

5. If the average bond length for a carbon-carbon bond is 1.54 Å and the average C—C—C bond angle is 112° in polyethylene, what is the maximum length of a polymer chain of polyethylene that contains 1×10^6 monomer units?

6. What are the essential differences between a *bulk polymerization* [polymethyl methacrylate and poly(ethylene phthalate)], a *solution polymerization* [*polystyrene*], and an *interfacial polymerization* [nylon 6.6]?

EXPERIMENTAL PROCEDURE*

Polymerization of Methyl Methacrylate (*Bulk Polymerization*). Place 8 g of freshly distilled methyl methacrylate in an 18 × 150-mm test tube, add 10 to 20 mg of benzoyl peroxide, and swirl the contents to

---

CAUTION: Benzoyl peroxide is explosive and may detonate upon heating or with friction. Weigh out the amount required on glassine paper; do not use a metal spatula.

---

dissolve the solid peroxide. Clamp the test tube in a beaker of boiling water, insert a thermometer into the test tube, and heat the solution until rapid polymerization occurs. Record any temperature changes that occur, but remove the thermometer from the test tube before the polymer solidifies. Allow the polymer to cool. Observe the behavior of a portion of the polymer upon reheating over a flame, and determine the solubility of the polymer in boiling acetone, cyclohexane, and methanol.

A sample of the methyl methacrylate polymer prepared above (or a commercial sample of the same polymer—*Leucite* or *Plexiglass*) may be depolymerized by heating. Place a 5-g sample of the methyl methacrylate polymer in the small-scale distillation apparatus (Figure 11.1) and immerse the collection tube in an ice bath. Heat the flask with a small luminous flame or heating mantel. At about 300°C the polymer softens and rapidly depolymerizes. Continue heating until only a small dark residue remains. Weigh and then redistill the liquid that condensed in the receiver. Identify this product from its physical or spectral properties and calculate the percentage yield.

Polymerization of Styrene (*Solution Polymerization*). Mix 1.0 g of

*Several of these experiments should be run simultaneously to make efficient use of laboratory time.

inhibitor-free styrene* and 6 ml of toluene in an 18 × 150-mm test tube. Add 40 to 50 mg of benzoyl peroxide [*CAUTION: See previous warning.*] and mix to dissolve the solid. Loosely cork the test tube with a one-hole stopper and clamp the tube in a water bath at 75 to 85°C. Heat the solution at this temperature for 1 to 2 hr, and then cool it to room temperature. Pour this cool solution into 15 ml of methanol in a small beaker with rapid stirring. Allow the precipitate to coagulate, and then decant the liquid from the solid. Wash the solid with an additional 15-ml portion of methanol. Decant the methanol, and allow the solid polymer to air dry on a piece of filter paper. Weigh the polymer and calculate a percentage yield. A polystyrene film for infrared analysis may be obtained by the solvent casting technique.

Place a small portion of the polystyrene polymer on a watch glass, and melt the polymer over low heat. Allow the melt to cool, and note the appearance of this material.

Preparation of Nylon 6.6 (Interfacial Polymerization). Into a 18 × 150-mm test tube add 8 ml of an aqueous solution that is 0.50*M* in 1,6-diaminohexane and 0.50*M* in sodium hydroxide. Carefully add 8 ml of a 0.30*M* solution of adipoyl chloride in cyclohexane to the test

---

CAUTION: Adipoyl chloride and 1,6-diaminohexane are both toxic and should neither be inhaled nor come in contact with your skin.

---

tube by slowly pouring the adipoyl chloride solution down the wall of the slightly tilted tube. Two layers should form, and there should be immediate formation of a polymer film at the liquid-liquid interface. Using a piece of steel wire with a hook bent into one end, hook the polymer mass near the center and slowly raise the solid polymer mass so that a continuous "rope" of polymer is formed. Transfer the polymer

---

CAUTION: Do not handle the polymer with bare hands. It may contain unreacted reagents.

---

*Commercial styrene has an alkylated phenol added to inhibit spontaneous polymerization. If this inhibitor has not been removed by your instructor, it may be removed by passing ~2 ml of styrene through a column of 2 g of dry alumina contained in a disposable pipet with a small cotton plug. Collect and use only the first milliliter of styrene that elutes from the column. A rubber bulb can be used on top of the pipet to provide moderate pressure to force the liquid through the alumina.

rope directly into a beaker containing 50 ml of 5% aqueous sodium bicarbonate. Rinse the rope several times with water, and then place the polymer on a piece of filter paper to dry. Vigorously stir the remainder of the reagents in the test tube to form additional polymer. Decant the cyclohexane and aqueous solution from the solid polymer, and discard the liquid in the waste container that is provided. Thoroughly wash the solid polymer with 5% aqueous sodium bicarbonate and then with water. Dry the polymer on a piece of filter paper. Weigh the two batches of dry polymer and calculate a percentage yield.

A small amount of this washed and dried polymer can be melted in a test tube by gentle heating. A nylon fiber can be drawn from the melt by touching the surface of the melted polymer with a clean glass rod and then slowly withdrawing the rod with the fiber attached.

Preparation of Polyesters. Place 3.0 g of phthalic anhydride in each of two 18 × 150-mm test tubes. To one tube add 1.3 g (1.2 ml) of ethylene glycol, and to the second add 1.3 g (1.0 ml) of glycerol. Clamp both tubes so that they may be heated with a flame simultaneously, and then heat the tubes gently to 150 to 180°C. Vapor will evolve during the heating process. Continue heating an additional 5 minutes, and gradually raise the temperature to ~200°C during this heating period. Allow the tubes to cool and then determine the physical properties (appearance, softening point, and solubilities) of the two polymers. Obtain an infrared spectrum of the polyester formed from phthalic anhydride and ethylene glycol.

Postlab Questions

1. Is the polymerization of methyl methacrylate endothermic or exothermic? Why must this reaction be heated?

2. What is the identity of the major product formed upon the thermal depolymerization of polymethyl methacrylate? How does this compound form?

3. Assign the major absorption peaks in the infrared spectrum of polystyrene (Figure 12.20) to the corresponding functional groups in the polymer.

4. Compare the appearance of the polystyrene polymer obtained by solvent precipitation with the appearance of the same polymer after it had been heated to its melting point. Is polystyrene a thermoplastic material?

5. Why did the nylon 6.6 polymer form only at the interface between the two liquid phases?

6. Compare the physical properties of the two polyesters that you prepared. Explain the differences that you observed based on the structures of the two polymers.

Experiment Thirty-nine

Multistep Organic Syntheses: Formation and Reactions of Diethyl α-Acetoglutarate

Diethyl α-acetoglutarate with its β-carbonyl and two carboxylate groups is useful in numerous syntheses of cyclic organic compounds. The methyl group of diethyl α-acetoglutarate is suitably positioned for intramolecular condensation that would result in the formation of a 6-membered ring product. In

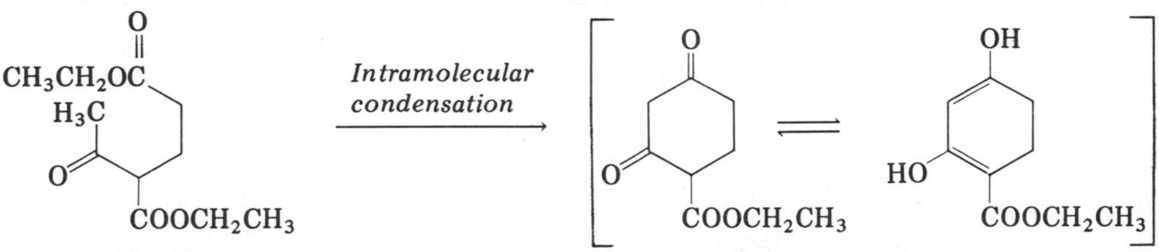

Diethyl α-acetoglutarate — *Tautomeric equilibrium of 1,3-cyclohexanedione derivative*

this series of experiments, you will utilize this capability for the preparation of a fused tricyclic organic compound (I). Careful examination of I will indicate

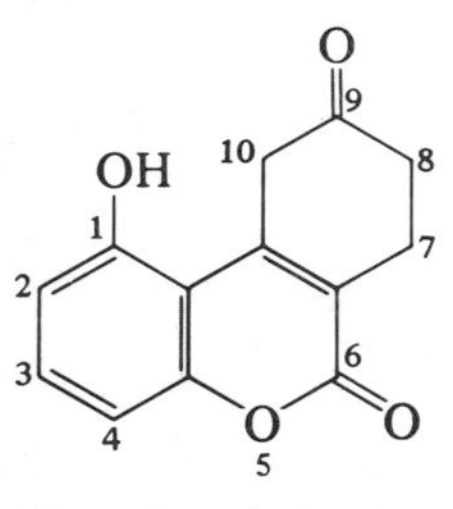

7,10-Dihydro-1-hydroxy-6H-dibenzo[b,d] *pyran-6,9(8H)-dione**

I

its construction from diethyl α-acetoglutarate and resorcinol.

*The systematic name for this compound is based on 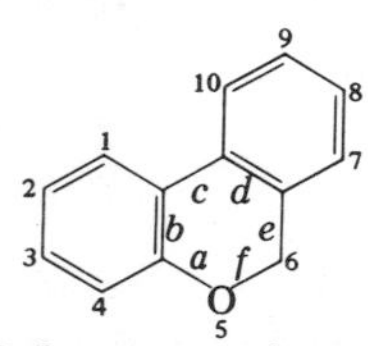 6H-Dibenzo[*b*,*d*]pyran

Ring fusion is distinguished by lettering the sides of the pyran ring with sequential letters (*a*, *b*, *c*, and so forth) and denoting the sides where fusion occurs by the letters in brackets.

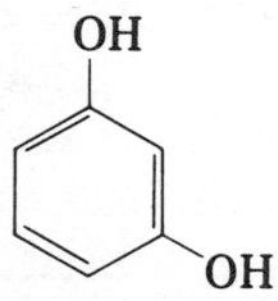

Resorcinol

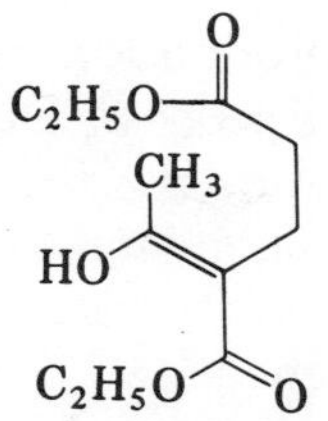

Diethyl α-acetoglutarate

The first step in the reaction sequence designed for the preparation of **I** is the synthesis of diethyl α-acetoglutarate by Michael addition of ethyl acetoacetate to ethyl acrylate:

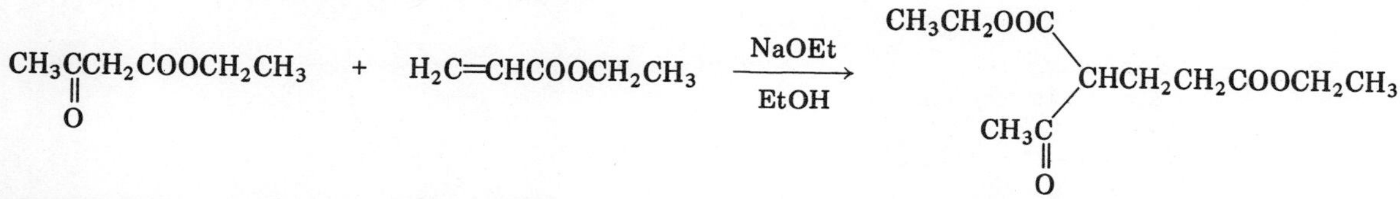

Ethyl acetoacetate
M.w. = 130.2 g/mol
b.p. = 180° C
$n_D^{20°C}$ = 1.4194

Ethyl acrylate
M.w. = 100.1 g/mol
b.p. = 100°C
$n_D^{20°C}$ = 1.405

Diethyl α-acetoglutarate
M.w. = 230.3 g/mol
b.p. = 271–222°C*
(110–112°C at 0.1 Torr)
$n_D^{19°C}$ = 1.4420

Ethyl acetoacetate is a relatively strong carbon acid (pK_a = 11). The stronger base, sodium ethoxide, abstracts a proton from the most acidic position of ethyl acetoacetate (see Experiment 29) to form the weaker conjugate base, sodium ethyl acetoacetate, and the weaker conjugate acid, ethanol (pK_a = 18). Conjugate addition of sodium ethyl acetoacetate to ethyl acrylate (*Michael reaction*) follows formation of this strong nucleophile. Proton transfer resulting

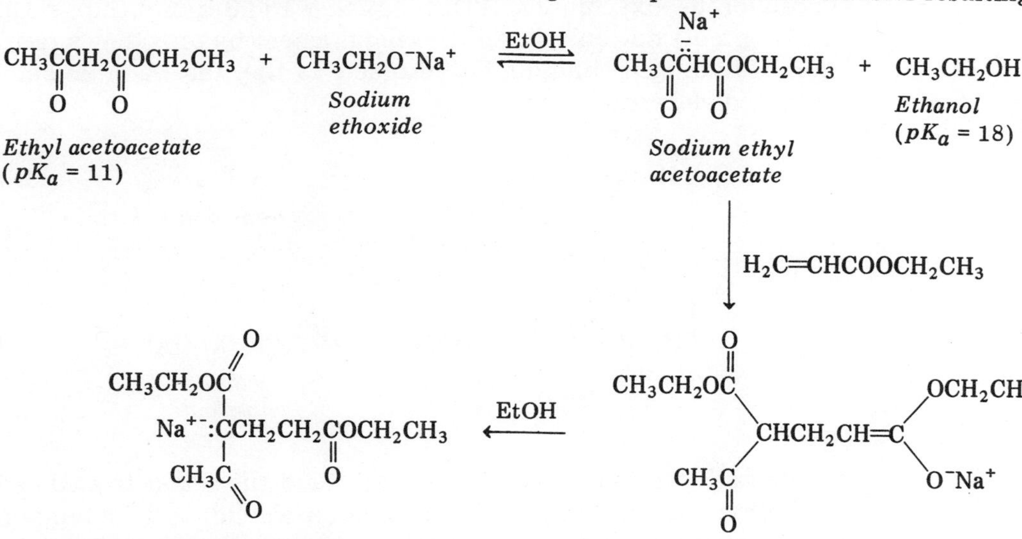

*Decomposes slowly at this temperature.

in the production of the sodium salt of diethyl α-acetoglutarate completes the reaction process. Diethyl α-acetoglutarate is isolated only after acidification of the reaction mixture.

The condensation of diethyl α-acetoglutarate with resorcinol (*von Pechmann condensation*) requires the use of a mild acid catalyst that will activate the β-ketoester for electrophilic substitution on the nucleophilic benzene ring.

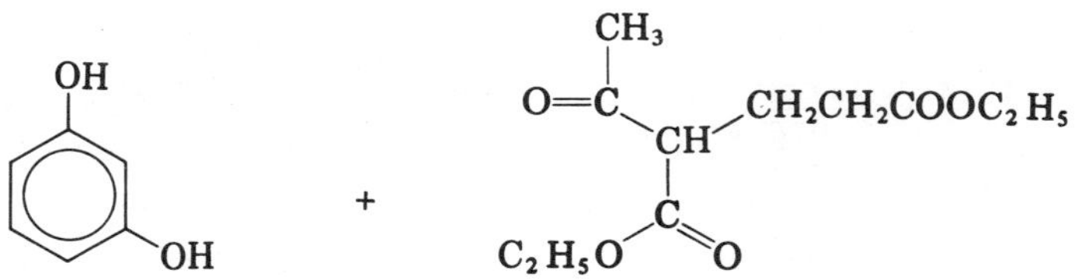

Resorcinol
M.w. = 110.1 g/mol
m.p. = 111°C
b.p. = 281°C

Diethyl α-acetoglutarate

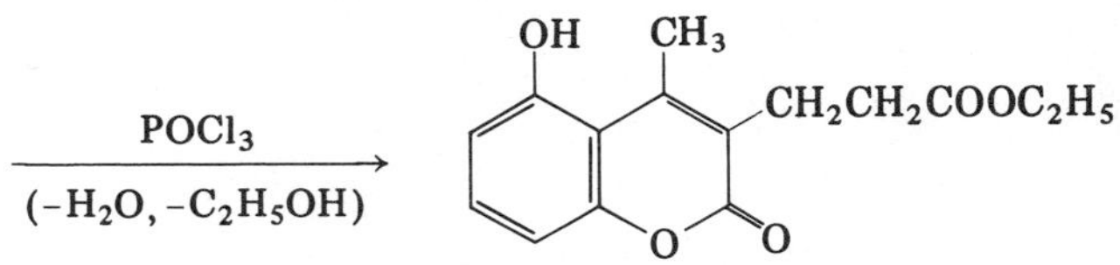

Ethyl 5-hydroxy-4-methylcoumarin-3-propionate
M.w. = 276.3 g/mol
m.p. = 108–110°C

Phosphorous oxychloride ($POCl_3$, M.w. = 153.3 g/mol, b.p. = 105°C, $d_{20°C}$ = 1.675 g/mol) not only fulfills this requirement by providing a regulated generation of hydrogen chloride, but also acts to trap the water and ethanol formed in this condensation process:

$$3\,H_2O + POCl_3 \longrightarrow H_3PO_4 + 3\,HCl$$

Phosphorus oxychloride → *Phosphoric acid*

$$3\,CH_3CH_2OH + POCl_3 \longrightarrow (CH_3CH_2O)_3PO + 3\,HCl$$

Phosphorus oxychloride → *Triethyl phosphate*

Condensation occurs between resorcinol and the acid-activated carbonyl group of diethyl α-acetoglutarate to produce an alcohol that undergoes elimination and intramolecular transesterification (see Experiment 29) under the acidic conditions employed for this reaction:

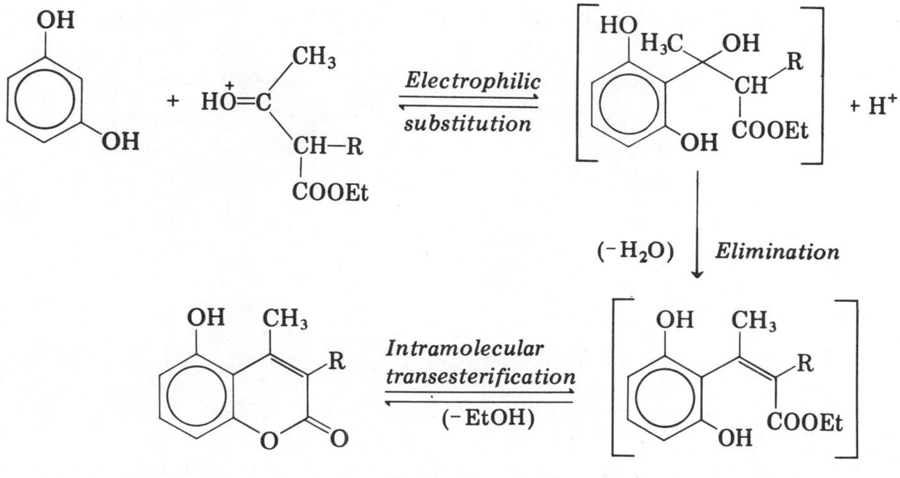

The von Pechmann condensation terminates at the coumarin stage so that a subsequent reaction must be employed to complete the synthesis of I.

The final reaction step in this experiment is an intramolecular condensation of ethyl 5-hydroxy-4-methylcoumarin-3-propionate. Hydrogens bonded to the

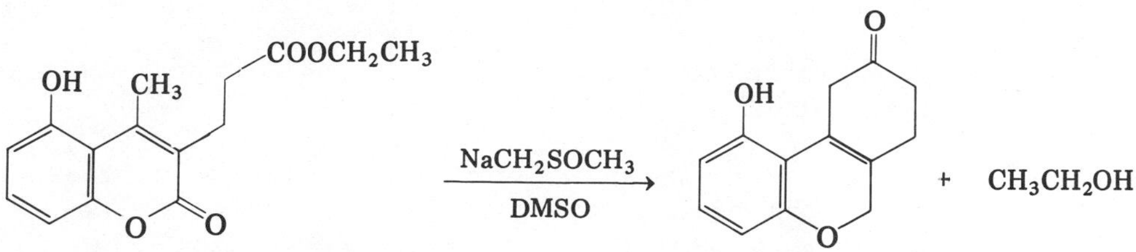

Ethyl 5-hydroxy-4-methylcoumarin-3-propionate

7,10-Dihydro-1-hydroxy-6H-dibenzo[b,d]*pyran-6,9(8H)-dione(I)*
M.w. = 230.3 g/mol
m.p. = 232–234°C

methyl group of this compound are uniquely acidic since proton removal produces a carbanion that is conjugated with the lactone carbonyl group:

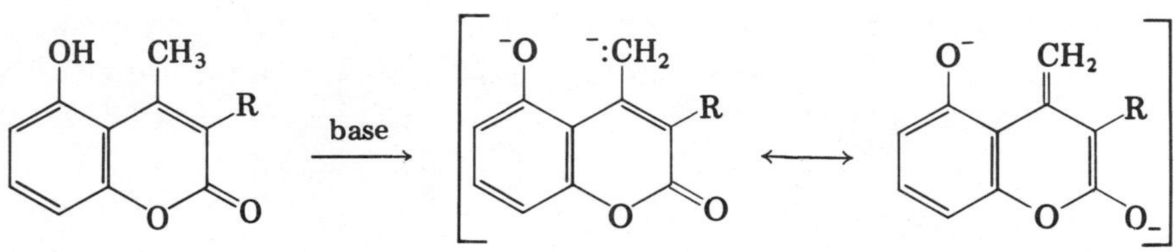

However, these hydrogens are only weakly acidic ($pK_a > 25$) and proton removal is not effective with the bases more commonly employed for condensation reactions. Consequently, the strong base sodium methylsulfinylmethylide (*dimsylsodium*), prepared by the reaction between sodium hydride

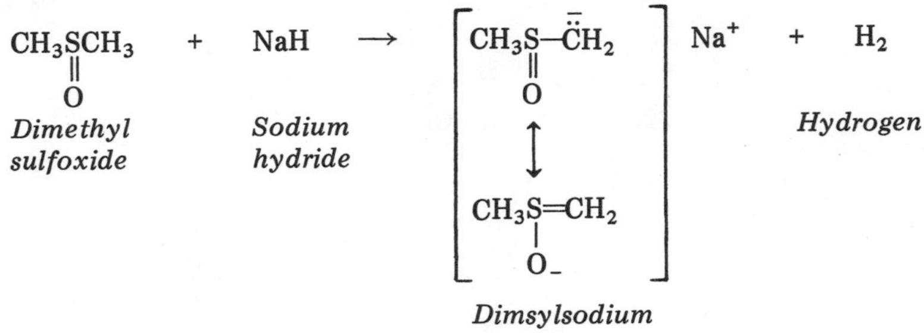

and dimethyl sulfoxide (DMSO, pK_a = 33.5), will be used in this experiment.

Prelab Questions

1. Obtain from the *Merck Index* the pertinent physical properties and toxicological data for ethyl acrylate, phosphorus oxychloride, and dimethyl sulfoxide.

2. Why is sodium ethoxide and not sodium hydroxide used for the synthesis of diethyl α-acetoglutarate?

3. Could a base such as sodium ethoxide have been used instead of phosphorus oxychloride for the synthesis of ethyl 5-hydroxy-4-methylcoumarin-3-propionate? Why?

4. What is the minimum number of molar equivalents of dimsylsodium required for the synthesis of I?

5. Outline the procedures for each step in this experiment. List each precaution for reference while you are performing this experiment. Provide flow diagrams of the purification procedures for each synthetic step in this experiment.

EXPERIMENTAL PROCEDURE

Preparation of Diethyl α-Acetoglutarate. Prepare an ethanolic solution of sodium ethoxide by slowing adding 1.2 g of freshly cut 2 × 2-mm pieces of sodium metal (0.052 mol) to 20 ml of absolute ethanol in a previously dried 100-ml round-bottom flask fitted with a dry reflux

---

CAUTION: Sodium metal reacts violently with water. When working with sodium metal, all skin contact with this reagent must be avoided.

---

condenser. (Sodium is a soft metal that can be cut with a spatula while holding the piece of sodium with forceps. Freshly cut sodium has a bright metallic sheen. Place the unused cuttings and residual sodium back into the container from which it was obtained. Use forceps for all transfers of sodium.) Sodium reacts vigorously with ethanol and forms white balls of the metal as the reaction proceeds. The spheres of sodium gradually dissipate with a mild exothermic reaction.

CAUTION: The reaction of sodium with ethanol produces hydrogen. Be certain that there are no flames in your vicinity. The preparation of sodium ethoxide should be performed in a well-ventilated area or in a fume hood.

When sodium metal is no longer visible in the ethanolic reaction solution, add 13.0 g of ethyl acetoacetate (0.10 mol) followed by 5.0 g of ethyl acrylate (0.050 mol) and heat the resulting solution to reflux on a steam bath. After refluxing for 2 hr, cool the reaction solution and pour the contents of the reaction flask into a separatory funnel that contains 60 ml of cold water. Wash the resulting mixture with three 30-ml portions of ether, dry the combined ether washings over anhydrous magnesium sulfate, and, after filtering, distill the ether from the solution at atmospheric pressure using a simple distillation apparatus. Dispose of the collected ether in the container provided for waste ether. Transfer the residue remaining after removal of ether to a side-arm distillation flask or to a 50-ml round-bottom flask and fractionally distill this residue under reduced pressure (Figure 14.2). Collect the fractions containing unreacted ethyl acetoacetate and diethyl α-acetoglutarate. (Diethyl α-acetoglutarate undergoes slight decomposition at its boiling point under atmospheric pressure.) Record the boiling point ranges for your distilled products and the pressure at which distillation occurred. Obtain the weight of diethyl α-acetoglutarate and calculate its percentage yield. The isolated product may be identified by ^{1}H NMR (Figure 39.1) and IR spectroscopy.

Preparation of Ethyl 5-Hydroxy-4-methylcoumarin-3-propionate. Add 5.0 g of diethyl α-acetoglutarate (0.022 mol) to a mixture of 2.2 g of resorcinol (0.020 mol) and 3.1 g (1.9 ml) of phosphorus oxychloride

---

CAUTION: Phosphorous oxychloride is corrosive and gives off irritating, toxic fumes. Measurement of this reagent and the operation of this reaction should be performed in the hood.

(0.02 mol) in a thoroughly dried 50-ml round-bottom flask fitted with

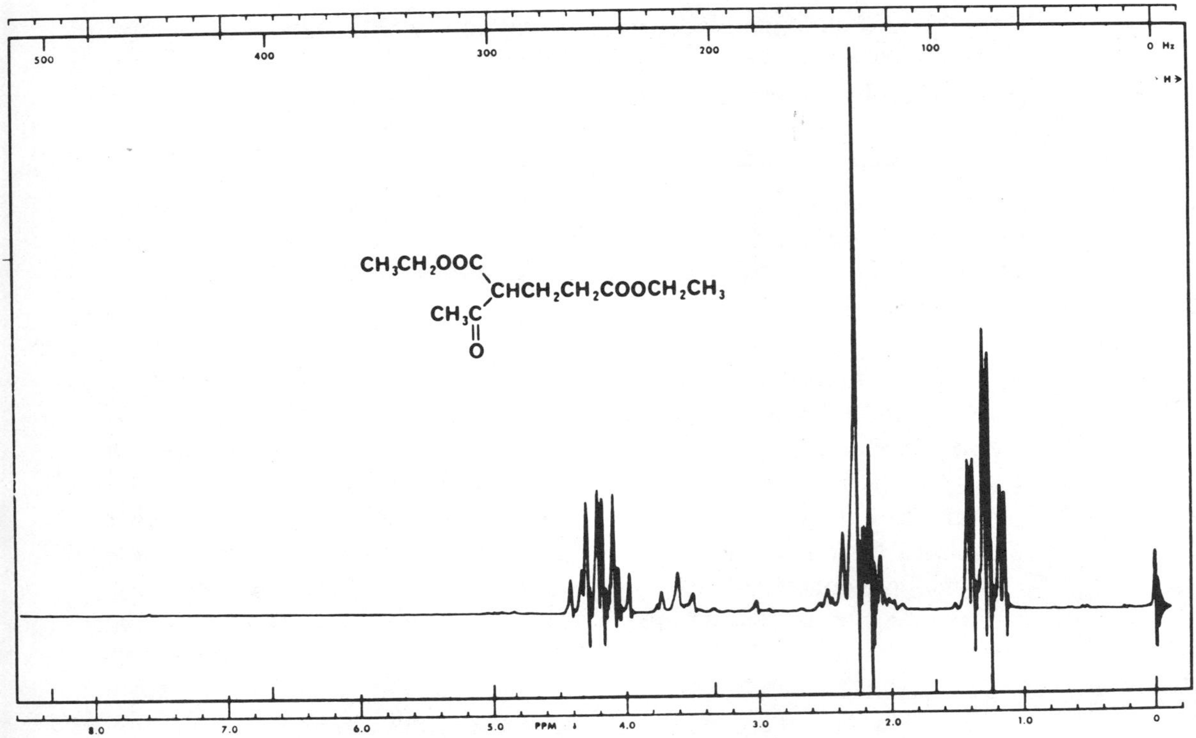

Figure 39.1 [1]*H NMR spectrum of diethyl α-acetoglutarate.*

a reflux condenser and protected from the atmosphere with a drying tube (containing anhydrous calcium chloride) or a piece of cotton. Mix the reactants thoroughly by swirling the contents of the reaction flask. If the reaction mixture becomes hot to your touch, cool the flask immediately in an ice bath. However, the reaction temperature should not be maintained below room temperature. Continue swirling the contents of the reaction flask for 60 minutes and, when the reaction mixture is at or below room temperature, remove the reflux condenser and place the drying tube or piece of cotton on the flask. Allow the reaction mixture to set until your next laboratory period (or for five days, whichever is greater). Dissolve the resulting dark solid in 80 ml of ether, wash the ether solution three times with 30-ml portions of water, and then dry the ether solution over anhydrous magnesium sulfate. Reduce the volume of ether by distillation in order to crystallize the product. If a crystalline solid does not form upon cooling the concentrated ether solution, add pentane as a cosolvent. If an oil forms upon cooling, add additional amounts of ether to redissolve the oil. Obtain the weight and melting point of your isolated product and calculate its percentage yield. The isolated product may be identified by ^{1}H NMR and IR spectroscopy.

Preparation of 7,10-Dihydro-1-hydroxy-6H-dibenzo[b,d]pyran-6,9-(8H)-dione (I). Place 2.40 g of a 50% mineral oil dispersion of sodium hydride (1.20 g NaH, 5.0 mmol) in a clean, dry, 50-ml round-bottom

---

CAUTION: Sodium hydride reacts violently with water. When working with sodium hydride, all skin contact with this reagent must be avoided.

---

flask. Wash the mineral oil from the sodium hydride by adding 10 ml of dry hexane to the dispersion, swirl the flask to ensure mixing, allow the sodium hydride to settle, and then decant the hexane from the sodium hydride. Repeat the washing procedure twice with 10-ml portions of dry hexane.

Add 2.8 g of ethyl 5-hydroxy-4-methylcoumarin-3-propionate (1.0 mmol) to the sodium hydride in the 50-ml round-bottom flask and mix the two reactants. Attach a reflux condenser to the round-bottom flask and cool the reaction flask to approximately 15°C in a cold water bath. Pour 10 ml of dry dimethyl sulfoxide* into the cooled reaction flask

---

CAUTION: The reaction of sodium hydride with dimethyl sulfoxide produces hydrogen. Be certain that there are no flames in your vicinity. This reaction should be performed in a well ventillated area or in a fume hood.

---

and, with swirling of the solution at regular intervals, maintain the reaction solution at 15°C for 1 hour. Allow the reaction temperature to rise to room temperature, replace the reflux condenser with a stopper, and store the reaction flask, preferably in a refrigerator, until your next laboratory period.

After warming to room temperature pour the reaction mixture into 100 ml of ice-cold 10% aqueous hydrochloric acid contained in a 250-ml Erlenmeyer flask. Thoroughly mix the solution at regular intervals over a 30-min period. Filter the solid obtained from the acidic solution by suction filtration and wash the solid with three 10-ml portions of water and then with 10-ml portions of a saturated sodium bicarbonate solution. Recrystallize the resulting solid from acetone. Obtain the

*Dry dimethyl sulfoxide is prepared by distilling the commercially available reagent (b.p. = 189° C, m.p. = 18° C) from calcium hydride (1 g CaH_2/50 ml DMSO) or by fractional distillation of dimethyl sulfoxide followed by drying with molecular sieves (D. R. Burfield and R. H. Smithers, *J. Org. Chem.* 43, 3966 (1978)).

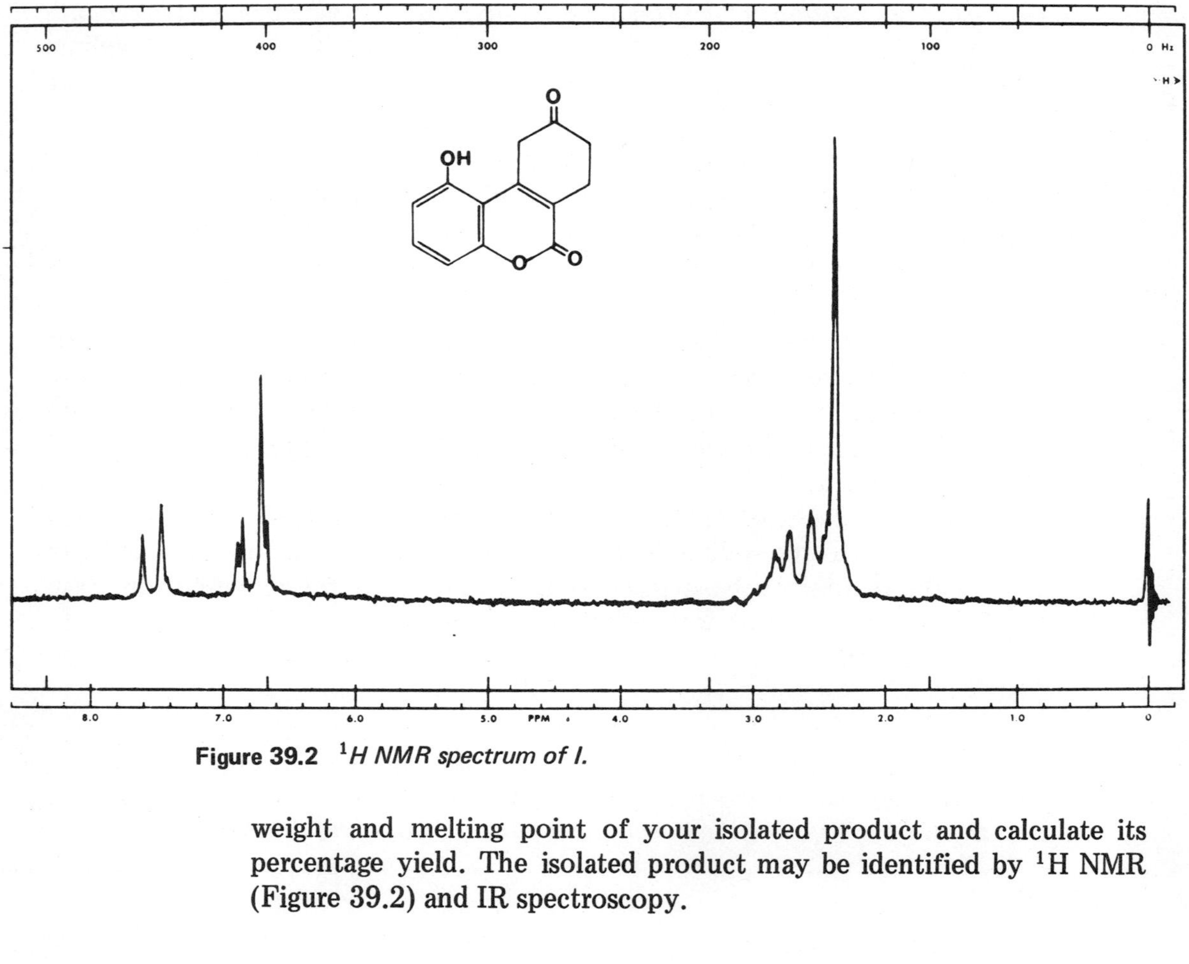

Figure 39.2 *¹H NMR spectrum of I.*

weight and melting point of your isolated product and calculate its percentage yield. The isolated product may be identified by ¹H NMR (Figure 39.2) and IR spectroscopy.

Postlab Questions

1. Why would the use of commercially available sodium methoxide or potassium *tert*-butoxide not be acceptable for the preparation of diethyl α-acetoglutarate in this experiment?

2. A twofold excess of ethyl acetoacetate is employed in the procedure for the preparation of diethyl α-acetoglutarate. Why is an excess of this reagent used? (*Hint:* When equimolar amounts of ethyl acetoacetate and ethyl acrylate are employed, the yield of diethyl α-acetoglutarate is substantially reduced. Why?)

3. If ethyl 5-hydroxy-4-methylcoumarin-3-propionate was contaminated with diethyl α-acetoglutarate, how could you separate these two compounds by simple extraction techniques? This procedure is particularly advantageous if you are not able to obtain a solid product in the procedure for the preparation of the coumarin derivative.

4. Calculate the overall yield of I from ethyl acrylate.

5. Assign the NMR absorptions of Figures 39.1 and 39.2 to the protons that produce these signals.

Experiment Forty

Use of the Organic Chemical Literature

The chemical literature makes available all of the collected knowledge that has been published concerning chemical compounds, chemical reactions, methods for chemical analyses, and chemical concepts and laws; furthermore, it details the progress of chemical discoveries. Through the use of this exceedingly valuable resource, the chemist can employ the results of prior investigations to design new experiments and develop new concepts. Proficient use of the chemical literature minimizes unnecessary repetition of experiments and allows a more rapid advance of chemical science.

In this experiment you will be asked by your instructor to use the chemical literature to plan the synthesis of an organic compound. Since the organic chemical literature is well organized, you will learn in a relatively short period of time whether or not a particular compound has been previously reported, the preferred method for synthesis of that compound if more than one synthetic method is available, and, if the compound is known, its physical and spectral properties. If the desired compound has not been previously reported, information about structurally similar compounds can often be obtained and employed for the design of a synthetic strategy for the unknown compound.

The literature of chemistry contains both *primary* and *secondary sources* of information. The primary sources are publications such as *research journals* and *patents* where chemists describe the original results of their research. The secondary sources include *review journals*, *series*, and *monographs* that review, cite, and evaluate results published in a specific area of chemistry, often over a limited time period. Other secondary sources of information include *handbooks* and *compilations* of physical and chemical properties. Since many different primary and secondary publications are available, *abstract journals* and comprehensive *indices* of the chemical literature are published to aid in searching the literature and to assist scientists in maintaining a current awareness of the developments in areas of interest.

Research Journals

Research journals comprise the most important source of original research reports. Publications in these journals are usually classified as *articles*, *notes*, or *communications.* Articles generally contain a brief historical background, a discussion of experimental results and detailed experimental procedures including physical properties for new compounds. Notes are usually shorter and of more limited scope than articles, but they also include detailed experimental procedures. Communications or letters are brief announcements of experimental results and do not contain the extensive discussion or detailed procedures found in articles or notes. Generally, communications and letters are used to rapidly announce the preliminary results of investigations.

Original manuscripts are usually evaluated prior to publication by other chemists (referees) familiar with the area of research described in the manuscript. However, on occasion incorrect data or conclusions are published; thus one should always read the literature critically and, if possible, compare results and conclusions from several different authors.

Several thousand primary journals are currently published each year. Examples of major journals that include coverage of organic chemistry are: *Journal of the American Chemical Society*, *Journal of Organic Chemistry*, *Journal of the Chemical Society: Perkin Transactions I and II*, *Chemische Berichte*, *Tetrahedron*, *Tetrahedron Letters*, *Synthesis*, and *Angewandte Chemie* (*International Edition*).

Patents

The chemical patent literature is also a major source of original chemical information. A patent is a government grant to an inventor for a specific period of time giving the inventor the exclusive right to manufacture or use an invention. According to the U.S. Constitution the purpose of a patent is: "to promote the progress of science and useful arts by securing for limited times to authors and inventors the exclusive rights to their respective writings and discoveries." It is estimated that up to 25% of the over 4,000,000 patents issued by the United States government are chemical patents.

Patent documents are primarily legal documents, but in order to obtain the exclusive rights granted by the government, the inventor must make full disclosure of his discovery in the patent document. This disclosure, which is designed "to promote the progress of science and useful arts," is what makes patent documents useful sources of information to other scientists. Patents tend to be most useful as sources of information to chemists working in industry and in highly technical fields of research. *Chemical Abstracts*, which is discussed in detail in a later section of this experiment, publishes abstracts of patents and indexes patents from the major industrial nations.

Review Journals and Series

The large amount of original information that is published each year in research journals and patents make periodic reviews of this information essential. A review will generally cover a limited research field and will summarize, evaluate, and give references to the original publications. However, a review does not necessarily include reference to every paper published in a given field during the period of time covered by the review. Most journals that publish reviews contain annual subject and author indexes. A very useful volume, *Index of Reviews in Organic Chemistry*, is published periodically by the Chemical Society, London. Examples of journals that publish reviews of interest to organic chemists include: *Accounts of Chemical Research, Angewandte Chemie (Internat. Ed.)*, *Journal of Organometallic Chemistry, Chemical Reviews, Quarterly Reviews*, and *Synthesis.*

Reviews are also available in volumes published as series. Often the titles of these series include the words, *Advances in . . .* , or *Progress in . . . .* Each volume generally contains several reviews on different subjects written by different authors. An index is usually included in each volume and often a cumulative index for the series is published after every five or ten volumes. Examples of series that contain reviews that are of interest to organic chemists include: *Organic Reactions, Progress in Organic Chemistry, Advances in Organic Chemistry: Methods and Results, Annual Reports on the Progress of Chemistry* (Section B), *Progress in Physical Organic Chemistry, Advances in Heterocyclic Chemistry*, and *Advances in Organometallic Chemistry.*

There are also several series that review methods and procedures for the synthesis of organic compounds:

Organic Syntheses is published annually. The first 49 volumes have been edited and republished in five collective volumes. A comprehensive index to all five collective volumes is also available. *Organic Syntheses* contains detailed procedures for the preparation of over 1000 compounds. Unlike most other sources, procedures are independently tested before publication. Apparatus, reaction conditions, and purification procedures are described. Many of the syntheses illustrate general methods which can be used in the preparation of related compounds.

Methoden der Organische Chemie (Houben-Weyl-Müller) is a multivolume series written in German that describes general laboratory procedures and reactions of classes of organic compounds. The coverage in this series is encyclopedic, but specific procedures are given for the preparation of representative compounds of each class.

Reagents for Organic Synthesis is a series that organizes synthetic reactions by the reagent used for the synthetic transformation. Information on preparation and purification of solvents and reagents is provided as are general procedures and specific examples of organic reactions.

Technique of Organic Chemistry is a multivolume series that has reviewed experimental techniques and physical methods in organic chemistry. In 1971 this series was combined with *Technique of Inorganic Chemistry*, and the combined series is now titled *Techniques of Chemistry.*

Annual Reports in Organic Synthesis, which has been published each year since 1970, contains summaries of new, synthetically useful organic reactions and methods reported during the preceding year in selected chemical journals. Reports are organized by reaction type in each volume.

Monographs

A large number of monographs (books) on a particular subject in organic chemistry are published each year. The most efficient method of locating a book on a subject of interest is to look in the subject index of the card catalog in the library. If a pertinent monograph is not located by this method, it is often useful to scan the titles of books on the shelves in the library. This task is aided by your knowledge of the particular organization of the library. Most university and technical libraries in the United States use the Library of Congress (L.C.) System to classify and organize books and periodicals. In the L.C. System all library materials are divided into subject fields with each designated by a letter of the alphabet. Letters of fields most often of interest to chemists are: Q, science; R, medicine; and T, technology. A second letter is used to further subdivide the major field divisions. Thus, in the Q section, books related to chemistry are found in the QD subsection. Other subsections of interest to the organic chemist include: QP, biochemistry and physiology; RM, drugs; RS, medicinal chemistry; and TP, chemical technology. Subsections are further divided into more specific areas by the addition of numbers to the two-letter designation. Information classified as organic chemistry is located in the QD 248 to QD 449 classifications.

Handbooks

For quick reference to physical properties of common compounds, several handbooks and compilations of data are available. A few of the most common volumes of this type are the following:

Handbook of Chemistry and Physics, published in yearly editions by the Chemical Rubber Co., contains data on the molecular weights, boiling points, melting points, densities, solubilities, and other properties of over 14,000 organic compounds.

Lange's Handbook of Chemistry, published by McGraw-Hill Book Co., is similar to the preceding handbook but covers only about 7000 organic compounds.

Handbook of Tables for the Identification of Organic Compounds, The Chemical Rubber Company, contains data on over 8000 organic compounds arranged by functional group and increasing boiling point or melting point.

Merck Index, Merck and Co., contains physical data on nearly 10,000 chemicals, drugs, and biological substances. A useful cross-index of common names and trade names and a formula index is included in this work. An index of organic name-reactions is also included.

Dictionary of Organic Compounds (4th edition and supplements), published by Oxford University Press, is a multivolume encyclopedia of physical and chemical properties of organic compounds and their derivatives. This reference work includes citations to the original literature.

Kirk-Othmer, Encyclopedia of Chemical Technology (2nd edition, 1972; 3rd edition publication started in 1978, John Wiley and Sons), is a very useful reference that includes information on organic compounds manufactured by chemical industry.

The Chemist's Companion (A. J. Gordon and R. A. Ford, John Wiley and Sons, 1972) is a handbook of practical chemical data, techniques, and references. Especially useful sections on spectroscopy, chromatography, and experimental techniques are included.

Beilstein's Handbuch der Organische Chemié.

This German reference work is really more than a handbook of organic compounds. It is the most comprehensive secondary source of information on organic compounds covering the literature prior to 1930. The original *Beilstein's Handbuch* (The *Hauptwerk*) was published in 31 volumes and covered all known organic compounds up to the year 1910. The first supplement (*Erstes Erganzungwerk*) extended this coverage to 1920, and the second supplement (*Zweites Erganzungwerk*) further extended the coverage to 1930. The third supplement, which extends coverage to 1950, and the fourth supplement, which extends coverage to 1960, are not yet complete.

The order in which compounds appear in *Beilstein* is systematically arranged by compound structure. In the *Hauptwerk* and the various supplements the organization and division of the volumes has remained consistent*: Volumes (*Band*) I–IV cover aliphatic and substituted aliphatic compounds; Volumes V–XVI cover carbocyclic and aromatic compounds and their functional derivatives; Volumes XVII–XXVII cover hetrocyclic compounds and functional derivatives of these compounds; Volumes XXVIII–XXIX are index volumes; and Volumes XXX and XXXI, which were only published in the *Hauptwerk*, cover natural substances not previously assigned in the first 27 volumes. Within each volume, compounds are separated into classes on the basis of structural complexity and functional groups. The major functional group categories are hydrocarbons (*Kohlenwasserstoffe*), alcohols (*Oxyverbindungen*), aldehydes

*A complete description of the organization of *Beilstein* can be found in *A Brief Introduction to the Use of Beilstein's Handbuch der Organischen Chemie*, 2nd ed., by E. H. Huntress, John Wiley and Sons, New York, 1938. A more recent publication by O. Weissbach, *A Manual for the Use of Beilstein's Handbuch der Organischen Chemie*, Springer-Verlag, New York, 1976, is similarly useful as an English-language guide to the *Hauptwerk* and its supplements.

and ketones (*Oxoverbindungen*), carboxylic acids (*Carbonsäuren*), sulfinic acids (*Sulfinsäuren*), sulfonic acids (*Sulfonsäuren*), amines (*Amine*), other nitrogen compounds, organophosphorus compounds, organic compounds of arsenic, antimony, and bismuth, and organometallic compounds. Compounds containing halogen atoms, nitro, nitroso, or azide groups are listed as a derivative under the parent compound. Acid chlorides, anhydrides, esters, primary amides, and nitriles are listed as derivatives of the parent carboxylic acid. Secondary and tertiary amides are classified as derivatives of the corresponding amine. Ethers, peroxides, thiols, thioethers, and sulfones are classified as derivatives of the corresponding alcohol.

Fortunately, it is not necessary to master *Beilstein's* system of organization to use this reference work. With a rudimentary knowledge of German nomenclature, it is possible to locate a compound in *Beilstein* using the formula index (*General Formelregister, Band XXIX*) for the second supplement. This index is cumulative and gives reference to the *Hauptwerk* and the first and second supplements.

If, for example, you wish to find 5-bromosalicylaldehyde (5-bromo-2-hydroxybenzaldehyde) in *Beilstein*, you would look under $C_7H_5BrO_2$ in the formula index to the second supplement. Among other compounds listed by name under this formula you find the following:

5-Brom-2-oxy-benzaldehyd 8, 54, II 45

This listing tells you that information on 5-bromo-2-hydroxybenzaldehyde can be found in Volume 8, page 54, of the *Hauptwerk* and in the same volume of the second (II) supplement on page 45. (No references to this compound are in the first supplement.) On these pages references to the preparation of this compound are cited and the physical properties reported for the compound are listed. The references cited in *Beilstein* can generally be assumed to cover all significant literature references prior to 1930.

Access to the third (and fourth) supplement volumes, which do not yet have a formula index, can also be made by the use of the page numbers found in the *Hauptwerk* and previous supplements of *Beilstein*. In the example, note that at the center of the top of the page in the second supplement volume where 5-bromo-2-hydroxybenzaldehyde is listed there is a reference to the page (54) in the *Hauptwerk* where this compound was originally listed. This practice of reference to the page where compounds appear (or would have appeared if not originally listed) in the *Hauptwerk* has been continued in later supplements. Thus, in Volume 8 of the third supplement, look for the reference numbers, H54, listed at the top-center of the page. Pages 183–187 all have this reference number since each compound listed on these pages would have appeared on page 54 of the *Hauptwerk* according to their place in the compound classification scheme. After looking through these pages, you will find the entry for 2-bromo-5-hydroxybenzaldehyde on page 185, and references to this compound are listed up to the year 1950. If you wanted details for the preparation of this compound you would look-up the original articles cited in this entry.

Chemical Abstracts

Chemical Abstracts has been publishing abstracts and indexes of information appearing in the chemical literature since 1907. It published over 400,000 abstracts in 1979, and its indexes are recognized as "the key to the world's chemical literature." Indexes are now published covering each issue of *Chemical Abstracts*, each volume (six months), and each collection period. Collective Indexes first covered ten-year periods, but after the Fifth Collective Index (1947–1956) collective indexes have been published covering each five-year period. The Ninth Collective Index, the last one to be published, covers the years 1972–1976. Only the collective indexes need to be consulted for the period prior to 1977.

Compounds are indexed both by molecular formula and by chemical name. Unless you are certain of the *Chemical Abstracts* preferred name, the Molecular Formula Index offers the quickest way of locating references to a specific chemical substance. Following the molecular formula in the Index are given the preferred index name and one or more abstract numbers where reference to that compound can be found. If more than one abstract reference is listed, it is usually faster to find the compound under the preferred name in the corresponding Subject Index since the Subject Index generally gives a short description of the content of the abstract.

Since 1972 the Subject Indexes to *Chemical Abstracts* have been divided into a Chemical Substance Index and a General Subject Index. The Chemical Substance Index contains references to specific compounds, and the General Subject Index contains references to classes of chemical compounds, experimental techniques, reactions, concepts, and other general chemical terms.

To illustrate the use of Chemical Abstracts, information on the synthesis of 5-bromosalicylaldehyde is found in the Ninth Collective Molecular Formula Index under the formula $C_7H_5BrO_2$. Among other entries listed with this formula is the following:

Benzaldehyde, 5-bromo-2-hydroxy- [*1761-61-1*]
See Chemical Subject Index

This entry gives the preferred name for this compound and, in brackets, the unique Registry Number for the compound. Since many abstracts cite this compound, you are referred to the Chemical Substance Index under the preferred name. Upon consulting the Chemical Substance Index, you learn that there are no papers reporting the preparation of this compound that were abstracted and indexed during this period. However, the following entry for this compound:

ir and NMR spectrum of, 82:30811a

informs you that abstract number 30811a in Volume 82 of *Chemical Abstracts* gives details on the infrared and nuclear magnetic resonance spectra of this compound. Since the 66th volume of *Chemical Abstracts*, the abstract reference numbers have been assigned in consecutive order as the abstracts appear in each separate volume. The letter following these numbers is a computer data check that is only meaningful to data processing equipment. In volumes prior to the 66th volume, the reference numbers refer to the column number or page number in the volume in which the abstract appears. The letter following the number corresponds to the position of the abstract down the column on the page. The Forward at the beginning of each volume of *Chemical Abstracts* contains a description of the basic abstract formats used and a list of abbreviations and symbols used in the abstracts. If the abstract indicates that the original paper contains information that is useful to you, you must then go to the original publication cited in the abstract for more detailed information. Since no reference to a preparation of 5-bromosalicylaldehyde was found in the Ninth Collective Index, you must now go on and search previous Collective Indexes of *Chemical Abstracts* to locate a synthesis of the desired compound.

Over the years the use of nonsystematic names in *Chemical Abstracts* has been greatly reduced, but in earlier indexes nonsystematic names were used frequently. Thus, in the Formula Index to the Fifth Cumulative Index entries under $C_7H_5BrO_2$ are found for both bromohydroxybenzaldehyde and bromosalicylaldehyde with no reference to a specific isomer. The 5-bromo-2-hydroxy isomer is listed as 5-bromosalicylaldehyde in the corresponding Fifth Collective Subject Index.

In the Fifth Collective Subject Index there are 20 different abstracts referenced that contain information on 5-bromosalicylaldehyde. The descriptors on the references inform you that most of the references contain information on biological properties or specific derivatives such as acylhydrazones of this compound, but two references are made directly to this compound and these abstracts should be examined. The first reference is listed as: "and derivs., 44:1452bh." Upon looking in Volume 44 of *Chemical Abstracts*, column number 1452, positions b and h in the column, you see that there are two different abstracts, each citing a different article, but written by the same authors, that describe the synthesis of 5-bromosalicylaldehyde by the direct reaction of bromine with salicylaldehyde. The yield of this reaction is reported to be "good." At this point you must go to the original journal article to find details of the experimental procedure. The second abstract citation found in the Fifth Collective Subject Index is listed as: "and derivs., bactericidal action of, and its metal complexes, 44:8999h." Upon looking up this abstract, you see that the original paper does not report the synthesis of 5-bromosalicylaldehyde and is not useful for your purposes.

If you only have a common name and no structural formula for a compound that you wish to look up in *Chemical Abstracts*, The Cumulative Index Guides to the Eighth and Ninth Collective Indexes should be consulted. These Guides list the common names with the corresponding preferred names and, when appropriate, give illustrative structural diagrams for compounds indexed. The Index Guide also gives detailed rules for nomenclature and for the use of the Collective Subject Indexes. The Index Guide does not give any abstract citations. Using the preferred chemical name, you must consult the Subject Index (Chemical Substance Index) for abstract citations.

In addition to the Subject Index and the Molecular Formula Index, each Volume and Cumulative Index of *Chemical Abstracts* has an Author Index and a Numerical Patent Index. The first pages of each of these indexes gives a detailed introduction to the use of the index.

Compendium of Synthetic Methods

Several review journals and series that list synthetic methods have been discussed, but there are also several books which are compilations of synthetic methods for organic compounds. It is useful to consult these reference sources when planning a synthesis or looking for examples of a chemical reaction. Some of the more important sources are:

Synthetic Organic Chemistry, by Wagner and Zook (John Wiley and Sons, 1953), describes methods for the preparation of the major classes of organic compounds. Tables listing numerous organic compounds are provided with yields, physical data, and references to the synthesis of each listed compound.

Survey of Organic Synthesis, Vols. 1 and 2, by Buhler and Pearson (Wiley-Interscience, 1970 and 1977), describes methods of synthesis for the main types of organic compounds and lists many specific examples with literature references.

The Compendium of Organic Synthetic Methods, Vols. 1, 2, and 3, (Wiley-Interscience, 1971, 1974, and 1977), is divided into sections giving most of the possible interconversions between the major functional groups.

Modern Synthetic Reactions, 2nd ed., by H. O. House (W. A. Benjamin Co., 1972), is really a textbook, but it contains so many useful examples of chemical reactions with references to the original literature that it is also an excellent reference book for synthetic methods.

Vogel's Textbook of Practical Organic Chemistry, 4th ed. (Longman, N.Y., 1978), contains representative procedures for the preparation of organic compounds. This reference text also gives practical descriptions of many advanced organic laboratory techniques.

Comprehensive Organic Chemistry (Pergamon Press, 1978), is a unique, comprehensive survey of the synthesis and reactions of organic compounds. Volume 6 contains independent formula, compound, author, reagent, and reaction indexes to the work and to the major synthetic literature.

Searching the Chemical Literature for Synthetic Methods

In the chemistry laboratory, it is frequently necessary to prepare a compound that cannot be obtained commercially or that is prohibitively expensive in the quantities necessary. Before starting any laboratory work on a synthesis, it is essential that you determine if the preparation of the desired compound has been previously reported in the literature. If a preparation is not described in the literature, then reports of the preparation of similar compounds must be examined and utilized as precedents for the synthesis of the desired compound. These reports, including the details of the experimental procedures, will then enable you to select the best laboratory procedure for the preparation of the desired compound.

A thorough search of the recent literature for preparative methods for a particular compound is accomplished by searching the Formula Indexes and Subject Indexes of *Chemical Abstracts* under listings of the molecular formula and the preferred name of the compound. Since the literature coverage by *Chemical Abstracts* did not begin until 1907 and was not comprehensive prior to about 1940, it is also necessary to search *Beilstein* to reliably gain access to the pre-1940 chemical literature. It should be emphasized that these secondary sources give only a brief summary of the experimental results and procedure and that the original reference should be consulted for a detailed description of the experimental apparatus, reaction conditions, and procedure.

In many situations, particularly for the synthesis of more common compounds, a thorough search of the literature is not required to find a useful synthetic procedure. Many of the compendia discussed in this experiment and *Organic Syntheses*, *Methoden der Organische Chemie*, and *Merck Index* contain preparations or references to preparations of organic compounds. The compendia *Synthetic Organic Chemistry* by Wagner and Zook and *Survey of Organic Synthesis* by Buethler and Pearson are particularly useful in locating references to preparative procedures for many mono- and some difunctional compounds. However, if a search of these secondary sources fails to produce a reference to a suitable preparative method for the desired compound, a thorough search of *Chemical Abstracts* and *Beilstein* must be undertaken before you can assume that the compound has not been previously prepared.

Evaluation of Synthetic Procedures

The procedure or procedures for the synthesis of a desired compound described in the literature must be evaluated on the basis of three primary criteria. First, the procedure must be evaluated in terms of the economy and yield of the process. The cost and availability of starting materials, the reported experimental yield of the procedure, and the estimated time required for the synthesis are all factors considered under this criterion.

The synthesis must also be practical to run on a laboratory scale. The literature procedure may describe a process requiring specially designed or industrial scale equipment that is not available or practical for a laboratory preparation. On the other hand, the procedure may be designed to work well on a very small scale, but the reaction may be impossible to run on a preparative scale. Many syntheses employing electrochemical, photochemical, or enzyme reactions may fall into this latter category.

The third criterion for evaluation is safety. The procedure must be safe to carry out in the laboratory. The skill and experience of the laboratory worker is an important consideration in any laboratory procedure. Fire, explosion, and toxic hazards of all chemicals and processes must be evaluated. Equipment such as fume hoods, dry boxes, or safety shields must be available to safely perform certain chemical reactions. Some reactions involving high pressures or hazardous chemicals require special equipment for their safe utilization. Not only must safety factors involved in handling reagents and in running reactions be considered, but safety factors associated with purification procedures and with disposing of chemical wastes must also be evaluated.

EXPERIMENTAL PROCEDURE

You will be asked to use the chemical literature to find a practical laboratory scale synthesis for an organic compound. The procedure may require one or more steps, but special apparatus not available to you in the organic chemistry laboratory should not be required. After you have found a practical procedure for the synthesis of the assigned compound, write a report describing your proposed synthesis that includes a complete experimental procedure for the synthesis on a scale which will provide 8 to 10 g of the final product. The report should also include: an introduction; equations for all reactions; a table of reagents and products that includes physical properties, amounts used, and costs; a purification scheme for each synthetic procedure; all safety precautions; and all relevant literature references. The experimental procedure must be sufficiently detailed that the preparation of the compound could be carried out by another person without reference to the original literature.

At your instructor's suggestion, this synthesis can then be carried out in the laboratory as an independent project.

Appendix A

Tables of Selected Organic Compounds with their Physical Constants and Derivatives

---

Abbreviations Used in the Tables

d used after a boiling or melting point indicates the compound decomposes at that temperature.

d_{20} the density in g/ml at 20°C.

n_D^{20} the refractive index at the sodium D line (589 nm) at 20°C.

subl indicates the compound sublimes.

General Comments Regarding the Tables

Compounds included in these tables are generally commercially available materials that have well-defined boiling points or melting points at atmospheric pressure. Systematic names are used for most compounds, but frequently used trivial names or alternative names are also given in parentheses.

If more than one melting point is reported in the literature for a derivative of a compound, the additional values are listed in parentheses after the selected value.

Table 1 Alkanes

Name of compound	B.p. (°C)	M.p. (°C)	n_D^{20}	d_{20} (g/ml)
Pentane	36		1.3575	0.626
Cyclopentane	49		1.4065	0.745
2,2-Dimethylbutane	50		1.3688	0.649
2,3-Dimethylbutane	58		1.3750	0.662
2-Methylpentane	60		1.3715	0.653
3-Methylpentane	63		1.3765	0.664
Hexane	69		1.3749	0.659
Methylcyclopentane	72		1.4097	0.749
2,2-Dimethylpentane	79		1.3822	0.674
2,4-Dimethylpentane	80		1.3815	0.673
Cyclohexane	81	6	1.4262	0.779
3,3-Dimethylpentane	86		1.3909	0.693
2,3-Dimethylpentane	89		1.3920	0.695
2-Methylhexane	90		1.3849	0.679
3-Methylhexane	92		1.3890	0.687
3-Ethylpentane	93		1.3934	0.698
Heptane	98		1.3876	0.684
2,2,4-Trimethylpentane	99		1.3915	0.692
Methylcyclohexane	101		1.4231	0.769
2,5-Dimethylhexane	109		1.3925	0.694
2-Methylheptane	118		1.3950	0.698
Cycloheptane	119		1.4449	0.827
trans-1,4-Dimethylcyclohexane	119		1.4209	0.763
1,1-Dimethylcyclohexane	120		1.4290	0.781
trans-1,2-Dimethylcyclohexane	123		1.4270	0.776
Octane	126		1.3974	0.702
2,2-Dimethylheptane	133		1.4016	0.710
Cyclooctane	151	13	1.4586	0.835
Nonane	151		1.4054	0.717
Isopropylcyclohexane	154		1.4410	0.802
1-Isopropyl-4-methycyclohexane	169		1.4375	0.792
Decane	174		1.4119	0.730
trans-Decahydronaphthalene (*trans*-decalin)	187		1.4695	0.869
Undecane	196		1.4172	0.740
Cyclodecane	201	10	1.4692	0.858
Dodecane	216		1.4216	0.749
Tridecane	235		1.4256	0.756
Tetradecane	254	6	1.4289	0.763
Pentadecane	271	10	1.4319	0.768
Hexadecane	287	18	1.4345	0.773
Heptadecane	302	22	1.4348	0.774
Octadecane	316	28	1.4191^{70}	0.775^{30}
Nonadecane	330	32	1.4211^{70}	0.779^{30}
Eicosane	343	37	1.4230^{70}	0.755^{70}
Norbornane		87 (subl)		
Adamantane		210 (subl)		

Table 2 Alkenes and Alkynes

Name of compound	B.p. (°C)	M.p. (°C)	n_D^{20}	d_{20} (g/ml)
1-Pentene	30		1.3715	0.640
2-Methyl-1-buten-3-yne	32		1.4140	0.680
2-Methyl-1,3-butadiene (isoprene)	34		1.4219	0.681
trans-2-Pentene	36		1.3793	0.648
2-Methyl-2-butene	38		1.3874	0.662
1-Pentyne	40		1.3852	0.690
1,3-Pentadiene (piperylene)	41		1.4309	0.680
Cyclopentene	44		1.4225	0.772
4-Methyl-1-pentene	54		1.3828	0.664
3-Methyl-1-pentene	54		1.3842	0.667
2-Pentyne	56		1.4039	0.711
1,5-Hexadiene	59		1.4042	0.692
1-Hexene	63		1.3879	0.673
2-Ethyl-1-butene	65		1.3969	0.689
trans-1,3-Hexadiene	65		1.4060	0.692
trans-3-Hexene	67		1.3943	0.677
trans-3-Methyl-2-pentene	68		1.4016	0.694
trans-2-Hexene	68		1.3935	0.678
2,3-Dimethyl-1,3-butadiene	69		1.4394	0.727
cis-3-Methyl-2-pentene	70		1.4045	0.700
1-Hexyne	71		1.3989	0.716
1,4-Hexadiene	72		1.4402	0.705
2,3-Dimethyl-2-butene	73		1.4122	0.708
trans-2-Methyl-1,3-pentadiene	76		1.446	0.719
2,4-Hexadiene	82		1.4529	0.720
3-Hexyne	82		1.4112	0.726
Cyclohexene	83		1.4465	0.811
2-Hexyne	84		1.4135	0.732
1-Heptene	94		1.3998	0.697
trans-2-Heptene	98		1.4045	0.701
1-Heptyne	100		1.4087	0.733
2,4,4-Trimethyl-1-pentene	101		1.4086	0.715
4-Methylcyclohexene	103		1.4414	0.795
2,4,4-Trimethyl-2-pentene	105		1.4160	0.722
Cycloheptene	114		1.4580	0.825
1,3,5-Cycloheptatriene	116		1.5222	0.888
1-Octene	121		1.4087	0.715
trans-4-Octene	122		1.4116	0.715
trans-2-Octene	125		1.4132	0.720
1-Octyne	126		1.4159	0.7461
4-Vinyl-1-cyclohexene	129		1.4640	0.832
2,5-Dimethyl-2,4-hexadiene	134	15	1.4796	0.765
1,3,5,7-Cyclooctatetraene	141		1.5290	0.920
Phenylacetylene (phenylethyne)	142		1.5485	0.928
Styrene (phenylethene)	145		1.5468	0.906
Cyclooctene	145		1.4693	0.846
1-Nonene	147		1.4160	0.730
1,5-Cyclooctadiene	149		1.4905	0.882
1-Nonyne	151		1.4217	0.757
α-Pinene	156		1.4560	0.860

Table 2 (Continued)

Name of compound	B.p. (°C)	M.p. (°C)	n_D^{20}	d_{20} (g/ml)
3-Phenylpropene (allylbenzene)	157		1.5132	0.892
d,l-Camphene	160	50		
β-Pinene	164		1.4782	0.869
α-Methylstyrene (2-phenylpropene)	165		1.5303	0.911
Myrcene	166		1.4722	0.798
β-Methylstyrene (1-phenylpropene)	170		1.5508	0.911
Dicyclopentadiene	170		1.5120	1.071
1-Decene	171		1.4215	0.740
1-Decyne	174		1.4265	0.766
Limonene	178		1.4743	0.841
Indene	182		1.5764	0.992
1-Dodecene	213		1.4300	0.758
1-Tetradecene	251	13	1.4363	0.771
1,2-Diphenylethyne (tolan)	298	62		
trans-1,2-Diphenylethene (stilbene)		124		

Table 3 Aromatic Hydrocarbons

Name of compound	M.p. (°C)	B.p. (°C)	n_D^{20}	Nitro derivative: Position	Nitro derivative: m.p. (°C)
Benzene	5	80	1.5011	1,3	89
Toluene		111	1.4961	2,4	70
Ethylbenzene		136	1.4959	2,4,6	37
p-Xylene	13	138	1.4958	2,3,5	139
m-Xylene		139	1.4972	2,4,6	183
o-Xylene		144	1.5054	4,5	118
Isopropylbenzene (cumene)		152	1.4915	2,4,6	109
n-Propylbenzene		159	1.4920	2,4	liquid
1,3,5-Trimethylbenzene (mesitylene)		165	1.4994	2,4 2,4,6	86 235
tert-Butylbenzene		169	1.4927	2,4 2,4,6	62 124
1,2,4-Trimethylbenzene		169	1.5048	3,5,6	185
1,2,3-Trimethylbenzene		176	1.5139		
Indane		177	1.5381	5	40
4-Isopropyltoluene		177	1.4909	2,6 2,3,6	54
Indene		182	1.5764		
1,2,3,5-Tetramethylbenzene (isodurene)		198	1.5125	4,6	181 (157)
1,3-Diisopropylbenzene		203	1.4883	4,6	77
1,2,3,4-Tetramethylbenzene		205	1.5201	5,6	176
1,3-Dimethyl-5-*tert*-butylbenzene		206	1.4958	2,4,6	107 (114)
1,2,3,4-Tetrahydronaphthalene (tetralin)		207	1.5413	5,7	95
1,4-Diisopropylbenzene		210	1.4898		
1-Phenylhexane		226	1.4864		

Table 3 (Continued)

Name of compound	M.p. (°C)	B.p. (°C)	n_D^{20}	Nitro derivative: Position	Nitro derivative: m.p. (°C)
Cyclohexylbenzene		236	1.5329		
1-Methylnaphthalene		245	1.6174	4	71
Diphenylmethane	27	264	1.5753	2,2′,4,4′	172
2-Methylnaphthalene	38	240		1	81
1,2-Diphenylethane	53	284		4,4′	180
				2,2′, 4,4′	169
Pentamethylbenzene	54	232		6	154
Biphenyl	69	254		4,4′	237
				2,2′,4,4′	150
1,2,4,5-Tetramethylbenzene (durene)	80	198		3,6	205
Naphthalene					
Triphenylmethane	92	358		4,4′,4″	206
Acenaphthene	96	278		5	101
Phenanthrene	101	340			
2,3-Dimethylnaphthalene	104	266			
2,6-Dimethylnaphthalene	111	262			
Fluorene	114	295		2	156
				2,7	199
trans-Stilbene	124				
1,4-Diphenyl-1,3-butadiene	152				
Anthracene	216				

Table 4a Alkyl Chlorides

Name of compound	B.p. (°C)	n_D^{20}	d_{20} (g/ml)
2-Chloropropane	36	1.3777	0.862
1-Chloro-1-propene	37	1.4054	0.935
Dichloromethane	41	1.4237	1.34
3-Chloro-1-propene (allyl chloride)	45	1.4157	0.938
1-Chloropropane	47	1.3879	0.891
trans-1,2-Dichloroethene	48	1.4454	1.26
2-Chloro-2-methylpropane (*tert*-butyl chloride)	51	1.3857	0.842
2-Chloro-1,3-butadiene	59	1.4583	0.958
Chloroform	61	1.446	1.50
2-Chlorobutane	68	1.3971	0.873
1-Chloro-2-methylpropane (isobutyl chloride)			
1,1,1-Trichloroethane			
Carbon tetrachloride	77	1.4630	1.60
1-Chlorobutane	78	1.4021	0.886
1-Chloro-2,2-dimethylpropane	84	1.4044	0.866
1,2-Dichloroethane	84	1.4443	1.256
2-Chloro-2-methylbutane	86	1.4055	0.865
1,2-Dichloropropane	96	1.4388	1.16
2-Chloropentane	97	1.4069	0.870

Table 4a (Continued)

Name of compound	B.p. (°C)	n_D^{20}	d_{20} (g/ml)
3-Chloropentane	98	1.4082	0.873
1-Chloro-3-methylbutane	101	1.4096	0.893
1-Chloropentane	108	1.4120	0.882
Chlorocyclopentane	115	1.4510	1.01
3-Chloro-3-methylpentane	116	1.421	0.890
1,1,2,2-Tetrachloroethene	121	1.5055	1.62
3-Chlorohexane	123	1.4163	0.870
2-Chlorohexane	125	1.4142	0.869
1,3-Dichloropropane	125	1.449	1.18
1-Chlorohexane	134	1.4196	0.878
Chlorocyclohexane	143	1.462	0.990
1-Chloro-3-bromopropane	143	1.4861	1.59
1-Chloroheptane	159	1.4256	0.876
Benzyl chloride	179	1.539	1.10
1-Chlorooctane	180	1.4305	0.874
2-Chloro-1-phenylethane (β-phenethyl chloride)	198	1.5276	1.07
1-Chloronoane	203	1.4345	0.872
α,α-Dichlorotoluene (benzal chloride)	205	1.5502	1.25
m-Chlorobenzyl chloride	216	1.5554	1.26
o-Chlorobenzyl chloride	217	1.5591	1.26
p-Chlorobenzyl chloride (m.p. = 30°C)	222		
1-Chlorodecane	223	1.4379	0.871

Table 4b Alkyl Bromides

Name of compound	B.p. (°C)	n_D^{20}	d_{20} (g/ml)
Bromoethane	38	1.4239	1.46
2-Bromopropane	60	1.4251	1.31
1-Bromopropane	71	1.4343	1.35
3-Bromo-1-propene (allyl bromide)	71	1.4655	1.40
2-Bromo-2-methylpropane (*tert*-butyl bromide)	73	1.4278	1.22
2-Bromobutane	91	1.4367	1.26
1-Bromo-2-methylpropane (isobutyl bromide)	93	1.4350	1.25
Dibromomethane	99	1.538	2.50
1-Bromobutane	102	1.4401	1.28
2-Bromopentane	117	1.4413	1.21
3-Bromopentane	119	1.4441	1.21
1-Bromo-3-methylbutane	120	1.4420	1.20
3-Bromo-3-methylpentane	130	1.4525	1.18
1-Bromopentane	130	1.4443	1.21
1,2-Dibromoethane	132	1.5379	2.18
Bromocyclopentane	137	1.489	1.39
1-Bromo-3-chloropropane	143	1.4861	1.59
Bromoform	151	1.598	2.89

Table 4b (Continued)

Name of compound	B.p. (°C)	n_D^{20}	d_{20} (g/ml)
1-Bromohexane	155	1.4475	1.17
Bromocyclohexane	165	1.495	1.34
1,3-Dibromopropane	168	1.523	1.98
1,4-Dibromobutane	198	1.5186	1.81
1-Bromooctane	201	1.4524	1.11
1-Bromononane	221	1.4542	1.10
1-Bromodecane	241	1.4557	1.07
Carbon tetrabromide	(m.p. = 90°C)		

Table 4c Alkyl Iodides

Name of compound	B.p. (°C)	n_D^{20}	d_{20} (g/ml)
Iodomethane	42	1.5308	2.28
Iodoethane	72	1.5133	1.93
2-Iodopropane	90	1.4997	1.70
3-Iodo-1-propene (allyl iodide)	102	1.5530	1.85
1-Iodopropane	103	1.5058	1.75
2-Iodobutane	118	1.4990	1.59
1-Iodo-2-methylpropane (isobutyl iodide)	120	1.4960	1.60
1-Iodobutane	131	1.5001	1.61
1-Iodopentane	157	1.4959	1.51
Iodocyclopentane	167	1.5447	1.71
Iodocyclohexane	179	1.547	1.62
1-Iodohexane	181	1.4928	1.44
Diiodomethane	181	1.7425	3.33
1-Iodoheptane	204	1.4904	1.38
1-Iodooctane	225	1.4885	1.33
Iodoform	(m.p. = 119°C)		

Table 5 Aryl Halides

Name of compound	B.p. (°C)	M.p. (°C)	n_D^{20}	Nitro derivative	
				Position	m.p. (°C)
Chlorobenzene	132		1.5241	2,4	52
Bromobenzene	156		1.5597	2,4	70
2-Chlorotoluene	159		1.5268	3,5	63
3-Chlorotoluene	162		1.521	4,6	91
4-Chlorotoluene	162	7	1.521	2	38
1,3-Dichlorobenzene	173		1.5459	4,6	103
1,2-Dichlorobenzene	180		1.5515	4,5	110
2-Bromotoluene	182		1.5565	3,5	82
3-Bromotoluene	184		1.551	4,6	103
4-Bromotoluene	184	28	1.5490	2	47
Iodobenzene	188		1.6200	4	171

Table 5 (Continued)

Name of compound	B.p. (°C)	M.p. (°C)	n_D^{20}	Nitro derivative Position	m.p. (°C)
2,6-Dichlorotoluene	199		1.5510	3	50
2,4-Dichlorotoluene	200		1.549	3,5	104
3,4-Dichlorotoluene	201		1.5472	2,6	91
3,5-Dichlorotoluene	201	26		2,6	99
3-Iodotoluene	204		1.6053	4,6	108
1,2,4-Trichlorobenzene	213	17	1.5671	5	56
1,2-Dibromobenzene	225	7	1.6155	4,5	114
1-Chloronaphthalene	259		1.633	4,5	180
1-Bromonaphthalene	281	6	1.658	4	85
3,5-Dichlorotoluene	201	26		2,6	99
4-Bromotoluene	184	28	1.5490	2	47
2,4,6-Trichlorotoluene		34		3	54
4-Iodotoluene	211	35			
1,2,3,4-Tetrachlorobenzene	275	46		5	64
				5,6	151
1,2,3,5-Tetrachlorobenzene	246	51		4	41
				4,6	162
1,4-Dichlorobenzene	173	53		2	54
				2,6	106
1,3,5-Trichlorobenzene	208	63		2	68
				2,4	131
1-Bromo-4-chlorobenzene	197	67		2	72
4-Chlorobiphenyl	293	77			
1-Chloroanthracene		81			
2,4,5-Trichlorotoluene		82		3	92
				3,6	227
Pentachlorobenzene	276	86		1	146
1,4-Dibromobenzene	219	89		2,5	84
1,3,5-Tribromobenzene	271	120			
1,4-Diiodobenzene	289	129		2,5	171
1,2,4,5-Tetrachlorobenzene	245	140		3	100
4,4′-Dichlorobiphenyl	315	149		2	102
				2,2′	138

Table 6a Alcohols—Liquid

Name of compound	B.p. (°C)	M.p. (°C)	n_D^{20}	Melting points of derivatives, (°C): 3,5-Dinitro-benzoate	4-Nitro-benzoate	α-Naphthyl-urethane
Methanol	65		1.3306	108	96	124
Ethanol	78		1.3610	93	57	79
2-Propanol	82		1.3793	123	110	106
2-Methyl-2-propanol (*tert*-butyl alcohol)	83	25	1.3838	142	116	101
3-Buten-2-ol	95		1.4137	54	43	
2-Propen-1-ol (allyl alcohol)	97		1.4135	50	28	108
1-Propanol	97		1.3850	74	35	80
2-Butanol	100		1.3950	76	25	97
2-Methyl-2-butanol	102		1.4052	116	85	72
2-Methyl-1-propanol (isobutyl alcohol)	108		1.3939	87	69	104
3-Buten-1-ol	113		1.4224	59		
3-Methyl-2-butanol	114		1.3973	76		109
3-Pentanol	116		1.4103	101	17	95
1-Butanol	117		1.3974	64	70	71
2-Pentanol	120		1.4060	62	17	75
3,3-Dimethyl-2-butanol	120		1.4148	107		
2,3-Dimethyl-2-butanol	120		1.4140	111	82	101
3-Methyl-3-pentanol	123		1.4166	96	67	84
2-Methyl-2-pentanol	123		1.4113	72	70	104
2-Methoxyethanol	125		1.4024		50	113
1-Chloro-2-propanol	127		1.4392	77		
2-Methyl-3-pentanol	128		1.4168	85		
2-Methyl-1-butanol	129		1.4107	70		82
2-Chloroethanol	130		1.4419	95		101
4-Methyl-2-pentanol	132		1.4011	65	26	88
3-Methyl-1-butanol	132		1.4085	61	21	68
2-Ethoxyethanol	135		1.4080	75		67
3-Hexanol	136		1.4159	97		72
2,2-Dimethyl-1-butanol	137		1.4208	51		81
1-Pentanol	138		1.4099	46	oil	68
2-Hexanol	139		1.4126	39	40	61
2,4-Dimethyl-3-pentanol	140		1.4226		155	99
Cyclopentanol	141		1.4530	115	62	118
4-Methyl-1-pentanol	153		1.4153	72		58
4-Heptanol	156		1.4205	64	35	80
1-Hexanol	157		1.4178	60	oil	62
2-Heptanol	159		1.4210	49		54
Cyclohexanol	161	25	1.4650^{25}	112	50	129
3-Chloro-1-propanol	161		1.4469	77	oil	76
Furfuryl alcohol	172		1.4863	81	76	130
1-Heptanol	177		1.4245	47	oil	62
2-Octanol	179		1.4265	32	28	63
2-Ethyl-1-hexanol	185		1.4328			60
1,2-Propanediol	187		1.4316		127	

Table 6a (Continued)

Name of compound	B.p. (°C)	M.p. (°C)	n_D^{20}	Melting points of derivatives, (°C): 3,5-Dinitro-benzoate	4-Nitro-benzoate	α-Naphthyl-urethane
1-Octanol	195		1.4293	62	17	67
1,2-Ethanediol	198		1.4319	169	140	176
2-Nonanol	198		1.4290	43		56
l-Linalool	199		1.4624		70	53
1-Phenylethanol	202	20	1.5244	95	43	106
Benzyl alcohol	206		1.5395	113	85	134
1-Nonanol	213		1.4311	52	60	66
1,3-Propanediol	215		1.4398	178	119	164
2-Phenylethanol	220		1.5240	108	62	119
Geraniol	230		1.4766	63	35	48
1-Decanol	231	7	1.4368	58	30	73
3-Phenyl-1-propanol	237		1.5356	92	47	

Table 6b Alcohols—Solid

Name of compound	M.p. (°C)	B.p. (°C)	Melting points of derivatives (°C): 3,5-Dinitro-benzoate	4-Nitro-benzoate	α-Naphthyl-urethane
1-Phenylethanol	20	202	95	43	106
1-Dodecanol	24	259	60	45	80
Cyclohexanol	25	161	113	50	129
2-Methyl-2-propanol	25	83	142	116	101
3-Nitrobenzyl alcohol[a]	27				
Cinnamyl alcohol	33	257	121	78	114
α-Terpineol (1-*p*-menthen-8-ol)	36	221	79	97	152
1-Tetradecanol	39	286	67	51	82
Pinacol[b]	43	172			
Menthol	44	216	153	61	126
1-Hexadecanol	50	344	66	58	82
2,2-Dimethyl-1-propanol (neopentyl alcohol)	52	113			100
Piperonyl alcohol[c]	58				
4-Methylbenzyl alcohol	60	217	118		
1-Octadecanol	60	332	77	64	
1,2-Diphenylethanol[d]	67				
Benzhydrol	68	299	141	132	139
2-Nitrobenzyl alcohol[e]	74				
2-Chlorobenzyl alcohol	74			94	
4-Chlorobenzyl alcohol	75				
4-Nitrobenzyl alcohol[f]	93		157		
Benzoin	137			123	140
Lanosterol[g]	140		201		

Table 6b (Continued)

Name of compound	M.p. (°C)	B.p. (°C)	Melting points of derivatives (°C): 3,5-Dinitrobenzoate	4-Nitrobenzoate	α-Naphthylurethane
Cholesterol[h]	148	360d	195	190	176
Triphenylmethanol[i]	161	380			
d-Borneol (*d*-2-camphanol)	208[j]	212	155	137	132
d,l-Isoborneol	212[j]		140	129	

[a]3-Nitrobenzyl alcohol: benzoate derivative, m.p. = 71°C.
[b]Pinacol: diacetate derivative, m.p. = 65°C.
[c]Piperonyl alcohol: benzoate derivative, m.p. = 66°C.
[d]1,2-Diphenylethanol: benzoate derivative, m.p. = 70°C.
[e]2-Nitrobenzyl alcohol: benzoate derivative, m.p. = 102°C.
[f]4-Nitrobenzyl alcohol: benzoate derivative, m.p. = 95°C.
[g]Lanosterol: benzoate derivative, m.p. = 191°C.
[h]Cholesterol: benzoate derivative, m.p. = 151°C.
[i]Triphenylmethanol: acetate derivative, m.p. = 88°C.
[j]Melting point in sealed tube.

Table 7 Phenols

Name of compound	M.p. (°C)	B.p. (°C)	Melting points of derivatives (°C): 3,5-Dinitrobenzoate	4-Nitrobenzoate	α-Naphthylurethane	Bromo derivative
2-Chlorophenol	7	176	143	115	120	76 *di*
Phenol	42	182	146	127	132	95 *tri*
2-Methylphenol (*o*-cresol)	31	191	138	94	142	56 *di*
2-Bromophenol	5	195			129	95 *tri*
Salicylaldehyde	2	197		128		
3-Methylphenol (*m*-cresol)	12	202	165	90	128	84 *tri*
4-Methylphenol (*p*-cresol)	35	202	188	98	146	198 *tetra*
2-Ethylphenol	—	207	108	57		
2,4-Dimethylphenol	27	212	164	105	135	
2-Hydroxyacetophenone[a]	28	215				
Methyl salicylate		224		128		
3-Methoxyphenol		243			129	104 *tri*
4-Allyl-2-methoxyphenol (eugenol)		255	131	81	122	118 *tetra*
2-Methoxy-4-propenylphenol (isoeugenol)		268	158	109	150	
2-Methoxyphenol	32	205	141	93	118	116 *tri*

Table 7 (Continued)

Name of compound	M.p. (°C)	B.p. (°C)	Melting points of derivatives (°C): 3,5-Dinitrobenzoate	4-Nitrobenzoate	α-Naphthylurethane	Bromo derivative
3-Bromophenol	32	236			108	
3-Chlorophenol	33	214	156	99	158	
4-Methylphenol (*p*-cresol)	35	202	188	98	146	198 *tetra*
2-Nitro-4-methylphenol	36		192			
2,4-Dibromophenol	36	238		183		95 *tri*
Phenol	42	182	146	127	132	95 *tri*
4-Chlorophenol	43	217	186	171	166	
2,4-Dichlorophenol[b]	43	209	142			68 *mono*
2-Nitrophenol	45	216	155	141	113	117 *di*
4-Ethylphenol	47	219	133	81	128	
4-Chloro-2-methylphenol	49	225				
2,6-Dimethylphenol	49	203	159		176	79 *mono*
5-Methyl-2-isopropylphenol (thymol)	49	233	103	70	160	55 *mono*
4-Methoxyphenol	55	244	166			
2,5-Dichlorophenol[c]	59	212				
3,4-Dimethylphenol	63	225	181		142	171 *tri*
4-Bromophenol	64	238	191	180	169	95 *tri*
4-Chloro-3-methylphenol	66				153	
2,6-Dichlorophenol[d]	67	219				
3,5-Dimethylphenol	68	219	195	109		166 *tri*
3,5-Dichlorophenol[e]	68	233				189 *tri*
3,4-Dichlorophenol	68	252				
2,4,5-Trichlorophenol[f]	68	248				
2,4,6-Trichlorophenol[g]	68	245		106		
2,4,6-Trimethylphenol	69	220				158 *di*
2,6-Di-*tert*-butyl-4-methylphenol	70	265				
2,5-Dimethylphenol	75	212	137	87	173	178 *tri*
8-Hydroxyquinoline	76			175		
4-Hydroxy-3-methoxybenzaldehyde (vanillin)	81	285				
1-Naphthol	94	280	217	143	152	105 *di*
2,3,5-Trimethylphenol	96	233				
2-Methyl-4-nitrophenol	96	190				
3-Nitrophenol	97		159	174	167	91 *di*
4-*tert*-Butylphenol	100	237			110	50
1,2-Dihydroxybenzene (catechol)	105	245	152 *di*	169 *di*	175	193 *tetra*
3,5-Dihydroxytoluene	106	290	190	214	160	104 *tri*
1,3-Dihydroxybenzene (resorcinol)	110	281	201	182 *di*		112 *tri*
2-Chloro-4-nitrophenol[h]	111					
4-Nitrophenol	114		188	159	151	145 *di*
2,4-Dinitrophenol	114			139		118 *mono*
4-Hydroxybenzaldehyde[i]	117					
1,3,5-Trihydroxybenzene (phloroglucinol dihydrate)	117			283		151 *tri*

Table 7 (Continued)

Name of compound	M.p. (°C)	B.p. (°C)	Melting points of derivatives (°C): 3,5-Dinitro-benzoate	4-Nitro-benzoate	α-Naphthyl-urethane	Bromo derivative
2,3,5,6-Tetramethylphenol	118	249				118 *mono*
2,4,6-Trinitrophenol (picric acid)	122					
2-Naphthol	123	286	210	169	157	84 *mono*
3-Methyl-4-nitrophenol[j]	129					
1,2,3-Trihydroxybenzene (pyrogallol)	133		205	230		158 *di*
2,4-Dihydroxyacetophenone[k]	147					
Salicylic Acid	158			205		
2,3-Dihydroxynaphthalene[l]	160					
1,4-Dihydroxybenzene (hydroquinone)	171	286	317	258		186 *di*
3,5-Dinitrosalicylic acid[m]	173					
2-Aminophenol[n]	174					
1,4-Dihydroxynaphthalene	176				220	
4-Aminophenol	184		178			
2,7-Dihydroxynaphthalene[o]	190					
Pentachlorophenol[p]	190					
3-Hydroxybenzoic acid[q]	200					
4-Hydroxybenzoic acid[r]	215					
1,3,5-Trihydroxybenzene (phloroglucinol anhydrous)	217		162	283		151 *tri*
1,5-Dihydroxynaphthalene[s]	265					
4,4′-Biphenol[t]	274					

[a]2-Hydroxyacetophenone: acetate derivative, m.p. = 89°C; semicarbazone derivative, m.p. = 210°C.
[b]2,4-Dichlorophenol: benzoate derivative, m.p. = 96°C.
[c]2,5-Dichlorophenol: benzoate derivative, m.p. = 69°C.
[d]2,6-Dichlorophenol: benzoate derivative, m.p. = 74°C.
[e]3,5-Dichlorophenol: benzoate derivative, m.p. = 55°C.
[f]2,4,5-Trichlorophenol: benzoate derivative, m.p. = 93°C.
[g]2,4,6-Trichlorophenol: benzoate derivative, m.p. = 75°C.
[h]2-Chloro-4-nitrophenol: acetate derivative, m.p. = 63°C.
[i]4-Hydroxybenzaldehyde: benzoate derivative, m.p. = 90°C; 2,4-dinitrophenylhydrazone derivative, m.p. = 270°C.
[j]3-Methyl-4-nitrophenol: benzoate derivative, m.p. = 74°C.
[k]2,4-Dihydroxyacetophenone: semicarbazone derivative, m.p. = 216°C.
[l]2,3-Dihydroxynaphthalene: dibenzoate derivative, m.p. = 235°C.
[m]3,5-Dinitrosalicylic acid: benzoate derivative, m.p. = 163°C.
[n]2-Aminophenol: benzoate derivative, m.p. = 175°C.
[o]2,7-Dihydroxynaphthalene: dibenzoate derivative, m.p. = 139°C.
[p]Pentachlorophenol: benzoate derivative, m.p. = 164°C.
[q]3-Hydroxybenzoic acid: amide derivative, m.p. = 170°C; anilide derivative, m.p. = 156°C.
[r]4-Hydroxybenzoic acid: amide derivative, m.p. = 162°C; anilide derivative, m.p. = 196°C.
[s]1,5-Dihydroxynaphthalene: dibenzoate derivative, m.p. = 235°C.
[t]4,4′-Biphenol: dibenzoate derivative, m.p. = 241°C.

Table 8 Ethers

Name of compound	B.p. (°C)	M.p. (°C)	n_D^{20}	d_{20} (g/ml)	Bromo derivative m.p. (°C)
Furan	31		1.4216	0.937	
Ethyl ether	35		1.3526	0.714	
Ethyl vinyl ether	36		1.3768	0.759	
Methyl *n*-propyl ether	39		1.3579	0.736	
Ethyl isopropyl ether	53		1.3698	0.721	
tert-Butyl methyl ether	55		1.3689	0.741	
Ethyl *n*-propyl ether	64		1.3695	0.739	
Tetrahydrofuran	66		1.4070	0.887	
Isopropyl ether	68		1.3688	0.726	
2-Methyltetrahydrofuran	79		1.4056	0.855	
Ethylene glycol dimethyl ether (1,2-dimethoxyethane)	85		1.3797	0.867	
3,4-Dihydropyran	86		1.4400	0.923	
Tetrahydropyran	88		1.4210	0.881	
n-Propyl ether	90		1.3883	0.747	
n-Butyl vinyl ether	94		1.4010	0.774	
1,4-Dioxane	101	12	1.4232	1.03	
β-Chloroethyl ethyl ether	107		1.411	0.989	
1,2-Epoxy-3-chloropropane	117		1.438	1.18	
Isobutyl ether	123			0.761	
n-Butyl ether	142		1.3989	0.768	
Anisole (methoxybenzene)	155		1.5221	0.994	61 *di*
Diethylene glycol dimethyl ether (dimethyl carbitol)	162		1.4099	0.944	
o-Methylanisole	171		1.5161	0.985	64 *mono*
Phenetole (ethoxybenzene)	172		1.5080	0.966	
p-Methylanisole	174		1.5112	0.970	
m-Methylanisole	176		1.5130	0.969	
2,2′-Dichloroethyl ether	178		1.4568	1.22	
n-Pentyl ether	188		1.416	0.783	
3-Chloroanisole	194		1.5362	1.16	
2-Chloroanisole	195		1.5445	1.12	
4-Chloroanisole	200		1.5358	1.16	
1,2-Dimethoxbenzene (veratrole)	207	22	1.5287	1.08	93 *di*
Butyl phenyl ether	210		1.4970	0.935	
1,3-Dimethoxybenzene	217		1.4233	1.055	140 *di*
n-Hexyl ether	229		1.4204	0.793	
Safrole	233	11	1.5383	1.100	108 *tri*
4-Propenylanisole (anethole)	235	22	1.5600	0.988	108 *tri*
2-Nitroanisole	277	10	1.562	1.25	
Benzyl ether	298	2	1.5610	1.04	108 *di*
Phenyl ether	258	28	1.5826	1.07	55 *di*
2-Ethoxynaphthalene	282	36	1.5975^{36}	1.061	66 *mono*
3-Nitroanisole	258	39		1.373	
1,2,3-Trimethoxybenzene	241	47			73 *tri*
4-Iodoanisole	240	52			
1,3,5-Trimethoxybenzene	255	53			130 *di*
4-Nitroanisole	274	54			
1,4-Dimethoxybenzene	213	56			142 *di*

Table 8 (Continued)

Name of compound	B.p. (°C)	M.p. (°C)	n_D^{20}	d_{20} (g/ml)	Bromo derivative m.p. (°C)
1,4-Diethoxybenzene	246	72			
2-Methoxynaphthalene	273	73			
1,2-Diphenoxyethane		98			135 *di*

Table 9a Aldehydes—Liquid

Name of compound	B.p. (°C)	Melting points of derivatives (°C) Semi-carbazone	2,4-Dinitro-phenyl-hydrazone	Phenyl-hydrazone	Oxime
Acetaldehyde	20	169	168	57	47
Propionaldehyde	48	89	150	oil	40
Glyoxal (m.p. = 15°C)	50	270	328	180	178
2-Propenal (acrolein)	52	171	165	51	
Isobutyraldehyde	64	126	187	oil	oil
2-Methyl-2-propenal (methacrolein)	68	198	206	74	
n-Butyraldehyde	75	96	123	95	oil
Trimethylacetaldehyde	75	190	210		41
Chloroacetaldehyde	86	148			oil
3-Methylbutanal	93	107	123	oil	49
Pentanal (valeraldehyde)	103		98		52
2-Butenal (crotonaldehyde)	104	199	190	56	119
2-Ethylbutanal	117	99	129		
4-Methylpentanal	121	127	99		oil
Paraldehyde (m.p. = 12°C)	125	169	168	57	47
Hexanal	131	106	104		51
5-Methylhexanal	144	117	117		
Heptanal	155	109	108		57
Furfural	162	202	230	97	92
Octanal	171	101	106		60
Benzaldehyde	179	222	237	158	35
Nonanal	185	100	100		64
Glutaraldehyde	189				178
Phenylethanal (m.p. = 34°C) (phenylacetaldehyde)	194	153	121	62	100
Salicylaldehyde	197	231	252	142	63
3-Methylbenzaldehyde (*m*-tolualdehyde)	199	204	195	91	60
2-Methylbenzaldehyde (*o*-tolualdehyde)	200	209	194	106	49
4-Methylbenzaldehyde (*p*-tolualdehyde)	205	234	234	113	80
Decanal	207	102	104		69
2-Chlorobenzaldehyde (m.p. = 11°C)	214	230	209	86	103
3-Chlorobenzaldehyde (m.p. = 17°C)	214	228	256	134	70(118)

Table 9a (Continued)

		Melting points of derivatives (°C)			
Name of compound	B.p. (°C)	Semi-carbazone	2,4-Dinitro-phenyl-hydrazone	Phenyl-hydrazone	Oxime
3-Methoxybenzaldehyde	230	233		76	40
3-Bromobenzaldehyde	234	228		141	72
2-Ethoxybenzaldehyde (m.p. = 20°C)	247	219			59
4-Methoxybenzaldehyde	248	210	254	120	64(133)
Cinnamaldehyde	252	216	255	168	65(138)

Table 9b Aldehydes—Solid

		Melting points of derivatives (°C)			
Name of compound	M.p. (°C)	Semi-carbazone	2,4-Dinitro-phenyl-hydrazone	Phenyl-hydrazone	Oxime
1-Naphthaldehyde (b.p. = 292°C)	34	221		80	98
Phenylethanal (b.p. = 195°C) (phenylacetaldehyde)	34	163	121	62	100
Piperonal	37	234	266	102	146
2-Methoxybenzaldehyde	39	215	254		92
4-Diethylaminobenzaldehyde	41	241		103	93
3,4-Dichlorobenzaldehyde	44		301		120
2-Nitrobenzaldehyde	44	256	265	156	102(154)
3,4-Dimethoxybenzaldehyde (veratraldehyde)	45	177	261	121	95
4-Chlorobenzaldehyde	48	233	254	127	110(146)
2,3-Dimethoxybenzaldehyde	54	231		138	99
4-Bromobenzaldehyde	57	229	128(257)	113	157(111)
3-Nitrobenzaldehyde	58	246	293	124	122
2-Naphthaldehyde	60	245	270	206	156
3,5-Dichlorobenzaldehyde	65			106	112
2,6-Dichlorobenzaldehyde	71				150
2,4-Dimethoxybenzaldehyde	71				106
4-Aminobenzaldehyde	72	153		156	124
2-Chloro-4-nitrobenzaldehyde	74	234	247d	154	
2,4-Dichlorobenzaldehyde	74				137
4-Dimethylaminobenzaldehyde	74	222	325	148	185
3,4,5-Trimethoxybenzaldehyde	78	219			84
2-Chloro-5-nitrobenzaldehyde	79		277		176
4-Hydroxy-3-methoxybenzaldehyde (vanillin)	81	230	271d	105	122
3,5-Dibromosalicylaldehyde	85				220
Isophthaldehyde (1,3-benzenedicarboxaldehyde)	89			242	180
3-Hydroxybenzaldehyde	104	198	257d	131	90
5-Bromosalicylaldehyde	106	297d			126
4-Nitrobenzaldehyde	106	221	322	159	133

Table 9b (Continued)

Name of compound	M.p. (°C)	Melting points of derivatives (°C): Semi-carbazone	2,4-Dinitro-phenyl-hydrazone	Phenyl-hydrazone	Oxime
4-Hydroxybenzaldehyde	116	224	271	184	72(112)
Terphthalaldehyde	116			278d	200
2,4,6-Trimethoxybenzaldehyde	118				203
5-Nitrosalicylaldehyde	126				218
2,4-Dihydroxybenzaldehyde	136	260d	286	158	192
3,4-Dihydroxybenzaldehyde	154	230d	275d	175d	157
3,5-Dihydroxybenzaldehyde	156	223			
Benzaldehyde-3-carboxylic acid	175	265		164	188d
Benzaldehyde-4-carboxylic acid	256			226	210

Table 10a Ketones—Liquid

Name of compound	B.p. (°C)	Melting points of derivatives (°C): Semi-carbazone	2,4-Dinitro-phenyl-hydrazone	Phenyl-hydrazone	Oxime
Acetone	56	190	128	42	59
3-Buten-2-one (methyl vinyl ketone)	81	141			
2-Butanone	82	136	117	oil	oil
3-Butyn-2-one	86		181		
3-Methyl-2-butanone	94	114	120	oil	oil
3-Methyl-3-buten-2-one	98	173	181		
Cyclobutanone	100		146		
3-Pentanone	102	139	156	oil	oil
2-Pentanone	102	112	144	oil	oil
1-Penten-3-one	103		129		
3,3-Dimethyl-2-butanone (pinacolone)	106	158	125	oil	79
1-Methoxy-2-propanone	115		163		
4-Methyl-2-pentanone	117	135	95		oil
3-Methyl-2-pentanone	118	95	71		oil
Chloroacetone	119	150	125		oil
3-Penten-2-one	122	142	155		
2,4-Dimethyl-3-pentanone	124	160	88		
3-Hexanone	125	113	130		oil
4,4-Dimethyl-2-pentanone	125		100		
2-Hexanone	128	125	106	oil	49
5-Hexen-2-one	130	102	108		oil
4-Methyl-3-penten-2-one (mesityl oxide)	130	164	206	142	49
Cyclopentanone	131	210	146	55	56
5-Methyl-3-hexanone	136	152			
2-Methyl-3-hexanone	136	119			

Table 10a (Continued)

Name of compound	B.p. (°C)	Melting points of derivatives (°C) Semi-carbazone	2,4-Dinitro-phenyl-hydrazone	Phenyl-hydrazone	Oxime
2,4-Pentanedione (acetylacetone)	139	209 *di*	209		149 *di*
4-Heptanone	144	132	75		oil
1-Hydroxy-2-propanone	146	196	129		
3-Heptanone	148	103			
2-Heptanone	151	123	89	207	
Cyclohexanone	156	167	162	82	91
2,3-Hexanedione	158				175 *di*
3,5-Dimethyl-4-heptanone	162	84			
2-Methylcyclohexanone	165	191	137	oil	43
2,6-Dimethyl-4-heptanone	168	126	66		
4-Octanone	170	96	41		
4-Methylcyclohexanone	171	199		109	39
2-Octanone	173	123	58		
2,2,6-Trimethylcyclohexanone	179	209	141		
Ethyl acetoacetate	181	133	93		
Cycloheptanone	182	163	148		23
5-Nonanone	186	90			
3-Nonanone	187	112			
2,5-Hexanedione (acetonylacetone)	194	224 *di*	257 *di*	120 *di*	137 *di*
2-Nonanone	195	119	56		
Acetophenone (m.p. = 20°C)	202	199	240	105	60
Menthone	209	189	146	53	59
2-Methylacetophenone	214	205	159		61
1,5,5-Trimethylcyclohexen-3-one (isophorone)	215	200		68	79
1-Phenyl-2-propanone (m.p. = 27°C)	216	200	156	87	70
Propiophenone (m.p. = 20°C)	220	174	191		54
3-Methylacetophenone	220	198	207		55
Isobutyrophenone	222	181	163	73	94
Pulegone	224	175	142		119
1-Phenyl-2-butanone	226	135			
6-Undecanone	228	oil			oil
3-Chloroacetophenone	228	232			88
2-Undecanone (m.p. = 12°C)	228	122	63		45
2,4-Dimethylacetophenone	228	187			63
2-Chloroacetophenone	229	160			113
n-Butyrophenone (m.p. = 12°C)	230	188	190		50
d-Carvone	230	163	191		73
4-Chloroacetophenone (m.p. = 12°C)	232	204	231	114	95
3,5-Dimethylacetophenone	237				114
2-Methoxyacetophenone	239	183			83
3-Methoxyacetophenone	240	196			
n-Valerophenone	248	160	166	162	52
2,5-Dichloroacetophenone (m.p. = 14°C)	251				130

Table 10b Ketones—Solid

Name of compound	M.p. (°C)	Melting points of derivatives (°C): Semi-carbazone	2,4-Dinitro-phenyl-hydrazone	Phenyl-hydrazone	Oxime
Acetophenone (b.p. = 202°C)	20	199	240	105	60
Propiophenone (b.p. = 220°C)	20	174	191		54
1-Phenyl-2-propanone (b.p. = 216°C)	27	200	156	87	70
4-Methylacetophenone (b.p. = 226°C)	28	205	260	97	88
2-Hydroxyacetophenone (b.p. = 215°C)	28	210	212	110	118
Phorone (b.p. = 198°C) (2,6-dimethyl-2,5-heptadien-4-one)	28	221	118		48
2,4-Dichloroacetophenone	34	208			148
4-Chloropropiophenone	36	176	223		63
4-Methoxyacetophenone (b.p. = 258°C)	38	198	220	142	87
2-Hydroxybenzophenone	39			155	143
2-Methoxybenzophenone	39		251		148
3-Bromopropiophenone	40	183			
4-Phenyl-3-buten-2-one (benzalacetone)	41	187	227	157	115
1-Indanone	42	233	258	131	146
8-Pentadecanone (b.p. = 178°C)	43				120
4-Bromopropiophenone	46	171			91
Benzophenone (b.p. = 306°C)	48	165	239	138	143
4-Bromoacetophenone (b.p. = 225°C)	51	208	230	126	128
3,4-Dimethoxyacetophenone	51	218	207	131	140
Methyl 2-naphthyl ketone	54	235	262	177	149
4-Methyl benzophenone	57	121	200	109	154
Benzalacetophenone (chalcone)	58	170	245	120	140(68)
α-Chloroacetophenone	59	156	214	212	89
Desoxybenzoin (benzyl phenyl ketone)	60	148	204	116	98
Benzoylacetone	61		151	150	
1,1-Diphenylacetone	61	170		131	165
4-Methoxybenzophenone	62		180	132	116 (140)
Cinnamalacetone	68	186	223	180	153
2,6-Dimethyl-1,4-benzoquinone	73				175
4-Chlorobenzophenone	78		185	106	105
1,4-Cyclohexanedione	79	220 *mono* 231 *di*	240		188
3-Nitroacetophenone	80	257	228	128	132
4-Nitroacetophenone	81		258	132	174
4-Bromobenzophenone	82	350	230	126	116 (110)
9-Fluorenone	83		284	152	196
4,4′-Dimethylbenzophenone	95	143	219	100	163
Benzil	95	243 *di*	189 *di*	235 *di*	237 *di*
3-Hydroxyacetophenone	96	195	257		
1,3-Cyclohexanedione	104				156

Table 10b (Continued)

Name of compound	M.p. (°C)	Melting points of derivatives (°C): Semi-carbazone	2,4-Dinitro-phenyl-hydrazone	Phenyl-hydrazone	Oxime
4-Hydroxyacetophenone	109	199	261	151	145
3,4-Dihydroxyacetophenone	116				184d
1,4-Benzoquinone	116	243	231 *di*	152	240
1,4-Naphthoquinone	125	247	278	206d	198
4-Hydroxybenzophenone	135	194	242	144	81(152)
Benzoin	137	206d	245	159 (106)	152 (99)
2,4-Dihydroxyacetophenone	147	218	208	157	200d
4,4′-Dichlorobenzophenone	148		241		135
Anthrone	154				
Xanthone	174			152	161
4,4-Bis(dimethylamino)benzophenone	177		274	175	233
Camphor (b.p. = 209°C)	178	248d	177	233	119
Isatin	203				200 (255)
4,4′-Dihydroxybenzophenone	210		192		
Ninhydrin	243			208	201
Acenaphthenequinone	261	271 *di*		219 *di*	223 *di*
9,10-Anthraquinone	285			183	224
Chloranil (tetrachloro-1,4-benzoquinone)	290			220	

Table 11a Amines—Liquid

Name of compound	B.p. (°C)	Melting points of derivatives (°C): Benzamide	Benzene-sulfon-amide	Phenyl-thiourea
Isopropylamine	33		26	101
Ethylmethylamine[a]	36			
tert-Butylamine	46	134		120
n-Propylamine	49	84	36	63
Diethylamine	56	42	42	34
sec-Butylamine	63	76	70	101
Isobutylamine	69	57	53	82
n-Butylamine	77	42		65
Diisopropylamine[b]	84		94	
Pyrrolidine[c]	89			
Triethylamine[d]	89			
2-Aminopentane[e]	92			
Isopentylamine	96			102
n-Pentylamine	104			69
Piperidine	106	48	93	101
Di-*n*-propylamine	110		51	69
Ethylenediamine	116	244 *di*	168 *di*	102
Pyridine[f]	116			

Table 11a (Continued)

Name of compound	B.p. (°C)	Melting points of derivatives (°C)		
		Benzamide	Benzene-sulfon-amide	Phenyl-thiourea
2-Methylpyridine (2-picoline)[g]	129			
Morpholine	130	75	118	136
n-Hexylamine	132	40	96	77
Cyclohexylamine	134	149	89	148
2-Dimethylaminoethyl alcohol[h]	135			
1,3-Diaminopropane	136	148 *di*	96	
Diisobutylamine	139		55	113
2,6-Dimethylpyridine (2,6-lutidine)[i]	143			
3-Methylpyridine (3-picoline)[j]	143			
4-Methylpyridine (4-picoline)[k]	146			
n-Heptylamine	156			75
Tri-*n*-propylamine[l]	157			
Di-*n*-butylamine	159			86
1,4-Diaminobutane (m.p. = 27°C)	159	177 *di*		168
2-Aminoethanol	171			138
2,4,6-Trimethylpyridine[m] (2,4,6-collidine)	172			
1,5-Diaminopentane	178	135 *di*	119	148
n-Octylamine (m.p. = 0°C)[n]	180			
Benzyldimethylamine[o]	181			
Benzylmethylamine[p]	181			
Aniline	184	160	112	54
Benzylamine	185	105	88	156
1-Amino-1-phenylethane (α-phenethylamine)	187	120		
N,*N*-Dimethylaniline[q]	193			
N-Methylaniline	196	63	79	87
2-Amino-1-phenylethane (β-phenethylamine)	198	116	69	135
2-Methylaniline (*o*-toluidine)	200	146	124	136
n-Nonylamine	201	49		
3-Methylaniline (*m*-toluidine)	203	125	95	94
N-Ethylaniline	205	60	oil	89
2-Chloroaniline	208	99	129	156
4-Methylbenzylamine	208	137		
Tri-*n*-butylamine[r]	211			
2,6-Dimethylaniline (m.p. = 11°C)	215	168		204
2,5-Dimethylaniline (m.p. = 14°C)	215	140	138	148
2,4-Dimethylaniline	216	192	130	152
N,*N*-Diethylaniline[s]	218			
3,5-Dimethylaniline	220	144		153
2,3-Dimethylaniline	221	189		
2-Methoxyaniline (*o*-anisidine)	225	60(84)	89	136
4-Isopropylaniline	225	162		
2,4,6-Trimethylaniline	229	204	137	193
3-Chloroaniline	230	119	121	124
Quinoline[t]	237			

Table 11a (Continued)

Name of compound	B.p. (°C)	Melting points of derivatives (°C) Benzamide	Benzenesulfonamide	Phenylthiourea
2-Chloro-6-methoxyaniline	246	135		
4-Ethoxyaniline (*p*-phenetidine)	248	173	143	136
3-Bromoaniline (m.p. = 18°C)	251	120(136)		143
3-Methoxyaniline (*m*-anisidine)[u]	251			
Dicyclohexylamine (m.p. = 20°C)	255	153		
Tri-*n*-pentylamine[v]	257			
Dibenzylamine	300	112	68	

[a]Ethylmethylamine: hydrochloride salt, m.p. = 129°C.
[b]Diisopropylamine: hydrochloride salt, m.p. = 216°C.
[c]Pyrrolidine: *p*-toluenesulfonamide derivative, m.p. = 123°C.
[d]Triethylamine: hydrochloride salt, m.p. = 245°C (subl); m.p. = 253°C (sealed tube).
[e]2-Aminopentane: hydrochloride salt, m.p. = 168°C.
[f]Pyridine: hydrochloride salt, m.p. = 82°C (hygroscopic).
[g]2-Methylpyridine: n_D^{20} 1.4957.
[h]2-Dimethylaminoethyl alcohol: n_D^{20} 1.4294.
[i]2,6-Dimethylpyridine: n_D^{20} 1.4953.
[j]3-Methylpyridine: n_D^{20} 1.5040.
[k]4-Methylpyridine: n_D^{20} 1.5037; m.p. = 4°C.
[l]Tripropylamine: hydrochloride salt, m.p. = 90°C (hygroscopic).
[m]2,4,6-Trimethylpyridine: n_D^{20} 1.4959.
[n]Octylamine: n_D^{20} 1.4924.
[o]Benzyldimethylamine: n_D^{20} 1.5011; hydrochloride salt, m.p. = 175°C.
[p]Benzylmethylamine: n_D^{20} 1.5224; hydrochloride salt, m.p. = 178°C.
[q]*N*,*N*-Diethylaniline: n_D^{20} 1.5582; hydrochloride salt, m.p. = 85–95°C (hygroscopic).
[r]Tributylamine: n_D^{20} 1.4297.
[s]*N*,*N*-Diethylaniline: n_D^{20} 1.5409.
[t]Quinoline: n_D^{20} 1.6268.
[u]3-Methoxyaniline: acetamide derivative, m.p. = 81°C; *p*-toluenesulfonamide derivative, m.p. = 68°C.
[v]Tripentylamine: n_D^{20} 1.4366.

Table 11b Amines—Solid

Name of compound	M.p. (°C)	Melting points of derivatives (°C) Benzamide	Benzenesulfonamide	Phenylthiourea
2-Bromoaniline	32	116		146(161)
3-Iodoaniline	33	157		
N-Benzylaniline	37	107	119	103
2,6-Dichloroaniline	39		157	
1,6-Diaminohexane	42	155 *di*	154 *di*	
4-Methylaniline (*p*-toluidine)	44	158	120	141

Table 11b (Continued)

Name of compound	M.p. (°C)	Melting points of derivatives (°C) Benzamide	Benzene-sulfon-amide	Phenyl-thiourea
3,4-Dimethylaniline	49	185	118	
2,5-Dichloroaniline	50	120		166
3,5-Dichloroaniline	50	147		
Indole	52	68	254	
Diphenylamine	53	180	124	152
2-Aminopyridine	57	165		
4-Methoxyaniline (*p*-anisidine)	58	154	95	157(171)
4-Iodoaniline	62	222		153
1,3-Diaminobenzene (*m*-phenylenediamine)	63	125 240 *di*	194	
2,4-Dichloroaniline	63	117	128	
4-Bromoaniline	66	204	134	148
2-Nitroaniline	71	110	104	
4-Chloroaniline	72	192	122	152
3,4-Dichloroaniline	72	144	130	
8-Hydroxyquinoline[a]	75			
4-Methyl-3-nitroaniline	78	172	160	171
2,4,6-Trichloroaniline	78	174	154	
2,4-Dibromoaniline	79	134		171
3,4-Diaminotoluene	89	264 *di*	179 *di*	
Tribenzylamine[b]	91			
2-Methyl-3-nitroaniline	92	168		
2-Methyl-6-nitroaniline	97	167		
2,4-Diaminotoluene	99	224 *di*	192 *di*	
1,2-Diaminobenzene (*o*-phenylenediamine)	102	301 *di*	185	
Piperazine	104	196 *di*	285 *di*	
2-Bromo-4-nitroaniline	105	160		
4-Aminoacetophenone	106	205	128	
2-Chloro-4-nitroaniline	107	161		
2-Methyl-5-nitroaniline	107	186	172	
3-Nitroaniline	114	157	136	160
4-Methyl-2-nitroaniline	115	148	102	
4-Chloro-2-nitroaniline	116	133		
2,4,6-Tribromoaniline	122	198		
Triphenylamine[c]	127			
2-Nitro-4-methoxyaniline	129	140		
2-Methyl-4-nitroaniline	130		158	
2-Methoxy-4-nitroaniline	139	150	181	
1,4-Diaminobenzene (*p*-phenylenediamine)	142	300 *di*	247 *di*	
4-Nitroaniline	147	199	139	
2-Aminobenzoic acid (anthranilic acid)	147	182	214	
4-Nitro-*N*-methylaniline	152	112	121	
4-Aminopyridine	159	202		
2-Hydroxyaniline	174	167	141	146

Table 11b (Continued)

		Melting points of derivatives (°C)		
Name of compound	M.p. (°C)	Benzamide	Benzene-sulfon-amide	Phenyl-thiourea
3-Aminobenzoic acid	174	113		
2,4-Dinitroaniline	180	202 (220)		
4-Hydroxyaniline (4-aminophenol)	184	216	125	150
4-Aminobenzoic acid	188	278	212	
2,4,6-Trinitroaniline	190	196	211	

[a]8-Hydroxyquinoline: hydrochloride salt (hydrate), m.p. = 235°C.
[b]Tribenzylamine: hydrochloride salt, m.p. = 228°C.
[c]Triphenylamine: hydrochloride salt, m.p. = 214°C.

Table 12 Nitriles[a]

Name of compound	B.p. (°C)	M.p. (°C)	n_D^{20}
Acrylonitrile	77		1.3911
Acetonitrile	81		1.3442
Propanenitrile	97		1.3659
2-Methylpropanenitrile	108		1.3720
Butanenitrile	117		1.3812
4-Methylbutanenitrile	130		1.3927
Pentanenitrile	141		1.3991
4-Methylpentanenitrile	155		1.4059
Hexanenitrile	165		1.4115
3-Chloropropanenitrile	178		1.4379
5-Methylhexanenitrile	180		
Heptanenitrile	183		1.4104
Benzonitrile	190		1.5289
4-Chlorobutanenitrile	196		1.4413
2-Methylbenzonitrile (2-tolunitrile)	205		1.5279
Octanenitrile	206		1.4200
Ethyl cyanoacetate	207		1.4179
3-Methylbenzonitrile (3-tolunitrile)	212		1.5256
Nonanenitrile	224		1.4252
Phenylacetonitrite (benzyl cyanide)	234		1.5211
Decanenitrile	245		1.4295
Dodecanenitrile	277		1.4358
Glutaronitrile (1,3-dicyanopropane)	286		1.4365
Cinnamonitrile	256	20	
4-Methylbenzonitrile (4-tolunitrile)	217	27	
Malononitrile	219	30	
4-Chlorobenzyl cyanide	267	30	
1-Cyanonaphthalene	299	34	
3-Bromobenzonitrile	225	38	
3-Chlorobenzonitrile		41	
2-Chlorobenzonitrile	232	43	

Table 12 (Continued)

Name of compound	B.p. (°C)	M.p. (°C)	n_D^{20}
4-Cyanobutanoic acid		45	
3-Cyanopropanoic acid		48	
3-Cyanopyridine		50	
2-Aminobenzonitrile	266	51	
2-Bromobenzonitrile	253	53	
2,4,6-Trimethylbenzonitrile		55	
Succinonitrile	267	57	
4-Methoxybenzonitrile	256	62	
3,5-Dichlorobenzonitrile		65	
Cyanoacetic acid		67	
3,4-Dichlorobenzonitrile		72	
Diphenylacetonitrile		75	
4-Cyanopyridine		78	
3-Cyanobenzaldehyde		80	
2-Chloro-6-methylbenzonitrile		82	
4-Aminobenzonitrile		86	
4-Chlorobenzonitrile	223	96	
2-Cyanophenol		98	
2,4-Dinitrobenzonitrile		104	
2-Nitrobenzonitrile		110	
4-Bromobenzonitrile	237	112	
4-Cyanophenol		113	
p-Nitrophenylacetonitrile		116	
3-Nitrobenzonitrile		118	
2,5-Dichlorobenzonitrile		130	
1,2-Dicyanobenzene		141	
2,6-Dinitrobenzonitrile		145	
4-Nitrobenzonitrile		147	
1,3-Dicyanobenzene		162	
2-Cyanobenzoic acid		187	
Tetracyanoethylene		198	
3-Cyanobenzoic acid		217	
4-Cyanobenzoic acid		219	

[a]See Table 14 for the physical properties and derivatives of the corresponding carboxylic acid formed by hydrolysis.

Table 13 Nitro Compounds[a]

Name of compound	B.p. (°C)	M.p. (°C)	n_D^{20}	Nitration product: Nitro positions	Nitration product: m.p. (°C)
Nitromethane	101		1.3817		
Nitroethane	115		1.3917		
2-Nitropropane	120		1.3944		
1-Nitropropane	131		1.4016		
2-Nitrobutane	140		1.4013		
1-Nitrobutane	153		1.4103		

Table 13 (Continued)

Name of compound	B.p. (°C)	M.p. (°C)	n_D^{20}	Nitration product: Nitro positions	Nitration product: m.p. (°C)
1-Nitropentane	173		1.4175		
Nitrobenzene	211		1.5562	1,3	90
2-Nitrotoluene	222		1.5474	2,4	71
1,3-Dimethyl-2-nitrobenzene	226	13		1,3,5	182
3-Nitrotoluene	233	16	1.5470	mixture	
1,4-Dimethyl-2-nitrobenzene	241		1.5414	1,2,4	139
1,3-Dimethyl-4-nitrobenzene	246	2	1.5497	1,3,5	182
1,2-Dimethyl-3-nitrobenzene	248	15	1.5434	1,2	82
2-Nitro-*p*-cymene (1-isopropyl-4-methyl-2-nitrobenzene)	264		1.5280	2,6	54
2-Nitroanisole	273	10	1.5620	2,4,6	68
2-Methyl-2-nitropropane	127	26	1.4036		
1,2-Dimethyl-4-nitrobenzene		31		1,2	82
2-Chloronitrobenzene	246	32		2,4	52
2,4-Dichloronitrobenzene	258	33			
2-Chloro-6-nitrotoluene	238	37		4,6	49
3-Nitroanisole		38		3,5	106
4-Chloro-2-nitrotoluene		38		2,6	77
2-Bromonitrobenzene		43		1,3	72
2,4,6-Trimethylnitrobenzene		44		1,3	86
3-Chloronitrobenzene		45		mixture	
4-Nitrotoluene	234	52		2,4	70
1-Chloro-2,4-dinitrobenzene		52		2,4,6	183
4-Nitroanisole		53		2,4	89
3-Nitrobromobenzene		56		3,4	59
1-Nitronaphthalene		57			
3,4-Dinitrotoluene		61			
2,6-Dinitrotoluene		66		2,4,6	82
2,4-Dinitrotoluene		70		2,4,6	82
3,5-Dimethylnitrobenzene		75			
4-Chloronitrobenzene	242	84		2,4	52
1,3-Dinitrobenzene		90			
1-Chloro-8-nitronaphthalene		94			
2,4-Dinitroanisole		95		2,4,6	68
4-Nitrobiphenyl		114		4,4′	240
1,2-Dinitrobenzene		118			
4-Nitrobromobenzene		126			
9-Nitroanthracene		146			
1,8-Dinitronaphthalene		170			
1,4-Dinitrobenzene		173			
1,5-Dinitronaphthalene		217			
4,4′-Dinitrobiphenyl		240			

[a]See Table 11 for derivatives of the corresponding amine obtained on reduction of the nitro group.

Table 14a Carboxylic Acids—Liquid

Name of compound	B.p. (°C)	n_D^{20}	Melting points of derivatives (°C)		
			Anilide	*p*-Toluidide	Amide
Formic acid	101	1.3714	50	53	
Acetic acid	118	1.3721	114	147	82
Propenoic acid (acrylic acid)	140	1.4224	105	141	85
Propionic acid	141	1.3861	106	126	81
Isobutyric acid	155	1.3920	105	107	129
2-Methylpropenoic acid (methacrylic acid)	163	1.4314			109
Butanoic acid	164	1.3983	96	75	115
3-Butenoic acid	164	1.4249	58		73
cis-2-Butenoic acid	169	1.4456	102	132	102
3-Methylbutanoic acid (isovaleric acid)	177	1.4043	110	107	137
3,3-Dimethylbutanoic acid	184	1.4096	132	134	132
2-Chloropropanoic acid	186	1.4345	92	124	80
Pentanoic acid (valeric acid)	186	1.4086	63	74	106
Dichloroacetic acid	194	1.4659	118	153	98
4-Methylpentanoic acid (isocaproic acid)	199	1.4144	112	63	121
2-Bromopropanoic acid	205d	1.4750	99	125	123
Hexanoic acid (caproic acid)	205	1.4163	95	75	100
2-Bromobutanoic acid	217d	1.4720	98	92	112
Heptanoic acid	223	1.4234	70	81	96
2-Ethylhexanoic acid	228	1.4241			102
Octanoic acid (m.p. = 16°C)	239	1.4268	57	70	110
Nonanoic acid (m.p. = 12°C)	255	1.4319	57	84	99
2-Phenylpropanoic acid	265	1.5225			92
Decanoic acid (m.p. = 32°C)	270		70	78	108
Undecanoic acid (m.p. = 30°C)	284		71	80	103
Oleic acid (m.p. = 16°C) (*cis*-9-octadecenoic acid)	d	1.4582	41	42	76

Table 14b Carboxylic Acids—Solid

Name of compound	M.p. (°C)	B.p. (°C)	Melting points of derivatives (°C)		
			Anilide	*p*-Toluidide	Amide
Undecylenic acid (10-undecenenoic acid)	25	275	67	68	87
2-Bromopropanoic acid	26	205d	99	125	123
Undecanoic acid	29	284	71	89	103
Cyclohexanecarboxylic acid	31	233	146		186
Decanoic acid	32	270	70	78	108
2-Oxobutanoic acid (α-ketobutyric acid)	32				117
Levulinic acid (4-oxopentanoic acid)	34	245	102	108	108d
Pivalic acid (trimethylacetic acid)	36	164	130	120	157

Table 14b (Continued)

Name of compound	M.p. (°C)	B.p. (°C)	Melting points of derivatives (°C) Anilide	*p*-Toluidide	Amide
Acetoacetic acid (3-oxobutanoic acid)	37	d	86	95	54
3-Chloropropanoic acid	42	204			101
2-Phenylbutanoic acid	42	270			85
Dodecanoic acid (lauric acid)	44	299	78	87	100
Tridecanoic acid	44	312	80	88	100
Hydrocinnamic acid (3-phenyl-propanoic acid)	48	280	98	135	105
Bromoacetic acid	50	208	131	91	91
4-Phenylbutanoic acid	52	290			84
Pentadecanoic acid	52		78		103
Tetradecanoic acid (myristic acid)	54		84	93	107
Trichloroacetic acid	58	198	97	113	141
5-Phenylpentanoic acid	60		90		109
Heptadecanoic acid	61				108
3-Bromopropanoic acid	61				111
Hexadecanoic acid (palmitic acid)	62		91	98	106
Chloroacetic acid	63	189	137	162	118
cis-2-Methyl-2-butenoic acid (tiglic acid)	65	199	77	71	76
Cyanoacetic acid	66		198		120
Benzoylformic acid	66				91
Octadecanoic acid (stearic acid)	70		96	102	109
trans-2-Butenoic acid (crotonic acid)	72	189	118	132	160
m-Methoxyphenylacetic acid	73				125
Phenylacetic acid	76		117	136	156
Glycolic acid (hydroxyacetic acid)	79		97	143	120
α-Methylcinnamic acid	81				128
Iodoacetic acid	83		144		95
p-Methoxyphenylacetic acid	87				189
o-Methylphenylacetic acid	90				161
2-Benzoylbenzoic acid hydrate	90		195		165
Citraconic acid (methylmaleic acid)	92d		170 *mono*	175 *di*	187 *di*
β-Chlorocrotonic acid (Z-3-Chloro-2-butenoic acid)	94		124		101
o-Chlorophenylacetic acid	95		139	170	175
p-Methylphenylacetic acid	95				185
3,4-Dimethoxyphenylacetic acid	95				147
Glutaric acid (pentanedioic acid)	98		224	218	176 *di*
3-Phenoxypropionic acid	98				119
Phenoxyacetic acid	99		99		102
Citric acid hydrate	100		192 *tri*	189 *tri*	210d *tri*
2-Methoxybenzoic acid (*o*-anisic acid)	101	200	131		129
Malic acid (2-hydroxybutanedioic acid)	101		197 *di*	207 *di*	157 *di*
Oxalic acid dihydrate	101		149 *mono* 254 *di*	169 *mono* 268 *di*	219 *mono* 419d *di*
o-Toluic acid (2-methylbenzoic acid)	104		125	144	143

Table 14b (Continued)

Name of compound	M.p. (°C)	B.p. (°C)	Melting points of derivatives (°C) Anilide	*p*-Toluidide	Amide
Heptanedioic acid (pimelic acid)	105		109 *mono*	206 *di*	175 *di*
Nonanedioic acid (azelaic acid)	107		108 *mono* 186 *di*	201 *di*	95 *mono* 172 *di*
3-Methoxybenzoic acid (*m*-anisic acid)	110				136
Ethylmalonic acid	111		150		214 *di*
m-Toluic acid	112		126	118	95
2-Phenylbenzoic acid	113				177
2-Phenoxybenzoic acid	113				131
p-Bromophenylacetic acid	114				194
2-Acetylbenzoic acid	115				116
Methylsuccinic acid	115		159 *mono* 200 *di*	164	225 *di*
β-Benzoylpropionic acid	116		150		146
2,6-Dimethylbenzoic acid	116				139
4-Isopropylbenzoic acid	117				133
Benzylmalonic acid	117d		217 *di*		225 *di*
Tropic acid (2-phenyl-3-hydroxy-propanoic acid)	117				169
d,l-Mandelic acid	118		152	172	132
m-Nitrophenylacetic acid	120				110
3-Furoic acid	121				169
Benzoic acid	122		160	158	130
Picric acid (see Table 7)	123				
2,4-Dimethylbenzoic acid	127		141		180
2-Benzoylbenzoic acid	127		195		165
Dodecanedioic acid	128		191 *di*	165 *di*	185 *di*
Maleic acid (*cis*-butenedioic acid)	130		187 *di*	142 *di*	173 *mono* 260 *di*
1-Naphthylacetic acid	132		155(160)		180
2,5-Dimethylbenzoic acid	132		140		186
m-Chlorocinnamic acid	133		135	142	76
2-Furoic acid	133		124	108	143
trans-Cinnamic acid	133		153	168	148
Decanedioic acid (sebacic acid)	134		202 *di*	201 *di*	210 *di*
Malonic acid (propanedioic acid)	135		230 *di*	253 *di*	170 *di*
O-Acetylsalicylic acid (aspirin)	135		136		138
1,3-Acetonedicarboxylic acid	135d		155 *di*		
Pyridine-2-carboxylic acid (picolinic acid)	137		76	104	107
Phenylpropynoic acid	137		126	142	100
Methylmalonic acid	138d		182	228 *di*	217
5-Chloro-2-nitrobenzoic acid	139		164		154
3-Nitrobenzoic acid	140		154	162	143
meso-Tartaric acid	140		194 *mono*		187 *di*
2-Chloro-4-nitrobenzoic acid	141		168		172
o-Nitrophenylacetic acid	141				161
2,4-Dichlorophenoxyacetic acid	141				130
4-Chloro-2-nitrobenzoic acid	142				172
2-Naphthylacetic acid	142				200

Table 14b (Continued)

Name of compound	M.p. (°C)	B.p. (°C)	Melting points of derivatives (°C)		
			Anilide	*p*-Toluidide	Amide
2-Chlorobenzoic acid	142		118	131	142
Octanedioic acid (suberic acid)	144		186 *di*	218 *di*	217 *di*
2,4,5-Trimethoxybenzoic acid	144		155		185
2,6-Dichlorobenzoic acid	144				202
o-Chlorophenoxyacetic acid	146		121		150
2-Nitrobenzoic acid	146		155		176
2-Aminobenzoic acid (anthranilic acid)	147		131	151	109
Diphenylacetic acid	148		180	173	168
p-Hydroxyphenylacetic acid	148				175
2-Bromobenzoic acid	150		141		155
Benzilic acid	150		175	190	154
Citric acid (anhydrous)	153		192 *tri*	189 *tri*	210d *tri*
p-Nitrophenylacetic acid	153		198	210	198
2,5-Dichlorobenzoic acid	153				155
Hexanedioic acid (adipic acid)	154		241 *di*	239	220 *di*
3-Bromobenzoic acid	155		136		155
2,4,6-Trimethylbenzoic acid	155				188
p-Chlorophenoxyacetic acid	156		125		133
Tartronic acid (hydroxymalonic acid)	157d				198 *di*
3-Chlorobenzoic acid	158		123		134
Salicylic acid (2-hydroxybenzoic acid)	158		136	156	142
1-Naphthoic acid	162		163		205
2-Iodobenzoic acid	162		141		110
4-Nitrophthalic acid	164		192	172 *mono*	200d
2,4-Dichlorobenzoic acid	164				194
3,4-Dinitrobenzoic acid	165		189		166
Itaconic acid (propene-2,3-dicarboxylic acid)	166		152 *mono* (190)		192 *di*
5-Bromosalicylic acid	165		222		232
2-Chloro-5-nitrobenzoic acid	165				178
Tricarballylic acid (1,2,3,-propane-tricarboxylic acid)	166		252 *tri*		207d *tri*
3,4-Dimethylbenzoic acid	166		104		130
3-Methylsalicylic acid	166		83		112
3,5-Dimethylbenzoic acid	166				133
Phenylsuccinic acid	167		175α 171β 222 *di*	175α 169β	159α 145β 211 *di*
d-Tartaric acid	170		180d *mono* 264d *di*		172 *mono* 196d *di*
3,4,5-Trimethoxybenzoic acid	171				177
5-Chlorosalicylic acid	172				227
3-Aminobenzoic acid	174		140		111
3,5-Dinitrosalicylic acid hydrate	174				181
Acetylenedicarboxylic acid	179				249d *di*
p-Toluic acid (4-methylbenzoic acid)	179	275	145	160	160

Table 14b (Continued)

Name of compound	M.p. (°C)	B.p. (°C)	Melting points of derivatives (°C)		
			Anilide	*p*-Toluidide	Amide
3,4-Dimethoxybenzoic acid	181		154		164
4-Chloro-3-nitrobenzoic acid	182		131		156
2,4-Dinitrobenzoic acid	183				203
4-Methoxybenzoic acid (*p*-anisic acid)	185		171	186	167
2-Naphthoic acid	185		171	192	192
3-Iodobenzoic acid	187				186
Coumarin-3-carboxylic acid	188		250		236
p-Nitrophenoxyacetic acid	188		170		158
d-Camphoric acid	188		204 *mono* 226 *di*	214α 196β	182 *mono* 192 *di*
4-Aminobenzoic acid	188				183
Succinic acid (butanedioic acid)	189		149 *mono* 230 *di*	180 *mono* 255 *di*	157 *mono* 260d *di*
Hippuric acid	190		208		183
Dimethylmalonic acid	193 subl				269 *di*
trans-Aconitic acid (1,2,3-propenetricarboxylic acid)	194d		189 *di*		>250 *tri*
4-Ethoxybenzoic acid	198		170		202
trans-m-Nitrocinnamic acid	199				196
3,4-Dihydroxybenzoic acid	200d		166		212
Fumaric acid	200 (*on slow heating*)		314 *di*		266 *di*
3-Hydroxybenzoic acid	200		157	163	170
2,5-Dihydroxybenzoic acid	204				218
d,l-Tartaric acid	204		236 *di*		226 *di*
2,3-Dihydroxybenzoic acid	204				175
3,5-Dinitrobenzoic acid	207		234	147	183
3,4-Dichlorobenzoic acid	209				168
Phthalic acid	210		253 *di*	201 *di*	220 *di*
o-Chlorocinnamic acid	212		176		168
2,4-Dihydroxybenzoic acid	213 (216d)		126		222
trans-o-Hydroxycinnamic acid	214d				209d
4-Hydroxybenzoic acid	215		197	204	162
3-Nitrophthalic acid	218		234 *di*	226 *di*	201d *di*
4-Cyanobenzoic acid	219		179		223
4-Phenylbenzoic acid	226				223
Piperonylic acid	229				169
5-Nitrosalicylic acid	230		224		225
3-Chloro-2-nitrobenzoic acid	235		186		
trans-o-Nitrocinnanic acid	240				185
4-Nitrobenzoic acid	241		211	204	201
4-Chlorobenzoic acid	243		194		179
4-Dimethylaminobenzoic acid	245		183		206
4-Bromobenzoic acid	251		197		190
3,4,5-Trihydroxybenzoic acid (gallic acid)	254d		207		189
4-Iodobenzoic acid	270		210		217

Table 14b (Continued)

Name of compound	M.p. (°C)	B.p. (°C)	Melting points of derivatives (°C)		
			Anilide	*p*-Toluidide	Amide
p-Nitrocinnamic acid	285				204
Fumaric acid	289 (*sealed tube*)		314 *di*		266 *di*
2-Bromobenzene-1,4-dicarboxylic acid	299				270 *di*
Terephthalic acid (1,4,-benzene-dicarboxylic acid)	300 subl		336 *di*		
Isophthalic acid (1,3-benzene-dicarboxylic acid)	348 subl				280 *di*

Table 15 Carboxylic Acid Chlorides

Name of compound	B.p. (°C)	M.p. (°C)	Corresponding carboxylic acid	
			b.p. (°C)	m.p. (°C)
Acetyl chloride	52		118	
Oxalyl chloride	64			101
Methyl chloroformate	72			
Propionyl chloride	80		141	
Isobutyryl chloride	92		155	
Ethyl chloroformate	93			
Methacrylyl chloride	95		163	
Butyryl chloride	102		164	
Chloroacetyl chloride	108		189	63
Methoxyacetyl chloride	113		204	
Isovaleryl chloride	115		177	
Trichloroacetyl chloride	118		197	58
trans-Crotonyl chloride	126		189	72
Pentanoyl chloride	126		186	
Isobutyl chloroformate	129			
Hexanoyl chloride	153		205	
Fumaryl chloride	162			289(200)
Cyclohexanecarboxylic acid chloride	184			31
Succinyl chloride	190d	20		189
Octanoyl chloride	196		239	16
Benzoyl chloride	197			122
Phenylacetyl chloride	210			76
Nonanoyl chloride	215		255	12
Glutaryl chloride	218			98
4-Chlorobenzoyl chloride	222			243
3-Chlorobenzoyl chloride	225			158
Phenoxyacetyl chloride	226			99
4-Methylbenzoyl chloride	226			179
2-Chlorobenzoyl chloride	238			142
Phthaloyl chloride	280	15		210
Adipoyl chloride	d			154
Sebacoyl chloride	d			134

Table 15 (Continued)

Name of compound	B.p. (°C)	M.p. (°C)	Corresponding carboxylic acid b.p. (°C)	Corresponding carboxylic acid m.p. (°C)
trans-Cinnamoyl chloride	258	35		133
4-Bromobenzoyl chloride	247	42		251
Isophthaloyl chloride	276	44		348
2,4-Dinitrobenzoyl chloride		46		183
3,5-Dinitrobenzoyl chloride		69		207
4-Nitrobenzoyl chloride		75		241
Terephthaloyl chloride		84		300

Table 16 Carboxylic Acid Anhydrides

Name of compound	B.p. (°C)	M.p. (°C)	Corresponding carboxylic acid b.p. (°C)	Corresponding carboxylic acid m.p. (°C)
Acetic anhydride	140		118	
Propionic anhydride	167		141	
Isobutyric anhydride	182		155	
Butyric anhydride	198		164	
Dichloroacetic anhydride	216d		194	
cis-1,2-Cyclohexandicarboxylic anhydride		34		192
Benzoic anhydride		42		122
Chloroacetic anhydride		46	189	63
Maleic anhydride		54		130
Glutaric anhydride		56		98
4-Methylbenzoic anhydride		95		179
Succinic anhydride		120		189
Phthalic anhydride		132		210
trans-1,2-cyclohexanedicarboxylic anhydride		147		230
3-Nitrophthalic anhydride		162		218
4-Nitrobenzoic anhydride		189		241
d,l-Camphoric anhydride		225		187
Tetrachlorophthalic anhydride		256		250d
1,8-Naphthalic anhydride		274		270d
Tetrabromophthalic anhydride		280		266

Table 17a Carboxylic Acid Esters—Liquid

Name of compound	B.p. (°C)	n_D^{20}
Methyl formate	31	1.3433
Ethyl formate	54	1.3597
Methyl acetate	57	1.3593
Isopropyl formate	71	1.3678
Vinyl acetate	73	1.3954

Table 17a (Continued)

Name of compound	B.p. (°C)	n_D^{20}
Ethyl acetate	77	1.3723
Methyl propionate	80	1.3779
Methyl acrylate	80	1.3984
Propyl formate	81	1.3779
Isopropyl acetate	90	1.3773
Methyl carbonate	91	1.3687
Methyl isobutyrate	93	1.3840
Isopropenyl acetate	94	1.4033
tert-Butyl acetate	98	1.3853
Methyl methacrylate	100	1.4142
Ethyl propionate	100	1.3853
Ethyl acrylate	101	1.4059
Propyl acetate	102	1.3847
Methyl butyrate	102	1.3879
Allyl acetate	104	1.4049
Ethyl isobutyrate	110	1.3903
Isopropyl propionate	110	1.3872
sec-Butyl acetate	112	1.3888
Methyl isovalerate	117	1.3927
Isobutyl acetate	117	1.3901
Ethyl pivalate (ethyl trimethylacetate)	118	1.3906
Methyl crotonate	119	1.4242
Ethyl butyrate	122	1.4000
Propyl propionate	123	1.3933
Butyl acetate	126	1.3961
Diethyl carbonate	127	1.3852
Methyl valerate	127	1.4003
Methyl methoxyacetate	130	1.3964
Methyl chloroacetate	131	1.4221
Ethyl isovalerate	135	1.4009
Methyl pyruvate	137	1.4065
Methyl α-hydroxyisobutyrate	137	1.4112
Ethyl crotonate	138	1.4252
3-Methylbutyl acetate (isoamyl acetate)	142	1.4003
Methyl lactate	145	1.4144
Ethyl chloroacetate	145	1.4227
Ethyl valerate	146	1.4009
Ethyl α-chloropropionate	146	1.4170
Diisopropyl carbonate	147	1.3932
Pentyl acetate	149	1.4031
Methyl hexanoate	151	1.4049
Cyclopentyl acetate	153	1.432
Ethyl lactate	154	1.4130
Ethyl pyruvate	155	1.4056
Ethyl dichloroacetate	158	1.4386
Ethyl α-bromopropionate	162	1.4490
Butyl butyrate	167	1.4075
Ethyl hexanoate	168	1.4073
Ethyl trichloroacetate	168	1.450
Methyl acetoacetate	170	1.4196

Table 17a (Continued)

Name of compound	B.p. (°C)	n_D^{20}
Hexyl acetate	172	1.4092
Methyl heptanoate	172	1.4152
Cyclohexyl acetate	175	1.4401
Furfuryl acetate	176	1.4627
Ethyl β-bromopropionate	179	1.4254
Ethyl acetoacetate	181	1.4198
Methyl furoate	181	1.4860
Dimethyl malonate	182	1.4140
Diethyl oxalate	185	1.4104
Ethyl δ-chlorobutyrate	186	1.4306
Ethyl heptanoate	187	1.4100
Ethylene glycol diacetate	190	1.4150
Heptyl acetate	192	1.4130
Methyl octanoate	193	1.4170
Dimethyl succinate (m.p. = 18°C)	196	1.4197
Ethyl cyclohexanecarboxylate	196	1.4501
Dimethyl methylsuccinate	196	1.4200
Phenyl acetate	197	1.5033
Diethyl malonate	199	1.4139
Methyl benzoate	199	1.5168
γ-Butyrolactone	204	1.4365
Dimethyl maleate	204	1.4416
Ethyl levulinate	206	1.4229
γ-Valerolactone	207	1.4330
Ethyl octanoate	208	1.4178
Octyl acetate	210	1.4190
Ethyl benzoate	212	1.5057
Dimethyl glutarate	215	1.4242
Methyl nonanoate	215	1.4214
Benzyl acetate	217	1.5200
Diethyl succinate	217	1.4198
Diethyl fumarate	218	1.4410
Methyl phenylacetate	220	1.5075
Diethyl maleate	223	1.4416
Methyl salicylate	224	1.5369
Methyl decanoate	225	1.4256
Ethyl phenylacetate	228	1.4980
Propyl benzoate	231	1.5014
Diethyl glutarate	234	1.4240
Ethyl salicylate	234	1.5296
Methyl β-phenylpropionate	238	1.503
Propylene carbonate	240	1.4210
Diethyl adipate	245	1.4277
Methyl undecylenate	248	1.4393
Diethyl pimelate	255	1.4298
Ethyl benzoylacetate	265	1.5498
Dimethyl suberate	268	1.4333
Ethyl cinnamate	271	1.5598
Methyl 2-nitrobenzoate	275	1.5350
Diethyl tartrate (m.p. = 18°C)	280	1.4468

Table 17a (Continued)

Name of compound	B.p. (°C)	n_D^{20}
Diethyl suberate	282	1.4324
Dimethyl phthalate	284	1.5138
Diethyl phthalate	290	1.5019
Ethyl 3-aminobenzoate	294	1.5608
Diethyl benzylmalonate	300	1.4868
Methyl myristate (m.p. = 18°C)	323 (295)	1.4362
Diisobutyl phthalate	327	1.4888
Dibutyl phthalate	340	1.4900

Table 17b Carboxylic Acid Esters—Solid

Name of compound	M.p. (°C)	B.p. (°C)
Dimethyl succinate	18	196
Methyl myristate	18	323
Diethyl tartarate	18	280
Benzyl benzoate	21	324
Methyl anthranilate (methyl 2-aminobenzoate)	24	300
Dimethyl sebacate	27	
Bornyl acetate	29	221
Methyl palmitate	30	
Ethyl 2-nitrobenzoate	30	275
Methyl 4-toluate	33	222
Ethyl stearate	33	
Ethyl 2-furoate	36	196
Methyl cinnamate	36	261
Ethylene carbonate	37	245
Ethyl mandelate	37	254
Dimethyl itaconate	39	
Methyl stearate	39	
Phenyl salicylate	42	
Diethyl terephthalate	44	302
Ethyl 3-nitrobenzoate	47	296
Dimethyl tartrate	49	280
1-Naphthyl acetate	49	
Methyl mandelate	53	250d
Dimethyl oxalate	54	
Ethyl 4-nitrobenzoate	56	186
Coumarin	67	290
Dimethyl isophthalate	68	
Phenyl benzoate	69	314
Methyl 3-hydroxybenzoate	70	
Diphenyl phthalate	74	
Diphenyl carbonate	78	306
Methyl 3-nitrobenzoate	78	279
Methyl 4-bromobenzoate	83	
Ethyl 4-aminobenzoate	90	
Dimethyl *d,l*-tartarate	90	282

Table 17b (Continued)

Name of compound	M.p. (°C)	B.p. (°C)
3-Carbethoxycoumarin	93	
Methyl 4-nitrobenzoate	96	
Propyl 4-hydroxybenzoate	96	
Dimethyl fumarate	102	193
Cholesteryl acetate	114	
Ethyl 4-hydroxybenzoate	116	
Hydroquinone diacetate	124	
Methyl 4-hydroxybenzoate	131	
Ethyl 4-nitrocinnamate	137	
Dimethyl terephthalate	141	
Propyl gallate	150	

Table 18a Carboxylic Acid Amides—Liquid

Name of compound	B.p. (°C)	n_D^{20}
N,N-Dimethylformamide	153	1.4305
N,N-Dimethylacetamide	165	1.4380
N,N-Diethylformamide	178	1.4321
N-Methylformamide	185	1.4319
N,N-Diethylacetamide	186	1.4374
Formamide	193(195d)	1.4472
N-Ethylformamide	199	1.4320
N-Methyl-2-pyrrolidinone	202	1.4684
N-Ethylacetamide	205	
N-Methylformanilide (m.p. = 13°C)	244	1.5593
2-Pyrrolidinone (γ-butyrolactam) (m.p. = 24°C)	250	1.4860

Table 18b Carboxylic Acid Amides—Solid

Name of compound	M.p. (°C)
N-Methylacetamide	31
δ-Valerolactam (2-piperidone)	39
Ethyl urethane	49
Formanilide	50
Methyl urethane	52
Phenyl urethane	53
N-Ethylacetanilide	54
Butyl urethane	54
Acetoacetamide	54
Propyl urethane	60
N-Benzyformamide	60
Pentananilide	63
Heptananilide	70
Decananilide	70
ε-Caprolactam	71
3-Butenamide	73
N,N-Diphenylformamide	73
Oleamide	76
N-Acetylacetamide	79
α-Chloropropionamide	80
Propionamide	81
N-Methylbenzamide	82
Acetamide	82
Acrylamide	85
Acetoacetanilide	86
3-Bromoacetanilide	87
2-Chloroacetanilide	88
N-Phenylmaleimide	91
Bromoactamide	91
2-Nitroacetanilide	92
Maleimide	93
Hexananilide	95
Iodoacetamide	95
Heptanamide	96
Butyranilide	96
m-Toluamide	97
Dichloroacetamide	98
Nonanamide	99
Hexanamide	100
N-Methylacetamide	102
Undecanamide	103
Isobutyranilide	105
Propionanilide	106
Hexadecanamide (palmitamide)	106
Pentanamide	106
Tetradecanamide (myristamide)	107
Decanamide	108
Heptadecanamide	108
Anthranilamide	109
Octadecanamide (stearamide)	109

Table 18b (Continued)

Name of compound	M.p. (°C)
Dodecanamide (lauramide)	110
Octanamide	110
β-Bromopropionamide	111
4-Aminobenzamide	114
Acetanilide	114
Butyramide	115
Methacrylamide	116
Chloroacetamide	118
Cyanoacetamide	120
Succinimide	126
Isobutyramide	129
2-Methoxybenzamide	129
Benzamide	130
2-Ethoxybenzamide	130
Urea	133
3-Chlorobenzamide	134
Phenacetin	134
3-Methylbutanamide	136
Salicylanilide	136
2-Chlorobenzamide	142
Salicylamide	142
o-Toluamide	143
3-Nitrobenzamide	143
Cinnamamide	148
Trimethylacetamide (pivalamide)	155
2,5-Dichlorobenzamide	155
3-Brombenzamide	155
Phenylacetamide	156
Succinic acid monoamide	157
N-(1-Naphthyl)acetamide	159
p-Toluamide	159
2-Bromobenzamide	161
4-Hydroxybenzamide	162
Benzanilide	163
3,4-Dimethoxybenzamide	164
4-Bromoacetanilide	167
4-Methoxybenzamide	167
4-Hydroxyacetanilide	169
3-Hydroxybenzamide	170
Malonamide (diamide)	170
N-Bromosuccinimide	173
2-Nitrobenzamide	176
4-Chloroacetanilide	179
4-Aminobenzamide	183
3,5-Dinitrobenzamide	183
2-Iodobenzamide	184
3-Iodobenzamide	186
4-Bromobenzamide	189
Biuret	192d
4-Nitrobenzamide	200

Table 18b (Continued)

Name of compound	M.p. (°C)
Isatin	201d
1-Naphthamide	202
2,6-Dichlorobenzamide	202
2,4-Dinitrobenzamide	203
4-Nitroacetanilide	215
4-Iodobenzamide	217
Hydantoin	218
Phthalamide (diamide)	220
2,4-Dihydroxybenzamide	222
Phthalimide	238
sym-Diphenylurea	240
Succinamide (diamide)	260d

Appendix B

Characteristic Infrared Absorptions for Organic Compounds

CHARACTERISTIC INFRARED ABSORPTIONS FOR ORGANIC COMPOUNDS

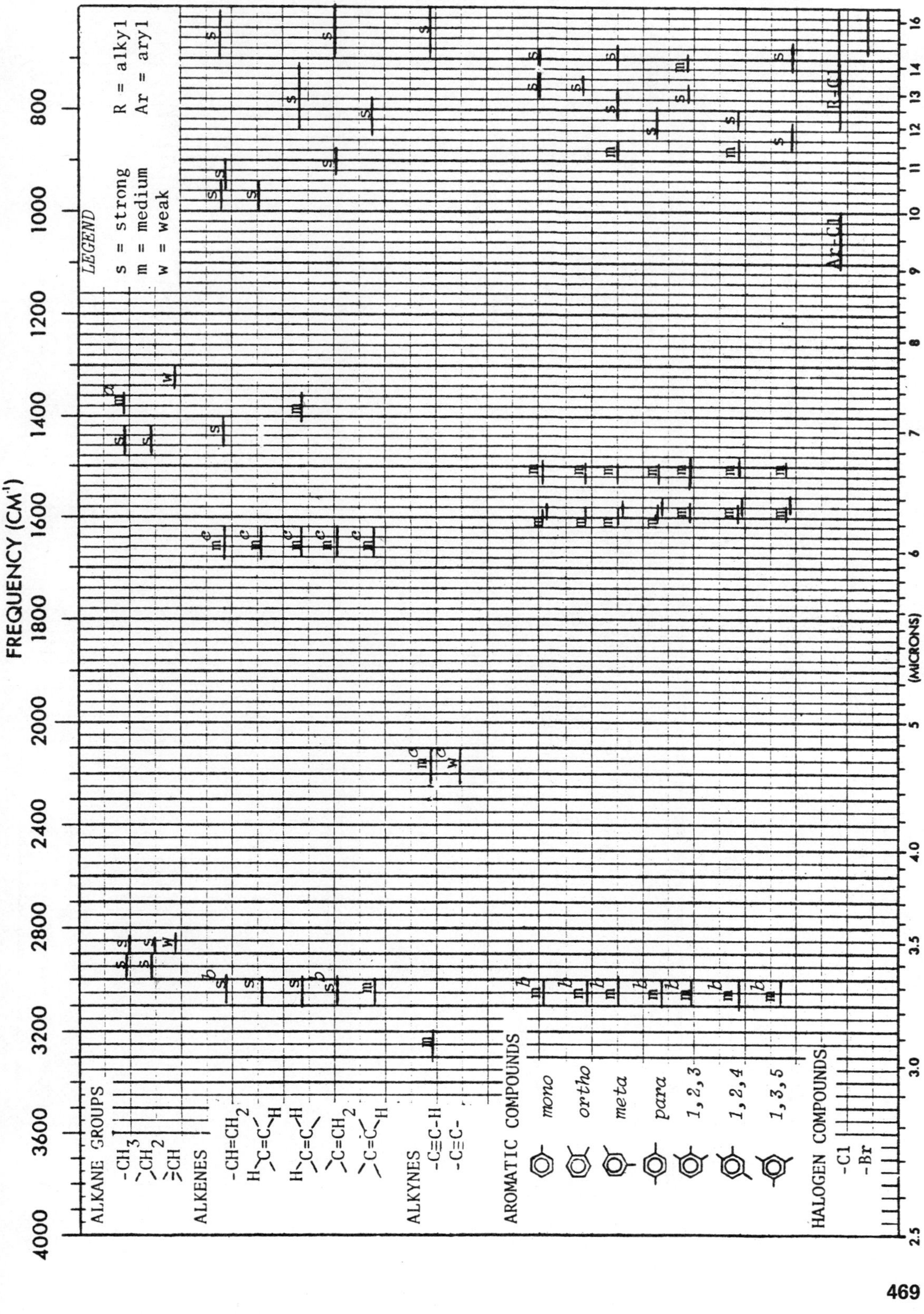

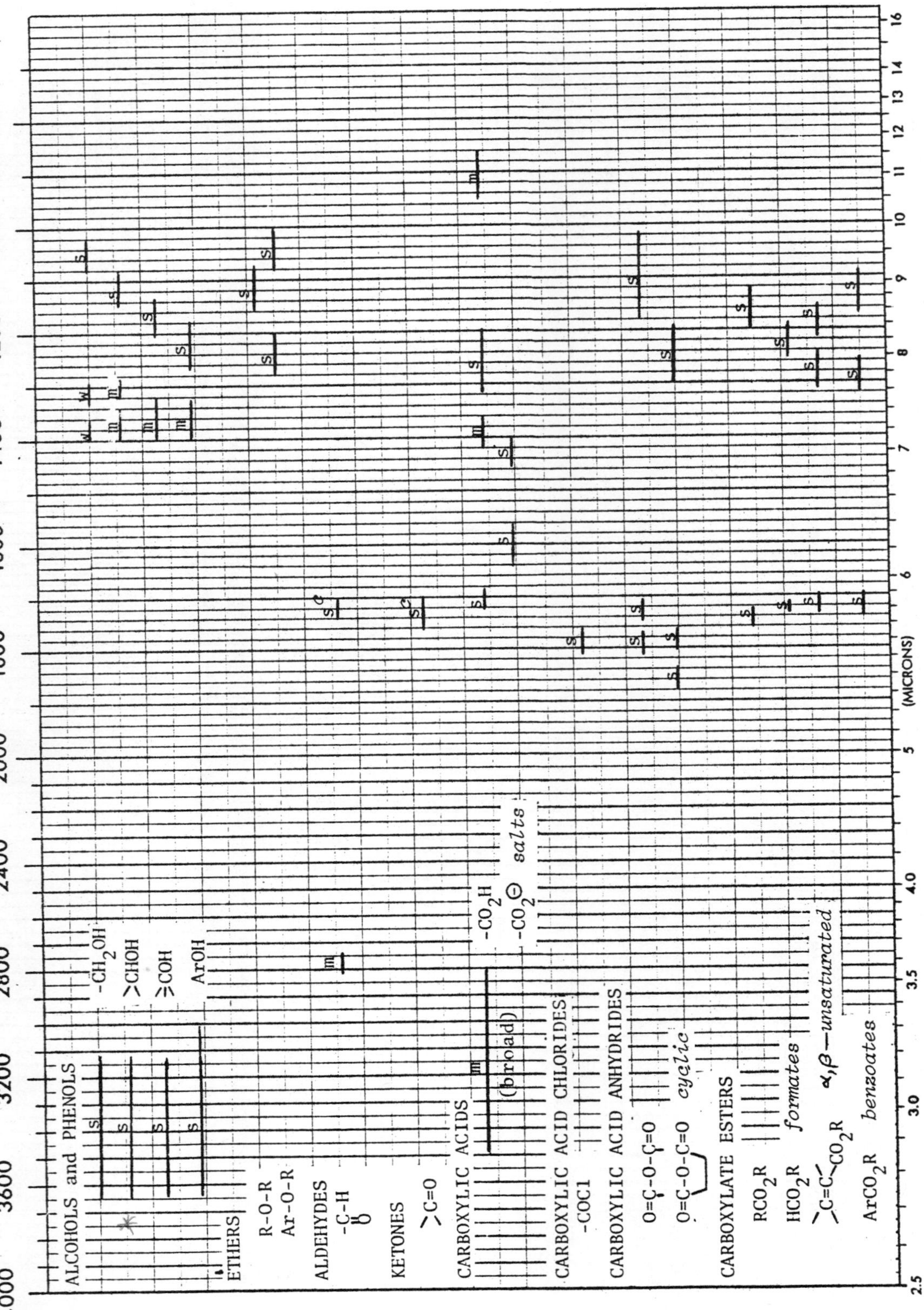
FREQUENCY (CM⁻¹)
4000 3600 3200 2800 2400 2000 1800 1600 1400 1200 1000 800
(MICRONS)
2.5 3.0 3.5 4.0 5 6 7 8 9 10 11 12 13 14 16
ALCOHOLS and PHENOLS
-CH2OH
>CHOH
≥COH
ArOH
ETHERS
R-O-R
Ar-O-R
ALDEHYDES
-C-H
KETONES
>C=O
CARBOXYLIC ACIDS
-CO2H
(broad)
-CO2⊖ salts
CARBOXYLIC ACID CHLORIDES
-COCl
CARBOXYLIC ACID ANHYDRIDES
O=C-O-C=O
O=C-O-C=O cyclic
CARBOXYLATE ESTERS
RCO2R
HCO2R formates
>C=C<CO2R α,β—unsaturated
ArCO2R benzoates
s
m
w

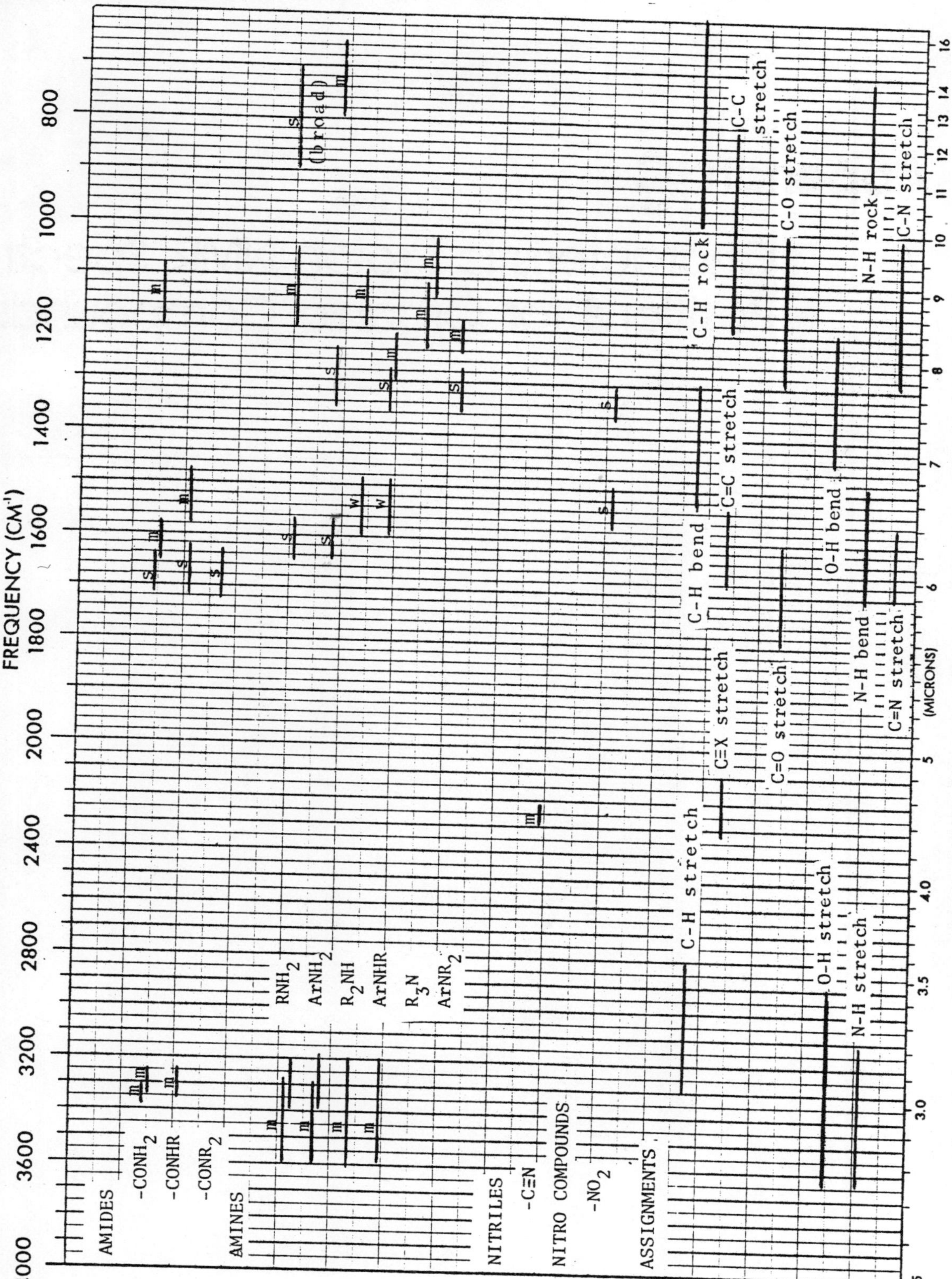

[a] An isopropyl group shows two absorption bands in this region.

[b] More than one absorption band may be observed in this region.

[c] Conjugation lowers the frequency range for this absorption band by 30 to 40 cm^{-1}.

Appendix C

Characteristic Proton NMR Spectral Positions for Organic Compounds

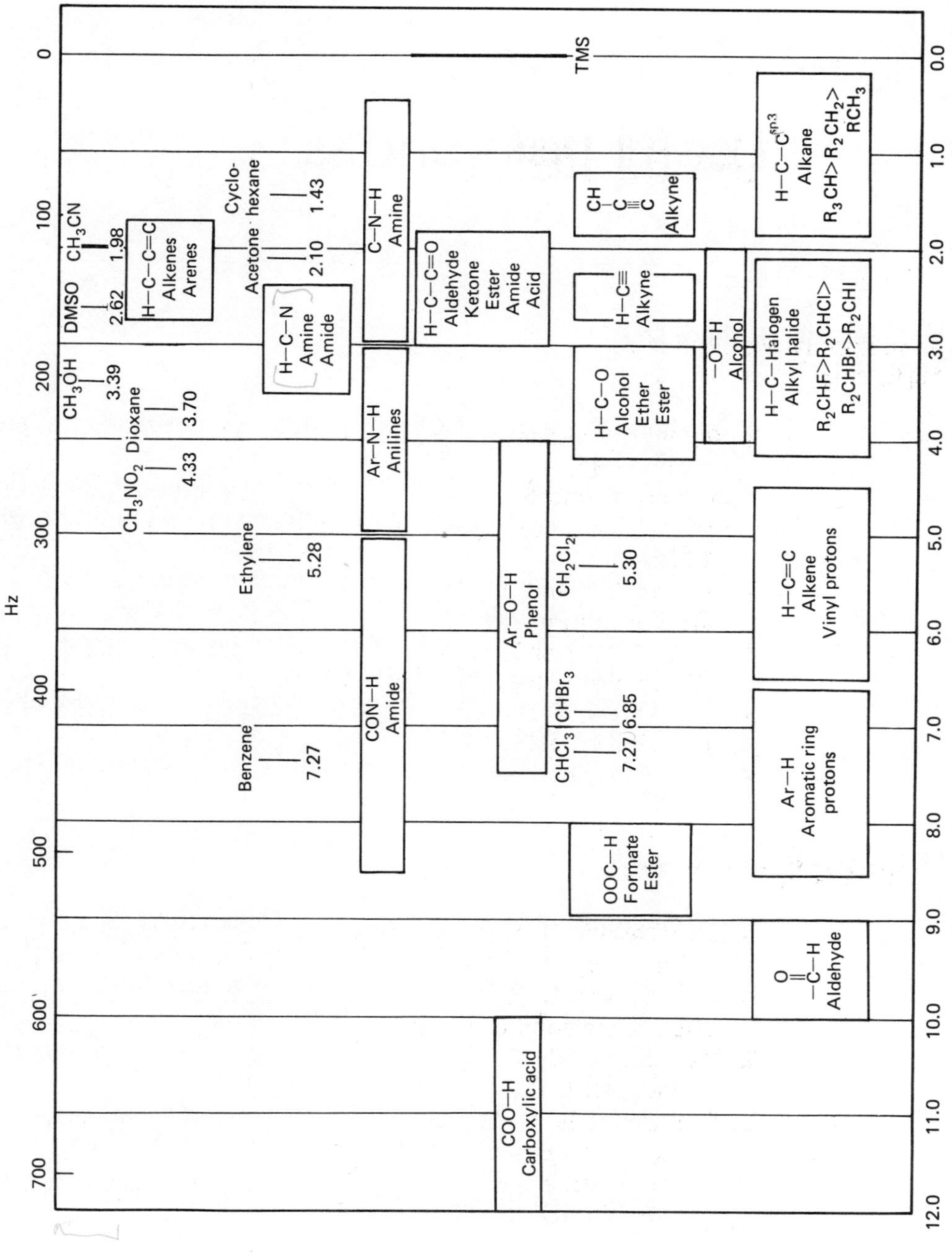
Hz
0
100
200
300
400
500
600
700
ppm δ
0.0
1.0
2.0
3.0
4.0
5.0
6.0
7.0
8.0
9.0
10.0
11.0
12.0
TMS
DMSO 2.62
CH_3CN 1.98
CH_3OH 3.39
Dioxane 3.70
CH_3NO_2 4.33
Cyclo-hexane 1.43
Acetone 2.10
Ethylene 5.28
Benzene 7.27
H—C—C=C
Alkenes
Arenes
H—C—N
Amine
Amide
C—N—H
Amine
Ar—N—H
Anilines
CON—H
Amide
H—C—C=O
Aldehyde
Ketone
Ester
Amide
Acid
Ar—O—H
Phenol
COO—H
Carboxylic acid
CH_2Cl_2 5.30
$CHBr_3$ 6.85
$CHCl_3$ 7.27
OOC—H
Formate
Ester
CH
C
C
Alkyne
H—C≡
Alkyne
H—C—O
Alcohol
Ether
Ester
—O—H
Alcohol
$H—C—C^{sp3}$
Alkane
$R_3CH > R_2CH_2 > RCH_3$
H—C—Halogen
Alkyl halide
$R_2CHF > R_2CHCl >$
$R_2CHBr > R_2CHI$
H—C=C
Alkene
Vinyl protons
Ar—H
Aromatic ring
protons
O
—C—H
Aldehyde

Useful References

Experimental Methods and Techniques

E. W. Berg, *Physical and Chemical Methods of Separation*, McGraw-Hill, New York, 1963.

A. J. Gordon and R. A. Ford, *The Chemist's Companion: A Handbook of Practical Data, Techniques, and References*, Wiley-Interscience, New York, 1972.

J. A. Riddick and W. B. Bunger, *Organic Solvents. Physical Properties and Methods of Purification*, 3rd ed., Wiley, New York, 1970.

F. L. J. Sixma and H. Wynberg, *A Manual of Physical Methods in Organic Chemistry*, Wiley, New York, 1964.

A. Weissberger, *Technique of Organic Chemistry*, 1st and 2nd eds., Interscience, New York, 1954, 1963.

K. B. Wiberg, *Laboratory Technique in Organic Chemistry*, McGraw-Hill, 1960.

Qualitative Organic Analysis

N. D. Cheronis, J. B. Entrikin, and E. M. Hodnett, *Semimicro Qualitative Organic Analysis*, 3rd ed., Interscience Publishers, New York, 1965.

D. J. Pasto and C. R. Johnson, *Organic Structure Determination*, Prentice-Hall, Inc., Englewood Cliffs, N.J., 1969.

Z. Rappaport, ed., *Handbook of Tables for Organic Compound Identification*, 3rd ed., CRC Press, Cleveland, Ohio, 1967.

R. L. Shriner, R. C. Fuson, and D. Y. Curtin, *The Systematic Identification of Organic Compounds*, 5th ed., Wiley, New York, 1964.

A. Vogel, *Vogel's Textbook of Practical Organic Chemistry*, 4th ed., Longman Group Limited, London, 1978.

Chromatography

J. M. Bobbitt, A. E. Schwarting, and R. J. Gritter, *Introduction to Chromatography*, Reinhold, New York, 1968.

J. M. Bobbitt, *Thin-Layer Chromatography*, Reinhold, New York, 1963.

R. C. Crippen, *Identification of Organic Compounds with the Aid of Gas Chromatography*, McGraw-Hill, New York, 1973.

D. W. Grant, *Gas-Liquid Chromatography*, Van Nostrand Reinhold, New York, 1971.

E. Heftmann, ed., *Chromatography: A Laboratory Handbook of Chromatographic and Electrophoretic Methods*, Van Nostrand Reinhold, New York, 1975.

J. J. Kirkland, ed., *Modern Practice of Liquid Chromatography*, Wiley-Interscience, New York, 1971.

H. M. McNair and E. J. Bonelli, *Basic Gas Chromatography*, Varian Aerograph, Walnut Creek, Calif., 1969.

H. Purnell, *New Developments in Gas Chromatography*, Wiley-Interscience, New York, 1973.

L. R. Snyder and J. J. Kirkland, *Introduction to Modern Liquid Chromatography*, Wiley-Interscience, New York, 1974.

Spectrometric Analyses

General Introduction to Spectrometric Methods and Analyses

J. R. Dyer, *Applications of Absorption Spectroscopy of Organic Compounds*, Prentice-Hall, Englewood Cliffs, N.J., 1965.

J. G. Grasselli, *Atlas of Spectral Data and Physical Constants for Organic Compounds*, CRC Press, West Palm Beach, Fla., 1975.

W. A. Guillory, *Introduction to Molecular Structure and Spectroscopy*, Allyn & Bacon, Boston, 1977.

J. B. Lambert, H. F. Shurvell, L. Verbit, R. G. Cooke, and G. H. Stout, *Organic Structural Analysis*, Macmillan, New York, 1976.

D. J. Pasto and C. R. Johnson, *Laboratory Text for Organic Chemistry*, Prentice-Hall, Englewood Cliffs, N.J., 1979.

J. W. Robinson, ed., *Handbook of Spectroscopy, Volume II*, CRC Press, West Palm Beach, Fla., 1974.

F. Scheinmann, ed., in *Introduction to Spectroscopic Methods for the Identification of Organic Compounds*, Pergamon Press, Oxford, 1973.

R. M. Silverstein, C. G. Bassler, and T. C. Merril, *Spectrometric Identification of Organic Compounds*, 3rd ed., Wiley, New York, 1974.

Infrared Spectroscopy

M. Avrum and Gh. D. Mateescu, *Infrared Spectroscopy*, Wiley, New York, 1972.

L. J. Bellamy, *The Infrared Spectra of Complex Organic Molecules*, 3rd ed., Wiley, New York, 1975.

R. T. Conley, *Infrared Spectroscopy*, 2nd ed., Allyn and Bacon, Boston, 1972.

K. Nakanishi and P. H. Solomon, *Infrared Absorption Spectroscopy*, Holden-Day, San Francisco, 1977.

C. J. Pouchert, *The Aldrich Library of Infrared Spectra*, 2nd ed., Aldrich Chemical Co., Milwaukee, Wisc., 1975.

Sadtler Standard Spectra, Sadtler Research Laboratories, Philadelphia.

Nuclear Magnetic Resonance Spectroscopy

F. A. Bovey, *Nuclear Magnetic Resonance Spectroscopy*, Academic Press, New York, 1969.

F. A. Bovey, *NMR Data Tables for Organic Compounds*, Interscience, New York, 1967.

W. Brugel, *Nuclear Magnetic Resonance Spectra and Chemical Structures*, Academic Press, New York, 1967.

High Resolution NMR Spectra Catalogue, Varian Associates, Palo Alto, Calif., Vol. 1, 1962, Vol. 2, 1963.

L. M. Jackman and S. Sternhell, *Application of Nuclear Magnetic Resonance Spectroscopy in Organic Chemistry*, 2nd ed., Pergamon Press, Elmsford, N.Y., 1969.

D. W. Mathieson, *Nuclear Magnetic Resonance for Organic Chemists*, Academic Press, New York, 1967.

C. J. Pouchert and J. R. Campbell, *The Aldrich Library of NMR Spectra*, Vols. I–XI, Aldrich Chemical Company, Milwaukee, Wisc., 1974–1975.

Sadtler NMR Spectra, Sadtler Research Laboratories, Philadelphia.

Ultraviolet-visible Spectroscopy

R. P. Bauman, *Absorption Spectroscopy*, Wiley, New York, 1962.

L. F. Fieser and M. Fieser, *Steroids*, Reinhold, New York, 1959, pp. 15–21.

K. Hiraya, *Handbook of Ultraviolet and Visible Absorption Spectra of Organic Compounds*, Plenum Press Data Division, New York, 1967.

H. H. Jaffe and M. Orchin, *Theory and Applications of Ultraviolet Spectroscopy*, Wiley, New York, 1962.

C. N. R. Rao, *Ultraviolet and Visible Spectroscopy*, 2nd ed., Butterworths, London, 1967.

Standard Ultraviolet Spectra, Sadtler Research Laboratories, Philadelphia.

R. G. White, *Handbook of Ultraviolet Methods*, Plenum Press, New York, 1965.

Specific Experimental Procedures

Experiment 14: W. S. Trahanovsky, L. B. Young, and G. L. Brown, "Oxidation of Organic Compounds with Cerium(IV). IV. Oxidation of Benzyl and Related Alcohols," *J. Org. Chem.*, **32,** 3865 (1967).

Experiment 17: M. F. Schlatter and R. D. Clark, "Orientation of *t*-Alkylation Products of Toluene and Ethylbenzene," *J. Amer. Chem. Soc.*, **75,** 361 (1953).

Experiment 19: I. B. Ruppel, Jr., F. L. Cuneo, and J. G. Krause, "Column and Thin-Layer Chromatography," *J. Chem. Ed.*, **48,** 635 (1971).

Experiment 20: H. Zollinger, *Azo and Diazo Chemistry*, Interscience, New York, 1961.

Experiment 27: W. P. Weber and G. W. Gokel, *Phase Transfer Catalysis in Organic Synthesis*, Springer-Verlag, New York, 1977.

Experiment 28: W. E. Kuhn, "2-Aminofluorene," *Organic Synthesis*, Coll. Vol. **II**, 448 (1943).

Experiment 29: E. C. Horning, M. G. Horning, and D. A. Dimmig, "3-Carbethoxycoumarin," *Organic Synthesis*, Coll. Vol. **III**, 1965 (1955).

Experiment 31: D. R. Paulson, A. L. Hartwig, and G. F. Moran, "The Ethylene Ketal Protecting Group in Organic Synthesis," *J. Chem. Ed.*, **50**, 216 (1973).

Experiment 32: M. P. Doyle and C. T. West, "Silane Reductions in Acidic Media. IV. Reductions of Alkyl-Substituted Cyclohexanones by Mono-, Di-, and Trialkylsilanes. Stereochemistry of Alcohol and Ether Formation," *J. Org. Chem.*, **40**, 3821 (1975).

Experiments 33, 34: M. P. Doyle, "A Spectrometric Study of the Oxidation of Alcohols by Cerium(IV)," *J. Chem. Educ.*, **51**, 131 (1974).

Experiment 35: G. Gorin, E. Fuchs, L. G. Butler, S. L. Chopra, and R. T. Hersh, "Some Properties of Urease," *Biochemistry*, **1**, 911 (1962). N. E. Dixon, G. Gazzola, J. J. Walters, R. L. Blakeley, and B. Zerner, "Inhibition of Jack Bean Urease by Acetohydroxamic Acid and by Phosphoramidate," *J. Amer. Chem. Soc.*, **97**, 4130 (1975). N. E. Dixon, C. Gazzola, R. L. Blakeley, and B. Zerner, "Jack Bean Urease. A Metalloenzyme. A Simple Biological Role for Nickel," *J. Amer. Chem. Soc.*, **97**, 4131 (1975).

Experiment 36: V. Hach, "A Most Simple Hydroboration Procedure," *Synthesis*, 340 (1974).

Experiment 37: M. C. Kloetzel, "The Diels-Alder Reaction with Maleic Anhydride," *Organic Reactions*, **IV**, 1 (1948).

Experiment 38: K. J. Saunders, *Organic Polymer Chemistry*, Chapman and Hall, London, 1973.

Experiment 39: K. E. Fahrenholtz, M. Lurie, and R. W. Kierstead, "The Total Synthesis of *d,l*-Δ^9-Tetrahydrocannabinol and Four of Its Isomers," *J. Amer. Chem. Soc.*, **89**, 5934 (1967).

Experiment 40: H. M. Woodburn, *Using the Chemical Literature*, Marcel Dekker, Inc., New York, 1974.

Index

Periodic Table of the Elements

Atomic number → 1 H 1.008 Hydrogen ← Atomic mass

PERIOD	IA	IIA	IIIB	IVB	VB	VIB	VIIB	VIIIB			IB	IIB	IIIA	IVA	VA	VIA	VIIA	O
1	1 H 1.008 Hydrogen																	2 He 4.003 Helium
2	3 Li 6.94 Lithium	4 Be 9.01 Beryllium											5 B 10.81 Boron	6 C 12.011 Carbon	7 N 14.01 Nitrogen	8 O 16.00 Oxygen	9 F 19.00 Fluorine	10 Ne 20.18 Neon
3	11 Na 22.99 Sodium	12 Mg 24.31 Magnesium											13 Al 26.98 Aluminum	14 Si 28.09 Silicon	15 P 30.97 Phosphorus	16 S 32.06 Sulfur	17 Cl 35.45 Chlorine	18 Ar 39.95 Argon
4	19 K 39.10 Potassium	20 Ca 40.08 Calcium	21 Sc 44.96 Scandium	22 Ti 47.90 Thallium	23 V 50.94 Vanadium	24 Cr 52.00 Chromium	25 Mn 54.94 Manganese	26 Fe 55.85 Iron	27 Co 58.83 Cobalt	28 Ni 58.71 Nickel	29 Cu 63.55 Copper	30 Zn 65.37 Zinc	31 Ga 69.72 Gallium	32 Ge 72.59 Germanium	33 As 74.92 Arsenic	34 Se 78.96 Selenium	35 Br 79.90 Bromine	36 Kr 83.80 Krypton
5	37 Rb 85.47 Rhodium	38 Sr 87.62 Strontium	39 Y 88.91 Yttrium	40 Zr 91.22 Zirconium	41 Nb 92.91 Niobium	42 Mo 95.94 Molyb- denum	43 Tc 98.91 Techne- tium	44 Ru 101.07 Ruthenium	45 Rh 102.91 Rhodium	46 Pd 106.4 Palladium	47 Ag 107.87 Silver	48 Cd 112.40 Cadmium	49 In 114.82 Indium	50 Sn 118.69 Tin	51 Sb 121.75 Antimony	52 Te 127.60 Tellurium	53 I 126.90 Iodine	54 Xe 131.30 Xenon
6	55 Cs 132.91 Cesium	56 Ba 137.34 Barium	57 *La 138.91 Lanthanum	72 Hf 178.49 Hafnium	73 Ta 180.95 Tantalum	74 W 183.95 Tungsten	75 Re 186.2 Rhenium	76 Os 190.2 Osmium	77 Ir 192.22 Iridium	78 Pt 195.09 Platinum	79 Au 196.97 Gold	80 Hg 200.59 Mercury	81 Ti 204.37 Titanium	82 Pb 207.2 Lead	83 Bi 208.98 Bismuth	84 Po (209) Polonium	85 At (210) Astatine	86 Rn (222) Radon
7	87 Fr (223) Francium	88 Ra 226.03 Radium	89 †Ac (227) Actinium	104 Rf (261) Ruther- fordium	105 Ha (262) Hahnium	106												

*Lanthanides	58 Ce 140.12 Cerium	59 Pr 140.91 Praseo- dymium	60 Nd 144.24 Neodym- ium	61 Pm (145) Prome- thium	62 Sm 150.4 Samarium	63 Eu 151.96 Europium	64 Gd 157.25 Gadolinium	65 Tb 158.93 Thorium	66 Dy 162.50 Dyspro- sium	67 Ho 164.93 Holmium	68 Er 167.26 Erbium	69 Tm 168.93 Thulium	70 Yb 173.04 Ytterbium	71 Lu 174.97 Lutetium
†Actinides	90 Th 232.04 Thorium	91 Pa 231.04 Protac- tinium	92 U 238.03 Uranium	93 Np 237.05 Neptunium	94 Pu (244) Plutonium	95 Am (243) Americium	96 Cm (247) Curium	97 Bk (249) Berkelium	98 Cf (249) Califor- nium	99 Es (254) Einstei- nium	100 Fm (257) Fermium	101 Md (258) Mendele- vium	102 No (259) Nobelium	103 Lr (260) Lawren- cium

Chemical Characterization Tests

Chemical Class	Characterization Test	Reagent	Page
All Classes	Solubility	Et_2O; H_2O; 10% HCl; 10% NaOH; H_2SO_4	82
All Classes	Qualitative elemental analysis		158
Alkene and Alkyne	Bromine addition	Br_2 in CCl_4	83
	Potassium permanganate oxidation	$KMnO_4$ in H_2O	85
Aromatic Hydrocarbon	Aluminum Chloride—chloroform	$AlCl_3/CHCl_3$	252
	Formaldehyde—sulfuric acid	H_2CO/H_2SO_4	253
Halohydrocarbon	Alcoholic silver nitrate	$AgNO_3$ in EtOH	130
	Beilstein flame	Cu wire	130
Alcohol	Ceric ammonium nitrate	$Ce(NH_4)_2(NO_3)_6$	163
	Lucas test*	$ZnCl_2/HCl$	164
	Chromic acid oxidation*	H_2CrO_4/H_2SO_4	163
	Iodoform formation*	I_2/NaOH	182
Phenol	Ceric ammonium nitrate	$Ce(NH_4)_2(NO_3)_6$	245
	Ferric chloride	$FeCl_3$ in H_2O	244
	Bromine water	Br_2 in H_2O	243
Aldehyde and Ketone	2,4-Dinitrophenylhydrazine	2,4-$(NO_2)_2C_6H_3NHNH_2$	179
	Chromic acid oxidation*	H_2CrO_4/H_2SO_4	181
	Iodoform formation*	I_2/NaOH	182
	Tollens' test*	$AgNO_3/NH_4OH$	181
Amine	Copper sulfate	$CuSO_4$ in H_2O	273
	Diazo coupling*	$NaNO_2$/HCl and β-Naphthol	245
	Hinsberg test*	$C_6H_5SO_2Cl$ in NaOH/H_2O	274
Nitrile	Hydroxamate	H_2NOH in $CH_3CH(OH)CH_2OH$	275
Nitro Compound	Ferrous hydroxide	$FeSO_4$/KOH	276
Carboxylic Acid	Neutralization equivalent	NaOH	289
Acid Chloride	Silver nitrate	$AgNO_3$ in EtOH	289
	Hydroxamate	H_2NOH/HCl in EtOH	288
Acid Anhydride	Hydroxamate	H_2NOH/HCl in EtOH	288
Ester	Hydroxamate	H_2NOH/HCl in EtOH	288
Amide	Hydroxamate	H_2NOH/KOH in $CH_3CH(OH)CH_2OH$	288

*Subclassification test.